Pathology of the Heart and Sudden Death in Forensic Medicine

Pathology of the Heart and Sudden Death in Forensic Medicine

Vittorio Fineschi
Giorgio Baroldi
and
Malcolm D. Silver

CRC Press
Taylor & Francis Group
Boca Raton London New York

CRC Press is an imprint of the
Taylor & Francis Group, an **informa** business

First published in paperback 2024

First published in 2006 by CRC Press
2385 NW Executive Center Drive, Suite 320, Boca Raton FL 33431

and by CRC Press
4 Park Square, Milton Park, Abingdon, Oxon, OX14 4RN

CRC Press is an imprint of Taylor & Francis Group, LLC

© 2006, 2024 Taylor & Francis Group, LLC

Library of Congress Card Number 2005055301

Library of Congress Cataloging-in-Publication Data

Pathology of the heart and sudden death in forensic investigations / [edited by] Vittorio Fineschi, Giorgio
   Baroldi, Malcolm D. Silver.
      p. cm.
   Includes bibliographical references and index.
   ISBN 0-8493-7048-5 (alk. paper)
   1. Forensic pathology. 2. Heart--Pathophysiology. 3. Cardiac arrest--Pathophysiology. 4. Sudden
death. I. Fineschi, Vittorio. II. Baroldi, Giorgio. III. Silver, Malcolm D. IV. Title.

RA1063.4.P38 2006
614'.1--dc22   2005055301

ISBN: 978-0-8493-7048-9 (hbk)
ISBN: 978-1-03-291895-2 (pbk)
ISBN: 978-0-429-24527-5 (ebk)

DOI: 10.1201/9781420006438

Visit the Taylor & Francis Web site at
http://www.taylorandfrancis.com

and the CRC Press Web site at
http://www.crcpress.com

Preface

In most jurisdictions of Western societies, sudden and unexpected deaths are mainly due to natural causes linked with heart diseases by an initiating factor triggering cardiac arrest. Advances in technology will procure a histo-ultrastructural view in living people. At present, however, the autopsy remains the gold standard; with the proviso that the diminishing requests for postmortem examination by clinicians may, in the future, confine autopsy investigation to forensic institutions. Science will continue to be nourished by well-suited hypotheses, but science advances on incontrovertible facts that, at present, only an autopsy can give.

In general, forensic pathology is mainly concerned with determining the role of trauma, intoxication, or poisoning in the cause of sudden and unexpected death as defined within legal boundaries. Its purpose is to settle whether death was due to natural causes or was a result of accident, suicide, or murder, and so help establish legal responsibility. So when performing an autopsy, the medical examiner and the pathologist should carefully investigate the inductive cause of sudden death, the history of disease, and the family history, in order to rule out the possibility of the above disorders.

With the marked advances in molecular techniques, DNA testing on peripheral blood and tissue has revolutionized the diagnosis of genetic causes of sudden death.

Although toxicological analyses are indispensable for final clarification of the cause of death, postmortem (external and internal) examination remains a fundamental challenge in forensic investigation. Because heart disease is responsible for most abusers' deaths, the study of macro and micro cardiac alterations is an essential step for the forensic pathologist.

That prompted us to write this book presenting the pathology of the heart/cardiovascular system with a forensic perspective. This is a book in which one can deliberate over recorded findings and then gain an understanding of how to make a critical, overall interpretation. This interpretation then guides the forensic pathologist in a court of law as the forensic pathologist takes on the role of interpreter of the sequence of events that led to death.

The recommendations presented in this book, along with others, are useful to the pathologist who will learn how to better collect and file the preliminary and necessary evidence needed for forensic judgment. In fact, it presumes an absolutely clear objectivity (supported by photographic, microfilm documents) and a willingness to investigate.

The harmonization of forensic autopsy rules is an essential initiative, not only for epidemiological and statistical aims, but also for unification in listing data and making that data available so that knowledge of the cause of death and the possible identification of that cause will be known. (There is also a project underway to develop a unique scheme for the autopsy protocol.) In addition to the general rules, this recommendation underlines particular ways of checking and drawing conclusions concerning the cardiovascular apparatus and, specifically, sudden death.

Acknowledgments

We thank the following for their collaboration in preparing this text: Mrs. Elisabetta Spagnolo and Miss Ilaria Citti, Institute of Clinical Physiology, National Research Council, Milan and Pisa, Italy.

We are grateful to the Monte dei Paschi Foundation for its financial support for research programs cited in this book.

The Editors

Vittorio Fineschi, M.D., Ph.D., received his basic medical degree from the University of Perugia, Italy, where he graduated in 1984. He undertook his postgraduate studies in forensic medicine at the University of Siena, Italy, in 1987 and received his clinical doctorate in legal medicine from the Universities of Siena and Perugia in 1990. He conducted his postdoctoral training in forensic pathology at the University of Siena ("Le Scotte" research institutes) in Siena.

Since 1992, Dr. Fineschi has worked continually with the Medical Examiner Department in Miami, Florida, where he conducted several research projects related to drug-abuse-related death and sudden cardiac death under the supervision of Dr. Joseph H. Davis.

Dr. Fineschi's research projects have been supported by a grant from major Italian agencies, including the National Research Council and the Monte dei Paschi of Siena Foundation.

Dr. Fineschi taught legal medicine in the graduate and postgraduate courses of the Medical School of the University of Siena (1990–1998). And, since 1999, he has been the Director of the Institute of Forensic Pathology at the University of Foggia, Italy, and Chairman of Legal Medicine at the University of Foggia, School of Medicine, in Italy. He is currently full professor of legal medicine at the University of Foggia, School of Medicine, in Italy.

Dr. Fineschi's main research interests are generally in forensic pathology, particularly in heroin- and cocaine-related death and cardiovascular pathologies related to sudden cardiac death. His primary research achievements are related to his studies on traumatic agents and their effects related to sudden cardiac death. He contributed greatly to the design of animal models currently used to perform *in vivo* physiology studies and oxidative stress studies in sudden cardiac death.

He is author and co-author of an extensive list of papers published in major scientific journals. Dr. Fineschi received from the Italian Society of Legal Medicine the National Award for the best scientific paper of the year in the years 1988, 1989, and 1993. Dr. Fineschi is a member of the advisory board of the ranking forensic journal *International Journal of Legal Medicine*; he is a Fellow of the American Academy of Forensic Sciences; and he was a founder of the Italian Society of Forensic Pathology.

Malcolm D. Silver, M.D., Ph.D., who has retired and become a full-time bird watcher, graduated in medicine from the University of Adelaide in South Australia and did post-graduate work both at McGill University and the Australian National University. Currently he is professor emeritus of the Department of Laboratory Medicine and Pathobiology at the University of Toronto (Ontario, Canada).

Formerly chair in pathology at both the University of Western Ontario (London, Ontario) and the University of Toronto and chief of pathology at University Hospital,

London (Ontario), and at the Toronto General Hospital, his areas of scientific interest ranged from platelet fine structure and behavior to cardiovascular diseases, with special interest in artificial heart valves, cardiomyopathies, and ischemic heart disease.

Dr. Silver is the author/co-author of more than 150 scientific articles and with others published two monographs. He edited/co-edited *Cardiovascular Pathology*, now in its third edition.

Giorgio Baroldi, M.D., Ph.D., FACC, FESC, is professor emeritus of pathology and cardiovascular pathology, Medical School, University of Milan, associate professor of the Institute of Clinical Physiology, National Research Council, Pisa and Milan, Italy. His main interest was and still is to correlate pathomorphology and clinical pattern in cardiac diseases. He started by demonstrating tridimensionally an extensive collateral or anastomotic network in the intramyocardial arterial system of the coronary circulation, with many facts supporting their ability to compensate for even numerous coronary obstructions and questioning the existence of a chronic ischemia. These data resulted in a Postdoctoral Research Fellowship Award FF452 of the National Health Institute, and he continued his research programs first at the Armed Forces Institute of Pathology, Washington, D.C., and after that at Banting Institute, Toronto, Canada; Division of Cardiology of Omaha, University of Nebraska; Institute of Cardiac Surgery Houston, Texas; and Istituto Patologhia, Università S. Paulo, Campo di Riberao Preto Brasile. From the demonstration that different types of damage of myocardial cells exist arose the concept that an adrenergic stress, namely, an excess of noradrenaline in the myocardial interstitium, may play an essential role in cardiac arrest after a cardiac infarct and sudden/unexpected death.

Dr. Baroldi received the Hektoen Silver Medal of the American Heart Association (San Francisco, CA, 1964), the Silver Medal of the College of American Pathologists (Chicago, 1965), and the Distinguished Achievement Award of the Society of Cardiovascular Pathology (New Orleans, LA, 2000). To date, he has published 269 papers, 4 monographs, and 11 chapters in textbooks.

Contributors

Thomas Bajanowski, M.D.
University of Essen
Essen, Germany

Giorgio Baroldi, M.D., Ph.D., FACC, FESC
University of Milan
Milan, Italy
and
Institute of Clinical Physiology
National Research Council
Pisa, Italy

Berndt Brinkmann, M.D.
University of Münster
Münster, Germany

Vittorio Fineschi, M.D., Ph.D.
University of Foggia
Italy

Kathryn A. Glatter, M.D.
University of California, Davis
Davis, California

Antonio L'Abbate, M.D., Ph.D.
University of Pisa
Pisa, Italy
and
Institute of Clinical Physiology
National Research Council
Pisa, Italy

Margherita Neri, M.D.
University of Foggia
Italy

Cristoforo Pomara, M.D.
University of Foggia
Italy

Alberto Repossini, M.D.
Cliniche Humanitas Gavazzeni
Bergamo, Italy

Irene Riezzo, M.D.
University of Foggia
Italy

Malcolm D. Silver, M.B.B.S., M.D., M.Sc., Ph.D., FRCPA, FRCPC
University of Toronto
Toronto, Ontario, Canada

Meredith M. Silver, M.B.B.S., M.Sc., FRCPA, FRCPC
University of Toronto
and
The Hospital for Sick Children
Toronto, Ontario, Canada

Emanuela Turillazzi, M.D., Ph.D.
University of Foggia
Italy

Contents

Introduction

1

GIORGIO BAROLDI AND VITTORIO FINESCHI

In general, forensic pathology is mainly concerned with determining the role of trauma, intoxication, or poisoning in the cause of sudden and unexpected death as defined within legal boundaries. Its purpose is to settle whether death was due to natural causes or was a result of accident, suicide, or murder, and so help establish legal responsibility.

In most jurisdictions of Western societies, sudden and unexpected deaths are mainly due to natural causes linked with heart diseases by an initiating factor triggering cardiac arrest. That prompted us to write this book, which presents the pathology of the heart and cardiovascular system with a forensic perspective. The aim is to provide a background of facts currently known or believed with a critical review to aid the reader in resolving cases for which he or she is responsible. In other words, we revisit pathologic findings versus etiopathogenetic hypotheses and theories and propose new ones substantiated by previously misinterpreted or ignored facts. In this book, detailed findings are presented along with overall interpretations to help the forensic pathologist in a court of law to act as interpreter of the sequence of events leading to death.

Sudden death due to cardiovascular causes is uncommon among the young, yet we provide insight as to cause. The incidence of cardiovascular disease increases with increasing age, and, as indicated above, it is now the leading cause of morbidity and death in Western societies.

In reaching conclusions, a forensic pathologist must correlate and discriminate between terminal events and pathological findings that denote preexisting cardiac disease. It is insufficient to rely upon the tendency to believe (erroneously) that current clinical imaging techniques are alone sufficient to provide evidence of the cause of a disease. Advances in technology will procure a histo-ultrastructural view in living people. At present, however, the autopsy remains the gold standard, with the proviso that the diminishing requests for postmortem examination by clinicians may, in the future, confine autopsy investigation to forensic institutions. Science will continue to be advanced by well-suited hypotheses, but at present, there are incontrovertible facts that only an autopsy can give. There is, therefore, a need to establish strict collaboration between clinicians and pathologists.

With this in mind, we determined the sequence of topics in this book: First, the functional anatomy of the heart and the method of its examination are discussed, then pathological changes and their frequency and extent in specific diseases are addressed, and a review of the etiopathogenetic significance of each morphopathologic finding is provided so as to aid in understanding coronary heart disease in general and sudden death in particular.

Functional Anatomy of the Heart

2

GIORGIO BAROLDI

Contents

Blood Vessels

The blood supply of the human heart in man has both an arterial system and a venous system joined through capillaries. To understand its complexity, one must keep in mind two points. First, extramural coronary arteries are filled during cardiac systole, as intramural vessels are compressed by the contracting myocardium with intramural flow occurring in diastole. This diphasic flow means a systolic increase in all types of wall stress, that is, radial-compressional, both circumferential and longitudinal-tensile stress, and shearing stress on the endothelium in relation to flow velocity and blood viscosity (Fry, 1969) in vessels prone to distension. The second point is the peculiar embriogenesis of cardiac blood vascular structures. In early ontogenesis (Grant, 1926; Grant and Regnier, 1926; Patten, 1968) and phylogenesis (Poupa and Ostdal, 1969; Poupa and Carlsten, 1973), the myocardium has a netlike disposition with large lacunae intercommunicating with each other and the heart cavities. No epicardial or intramural blood vessels are present (Figure 1A). At a later stage, the outer cardiac wall becomes compacted with a need for epicardial arterial vessels, while the inner part maintains its lacunar architecture (Figure 1B). Finally most

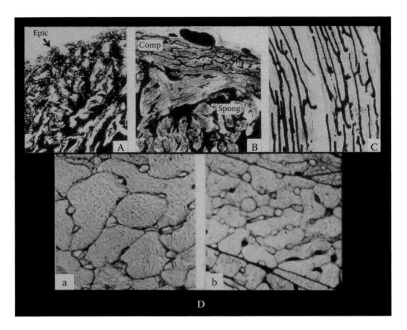

Figure 1 Different types of blood supply in the ventricular wall in vertebrates. (India-ink-injected specimens.) (A) Wall of the ventricle in a frog (*R. temporaria*). The lacunary type of blood supply occurs in many fishes (e.g., cod, platfish, etc.) and in amphibia. Blood-filled lacunary spaces (black) reach to the outer surface of the ventricle (epic). (B) Wall of the ventricle in a turtle (*Testudo horsfieldi*). The outer compact layer of the myocardium (comp) is supplied by coronary capillaries. The inner core of the ventricle is spongious (sponge) supplied by blood circulating in the lacunary spaces. Capillaries are extremely scarce in this muscular compartment. Quantitative relations of both muscular compartments are age and body-weight dependent. The installation and evolution of the coronary terminal vascular bed is closely related to the development of the outer compact muscular layer. This transitory type of myocardial blood supply occurs in some fishes (e.g., salmon, trout, mackerel, tuna, etc.) and in reptiles. (C) The wall of the ventricle in an albino rat (*R. rattus*). This type occurs in adult birds and mammals. The density of the capillaries depends on the physical activity of the respective species. (D) The capillary density in relation to physical activity. Left ventricle wall, with capillaries stained by periodic acid Schiff (PAS), of the (a) mouse (strain B 10 LP) and (b) flying mammal (brown bat, *Myotis myotis*) under the same magnification (630×). The body weight of both of these species was nearly the same (bat: 20 ± 0.7; mouse: 28 ± 0.5 g), but the relative heart weight was three times greater in the bat (1.01 ± 0.02; mouse: 0.35 ± 0.02). Cardiac cells/mm^2: bat, 3.328 ± 90; mouse, 2.578 ± 26. Capillaries mm^2: bat, 3.717 ± 103; mouse, 2.739 ± 58. Analogous differences were found between athletic and nonathletic species (hare and rabbit) and wild and domesticated species (wild rat and laboratory albino rat). (Courtesy of Dr. Otahar Poupa, Department of Clinical Physiology, University of Goteborg, Sweden.)

of the myocardium becomes compacted and is supplied by vessels, while the lacunar aspect is limited to the subendocardial layer among the columnae carneae (Figure 1C). This myocardial-vasculature molding in the presence of a rhythmic contraction may explain the final architecture of the coronary tree, some anomalies, and its peculiar flow dynamics.

Arteries

The heart has both an extramural and an intramural arterial system (Table 1). Their totally different environments give reason for their functional and pathological dissimilarities.

Table 1 Characteristics of Normal Coronary Arteries

Vessels	Average		Average Number of Branches[a] (Ventricles)				Annotations
	Present (%)	Diameter (mm)	Anterior (VS)	Posterior (VS)	Anterior (VD)	Posterior (VD)	
Extramural							
Left coronary artery	99	4.0 (2.0–5.5)	2.3 (2–4)	—	—	—	Small aortic and atrial branches
Anterior descending	100	3.6 (2.0–5.0)	4.1 (2–9)	—	1.5 (0–3)	—	36% ends of the apex; 64% in post interv. groove
Diagonal	33	2.0 (0.5–2.5)	2.5 (2–3)	—	—	—	9% hypoplastic
Circumflex	100	3.0 (1.5–5.0)	2.0 (0–5)	1.8 (0–6)	—	—	14% ends at post. interv. groove, 7% beyond it
Marginal	91	2.2 (1.0–3.0)	1.3 (0–3)	1.4 (0–5)	—	—	80% ends at apex
Posterior descending	8	2.1 (1.0–3.0)	—	0.5 (0–3)	—	2.0 (0–4)	90% ends in post. interv. groove
Right coronary artery	100	3.2 (1.5–5.5)	—	2.3 (0–6)	2. 0 (0–6)	2.3 (0–5)	10% before, 9% at, 81% beyond post. interv. groove
Marginal	100	1.7 (1.0–2.5)	—	—	1.3 (0–5)	1.0 (0–4)	93% ends at apex
Posterior descending	77	2.1 (1.0–3.0)	—	0.5 (0–3)	—	2.0 (0–4)	15% from both coronary arteries
Third coronary artery	46	1.1 (0.7–2.0)	—	—	—	—	64% independent; 36% common ostium; 7% ends at apex, 57% at 2/3; 33% at 1/3; 24% at 1/6 of right anterior wall

Table 1 (continued) Characteristics of Normal Coronary Arteries

Vessels	Average Present (%)	Average Diameter (mm)	Interventricular Septum Average Anterior (VS)	Interventricular Septum Average Posterior (VS)	Number of Branches[a] Anterior (VD)	Number of Branches[a] Posterior (VD)	Annotations
Anterior septal branches from anterior descending	100	1.0 (0.5–2.5)		13 (6–23)			Divided in superior-median-inferior apical
Posterior septal branches from posterior descending	100	0.7 (0.3–0.9)		11 (5–20)			Other septal branches may originate from adjacent vessels, left main trunk
Posterior septal branches from ascending posterior portion of anterior descending	64	0.4 (0.3–0.7)		6 (0–11)			
				Atria			
Main atrial branch	99	About 1		>30			May originate anteriorly, marginally, posteriorly
Minor atrial branches	100	<1		<30			May originate anteriorly, marginally, posteriorly

Note: LV, left vetricle; RV, right ventricle.

[a] In counting the ramifications, only branches >1 mm in diameter were considered.

Extramural Coronary Arteries

The *extramural coronary arterial* system presents vessels that normally run on the surface of the heart covered by epicardium and surrounded by fat tissue. The coronary arteries arise from the aorta usually from two ostia. In a heart removed from the body, the *left coronary artery* takes origin from the left anterior sinus of Valsalva and the *right coronary artery* from the right anterior sinus. The ostia are, in general, located at the same level as the free margins of the aortic cusps. According to Banchi (1904), the left ostium has such a location or one immediately below that level in 66% of cases or immediately above it in 34%. For the right ostium, the figures are 81% and 19%, respectively.

A third ostium marking the origin of a *third coronary* or *conus artery* is found in 33 to 51% of hearts (Schlesinger et al., 1949). It arises from a separated ostium in 64% of instances, shares the ostium of the right coronary artery in 36%, may be doubled in 8% of cases, and is tripled in 1% (Baroldi and Scomazzoni, 1967). A last consideration concerns the possibility that a third coronary artery can be missed by selective coronary cineangiography *in vivo* or by inaccurate inspection postmortem. A prominent third coronary artery could lead to an erroneous diagnosis of a hypoplastic right coronary artery.

The *left coronary artery* (LCA) has a short (3 to 23 mm; mean 13.5 mm) course and is rarely absent. Save for a few small aortic and atrial ramifications, and an occasional large branch ("interaortic-pulmonalis" [Crainicianu, 1922] or "sinus node artery" [James, 1961]), the LCA divides into the *left anterior descending* (LAD) and *left circumflex* (LCX) branches. In 33% of hearts, a third branch, the *diagonal branch*, exists.

The LAD continues the course of the main trunk in the anterior interventricular groove, extending to the heart's apex or frequently recurring around it and (64%) ascending one fourth or more in the posterior interventricular groove ("ramus recurrrens" [Mouchet, 1933]). In 5% of cases, the LAD divides into equal, parallel vessels and in its course gives off from zero to three (average 1.5) ramifications to the anterior surface of the right ventricle and from two to nine (average four) to the left ventricle. In 72% of cases, a conspicuous branch is distributed to the pulmonary conal region ("left adipose artery" [Banchi, 1904]).

The LCX follows the left atrioventricular groove, ending at different points in the latero-posterior tract of this groove. Its main branch, the left *marginal branch*, is present in 90% of hearts. Other atrial and ventricular minor ramifications exist. In 34% of cases, the LCX either does not follow (9%) or follows only in part the left atrioventricular groove, assuming an oblique course that crosses the posterior left ventricular wall in the direction of the cardiac apex. In these cases, the marginal branch is absent or hypoplastic.

The *right coronary artery* (RCA) passes posterior to the pulmonary artery and runs in the right atrioventricular groove, giving origin to atrial and ventricular branches. Anteriorly, its course is from its origin to the right margin of the heart where the *right marginal branch* originates. Its posterior course is from the latter margin to different levels in the posterior interventricular groove (see below), where, in 77% of instances, the RCA generates the *posterior descending branch*. Rarely, the posterior segment follows a diagonal course and crosses the posterior right ventricular surface (Figure 2).

Several authors (Banchi, 1904; Crainicianu, 1922; Gross, 1921; Mouchet, 1933; Schlesinger, 1940; Schoenmackers, 1958) proposed classifications of the extramural coronary arteries based on the different distributions of the posterior ventricular supply. The most popular was that of Schlesinger (1940), who recognized the following three types of coronary circulation:

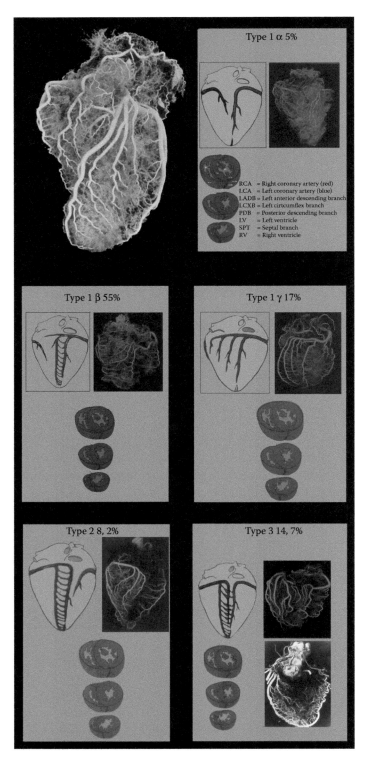

Figure 2 Plastic cast of the coronary arteries in a normal adult heart. Anterior view (A): (1) left main trunk (LM); (2) left anterior descending branch (LAD); (3) left circumflex (LCX); (4) diagonal branch (DB); (5) right coronary artery (RCA). Posterior view (B): five different types of posterior anatomical distribution of LCX (3) and RCA (5) and origin of the posterior descending (PD) branch (6). In all of these conditions, the left coronary artery predominates in terms of the amount of vascularized cardiac mass. The posterior recurrency of the LAD implies that the apex and varying portion of the inferior interventricular septum are nourished by this branch. In one case (C), the LAD reascended the whole posterior interventricular wall, with the interventricular septum being totally vascularized by LAD. In this plastic cast, other ramifications were removed to show the septal vascularization.

"Right preponderance" or type I (48% of cases): The right coronary artery supplies the right ventricle, the posterior half of the septum via the posterior descending branch, and the left posterior ventricle.

"Balanced" or type II (34%): The two ventricles are vascularized by the corresponding artery, with a posterior descending branch originating from the right coronary artery.

"Left preponderance" or type III (18%): The posterior descending is furnished by the left circumflex branch or two posterior descending branches, one from the RCA and one from the LCX.

This author claimed a sex difference and a relationship with coronary pathology related to the types of anatomical distribution, an opinion not confirmed by others (Baroldi and Silver, 1995; Baroldi and Scomazzoni, 1967).

By tridimensional plastic casts of the coronary arteries injected through the aorta — and therefore with reproduction of luminal ostia — we adopted the following classification (Baroldi et al., 1967):

Type I (77%; men 84%, women 68%): The right coronary artery supplies the posterior one third of the interventricular septum and, to a variable extent, the posterior left ventricle according to these subtypes:

Type Ia (5%): The RCA ends as a posterior descending branch without prominent branches to the left ventricle. This subtype is usually associated with a diagonal disposition of the LCX.

Type Ib (55%): The RCA terminates between the posterior interventricular groove, and the left cardiac margin supplies about one half of the posterior left ventricle.

Type Ic (13%): The RCA ends at the left margin, supplying the whole posterior left ventricular wall.

Type II (8%; men 6%, women 11%): The posterior descending branch arises from the LCX that supplies the whole posterior left ventricular wall, and the interventricular septum is entirely vascularized by the LCA through its two main branches.

Type III (15%; men 10%, women 21%): Two posterior descending branches exist, one from the RCA and the other from the LCX.

In our study, the misleading terms "balanced" or "preponderant" supply were avoided, because they give a false impression that the right coronary artery may predominate or be equal to the left coronary artery in the vascularization of the total ventricular–interventricular septal mass; it should be clear that the left coronary artery always predominates (Figure 2).

The diameters of the coronary arteries (Table 2) and their main branches vary according to their lengths and are related to the type of anatomical distribution and absence or presence of a third coronary artery. The RCA at its origin averages 3.8 or 3 mm in the absence or presence, respectively, of the latter vessel.

In general, if one "hyperplastic" coronary artery has a large extent of distribution, the other "hypoplastic" will have a small one so that myocardial territories do not lack normal vascularization. Thus, the diagnosis of a congenital hypoplasia of a coronary artery to explain cardiac dysfunction seems to be inappropriate (see Chapter 10).

Table 2 Diameter (ø) of Normal Coronary Arteries in Different Types of Distribution

Type	Ø LCA (mm)	Ø RCA (mm)	> LCA (%)	> RCA (%)	Equal (%)
I	2.5–5.5; average 4	2.5–5.5; average 3.6	60	20	20
II	2.5–5.0; average 4	2.0–3.3; average 2.6	83	No	17
III	3.5–5.0; average 4	1.5–5.0; average 3.5	50	33	17
Total	2.0–5.5; average 4	1.5–5.5; average 3.2	65	17	18

Notes: In the presence of a third coronary artery (RCA), average Ø: 3.00 mm; in absence, 3.8 mm. RCA, right coronary artery; LCA, left coronary artery.

Extramural coronary arteries and their branches may, occasionally, be covered for a variable distance by a layer or bridges or loops of myocardium (Crainicianu, 1922; Spalteholz, 1924). Such "mural coronary arteries" (Geiringer, 1951) were associated with 22% of the left anterior descending branch in hearts examined by Geiringer (1951) to less than 5% and were rarely seen in other branches (Edwards et al., 1956). Polacek (1961) described "myocardial bridges" in 86% of ventricular arterial vessels and "myocardial loops" at the atrioventricular groove level. In animals like the rabbit or rat, the coronary arteries are totally "mural." In man, their presence must be considered in interpreting systolic occlusion of coronary arteries (see below).

The extramural coronary arteries are muscular vessels lined by endothelium. They have a well-developed media limited by external and internal elastic membranes. Most smooth muscle cells in it have a circular disposition. However, some layers of these cells have a helical arrangement (Ahmed, 1969), a morphology also described in canine coronary arteries (Boucek et al., 1963). The adventitia is mainly formed by loose collagen, while the intima has a peculiar structure not seen in other arteries of similar size. It is characterized by an intimal thickening, found at birth in nodular form (Bork, 1926; Dock, 1946). This progressively extends with age. First exhaustively described by Wolkoff (1929), the thickening presents two layers more or less separated by an elastic lamina and is recognizable in the first decade of life. The external "elastic-muscular" layer is formed by a splitting of the internal elastic membrane with smooth cell proliferation through fenestrations in the latter. The internal "elastic-hyperplastic" layer is formed by elastic fibers originating from the separating lamina. In the first decade, the intimal thickness does not exceed that of the media being more prominent in males (Dock, 1946; Moon, 1952) and at branching sites without involvement of the whole intimal surface and with a great variability in different subjects and different segments of the same vessel. With aging, the elastic-hyperplastic layers progress in width and transform into a uniform fibrous layer approximately in the second decade, becoming prominent in the fourth decade (Figure 3). At that time, intimal thickness may exceed medial width (Wolkoff, 1929; Vlodaver and Neufeld, 1967, 1968). In this proliferative molding of the coronary intima, the smooth muscle cell as a multifunctional mesenchymal cell capable of proliferation, migration, and contraction, and able to synthesize ground substance, elastin, collagen, and basement membrane (Wissler, 1967), has a predominant role. *In vitro* culture of human smooth muscle cells

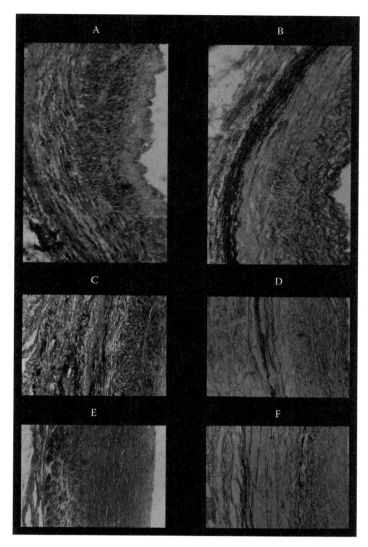

Figure 3 Coronary physiologic intimal thickening. This change starts as nodular (already visible at birth at the site of vessel bifurcation) smooth muscle cells (A) and elastic fibrils (B) hyperplasia, which in the second decade is diffuse to the whole intimal surface of all extramural arterial vessels. With aging, there is a progressive increase of fibrous tissue that substitutes for myoelastic tissue (C,D) with final, total, anelastic fibrosis (E,F). Arteriosclerosis is distinct from atherosclerosis.

shows that they may produce C-reactive protein — a possible linkage with atherogenesis (Calabrò et al., 2003).

The relationship of this *physiologic intimal thickening* to coronary pathology will be discussed later (see Chapter on coronary atherosclerosis). Here, we indicate that this phenomenon, found in all adults, is a normal response ("postnatal vasogenesis" [Vlodaver and Neufeld, 1967]) likely linked to the particular flow dynamics of the extramural coronary arterial system. In other arteries, for example, the carotid artery wall thickening

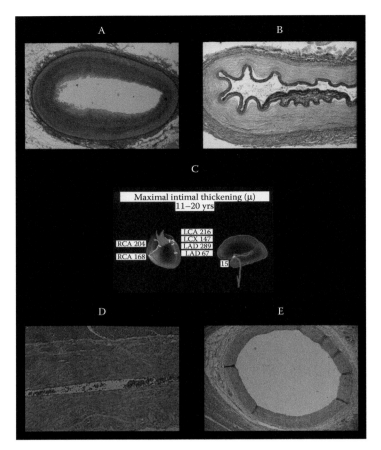

Figure 4   Comparison between the intimal thickening of the left anterior descending (LAD) artery (A) and the middle cerebral artery (B) of the same 18-year-old subject. In the latter artery, the intimal thickening is minimal in contrast to that of LAD, which is circumferential with a thickness greater than the tunica media. (C) Difference in maximal thickness in microns found in main coronary arteries and branches in respect to the middle cerebral artery. (D) Absence of intimal thickening in the LAD of dogs, despite an identical morpho-function. This suggests a possible role of the neurogenic control of coronary arteries in humans. On the other hand, the absence of intimal thickening in the mural tract of coronary arterial vessels (E) emphasizes the role of systolic dynamic stresses on the arterial wall free to expand versus those protected by encircling contracted myocardium.

regresses following local pulse pressure reduction by antihypertensive therapy (Bontonyric et al., 2000). In contrast, a physiologic intimal thickening is minimal in other muscular arteries, such as the middle cerebral artery, or is absent in extramural coronary arteries incorporated into the myocardium (Figure 4) and in intramural arteries where the systolic expansion of their wall is counteracted by the contracted myobridges or myocardium, respectively. At a critical point of these stresses, modification of the wall structure is expected (Langille, 1991).

Flow dynamics is an important component in the genesis of human coronary intimal thickening. However, its absence in the coronary arteries of dogs with a similar anatomic disposition (Figure 4) may indicate a more intense neurogenic–adrenergic control on human arterial wall tone. This may play a prominent role (Baroldi and Silver, 1995) and could explain divergencies among subjects, different ethnic groups (Vlodaver et al., 1969), and species with identical coronary arterial anatomies.

Intramural Coronary Arteries

The *intramural* or *intramyocardial coronary arterial system* is formed by ventricular, septal, and atrial secondary branches originating from subepicardial arterial vessels.

In the left ventricle, they arise at a right angle to the extramural arteries and assume a perpendicular course within the cardiac wall. They may cross the whole wall or branch at different levels as straight or "branching" types (Farrer-Brown, 1968). Most terminate in the subendocardial region as ramifications parallel to the subendocardial layer. Larger and longer straight branches supply the papillary muscles, assuming various patterns of distribution according to papillary muscle anatomy (Ranganathan and Burch, 1969). The source of individual papillary muscle blood supply depends upon the type of coronary artery distribution.

In the right ventricle, intramural branches ramify in a more oblique and transverse direction. The pattern is similar to that seen in atrial vascularization and relates to the thinness of the cardiac wall in these regions. Two main *atrial branches* should be mentioned: the branch to the *sinus node* and that to the *atrioventricular node*, both arising from the RCA. In contrast, intramural arteries within the interventricular septum show a peculiar disposition with parallel antero-posterior branches arising from the extramural anterior and posterior descending branches. They have an oblique or horizontal or vertical direction from the proximal to the distal portion of the septum (Figure 2). These branches are more numerous in the left than in the right part of the latter (Farrer-Brown and Rowles, 1969). Furthermore, the distal-apical portion of the interventricular septum is vascularized by branches originating from the reascending tract of the left anterior descending branch.

The arterial and arteriolar vessels have the same histological structure seen in comparable vessels of other organs, namely, a well-defined pluristratified tunica media circumscribed by outer and inner elastic laminae, a thin endothelial intima, and a loose adventitia.

Small blood vessels may be found in the basal third of both atrioventricular and semilunar valves, their number increasing if a valve thickens pathologically. They also develop in thickened chordae tendineae (Lautsch, 1971).

Terminal Vascular Bed

The increasing branching and diminishing caliber of arterial vessels implies a progressive loss of wall structures from the complexity previously described for extramural arteries to the simplicity of capillaries reduced to a thin endothelial layer. In this process, several types of vascular structures can be recognized:

1. The *smallest* or *terminal arterioles* that are formed by a monolayered media and an endothelium
2. The *metarterioles* with monolayered smooth muscle cell sphincters alternating with "naked" endothelial tracts
3. The *precapillaries* that have a capillary-like structure with a single myosphincter at their origin (Provenza and Scherlis, 1959)
4. The "*true*" *capillaries* that are simple endothelial tubes about 8 μm in diameter (Figure 5)

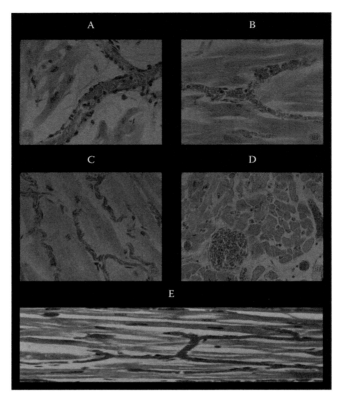

Figure 5   Intramural arterial system: (A) arteriole with monolayered tunica media; (B) meta-arteriole with alternated myosphincters; and capillary network (C) longitudinal and (D) transverse sections with (E) anastomotic connections.

Heart capillaries show marked histologic differences from those in skeletal muscles. They are more numerous, have a greater amount of enzyme-bearing mitochondria and nerve axons near or contacting their wall, and have more protoplasmic villi projecting into the lumen. According to Kisch (1963), cardiac capillaries may be formed by a spiral sequence of two or more endothelial cells (*metacapillaries*) or one embracing only the total lumen (*protocapillaries*).

In general, the capillaries run parallel to the myocardial cells with frequent branching in any direction. It is claimed that, on average, each myocardial cell is furnished by a capillary (Wearn, 1928) or is surrounded by four capillaries (Ludwig, 1971). The capillary density in man and guinea pig is 2000 capillaries/mm^2, while it is 3500/mm^2 in hares, and 3700/mm^2 in bats (Rakusan, 1971); capillary density, being dependent upon physical activity, is higher in more active species and wild versus domesticated animals of the same species (Poupa et al., 1970; Poupa and Carlsten, 1973). Whether regional variations of capillary density occur in different cardiac areas and in different cardiac wall layers is still controversial (see below). The anastomotic vessels, to be described next, can be included in the terminal vascular bed.

Coronary Arterial Anastomotic or Collateral System

The terms coronary artery *anastomoses* or *collaterals* indicate vessels that join different coronary arteries and their branches. Many authors (Banchi, 1904; Blumgart et al., 1940;

James, 1961; Spalteholz, 1924; Spateholz and Hockrein, 1931; for reviews of the literature, see Baroldi and Scomazzoni, 1967; Baroldi and Silver, 1995) studied them using different methods of investigation, often with contradictory or inconclusive results. Indeed, their existence has been questioned, the persistent assumption being that they do not exist or are poorly developed in man, with the coronary arteries being physiologically terminal arteries (Helfant et al., 1970). Whether or not this point is true is fundamental in the functional interpretation of a coronary arterial obstruction.

In our experience, using tridimensional plastic casts of the aorta and coronary system of normal human hearts, innumerable anastomoses were demonstrated at many levels along the courses of all intramyocardial branches (Figure 6). In contrast, only occasional collaterals were found to join extramural arteries on the heart surface, a finding that is normal in canine hearts (Baroldi and Scomazzoni, 1967). Intramyocardial collaterals can be distinguished as *homocoronary*, which connect branches of the same coronary artery, and *intercoronary*, which join branches of different coronary arteries. Their distinction is related to the anatomical distribution of the main vessels. For instance, septal collaterals may be homo- or intercoronary according to the origin of the posterior descending branch. The diameter of these collaterals, already present at birth and measured on plastic casts, ranged from less than 20 to 350 μm. In general, they ran parallel to myocardial bundles. In the subendocardium and atria, they had a mesh-like appearance.

Collaterals frequently present a corkscrew aspect, likely due to their adaptation to myocardial contraction (Figure 6). Due to their extremely large number, it is practically impossible to quantify their frequency. For comparative purposes, we adopted an *anastomotic index* (maximal collateral diameter + average collateral diameter × collateral frequency/100) that in normal hearts ranged from 3.4 to 6.2 with a mean value of 4.7. These normal collaterals have a histological structure identical to capillaries. Because the latter form an intercommunicating reticulum, we concluded that the whole intramyocardial arterial and capillary system is a widespread network, and that coronary arteries, at least from the anatomical standpoint, are not end arteries.

Other types of arterial collaterals exist:

1. *Extracardiac* arterials between atrial branches and the aortic and pulmonary vasa-vasorum (in turn connecting with pericardial, mediastinal, and diaphragmatic arterial vessels), particularly with bronchial arteries (Moberg, 1968)
2. *Arterioluminal vessels* that join the intramural arterial system with the ventricular heart cavities, mainly the left. The latter were described by Wearn et al. (1933) and distinguished into *arterioluminal vessels* directly opening within the intraventricular heart cavities, and *arteriosinusoidal vessels* formed by an arteriole that transforms into a 50 to 250 μm diameter sinusoid that communicates with the ventricular cavity. In our experience, arterioluminals had a frequency of less than 15 vessels in 81% of hearts and more than 15 vessels (maximum 20) in 5% of hearts, being totally absent in the remaining 14% of hearts.

Veins

Coronary veins originate from the capillary network draining into venules and then, in turn, converge into larger venous channels. Like coronary arteries, the venous system can be divided into *extramural* and *intramural veins* (Figure 7).

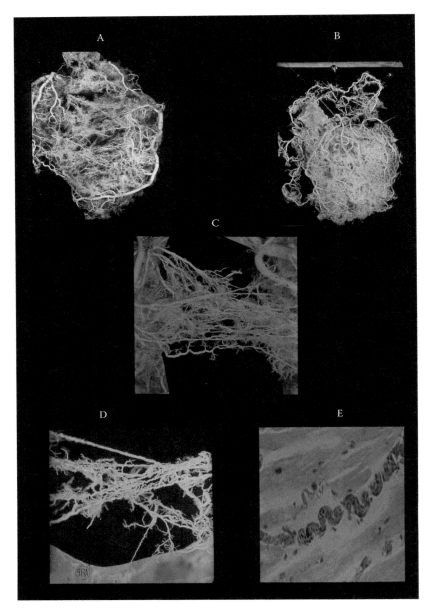

Figure 6 Coronary anastomoses or collaterals: (A) intercoronary ventricular, (B) atrial, and (C) homocoronary anastomoses. Note the innumerable collaterals joining different branches at any level of their course. They frequently have a corkscrew aspect (D) visible also histologically (E), as an adaptation to the cardiac contraction–relaxation cycle.

The *venous sinus system* includes all veins that drain into the coronary sinus and the *anterior cardiac venous system* formed by the right veins that drain directly into the right atrium or into the small cardiac vein (Baroldi and Scomazzoni, 1967).

The *coronary sinus* is a short, large diameter vein that runs in the posterior left atrioventricular groove from the left cardiac margin to its ostium in the right atrium distal and medial to the orifice of the superior vena cava. Mechanik (1934) described a ring of atrial myocardial cells around the coronary sinus capable of preventing a systolic backflow. The

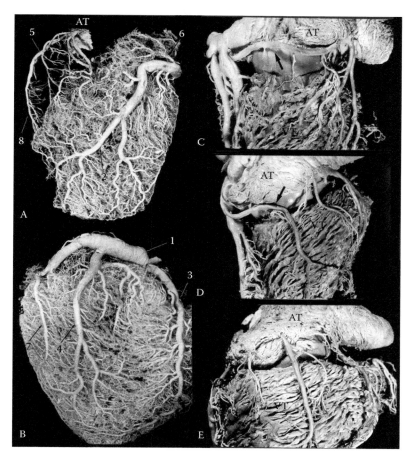

Figure 7   Cardiac veins: (A) anterior and (B) posterior view: 1: coronary sinus, 2: great cardiac vein, 3: middle vein, 4: left posterior veins, 5: atrial vein, 6: anterior cardiac veins, 7: left marginal vein, 8: right marginal vein. Different aspect of the small vein (arrow) draining (C) in the coronary sinus and right atrium (AT); or only coronary sinus (D) or atrium, the small vein being rudimentary.

incomplete valve of Thebesius limits its atrial ostium, while the valve of Vieussens separates it from the *great cardiac vein* that runs in the anterior interventricular groove and drains right and left venous branches, among which the main vessels are the *left marginal vein*, the *middle vein* running in the posterior interventricular groove, and the *posterior left ventricle veins*. The characteristics of these veins are given in Table 3 in which the *right anterior cardiac veins*, including the *right marginal vein* and the *small vein*, when it exists, are also reported. The latter is a rudimentary vessel that may or may not drain into the coronary sinus and may or may not collect the right anterior veins. Finally, *atrial* and *septal veins* and *venous-luminal* or *thebesian* vessels that open directly into the cardiac cavities exist. It must be stressed that innumerable venous anastomoses are present everywhere and are divided into *intersystemic, homosystemic,* and *homovessel venous anastomoses* (Mechanik, 1934).

A juxtaposed course of coronary arteries and veins is limited to only a few vessels. For example, the coronary sinus and left circumflex branch, the vena magna and left anterior descending branch, and the middle vein and posterior descending branch have more or

Table 3 Normal Characteristics of Extramural Cardiac Veins

Vessel	Present (%)	Average (Ø mm)	Average Number of Branches			
			LV Anterior	LV Posterior	RV Anterior	RV Posterior
Ventricles						
Coronary sinus	100	9 (4–15)	—	2.5 (1–4)	—	1 (in 60%)
Great vein	100	6 (4–9)	9 (2–12)	—	4 (0–9)	—
Left marginal	95	1 (1–2)	—	4 (0–12)	—	—
Middle vein	100	5 (3–7)	—	6 (2–12)	—	4 (1–8)
Left posterior	76	5 (3–5)	—	—	—	—
Small vein	67	2 (1–2)	—	—	1 (0–3)	—
Anterior veins	100	1	—	—	—	—
Right marginal	100	1	—	—	—	—
Atria						
Left oblique	94	1				
Anterior/posterior	99	1				
Interventricular septum						
Anterior	100	1	10 (4–22)	—	—	—
Posterior	100	1	—	11 (0–24)	—	—

Note: Only vessels with a diameter (Ø) equal to or greater than 1 mm were considered.

less parallel courses. Other venous vessels are not closely associated with arteries in their courses. True venous rings around an arterial vessel are occasionally seen. The claim that the venous vasculature is twice that of the arterial one (Truex and Angulo, 1952) is only partially true. According to our plastic cast study, this claim seems to be correct for superficial vessels only. On the heart's surface, veins apparently prevail both in number and diameter, likely due to the different types of flow in the two systems: the arteries supplying the myocardium penetrate into and extensively ramify within it, while the venous drainage at a low pressure converges toward the surface.

Arteriovenous Anastomoses

The existence of *arteriovenous anastomoses* is still controversial. Postmortem injection of calibrated glass spheres into the coronary arteries produced spheres of 70 to 170 μm and 80 to 360 μm in diameter in the coronary sinuses of both normal hearts and hearts with obstructive coronary disease (Prinzmetal et al., 1947; Aho, 1950). Many intracoronary injection studies using different substances including plastic material were unable to confirm the presence of such anastomoses. In our plastic cast study of more than 500 normal and diseased hearts, vessels 20 μm in diameter were injected. Thus, it is difficult to understand why we could not demonstrate these vessels with their reported large diameter. On this subject, mention must be made of the "hybrid vessels" described by Bucher (1947) and occasionally seen in the papillary muscles and columnae carneae of normal hearts. They show a hyperplastic media with longitudinal bundles of smooth muscle cells and a progressive increase of substitutive fibrous tissue. Postmortem arterial injection by radiopaque material confirmed their arterial nature. Their morphology is likely an adaptation to their location. This is a finding to be considered in endomyocardial biopsies (Figure 8; Baroldi, 1991a).

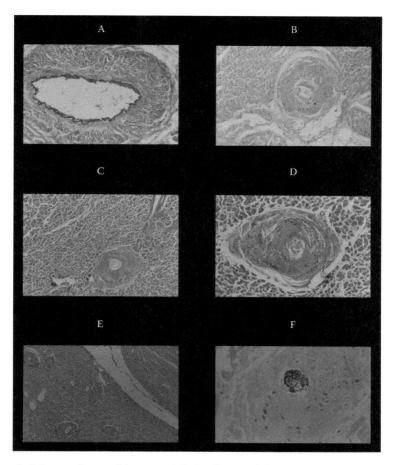

Figure 8 Medial hyperplasia obliterans ("hybrid vessel"). Localized within the trabeculae carneae, papillary muscles, and rarely in the interventricular septum, they are characterized by hyperplasia of myocells of the tunica media, forming a longitudinal bundle without early intimal thickening (A). This process progresses with fibrous transformation and lumen reduction (B through D) also visible in normal hearts of subjects who succumbed to accidental death (E). The surrounding myocardium is normal. The arterial nature of these vessels is demonstrated by selective postmortem injection of radiopaque material filling the lumen (F).

Physiological Structural Vascular Adaptation

In the absence of disease, coronary arteries and veins undergo structural changes that are an adaptation to an increased or reduced myocardial mass (Figure 9). With cardiac hypertrophy, extramural vessels show an increased length and diameter proportional to the tridimensional increase of cardiac mass, this occurring without evidence of new extramural vessel formation. Similarly, an increase in the length and diameter of intramural vessels, including collaterals, occurs. These changes are particularly evident in cor pulmonale, where the right ventricle may be as thick as the left one with a dramatic increase in vessel length and diameter. This vascular adaptation implies a hyperplasia of all cellular components of the arterial wall. In contrast, in cardiac atrophy, extramural arteries and branches do not change their caliber but become markedly tortuous as an adaptation to a reduced

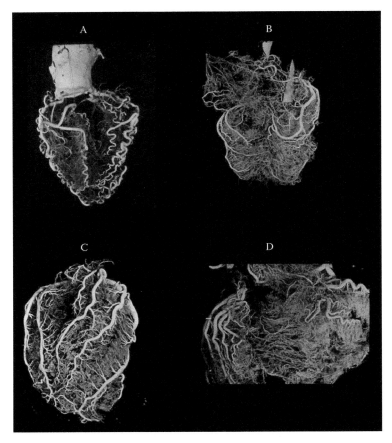

Figure 9 Vessel changes in relation to modification of the cardiac mass. (A) Atrophic heart with acquired serpentoid form of extramural vessels due to cardiac mass reduction and minor intramural vascularity. The contrary is seen in cardiac hypertrophy (B), in which the extramural arteries increase in length and diameter (but not in number) to adapt to the greater myocardial mass. Similarly, the same enlargement is seen in the intramural branches. Cor pulmonale, a condition in which the right ventricle may become greater than the left, is an extreme example of adaptation of the extramural (C) and intramural arteries, including collaterals (D). No histologic evidence of new vessel formation exists. The cardiac veins show similar behavior.

cardiac mass. In these circumstances, the intramyocardial vascularity is apparently diminished, and collaterals have smaller diameters. The relationship between vascular rearrangement and the metabolic needs of the myocardial mass will be discussed subsequently (see Chapter 5).

Functional Anatomy of Coronary Vessels

Normal cardiac metabolism for a rapid anabolic–catabolic cycle is maintained by blood flow. Without a vascular system, Rushmer (1963, p. 2) calculated that "by random movement a molecule of water could theoretically pass from a man's head to his toe unassisted by circulation or flow currents, but this would require more than 100 years." The arteries deliver oxygen and metabolites, and the veins remove metabolic waste products, with exchange taking place mainly in the capillary bed and in any area covered by an endothelial lamina. This is the reason why the whole collateral system formed by capillary-like structures must be included in the terminal bed. This means that any geometric one-to-one or one-to-four myocardial cell/capillary relationship has little meaning because, in reality, each cell is surrounded by an interstitium bounded by a myriad of capillary vessels. Therefore, any calculation that considers the diameter of the myocyte and its distance from a capillary to evaluate the degree of exchange is unreliable.

The nutrient flow is modulated by myocardial demand through a complex and not completely understood neurohormonal — including endothelial derived factors — regulatory or autoregulatory mechanisms that act on vessels equipped with a muscular coat. The demonstration of contractile proteins in endothelial cells and the presence of pericytes around capillaries also raise the question of a possible capillary regulation of flow. However, the endothelial contractile proteins likely control permeability by adjusting the openings at interendothelial cell junctions (Becker and Murphy, 1969). Indeed, the control of endothelial permeability by nerves, hormones, and drugs (Rodbard, 1965) is the other determining factor in metabolic exchange.

In reality, transcapillary exchange needs various, often interdependent factors to determine (1) the effective capillary or filtration pressure based on the difference between capillary and tissue pressure; (2) the effective plasma osmotic pressure, that is, the difference between plasma and tissue osmotic pressure; and (3) the capillary diffusion capacity already discussed above. Such diffusion, which depends on the grade of permeability to the solute, the surface area available for its transfer, and blood flow velocity, varies in different tissues being, for instance, 8 to 10 times greater in the myocardium than in skeletal muscle (Renkin, 1967). Capillary and precapillary structural variations, in terms of the thickening of the basement membrane, fenestrations or pores, degree of pinocytosis, status of perivascular spaces, and density of pericytes, in different organs and tissues suggested a morphologic classification (Bennet et al., 1959) that could explain diffusion variability in different territories (Rushmer, 1963).

The rhythmic coronary flow related to the myocardial contraction–relaxation cycle was mentioned briefly. More precisely, one must consider the homogeneity of flow distribution in different cardiac areas in relation to contraction and vascular disposition. Major divergencies exist between atrial/right ventricle and left ventricle and, in the latter, between subepicardial and subendocardial myocardium. Long ago, it was assumed that during systole a transmural pressure gradient existed, decreasing from deeper to superficial layers of the left cardiac wall, and that this resulted in a complete systolic occlusion of the deeper coronary vessels (Johnson and Di Palma, 1939). Some authors have shown poor perfusion of the subendocardium, explaining its propensity to ischemia (Honig et al., 1967), while others demonstrated the opposite (Fortuin et al., 1968), with a transmural flow homogeneity maintained by vasomotor tone control, by which the systolic flow reduction is counterbalanced by an autoregulatory diastolic dilatation (Moir, 1972). Furthermore, Gregg (1968) showed that with a low coronary perfusion pressure, subendocardial Rb[86] uptake can fall to 39 to 85% of the subepicardial uptake without evidence of left ventricular failure or electrographic signs of ischemia.

Unfortunately, all physiologic studies are done in animals, and the findings in man — even with the more sophisticated up-to-date techniques available today — are more speculative than objective, lacking a control in the normal population. Here it is important to recall the list of parameters involved in coronary flow and myocardial metabolism, such as "structure, function, tissue pressure, nutrient blood flow, capillary exchange, oxygen consumption, pO_2, aerobic and anaerobic metabolism, various substances such as myo-globin, glycogen, phosphorilase, nucleotides (in smooth muscle), various enzymes, capil-lary density and patency, intercapillary distance, smooth muscle susceptibility and reactivity in precapillary sphincters and other arterioles to effective stimuli such as potas-sium, calcium, pO_2, catecholamines and possibly others" mentioned by Gregg (1950, 1968). This is a long list that is still valid, even if others were added by more recent contributions (see below). What is important to stress is the role of an extravascular component of the coronary resistance as shown by an increased coronary blood flow immediately upon vagal cardiac arrest (Sabiston and Gregg, 1957). Furthermore, with heavy exercise, the rate of flow in late systole can be very high and therefore not explicable by a radial expansion of subepicardial arterial vessels. It is likely that much of this systolic flow penetrates the myocardium to an unknown depth (Gregg, 1950), a finding that emphasizes the complexity of the arterial coronary flow in relation to cardiac metabolism. Using high-speed cinema-tography and transillumination in the atria and superficial layer of ventricular myocardium from turtles and dogs, Tillmanns et al. (1974) demonstrated a diameter reduction of all vessels in systole with a reduction in red blood cell velocity in arterioles in contrast to an increase in capillaries and venules; a reverse pattern was seen in diastole. This important experiment showed that the shift in flow pattern occurs at the transition from arterioles to capillaries — a fact not yet confirmed in man.

A last point is the sensitivity of vascular structures to hemodynamic changes, partic-ularly in the coronary system with its complex functional anatomy (Rodbard, 1956, 1971). Already mentioned is the intimal thickening noted in these vessels, related, at least in part, to the different types of wall stresses. Other changes occur in the media by hyperplasia and hypertrophy of smooth muscle cells. Little do we know about nerves and lymphatics. However, an impressive example of structural plasticity to dynamics is given by the changed shape and orientation of endothelial cells according to flow direction. When the latter is changed experimentally, nuclear reorientation occurs (Flaherty et al., 1972; Langille, 1991).

Myocardium

The heart is a muscular pump formed by right- and left-sided cylindrical muscles that are internally and externally wrapped by another two helicoidal muscles anchored to a fibrous framework that surrounds valves and large vessels. The anatomic disposition of these muscles favors a maximal ejection by concentric shortening associated with helicoidal twisting. The old belief that the myocardium was an anatomical syncytium was revised following the ultrastructural demonstration of cylindric, 50 to 100 μm long and 10 to 20 μm wide elements separated by, and connected with, intercalated discs. The myocardial cells or myocytes have one central nucleus, rarely two or more nuclei, and a contractile sarcoplasmic structure, formed by myofibrils.

The myocardial cell (or myocyte or myocell; Figure 10) is covered by a *sarcolemma* that has an extracellular or *basement membrane* connected with the interstitium and vessels by a *collagen matrix* or fibrillar collagen network (Weber 1989), and a *surface structure* that has a typical three-layered unit membrane formed by a bimolecular hydrophobic lipid median layer and two external protein plus hydrophilic lipid layers. These lipoprotein structures are the equivalent of metal particles in a magnetic field. At rest, their negative charge is internal and the positive one external to the cell ("polarization"). Removal of the

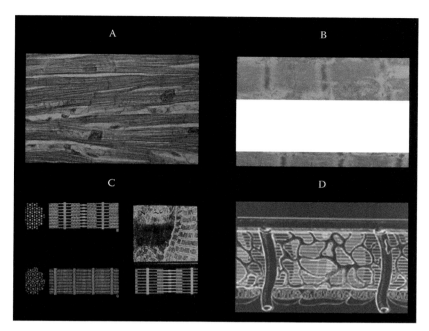

Figure 10 Functional morphology of the myocardium. (A) A histological view of normal myocardium in longitudinal section. Each myocell is formed by many myofibrils subdivided in sarcomeres (B) in registered order, resulting in physiological bands (C) that change according to functional cycle by the sliding of thin, actin filaments on the thick, myosin filaments. The sarcomere is the contractile functional unit included within two Z lines, with myofibrils enmeshed in the network of the sarcoplasmic reticulum (storage of Ca++) and between mitochondria (D). In (C), a sarcomere is shown in distention (a) with evident I bands that disappear in contraction (b). An ultrastructural imaging of a normal myocell adjacent to a myocell with a pathological contraction band is also shown.

field by a shift in the intracellular potential toward zero depolarizes the molecular structures that now orient in a random fashion. Depolarization occurs in systole when the myocardial cell is electrically, biochemically, mechanically, or automatically excited with changes in membrane permeability resulting in a passive egress of K+ and entry of Na+ in relation to their intra- and extracellular concentration gradients. The contrary happens in diastolic repolarization, in which, however, ionic transport is not passive but metabolically active, K+ and Na+ being pumped intra- and extracellular, respectively (Hecht, 1968). Furthermore, vesicles crossing the sarcolemma are an expression of pinocytosis, namely, the transport of all substances needed for myocardial metabolism and function. The intercalated disc is the end-to-end limit between two sequential cells, has a serpiginous course, and is formed by the two surface membranes of joined myocardial cells separated by an intercellular space. Electrical impendance across intercalated discs is low, permitting a rapid electrical excitation from cell to cell. The myocardium, if not an anatomic, is a functional syncytium. Another main myocellular component is the sarcoplasmic reticulum formed by a transverse or T system resulting from invagination of the surface membrane (Spiro et al., 1968) or of the entire sarcolemma (Leyton and Sonnenblick, 1969) at the level of the Z lines (intermediary vesicles), and a longitudinal system consisting of a network of channels disposed between the Z lines and intertwining between myofibrils. The triad complex, that is, one T vesicle and two longitudinal tubules, is visible at the Z-line level.

The role of the sarcoplasmic reticulum is to accumulate Ca^{++} ions on which the contraction–relaxation cycle depends. *Mitochondria* are extremely numerous, almost cylindrical bodies, forming from 30 to 50% of the cell volume. This richness is related to the continuous contractile function of the myocardium because the mitochondria transform the energy contained in carbohydrates, lipids, and amino acids to adenosine triphosphate (ATP) through oxidative phosphorylation. The close contiguity of mitochondria with myofibrils favors a rapid transfer of the high-energy compounds used for contraction.

The functioning machinery of the myocardial cell is the *contractile apparatus* that consists of bundles of longitudinal myofibrils. Each myofibril is subdivided into 20 to 50 subunits, the *sarcomeres*, which are the true functional units. They are limited by two Z lines and formed by 1 µm long *thin* or *actin* filaments, attached to the Z lines and separated centrally, and *thick* or *myosin* (L-meromyosin) *filaments* not affixed to the Z lines. They are 1.5 µm long and are arranged in the central part of the sarcomeres between and parallel to the actin filaments. Both actin and myosin filaments have corresponding interdigitations (*tropo-myosin-troponin* and *H-meromyosin digitations*, respectively) that are the active sites of the biochemical hinge that regulates the contraction–relaxation cycle. The latter is achieved by a back-and-forth movement of actin filaments. They penetrate the other half of the sarcomere by sliding on the thick filaments (sliding theory of contraction). The tropo-myosin-troponin complex inhibits contraction (diastole), which is promoted by Ca^{++} binding to troponin (systole). This needs a rhythmic to-and-fro pumping of Ca^{++} from the sarcoplasmic reticulum to myofibrils and vice versa. All myofibrils are in registered order, thus giving the myocardial cell its regular, characteristic cross-striations or *physiological bands*, which vary in relation to the contractile cycle. In relaxation, two clear I bands of actin filaments are visible at each site of a Z line. Internal to the I bands are two more dense S bands formed by both actin and myosin filaments. In cross-section, one thick filament can be seen to be surrounded by six thin filaments in hexagonal order. In the central part of the sarcomere, a band of thick filaments only forms the H-L-M complex (Figure 10). This complex and the S bands constitute the A band. In contraction, I and H bands disappear, with the actin filaments of one site penetrating totally in the other site of the sarcomere. In cross section, the number of actin filaments can be seen to double. The length of a sarcomere normally oscillates from a maximum of 2.4 µm in relaxation to a minimum of 1.5 µm in contraction, with a range from 1.86 to 1.95 µm in systole and from 2.05 to 2.15 µm in diastole (Spiro et al., 1968).

Myocardial Interstitium

The space between myocardial cells, the myocardial interstitium, contains structures such as vessels, including lymphatics, nerves, collagen matrix, proteoglycan ground substance, fibroblasts, monocytes, and possible fat tissue. In this complex compartment, myofibrillar bridges connecting adjacent myocardial cells, which convert the myocardium into a cellular network, have to be considered.

In general, the interstitium has many functions according to the diversified structures it contains: anabolic–catabolic turnover of myocardial cells by blood flow; drainage of interstitial fluid by lymphatics; nerve control of cardiac function; defense against noxious agents; structural support to the contracting myocardium; and proteoglycan lubricant production helpful for contractile function.

Fibrillar Collagen Network or Matrix

The collagen network (Weber, 1989) formed by type I (or thick) and type II (or thin) fibrils produced by fibroblasts connects cardiac valves and chordae tendineae with the subendothelial *epimysium* of both epicardium and endocardium, which in turn arborizes by tendon-like fibrils in a collagen weave or *perimysium*. The latter aggregates bundles of myocardial cells, while the *endomysium* joins by strands adjacent cells and by struts adjacent myofibrils within myocardial cells. The endomysial fibrils also connect the latter with related vascular elements and insert into its basal lamina at Z-line level with the intramyocellular cytoskeleton.

This fibrillar collagen network that forms the "skeleton" of the heart supports myocardial cell alignment, and by its tensile strength and elastic pliancy helps maintain myocardial shape and thickness as well as its passive and active stiffness and prevents deformation or rupture of myocardial cells. Acquired or congenital alterations of this structure can result in cardiac dysfunction by architectonic changes of the fibrillar network due to impaired collagen synthesis, giving rise to "interstitial heart disease" (Weber, 1989).

Cardiac Lymphatic Vessels

In the human heart, three interrelated plexuses of lymphatics exist at the subendocardial, intramural, and subepicardial levels. From these arise the *anterior middle lymphatic* vessel that runs parallel and superficial to the LAD, the *right anterior ventricular* lymphatic vessel that follows the atrioventricular sulcus, and the *posterior ventricular* lymphatic vessel that parallels the posterior descending arterial coronary branch. They all join at the pulmonary trunk and form a common lymphatic trunk that drains into a lymphatic mediastinal plexus through mediastinal lymph node transfer with a terminal discharge into the thoracic duct. The function of this system is to absorb and transfer to the blood from the interstitium proteins, water, and electrolytes; control tissue volume and pressure; remove debris following tissue injury; and limit infection (Miller, 1982).

Cardiac Nerves

The autonomic nervous system supplies the heart with parasympathetic and sympathetic fibers from the vagus and phrenic, cervical and thoracic afferent and efferent fibers and related ganglia. These form superficial (on the aorta) and deep (between aorta and trachea) cardiac plexuses that extend to right and left coronary artery nervous plexuses (Armour, 1999). Nerve ganglia are frequently seen histologically in the subendocardial atrial myocardium and around the main extramural coronary arteries. The sinus node and atrioventricular node have both adrenergic and vagal innervation. Ultrastructurally, unmyelinated nerve fibers are seen in cardiac muscle close to blood vessels, being distinguished into autonomic, both adrenergic and cholinergic fibers, and sensory fibers. In cholinergic fibers, varicosities contain agranular vesicles and a few mitochondria, while in adrenergic fibers the vesicles are granular. Sensory fibers have a larger diameter (1.5 to 3 μm) with many mitochondria, few granular vesicles, glycogen rosettes, and lamellar bodies. They are located around vessels and surrounded by Schwann cells. According to the distance between the vesicle-containing nerve terminal and myocardial cell, three types of neuromuscular contiguity are recognized: (1) with a gap of 100 μm or more, (2) 50 μm, and (3) 20 μm or less. In the last type, the basement membrane of the axon is continuous with that of a myocardial cell. These three types are not synaptic structures. The neurotransmitter is released into the interstitial space, with varying degrees in proximity to the sarcolemma (Ferrans and Rodriguez, 1991).

The autonomic cardiac nerves regulate the rhythm of cardiac contraction by influencing the anatomical pacemaker, the sympathetic fibers accelerating and the parasympathetic fibers slowing the rhythm. Furthermore, intramyocellular interstitial nervous fibers may act on the resistance vessels (vasoconstriction) or on myocardial contractility through acetylcholine, catecholamines, and neuropeptide Y (Marron et al., 1995).

Heart Examination

3

GIORGIO BAROLDI AND VITTORIO FINESCHI

Contents

Gross Examination

Among the many methods used to examine the heart anatomically (Reiner, 1968; Silver and Freedom, 1991), we need to adopt one that in performance is both compatible with the time available for a routine autopsy and provides sufficient information for a correct diagnosis based on a solid morphofunctional background.

The usual method is Virchow's method modified by Prausnitz. Dissection follows blood flow direction from the venae cavae, cutting the lateral margin of the right ventricle, its pulmonary cone, and its pulmonary artery. Then, from the left atrial veins and in sequence the left atrium, left ventricular margin, left outflow tract, and the aorta are opened. The past bad habit of examining coronary arteries either by stripping them from the heart surface or by opening them longitudinally using scissors — thereby, respectively, losing any possibility of controlling the structural relationship between coronary wall and subepicardium and of measuring lumen and plaque variables because of traumatic plaque breakage — was replaced by a cross-section of these vessels at intervals of 3 mm along their course.

More sophisticated and time-consuming methods of study can be used for coronary artery changes in relation to other variables such as postmortem coronary injection of plaster (Banchi, 1904), radiopaque, eventually colored, material in the unopened (Gross, 1921), cleared (Spalteholz, 1924), or opened (enrolled) heart (Schlesinger, 1938) or plastic substance plus corrosion of cardiac tissue (James, 1961) and tissue sampling before corrosion (Baroldi and Scomazzoni, 1967c). Furthermore, by comparing clinical imaging, for example, echocardiographic imaging, different plane dissections can be adopted. Each method has its advantages and disadvantages, and not one may evaluate all variables we need to know. For the best quantitation of coronary artery and myocardial variables, we found

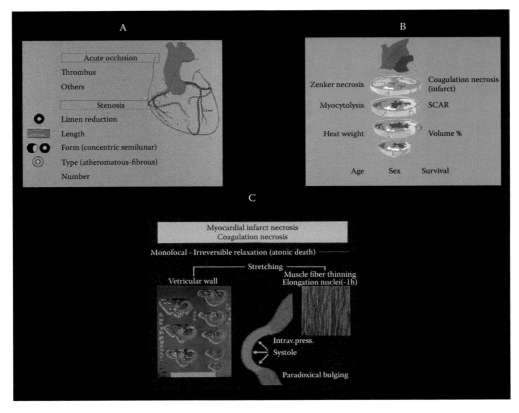

Figure 11 Examination of the heart: (A) cross-sectioning of extramural coronary arteries and (B) slicing of the heart. (C) An example is as follows: a case of an acute myocardial infarct of the left ventricular posterior-lateral wall and posterior interventricular septum. The necrotic myocardial cells in irreversible relaxation are hyperdistended by the intraventricular pressure.

that the type of examination adopted for the study of large populations (Baroldi et al., 1974b) permits a valid definition of most morphofunctional parameters and correlation with the clinical records; particularly today, because heart disease is the major cause of death in Western countries and is responsible for many forensic controversies. The following method is achievable even in a busy service and guarantees information for a correct diagnosis and research (Figure 11).

The heart, removed from the body by careful sectioning of all connecting arterial and venous vessels, is washed to free all cavities of blood, is weighed, and its surface is inspected. Then, the whole intact heart is suspended for 24 h in a large container filled with 10% buffered formalin. The mild fixation hardens the tissue and facilitates precise cutting of the organ. First, coronary arteries and their main extramural branches are cross-sectioned at 3 mm intervals along their whole course. Each 3 mm segment is removed together with the surrounding epimyocardial tissue to preserve any anatomical relationship. Before sectioning, when needed, the vessels are decalcified. Subsequently, the whole heart is cross-sectioned by a machine from apex to base in 1 cm thick slices, parallel to the atrioventricular groove. The last basal slice includes the tip of the left papillary muscle. All heart slices and coronary segments are placed according to their anatomical disposition and are color photographed with a metric scale (Figure 11). This permits calculation of gross data, such as cavity volume and thickness of the cardiac walls, so as to planimetrically measure any gross pathological changes in relation to the whole cardiac area. Careful inspection of the atria, cardiac valves, and any other anatomical structure can be done easily. When a heart has been surgically excised at transplantation, it is missing its atria. For comparative purposes, its weight can be adjusted by adding the theoretical atrial weight using the following formula: actual heart weight × 100/75 (Reiner, 1968).

Tissue Sampling

A systematic sampling for histological and immunohistochemical examination is needed:

1. The first segment of coronary arteries and main extramural branches and any other segment with pathological changes are processed for histology, any heavily calcified coronary tract being previously decalcified; up to four slices of coronary arteries marked for orientation can be included in a histologic block. For a correct evaluation of adventitial and periadventitial structures, the coronary samples must contain epicardium and underlying myocardium.
2. A central heart slice is sampled at the anterior, lateral, and posterior walls of the left and right ventricle, and at the anterior and posterior interventricular septum — as well as any pathological area seen grossly — by 2 cm wide sections of the whole wall thickness, including epicardium and endocardium. When a more complete examination is required, two heart slices, one superior and one inferior, are similarly processed. All samples are first placed in 10% buffered formalin to complete fixation. The uncut atriovalvular part and remaining tissue are preserved in a closed plastic bag containing a small amount of formalin, for any further quantitative study of the sinus node, conduction system, atria, valves, and so forth.

Quantification of Lesions

In general, postmortem diagnosis and related cause of death are based on a qualitative definition of lesions rather than on their extent. However, only the latter helps give exact morphofunctional meaning. For instance, an infarct involving 10% of the total myocardial mass may not, per se, be the cause of a cardiac arrest. Rather, it may be linked to other concomitant factors (see below). In other words, only correct morphologic quantification can interpret dysfunction.

Many methods have been employed to quantify lesions, but they are rarely used in routine diagnosis or in most research projects. Pathological conclusions are limited to the quality of a lesion, which is often imprecisely determined. We describe the methods of quantification of the main myocardial and coronary artery variables we use (Baroldi and Silver, 1995, 2004; Baroldi et al., 2003; Fineschi et al., 2004):

1. Any gross lesion, for example, an acute infarct or a scar, is measured by a planimeter or by a video system on the photographed heart slices, and its size is referred to in percent of the total cardiac area. On the same photographs, the thickness of cardiac walls and the volume of cardiac cavities are estimated. All quantified lesions are examined histologically.

2. The length of a coronary atherosclerotic plaque is measured over 3 mm coronary segments, while lumen reduction is calculated in each histologic section by assessing the average diameter of the residual lumen in percent related to the normal diameter measured on casts of that vessel preserved in maximal dilation in normal hearts (Baroldi, 1965). In fact, enlargement of the atherosclerotic plaque makes unreliable any lumen valuation in percent of the cross-sectional area included within the internal elastic membrane (Roberts and Jones, 1979; Farb et al., 1995). A normal or mild lumen reduction may be interpreted as severe stenosis. Keep in mind that when it occurs, enlargement of an atherosclerotic plaque may or may not be a compensatory remodeling phenomenon (Glagov et al., 1987). As a matter of fact, ischemic heart disease patients show at their first fatal episode a severe lumen reduction ($> 70\%$ lumen diameter or 90% luminal area) in one or all main coronary arteries. Finally, in the histologic section, all plaque variables can be measured in percent of the total plaque area.

3. The area of each histologic myocardial section is measured in millimeters by an image analysis system (for instance Vidas, Zeiss), and the number of pathological foci and total number of pathologic elements can be referred in square millimeters to the total area examined histologically. The extent of fibrous tissue is calculated by an orthogonally bisected ocular in percent of the total histologic area. This is the only reliable method because myocardial fibrosis often transforms into adipose tissue (Baroldi et al., 1997c).

In conclusion, what is needed is a uniform method of heart examination adopted by all pathological services, particularly forensic institutions, in order to have a constant approach in solving diagnostic problems (Baroldi, 1993). Equally, data must be fully recorded to an exact protocol (see annexed card). Data collected in this manner at our institutes provided an information bank for research and for detecting trends that may affect public health.

Examination of the conduction system is particulary important in some cases of sudden death, especially nowadays with ablation procedures. A complete study demands serial section of the sinus, atrioventricular node, and His' bundle — an expensive and time-consuming task.

In our study on sudden coronary death (Baroldi et al., 1979), we sampled the sinus node and conduction system according to Lev's method (1951, 1954). The various blocks were embedded in paraffin, and a few histological sections per block were examined. This procedure was considered sufficient for seeing any direct relation between coronary sudden death and histologic findings of the conduction system (see Chapter 7).

Basic Pathophysiological Changes

4

GIORGIO BAROLDI

Contents

Myocardium

Pathophysiology of Contraction and Related Myocardial Injuries

The myocardial cycle of contraction–relaxation can be interrupted acutely in irreversible contraction or relaxation or chronically by a progressive loss of function. Each of these three dysfunctional states shows a clear-cut, pathognomonic structural aspect.

Irreversible Relaxation (Atonic Death)

The loss of contraction of a myocardial region is the first functional change following a myocardial infarct (*atonic death*). It occurs within a few seconds. Consequently, the tissue in flaccid paralysis (Figure 11) is extruded by intraventricular pressure (paradoxical bulging), giving reason for early cardiac wall and myocellular stretching, shown by prominent I bands demonstrated electron microscopically (Korb and Totovic, 1969) and by histology (Hort, 1968). This causes extravascular compression of intramural vessels with blockage of intramyocardial blood flow, secondary wall degeneration, and platelet-fibrin thrombosis.

This necrosis is monofocal with a size ranging from less than 10% to more than 50% of the left ventricular wall and of the interventricular septum. Rare in the right ventricle (Blumgart et al., 1940; Wade, 1959) and in cor pulmonale, even in the presence of severe right coronary atherosclerosis (Baroldi, 1971), this form of myocardial necrosis is also rarely found in the conduction system (Ekelund et al., 1972) and atria (McCain et al., 1950; Wartman and Souders, 1950). When present at these sites, it is usually an expansion of a left ventricular or septal necrosis. In 40% of cases, the necrosis has an anterior or anteroseptal left ventricular location, in 28% posterior or posteroseptal, and anteroposterior in 32%. It is transmural in 57%, involves the inner two thirds of the ventricular or septal wall in 31% and the inner one third in 12% of cases (Baroldi and Silver, 1995). Thinning of the cardiac wall with a structurally normal myocardium is the first macroscopic change occurring in a few hours in relation to the size and wall extent of an infarct. At 10 to 15 h, the myocardium becomes pale. Subsequently, it assumes first a yellow color, being limited by a thin red rim (36 h), and then an increasing pale aspect until a scar is formed (30 days).

The earliest histologic signs are visible within 30 min of infarct onset and consist of mild myofiber eosinophilia and elongation of sarcomeres and nuclei. The latter show peripheral chromatin clumping followed by a progressive fading with their disappearance within 10 to 15 days (Figure 12). Other typical changes occur after 6 to 8 h with a margination of polymorphonuclear (PMN) leukocytes in vessels at the periphery of the necrotic zone followed by an infiltration of these elements, without fibrin or hemorrhage, into the dead tissue. A crowd of PMN is visible along a line between infiltrated and noninfiltrated necrotic myocardium in large areas of necrosis. The impression is that this PMN vallum, with an abscess-like aspect, represents a blockage of PMN infiltration likely due to a maximal stretching of the central portion of necrotic myocardium. PMNs disappear by lysis within 5 to 7 days, according to the size of the necrosis, without evidence of related myocellular change. Again at the periphery of the necrotic myocardium, a repair process starts at 1 week by macrophagic digestion of tissue within the sarcolemmal sheets and progressive collagenization to form a dense scar. This is not, as is generally assumed, repair by granulation tissue (Barrie and Urback, 1957; Baroldi and Silver, 1975). By postmortem intra-aortic injection of plastic substances or radiopaque material, even in the early phase, the mestruum did not penetrate into the acute necrotic tissue, showing an avascular intramyocardial area (Figure 13). Another finding, often neglected, is a severe intimal thickening of the small arteries and arterioles around the healing of necrotic tissue (Figure 13).

The timings of the evolutive phases of this lesion are reported in Table 4. Several important morphofunctional events during the evolution of this form of necrosis were ignored in earlier studies of myocardial infarct (Mallory et al., 1939; Lodge-Patch, 1951).

Finally, it is worth mentioning that the registered order of sarcomeres (Figure 12) can be maintained in the last remnants of necrotic myocardium within scar tissue; and it should

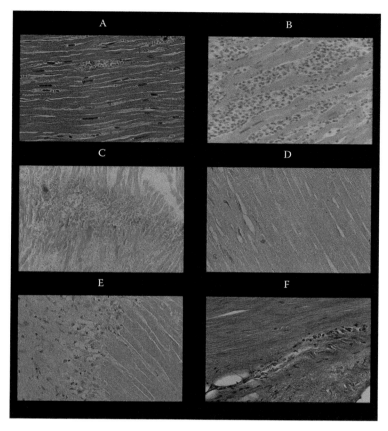

Figure 12 Infarct necrosis: The first change is loss of contraction with stretching of the myocardium in flaccid paralysis, (see Figure 11) resulting in a very early elongation of sarcomeres and nuclei (A) already visible within 30 min in experimental infarction. (B) Polymorphonuclear leukocyte infiltration from the periphery of the infarct after 6 to 8 h. In the largest infarcts, this infiltration arrests, along a line (maximal myocardial stretching in central part of infarct?) with occasional abscess-like formation (C). This infiltration disappears by lysis of the leuko-cytes, without evidence of myocellular colliquation or destruction (D). The myocardial cells maintain their sarcomeric registered order even in the terminal healing phase. The repair process is carried out by macrophagic digestion (E), and not by granulation tissue, ending in a compact and dense scar (F).

be emphasized that the absence of interstitial exudation, hemorrhage, and myocardial cell vacuolization at any evolutive stage of this form of myocardial necrosis is pathognomonic of a myocardial infarction. Its correct term is *infarct necrosis* rather than the too often used "coagulation necrosis" derived in the past from a false impression gained from gross examination. In reality, a coagulative process never occurs.

Irreversible Contraction (Tetanic Death)

In this condition, the myocardial cell arrests in irreversible hypercontraction (*tetanic death*), as shown by the marked shortness of sarcomeres with a length much less than that observed in normal contraction and with a characteristic anomalous, extreme thickening of Z lines. In its early phase, the lesion cannot be detected by the naked eye or by gross

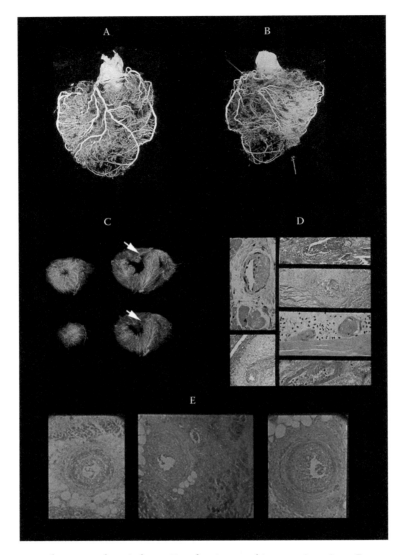

Figure 13 Avascular area of an infarct: By plastic cast (A: anterior view, B: posterior view) or postmortem angiogram (C), the infarcted zone (arrow) lacks intramural vessel injection (avascular area). Stretching of the necrotic myocardium and secondary vascular damage with wall degeneration and thrombosis (D) explain this vascular sequestration that occurs in the early phase. This may indicate a blockage without possibility of therapeutical intervention via blood flow within the infarcted myocardium. Note that the avascular area in this acute myocardial infarction case documented histologically, is supplied from the left anterior descending branch without evidence of occlusion or severe stenosis. The occluded vessel (arrow) was (B) the right coronary artery, the distal part of which was filled by numerous anastomoses. No myocardial damage was seen in its territory. By dissection conducted even by an expert pathologist, the diagnosis could be of myocardial infarction following occlusion of the right coronary artery. (E) Obliterative intimal hyperplasia in arterioles around a 7-day-old infarct with early repair process.

staining methods. Later, when myocardial cells coagulate and break with initial macrophagic repair, the myocardium shows focal discoloration and depressions on its cut surface. In the healing/healed phase, foci of fibrous tissue are seen.

Table 4 Timing of Evolving Infarct Necrosis

	Mallory et al., 1939	Lodge-Patch, 1951	Baroldi and Silver, 1995
Edema	—	2	—
Myocell nuclei disappearance	—	4	10
Myocell necrosis	20/180	6/28	1
Phagocytic remotion	5/180	6/28	7
Polymorphonuclear infiltration	6–24/180	6/10	6/7
Basophilic ground substance	—	20/14	—
Pigmented macrophages	6/365	3/365	7/365
Lympho-plasma-cells	6/60	5/180	In scar
Eosinophils	6/28	10/30	—
Fibroblasts	4/90	4/60	In scar
Collagen	14/365	10/365	In scar
Angiogenesis	5/90	4/90	No
Endoarteritis obliterans	—	—	7
Pathological contraction bands	+	—	<1
Apoptosis[a]	—	—	No

Note: Italics = days, Roman = hours.

[a] Diagnosis based on apoptotic bodies within and around infarct necrosis and normal myocardium.

The first histologic change, visible within 10 min of onset, is an intense hypereosinophilia of the hypercontracted myocardial cells with rhexis of the myofibrillar apparatus into cross-fiber, anomalous, and irregular or *pathological bands*. The latter are formed by segments of hypercontracted sarcomeres with scalloped sarcolemma. Normal cells around hypercontracted ones assume a wavy appearance. The spaces between bands are filled by mitochondria. No evidence of platelet aggregation or other vessel changes or of interstitial or sarcolemmal alterations exist. This necrosis is, in general, plurifocal, formed by foci ranging from one to thousands of myocardial cells and is found in any cardiac region (Baroldi and Silver, 1995; Baroldi et al., 2001a,c, 2003).

Two patterns of this lesion can be recognized. One corresponds to fragmentation of the whole myocell (*pancellular lesion*), which ranges from early breakdown in pathological bands to a total granular disruption (myofibrillar degeneration) (Figure 14). This myocardial cell destruction is likely due to the action of the contracting myocardium on these rigid elements in tetany. Repair of the pancellular lesion is by macrophagic digestion of all structures within the sarcolemmal tubes (*alveolar pattern*) followed by a progressive collagenization (Figure 15).

The second pattern, associated with the previous one, is characterized by a unique band of 10 to 20 hypercontracted sarcomeres close to the intercalated disc (*paradiscal lesion*). This band does not show rhexis of myofibrils and may assume a dark, dense, ultrastructural aspect or a pale, clear one, with very thin Z lines and myofibrils, and mitochondria "squeezed" in the normal portion of the myocyte (Figure 16). The paradiscal lesion does not show any macrophagic infiltrates. Two possibilities may explain this. Either the paradiscal lesion transforms in the pancellular one, or if it is a reversible change, the pale aspect represents a rebuilding of new sarcomeres because the major portion of the cell is normal and maintains a normal function without myofibrillar rhexis at the hypercontracted band level.

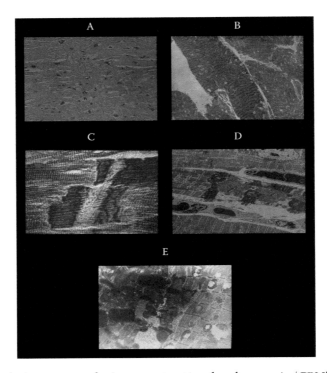

Figure 14 Coagulative myocytolysis or contraction band necrosis (CBN) or catecholamine necrosis. Pancellular lesion involving the whole myocardial cell. (A) Histological view of a CBN focus. (B) Ultrastructural hypercontraction with extremely short sarcomeres and highly thickened Z lines and minor focal myofibrillar rhexis. (C) rupture of a hypercontracted myocell. Electron micrograph view of pathological bands formed by segments of hypercontracted (D) and coagulated sarcomeres (intravenous infusion of catecholamines in dogs).

Both of these lesions are produced by an intravenous infusion of catecholamines (Todd et al., 1985a) and are found in human cases of pheochromocytoma (Figure 15). In animals, they can be prevented if beta-blocking agents are administered before catecholamine infusion.

In the literature, many terms were proposed for this lesion, including infarct-like necrosis, microinfarcts, focal myocytolysis, focal myocarditis, myocytolysis with major contraction bands, myofibrillar degeneration (Reichenbach and Benditt, 1969, 1970), and zonal lesion for paradiscal bands, described in canine experimental hemorrhagic shock (Martin and Hackel, 1963, 1966) and prevented by a beta-blocker (Entman et al., 1967). Overall, the term *contraction band necrosis* is used commonly. We proposed (Baroldi, 1975) the terms *coagulative myocytolysis* or the old equivalent German term of *Zenker necrosis*, because different patterns of pathological bands exist, or more directly, *catecholamine necrosis*. *Myocytolysis* is an old term that expresses myofibrillar damage (Schlesinger and Reiner, 1955), and coagulative refers to coagulation as the end result of these specific sarcomeric bands.

Other Types of Pathological Contraction Bands

A layer of hypercontraction with extremely short sarcomeres and thickened Z line without myofibrillar rhexis develops along the cutting margin of living myocardium. The layer

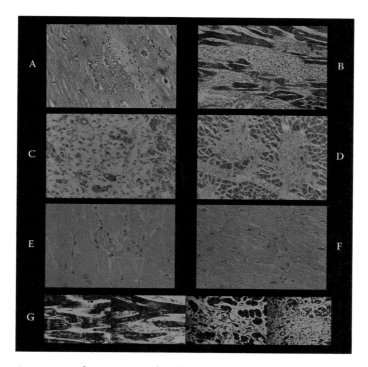

Figure 15   Repair process of contraction band necrosis (CBN). (A) Early monocyte infiltration that later becomes extensive, especially in large necrotic foci (B). It can be misinterpreted as lymphocytic myocarditis. This macrophagic reaction results in empty sarcolemmal tubes with numerous macrophages often loaded with lipofuscin and normal intramural vessels (C). The end result is a focal or plurifocal or confluent fibrosis (D). Microfocal fibrosis as result of necrosis of few myocells (E) can be confused with proliferation of the collagen matrix. (F) Waviness of normal myofiber around hypercontracted elements. (G) All stages of CBN in a case of human pheochromocytoma.

ranges from 0.2 to 0.5 cm thick (Todd et al., 1985a). This *cutting-edge lesion* seen in hearts excised at transplantation and at the edge of myocardial biopsies has the same histologic morphology as early pathologic contraction bands with the absence of myofibrillar rhexis. Their location and associated special circumstance of formation are so typical that they must not be confused with the lesions described above (Figure 17A and Figure 17B).

Another form of diffuse pathological contraction bands is that observed in experimental *reperfusion* or *reflow necrosis* when, after more than 20 min of ischemia, blood flow is restored. In this condition, pathological contraction bands, identical to those produced by catecholamine infusion, are associated with extensive interstitial hemorrhage and, as shown experimentally, with malignant arrhythmias/ventricular fibrillation. First described in an experimental model (Reimer et al., 1977), the condition has an equivalent in humans following heart surgery (concentric hemorrhagic necrosis, Gotlieb et al., 1977a) or reanimation maneuvers (Figure 17C). It is also seen, if rarely, when bypass surgery is done to treat an acute myocardial infarct.

Progressive Loss of Contractile Function (Failing Death)

Usually no gross change is obvious for this third morphofunctional myocardial injury if extensive areas of subendocardial myocardium are pale.

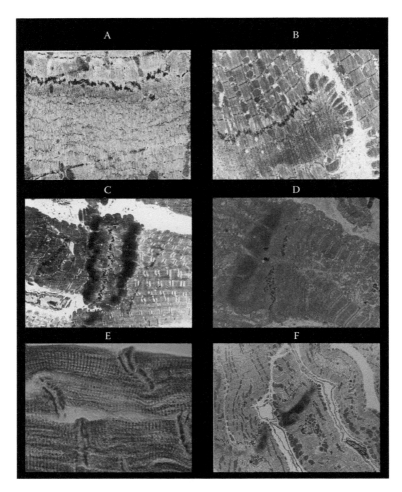

Figure 16 Contraction band necrosis (CBN). Paradiscal lesion. Always associated with the pancellular lesion, the paradiscal lesion is already visible in experimental intravenous infusion of catecholamines within 5 min (pancellular within 10 min). It is formed by a unique band of hypercontraction involving 10 to 15 sarcomeres adjacent to an intercalated disc. The major part of the myocell is normal, and this lesion shows ultrastructurally (A) a clear aspect without rhexis, thin myofibrils and Z lines (rebuilding of normal sarcomeres?) or as a band with different grades of density (B through D), often involving two myocells (C). The dense band can be seen histologically (E). A hypercontracted center (F) induces the waviness of normal adjacent myocells seen by electron micrography.

    The lesion consists of a progressive loss of myofibrils resulting in vacuolization of myocardial cells. It starts around a normal nucleus. The space, empty of myofibrils, is filled by small mitochondria and edema fluid (Figure 18). The impression is that of an increasing colliquation (*colliquative myocytolysis*) with the disappearance of myofibrils and without their resynthesis. The process is not associated with any inflammatory cellular or interstitial reaction. This injury is mainly found in congestive heart failure, irrespective of its cause (Baroldi et al., 1998c).

    Quantification of this lesion is difficult. In fact, the disappearance of myofibrils is a process in which the earliest phases are poorly measurable. Only when a partial or total

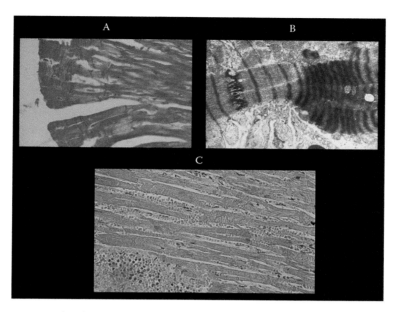

Figure 17 Cutting-edge lesion that involves a layer of 0.2 to 0.5 mm along the cut margin of a living myocardium (biopsies, surgical samples, heart excised at transplantation). (A) Histological aspect in heart excised at transplantation and (B) ultrastructural pattern in dog. (C) Reflow or reperfusion necrosis characterized by contraction band necrosis (CBN) plus massive interstitial hemorrhage never seen in other human and experimental conditions.

disappearance of myofibrils has occurred can a semiquantitative valuation be done. In our study, we adopted a grading of this pattern seen in the internal half of the left ventricular wall where prominent lesions are more frequent: grade 1 corresponds to occasional empty edematous myocells or to a small group of them, grade 2 is when less than 50% are seen, and grade 3 is when more than 50% of affected myocardial cells are seen in each histological section.

Etiopathogenesis and Pathophysiologic Meaning of Myocardial Dysfunctional Injuries

Myocardial infarct necrosis is caused by a reduction below a critical point of the nutrient blood flow. The different causes and mechanisms that may be responsible are, at present, under investigation and discussion (see Chapter 10). The common effect of pathological blood flow reduction is intramyocardial cell acidosis with irreversible myocardial relaxation, Ca^{++} being displaced from troponin and a consequent inability of the myocyte to contract (Katz, 1988; Opie, 1993).

In contrast, myocardial cell hypercontraction is associated with intracellular alkalosis, with rapid loss of adenosine triphosphate (ATP) and lack of energy to remove Ca^{++} from troponin (Meerson, 1969) or massive intramyocellular Ca^{++} influx (Fleckenstein et al., 1975). This form of injury is typical of catecholamine myotoxicity (Todd et al., 1985a,b) through free-radical-mediated lipid peroxidation (Mak and Weiglicki, 1988; Fineschi et al., 2000), and oxidative stress occurs after reperfusion in man (Ferrari et al., 1990). In this context, the calcium paradox phenomenon, that is, contraction band necrosis resulting

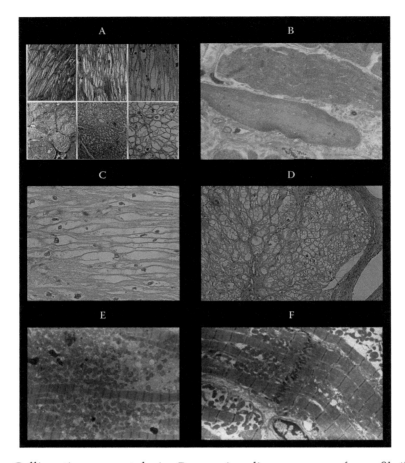

Figure 18 Colliquative myocytolysis. Progressive disappearance of myofibrils (A) with intramyocardial edema (B) resulting in an empty sarcolemmal tube seen in longitudinal (C) and transverse sections (D). Note the absence of any type of reaction. (E) Ultrastructural view of an edematous myocell filled by mitochondria and few contracted myofibrils in contrast with a normal contracted myocell (F) of the same case with congestive heart failure.

from perfusion of calcium-free solution followed by calcium containing solution (Zimmerman et al., 1967), is worth mentioning.

In heart failure, the sarcotubular system and mitochondria have a reduced capability to bind Ca^{++} (Bing et al., 1974) associated with a loss of K^+ and intracellular retention of Na^+. Lysis of myofibrils was obtained by prolonged beta-blocking (Sun et al., 1967), hypokalemia (Emberson and Muir, 1969), and hypocalcemia (Weiss et al., 1966).

The same myocardial damage may have different causes, but such damage always has a unique morphologic pattern and pathogenic mechanism. As shown, the three forms of myocardial injuries (Table 5) have totally different structural, dysfunctional, and biochemical disorders, with no possibility that one can be confused with the other. This means that at any time they are found together, we need to define their clinicopathological correlation and significance in the natural history of a disease, particularly in coronary heart disease (see Chapter 5, section entitled "Coronary Heart Disease").

Table 5   Histologic Patterns in Different Forms of Myocardial Damage According to Contractile Cycle and Coronary Heart Disease

Myocardium	Infarct Necrosis	Coagulative Myocytolysis (Contraction Band or Catecholamine Necrosis)	Colliquative Myocytolysis (Myofibrillolysis)
Functional status contraction cycle	Irreversibile relaxation ("atonic" death) plus stretching by intraventricular pressure within 1 h	Irreversibile hypercontraction ("tetanic" death) — crossband formation in 5 to 10 min	Progressive lost of function ("failing" death)
Myocardial cell	Early thinning	Early increase in diameter	Increasing edema-vacuolization
Nucleus	Elongation-pyknosis-progressive karyolysis	Normal	Normal
Myofibrils	Elongated sarcomeres in normal registered order, even in late stage	Rhexis-anomalous irregular crossband formations (coagulation of hypercontracted sarcomeres)	Progressive disappearance, "empty cell" (colliquation)
Vessels	Secondary wall degeneration and thrombosis during polymorphonuclear infiltration	Normal	Normal
Infiltration	Massive polymorphonuclear infiltration beginning 6 to 8 h and lysis within 5 to 6 days	No early cellular infiltrates; possible late lymphocytes	No infiltrates
Extension location	In general, unique massive focus of different size; subendocardial to transmural	Multiple (mono- or pluricellular) disseminated or confluent foci of different size in any area	Focal progressively spreading
Irreversible in	At least 20 to 60 min	A few minutes	?
Healing	Removal by macrophages; collagenization of empty sarcolemmal tubes	Removal by macrophages; sarcolemmal tube collagenization	Never documented
Frequency in coronary heart disease: acute infarct	100%	100% at outer limit of early infarct; 85% in myocardium elsewhere	38% subendoperivascular
Sudden/unexpected death	17% histologically demonstrated	72% (unique demonstrable lesion), 86% (including cases with infarct)	8% subendocardial plus fibrosis

Aging of Myocardial Necroses

Infarct Necrosis

A guide to aging in infarct necrosis proceeding to healing was discussed previously (Table 4). In general, no pathognomonic histologic sign is visible within 6 to 8 h of survival after onset when PMN infiltration starts. Many attempts to define early infarct necrosis have been unsuccessful. By incubating heart slices in a solution of nitroblue tetrazolium

(NBT), the normal myocardium stains dark, because NBT is reduced by dehydrogenase enzymes to formazan in contrast to unstained necrotic areas or scar tissue, which are depleted of these enzymes (Nachlas and Shnitka, 1963). This method becomes positive approximately 5 h from infarct onset. Even a postmortem intracoronary injection of NBT (Feldman et al., 1976) could not detect an earlier infarct. This is also so for many other substances, including glycogen, myoglobin, intracellular diffusion of IgG (Kent, 1982), fibrinogen complement C5b-9 (Brinkmann et al., 1993; Thomsen and Held, 1995), caeruloplasmin, C-reactive protein (Leadbeatter and Nawman, 1990), and the cytoskeletal proteins vinculin, desmin, and α-actin (Zhang and Riddlck, 1996). Any loss of these substances is a nonspecific sign possibly linked with terminal events. Similarly, the wavy fibers considered ischemic by Bouchardy and Majno (1971–1972) are not specific and are often found around nonischemic hypercontracted elements (see above). In our opinion, the stretching of mildly eosinophilic myocardial cells with elongated sarcomeres and nuclei already seen within 1 h in experimental and human infarct is a reliable early sign. This stretching, even if nonspecific, because it may occur in other conditions (e.g., surviving myocardial cells in a fibrous aneurysm wall) may be useful at least in a negative sense: if not seen, one may exclude an early (30 to 60 min) infarct.

A comment seems appropriate here on the clinical use of serum CK-MB enzyme levels or those of any other substance, for example, troponin and myoglobin (Werf, 1996), to measure *infarct size* for diagnostic and prognostic purposes (Erhardt, 1996). Because of early sequestration, with a lack of flow in infarcted myocardium, the latter may express not infarct size but the extent of the surrounding contraction band necrosis secondary to adrenergic damage (see below). Comparing postinfarct scintigrams by technetium-99 stannous pyrophosphates with histology, the positive zone *in vivo* corresponded to myocytolytic degeneration (Buja et al., 1977). Serum tumor necrosis factor alpha (Hirschl et al., 1996), or heart troponin I (Antman et al., 1996), or T troponin (Ohman et al., 1996) levels come with the same criticism. The number of severe coronary artery stenoses with angiographic imaging of thrombi did not correlate with T troponin levels, suggesting that another nonischemic lesion was present (Buja et al., 1977). Indeed, further analyses are needed to define the relationship between clinical substances released into the blood and the different forms of myocardial necrosis described above.

Catecholamine Necrosis

More difficult and incomplete is the timing of this necrosis in man. According to experimental isoproterenol and noradrenaline intravenous infusion in the dog, the lesion is detectable within 5 to 10 min (Todd et al., 1985). In the rat, the infarct-like lesion following subcutaneous isoproterenol injection was grossly demarcated in 48 h. The removal of necrotic myofibers by histiocytes leaving empty sarcolemmal sheaths began at 72 h. Thereafter, at 7 days, there was a granulomatous reaction by histiocytes, fibroblasts, Anitschkow myocytes, and Aschoff giant cells, with progressive reticulin and collagen fiber formation ending in a fibrous scar containing spindle-shaped fibroblasts, pigment-laden macrophages, and lymphocytes within 2 to 5 weeks. An acellular scar was present within 12 to 16 weeks (Rona et al., 1959; Rona and Kahn, 1969).

Failing Necrosis

As far as colliquative myocytolysis, reactive or repair processes were never seen in our studies. We emphasize that the association in coronary heart disease of these three forms of myocardial injury requires their morphologic distinction (Table 5) if a correct diagnosis, particularly from a forensic perspective, is required.

Meaning of Different Forms of Structural Myocardial Injury

Often in biomedical science, experiments may mislead our understanding. From the earliest experiments in dogs and rabbits (Cohnheim and von Schulthess-Rechberg, 1881) to more recent ones (Murry et al., 1986; Kelley et al., 1999; Lameris et al., 2000), many contributions have furthered our knowledge. Nevertheless, an experimental model that exactly mimics the natural history of human coronary heart disease is not yet available. Until now, experimental coronary artery occlusion, whether temporary or permanent, is performed on healthy coronary arteries of healthy hearts, while in human coronary heart disease, an occlusion occurs in severely stenosed vessels already bypassed by collateral flow. Furthermore, when a severe stenosis is experimentally or clinically occluded, there is no infarct or dysfunction (see below). The size of an infarct obtained by experimental coronary occlusion is reproducible in each species but varies among different animals because of differing coronary anatomy and collateral location. In contrast, in men and women, infarct size is not stable but ranges from less than 10% to more than 50% of left ventricular mass. What needs to be stressed here is the present controversy in defining ischemic necrosis, mainly developed from unreliable models of experimental temporary occlusion. In the dog, a subendocardial infarct is fully established within 1 h of coronary occlusion, while reopening the vessel within 20 min prevents infarction (Jennings et al., 1969). By varying the periods of occlusion and reflow or by occlusion lasting for 4 days, an infarct becomes transmural. This process was defined as the *wavefront phenomenon* and consists in a progressive expansion of myocardial necrosis from the endocardium to the subepicardium according to the duration of temporary (40 min, 3 and 6 h occlusion followed by 2 to 4 h reperfusion) or permanent (24 and 96 h) coronary occlusion in the dog. Three different zones were described: (1) a central region with myofibrillar relaxation without hemorrhage and inflammation; (2) an hemorrhagic midzone with contraction band necrosis particularly prominent after reperfusion; and (3) a peripheral, nonhemorrhagic zone in which inflammation, phagocytosis, and infarct repair occurred relatively quickly. By comparing different timings with and without reperfusion, the amount of viable myocardium at risk and salvaged by reperfusion was 55% after 40 min occlusion, 33% after 3 h, and 16% after 6 h, with a maximum damage (85%) after 96 h of permanent occlusion. Reperfusion after 40 min was associated with a high incidence of ventricular fibrillation (41%) in contrast to reflow after 3 or 6 h. Late death was infrequent in reperfused animals, while it happened in 10 of 29 dogs with permanent occlusion (Reimer et al., 1977). Protection by propranolol after temporary occlusion (Reimer et al., 1976) and overestimation of necrosis size and underestimation of collateral flow because of edema, hemorrhage, and acute inflammation (Reimer et al., 1977) were other important findings.

These experiments, and many similar ones, have had a major influence in relation to the theoretical and practical approach to save human myocardium at risk using a variety of pharmacologic and hemodynamic surgical procedures during the acute phase of coronary syndromes. The experimental experiences made a fundamental contribution, even through they can be criticized. First, the dog model has a totally different coronary collateral system that includes subepicardial anastomoses able to compensate immediately for an acute occlusion because they are not compressed by stretched ischemic myocardium, and thus, the timing of the "wavefront" becomes questionable when referring to acute myocardial infarction in humans. Second, the percentage of necrosis in a dog experiment is related to the posterior papillary muscle that is considered a "reliable index of overall left ventricular necrosis in this model," (Reimer et al., 1977) despite a strong difference in vascularization and function in respect to the intraventricular pressure acting on the dead myocardium of the free wall; the "central zone" likely being more extensive in humans. Third, there is interference of reperfusion with the formation of a hemorrhagic midzone with contraction band necrosis.

No matter if consequent to a presumed, but never morphologically proven, borderline ischemia or postspasm reflow, wavefront necrosis was never observed in 100 Italian (Baroldi et al., 1974b) and 100 Canadian (Silver et al., 1980a) fatal acute infarcts of different age or in 208 sudden and unexpected coronary deaths (Baroldi et al., 1979). In the absence of myocardial wall rupture, the infarct was always anemic, an observation confirmed by others (Oliva et al., 1993), without histologic changes that could support any concept of an expansion of an original infarcted area. Rather, an infarct was always associated with peripheral nonhemorrhagic contraction band necrosis. This means that hemorrhage, and therefore reperfusion, does not occur in the natural history of acute myocardial infarct in man. In our studies, we quantified the size of a myocardial infarct only, without measuring the variable extent of the associated contraction band necrosis that appeared as a more or less wide layer and numerous scattered foci of different dimension, often confluent, in continuity with infarct necrosis or in nonischemic myocardium (Figure 19), a wavefront that we interpret as being due to adrenergic myotoxicity and not to borderline ischemia or postspasm reflow.

The innumerable reflow procedures done in the past 40 years in humans never induced a 41% fatal malignant arrhythmia rate as in experimental reperfusion. Do these procedures

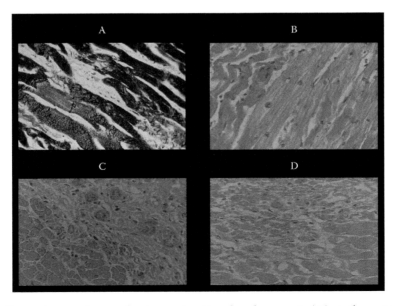

Figure 19 Extensive nonhemorrhagic contraction band necrosis (A) at the external layer of an acute myocardial infarct or focal in normal myocardium around the infarct (B). In a recent infarct, already encircled by a fibrous ring, internally there is typical infarct necrosis (C) and externally (D) an early nonhemorrhagic contraction band necrosis (CBN).

really revascularize a hypothetically ischemic myocardium? In human pathology reflow, necrosis was described following heart surgery with cardiopulmonary bypass or resuscitation maneuvers (see above). Hemorrhage within an infarction is reported in patients treated with selective thrombolysis (Fujiwara et al., 1986). In 19 acute infarct patients treated by fibrinolytic therapy or percutaneous transluminal coronary angioplasty or both, an infarct plus hemorrhage was present in all with fibrinolytic therapy whether or not balloon angioplasty had been done, while no hemorrhage was seen in infarcts after angioplasty alone. Intimal and medial hemorrhage plus plaque cracks and fractures were present in patients with pharmacologic and mechanical intervention, in contrast to those treated by balloon angioplasty in whom plaque cracks and fractures were free of intimal and medial hemorrhage. The conclusion was that plaque and infarct hemorrhage, the latter also extending into noninfarcted myocardium and not confined only in the median zone of infarct necrosis as in experimental reperfusion damage, was related to pharmacologic effects rather than to reperfusion, per se (Waller et al., 1987). The frequency of hemorrhage in 99 acute infarcts treated with fibrinolytic agent was 43% (Waller, 1991). The "hemorrhagic contraction band infarct" after vein graft surgery (Lie et al., 1978) in reality seems a reflow necrosis, and not infarct necrosis, following cardiopulmonary bypass surgery. The frequent elevation of creatine kinase and its MB subfraction (CK-MB) plus electrocardiogram (ECG) alterations after revascularization procedures (Calif et al., 1998) may represent this type of myocardial necrosis (see below), rather than any caused by diffuse coronary atherosclerosis (Kini et al., 1999).

There seems to be reluctance by pathologists and clinicians to differentiate forms of myocardial necrosis according to the contractile cycle. Rather, lesions of different morphology and cause are linked under the umbrella of ischemic injury. At present, the opinion of some authors is that contraction band necrosis is an ischemic lesion likely due to a temporary spasm followed by reperfusion (Factor and Bache, 1994). What needs to be stressed is that contraction band necrosis in many human conditions is not a hemorrhagic and ischemic lesion but is caused by catecholamines, as is shown experimentally by their infusion (Todd et al., 1985a,b) and the prevention of this form of necrosis and associated ventricular fibrillation by beta-blockers (Baroldi et al., 1977; Clusin et al., 1982; Anderson et al., 1983). Recent studies in the dog confirmed this. A progressively increasing extent $\times 100$ mm^2 of the lesion paralleled coronary occlusion that lasted 18, 20, 40, or 60 min with a maximum after conditioning (10 min occlusion plus 5 min reflow $\times$ four times) not only in ischemic but also in nonischemic myocardium without relation to blood flow measured by microspheres. Both pathological contraction bands and ventricular fibrillation were prevented by a beta-blocker. An increase of norepinephrine in the interstitial fluid of ischemic myocardium was documented (Lameris et al., 2000). Therefore, the reduction of damage observed in experimental coronary occlusion by transient ischemia (Sommers and Jennings, 1972), denervation (Jones et al., 1978), lidocaine (Nasser et al., 1980), superoxide dismutase (Przyklenk et al., 1986), or preconditioning (Przyklenk et al., 1993) seems antiadrenergic stress related. Beta-blockers infused into a conscious dog with coronary narrowing improved myocardial dysfunction and abnormal flow patterns (Tomoike et al., 1978).

Using a quantitative systemic study, the frequency and extent of nonhemorrhagic contraction band necrosis was calculated for the first time in different unrelated human pathological and normal conditions, attempting to delineate the natural history of this lesion (Baroldi et al., 2001a) (Table 6). In summary, its frequency and extent were practically

Table 6 Frequency Percentage and Extent of Different Morphofunctional Myocardial Cell Necrosis in Various Conditions

Disease	Acute Infarct	Sudden/Unexpected Death — Coronary	Coronary	Chagas'	Brain Hemorrhage	Brain Hemorrhage	Transplanted Hearts	Acquired Immunodeficiency	Congestive Heart Failure	Cocaine	Carbon Monoxide	Electrocution	Head Trauma	Head Trauma
Number of cases	200	13	12	34	14	13	46	38	144	26	26	21	26	19
Resuscitation	+	No	+	No	+	+	?	+	No	No	No	No	No	No
Survival	12 h–25 d	Min	<1 h	Min	<1 d	>1 d	1 d–yr	Months	—	5–12 h	>5 h	Min	<1 h	>1 h
Infarct necrosis														
Present (%)	100	23	8	—	—	—	15	—	3	—	—	—	—	—
Size (%)	10–50	10–30	20	—	—	—	10–30	—	10–30[a]	—	—	—	—	—
Catecholamine necrosis														
Present (%)	100	69	83	100	86	92	85	66	87	42	11	5	4	42
Foci	>30	9±6	39±47	3±5	16±21	37±42	36±55	4±11	2±4	4±4	1±1	8	0.5	12±18
Myocells 100 mm^2	>2000	102±143	241±196	26±56	26±29	108±134	262±293	13±27	11±21	11±14	5±4	46	35	21±23
Crossband	100	22	10	24	75	33	36	76	52	100	100	100	100	100
Alveolar	50	67	70	47	25	50	44	24	20	—	—	—	—	—
Healing	50	11	20	29	—	17	20	—	28	—	—	—	—	—
Colliquative myocytolysis														
Present (%)	43	—	—	—	—	4	28	—	97	—	—	—	—	—
<25	—	—	—	—	—	4	28	—	32	—	—	—	—	—
25–50	100	—	—	—	—	—	—	—	58	—	—	—	—	—
>50	—	—	—	—	—	—	—	—	10	—	—	—	—	—
Myocellular fibrosis														
10%	27	46	50	53	8	8	17	—	64	—	—	—	—	—
Fibrous index[c]	—	12±1	—	7±1	0.4±0.6	1±1	1±1	0.04±1	—[b]	—	—	—	—	—

a Only in coronary heart disease group.

b 17 ± 8 coronary heart disease, 2 ± 3 dilated cardiomyopathy, and 4 ± 4 chronic valvulopathy. All 144 hearts were excised at transplantation.

c Fibrous index: total fibrosis area/total histologic area in mm^2 × 100.

absent in people dying instantaneously or rapidly in less than 5 min from a variety of causes, with a progressive significant increase of lesions in other groups, with a maximum in acute infarct cases, transplanted hearts, in sudden and unexpected coronary and Chagas' death, intracranial hemorrhage, and with higher values in longer survivors. The greater extent in sudden and unexpected coronary cases is also likely due to a longer survival, rather than to resuscitation therapy. In fact, the same extent of early contraction band lesions was observed with and without the latter, older stages of the lesion being present in most cases. This indicates a possible increased adrenergic intramyocardial traffic in the agonal phase to stimulate the cardiac pump, independent of the cause of death.

According to these data, the threshold to document adrenergic stress in the natural history of any condition, seems to be around 37 ± 7 SE of foci of contraction band necrosis and 322 ± 99 SE $\times 100$ mm^2 of myocardial cells showing the lesion, that is, the mean values found in the anterior left ventricle of SD, transplanted, brain hemorrhage heart groups vs. normal controls with 7 ± 2 and 19 ± 5, respectively, and the presence of all stages of the lesion, from early hypercontraction to contraction bands, alveolar and healing–healed changes. Catecholamine necrosis, except around an infarct, shows a minimal myocellular loss that would seem unable, per se, to alter pump function. However, its presence is an important signal of sympathetic overdrive with possible arrhythmogenic supersensitivity (Inoue and Zipes, 1987) in the natural history of a disease. This view about the role of catecholamines in causing this lesion is in opposition to other hypotheses, such as ischemia, or the vague increased oxygen consumption concept. The opinion is based on the occurrence of contraction band necrosis any time there is an excessive and acute increase of heart work as occurs in many experimental conditions, for example, acute coronary occlusion (see above), acute severe aortic stenosis (Vitali-Mazza et al., 1972; Meerson, 1969), catecholamine infusion (Todd et al., 1985), and human conditions (e.g., pheochromocytoma, acute infarct). There is need for similar study in other nosologic patterns where a "sympathetic storm" may have a role in, for instance, coronary heart disease (Lesch and Kehoe, 1984), intracoronary embolization (Sabbah et al., 1992), hypertrophic cardiomyopathy (Anderson et al., 1983; Hecht et al., 1993); myxomatous mitral valve (Pasternac et al., 1982; Chesler et al., 1983), systemic amyloidosis (Falk et al., 1984), congestive heart failure (Packer, 1985; Parmley, 1987), cardiomyopathies (Brandenburg, 1985; Goodwin and Krikler, 1976, 1978; Hecht et al., 1993; Marcus et al., 1982; Maron et al., 1978; Oparil, 1985). radiation-induced coronary obstruction (Angelini et al., 1985), and systemic sclerosis (Bulkley et al., 1978).

In this context, it is worth mentioning that contraction band necrosis is practically absent in fatal carbon monoxide poisoning but becomes evident after reoxygenation by breathing air in preterminal rats. Its arrhythmogenic effect is prevented by a beta-blocker (Fineschi et al., 2000). This is a fact to be considered in correct forensic diagnosis and in resuscitation.

Finally, the catecholamine damage observed in hearts with irreversible congestive heart failure (CHF; Baroldi et al., 1998c) confirms the sympathetic overactivity recognized in this condition (Hartmann et al., 1996; Cohn et al., 2000; Liggett et al., 2000), with the interesting morphologic observation of a sympathetic hyperinnervation in congestive heart failure with a history of spontaneous ventricular arrhythmia (Cao et al., 2000).

The other form of myocellular morphofunctional damage, colliquative myocytolysis (even if not widely accepted; Karch et al., 1986), is an important marker of acute and chronic cardiac pump failure (Table 7). In acute myocardial infarcts, it can be seen in the subendocardial and perivascular myocardial layer preserved by infarct necrosis (Figure 20).

Table 7   Distribution of Different Grades of Colliquative Myocytolysis

| Disease | Total Cases | Myocytolysis (Grades[a]) | | | |
		0	1	2	3
Congestive heart failure[b]	144	5	44	81	14
Coronary heart disease	68	1	14	39	9
Dilated cardiomyopathy	63	3	28	28	4
Valvulopathy	18	1	2	14	1
Sudden/unexpected death[c]					
Coronary	34	28	6	—	—
Chagas'[d]	—	—	—	—	—
Acquired immunodeficiency syndrome	38	33	3	2	—
Acute intracranial hemorrhage[e]	27	26	—	1	—
Head trauma in otherwise healthy individuals	45	45	—	—	—

[a] For grading, see text. Congestive heart failure cases versus other groups, $p < 0.0001$. No relation was found between grade of colliquative myocytolysis and clinical cardiac dilatation (see Baroldi et al., 1998c).
[b] Hearts excised at transplantation.
[c] In apparently healthy subjects.
[d] Serum-positive for Chagas' disease.
[e] Aneurysm rupture in the absence of cardiac disease.

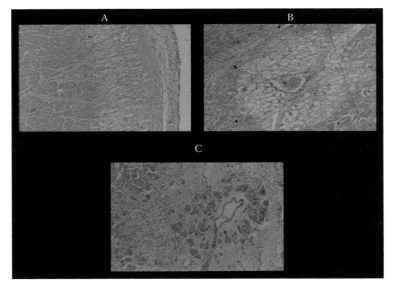

Figure 20   Colliquative myocytolysis associated with acute myocardial infarct. The lesion is confined in layers of the subendocardial myocardium (A) or around functioning vessels (B). These layers are generally preserved by the infarct necrosis, as shown (C) in a perivascular myocardial layer around a vessel in an old infarct without congestive heart failure.

An expression, more than a cause of the insufficiency, it has no relation with ischemia as is generally believed and suggested by an endomyocardial biopsy study (Clausell et al., 1996). Nevertheless, myocardial vacuolization may have different causes, for example, storage diseases or parasitosis (Figure 21), or may be an artifact following histologic processing or due to autolysis. We have never seen vacuolization of myocardial cells and it was not reported by others in the very early phase of human and experimental myocardial

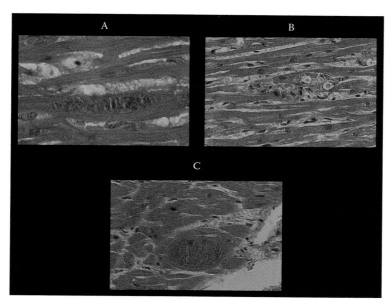

Figure 21   Parasitosis: (A) toxoplasma, (B) cryptococcus, and (C) sarcosporidium invading the myocells without inflammation.

infarction, particularly in the prenecrotic stage following temporary (plus reflow) or permanent coronary occlusion. In congestive heart failure, one cannot exclude that blood flow reduction in this condition (Parodi et al., 1993) is consequent to the reduced work or contractility of the failing myocardium. In the latter there is a lack of stretching of failing myocardial cells, confirmed by electron microscopy in hearts excised at transplantation (Figure 18); a reduced compliance of contractility due to a reduced capability to relax (Piper et al., 2000); variations in Ca^{++} content (Bakker et al., 1995) and modification of myofibrillar sensitivity to Ca^{++} (Michele et al., 1999); a dependence of systolic and end diastolic velocity on both number of myocytes and density of myocardial beta-adrenergic receptors (Shan et al., 2000b); and an increased length of cultured myocardial cells from ischemic hearts, despite a normal sarcomere length with a greater myocell length–width ratio (Gerdes et al., 1992). All these data suggest that failing hypertrophy of any nature, including that in dilated cardiomyopathy, is an anomalous hypertrophy that could result from an intrinsic biochemical disorder of the myocyte unable to relax (adrenergic over-activity) with impairment of the Ca^{++} pump to remove it from the troponine–tropomyosine complex. In other words, this is an "abortive" hypertrophy with a sarcomerogenesis limited to a longitudinal sarcomeric apposition without three-dimensional myofibrillogenesis, which results in an increase in length and not in volume, with a maintained inborn incapacity for relaxation (Baroldi et al., 1998c). This view suggests the use of new therapeutical approaches (Bristow and Gilbert, 1995; Barry, 1999) by gene therapy (Del Monte et al., 1999), with the annotation that this abortive hypertrophy is reversible as demonstrated by functional recovery following admistration of beta-blocking agents and after surgical repair of insufficient hearts with congenital malformations; and the adrenergic overstimulation is due to blood-derived catecholamines, because the related focal necrosis is found in the whole myocardium (Baroldi et al., 1998c), as occurs in pheochromocytoma and in experimental intravenous infusion (Todd et al., 1985a).

Reversibility of Lesions Related to Myocardial Dyssynergy

Any myocardial lesion, including the three forms under discussion, has an initial reversible phase that is often extremely difficult, if not impossible, to diagnose morphologically.

The reversibility or temporary asynergy of noncontracting myocardium is a well-known clinical phenomenon. However, its morphologic counterpart is still a matter of speculation. In general, a contractile dysfunction, that is, *asynergy* or *dyssynergy*, can be *global* when the whole heart contracts poorly or *regional* or *local* when limited to one area surrounded by normally contracting myocardium. A further distinction concerns its *permanent* or *temporary* duration that is defined as *hypokinesis* when there is a reduced contractility, *akinesis* when a total loss of contraction occurs, and *dyskinesis* when there is paradoxical systolic bulging of the cardiac wall. Possible recovery of function was recognized long ago, but only recently has the concept of a nonfunctioning but *viable myocardium* become popular in the attempts to reestablish function therapeutically (Pierard et al., 1990). Using dobutamine infusion, it was possible to predict a functional recovery of 198 out of 205 cardiac segments shown by echocardiography before and after surgical revascularization (La Canna et al., 1994).

Two forms of reversible myocardial asynergy were distinguished. One type, the *hibernating* or *hibernated myocardium*, is defined as "a state of persistently impaired myocardial left ventricular function at rest due to reduced coronary blood flow that can be partially or completely restored to normal if the myocardial oxygen supply/demand relationship is favorably altered either by improving blood flow and/or reducing demand" (Rahimtoola, 1989, p. 223). In other words, the myocardium seems capable of preserving its structure by abolishing its function in the presence of chronic ischemia. It is not clear, however, how an already nonfunctioning myocardium can reduce its demand and whether or not chronic ischemia exists (see below). The other type is the *stunned myocardium*, the temporary asynergy that may last hours, days, or weeks following reflow after temporary ischemia, as described mainly in experimental models (Braunwald and Kloner, 1982; Kloner et al., 1989; Taylor et al., 1992b) and in man (Kloner et al., 1990; Bax et al., 2001; Sicari et al., 2003). Spontaneous functional recovery was seen in the presence of identical myocardial blood flow in dysfunctional and normally contracting segments, that is, a perfusion–contraction mismatch in unstable angina (Sambuceti et al., 1998; Gerber et al., 1999), but this was not confirmed by other authors (Camici et al., 1997). In both these reversible conditions, histological changes are obscure, because by definition, a viable myocardium maintains its structure, the impairment being at the molecular or ion level. In the few reports dealing with biopsy sampling, the observed changes were nonspecific (Schroder et al., 1988; Zhao et al., 1987; Flameng et al., 1987; Vanoverschelde et al., 1993; Borges et al., 1993; Kloner et al., 1998), with time-dependent deterioration and fibrosis both impairing recovery after revascularization (Schwartz et al., 1998a), hibernation being an incomplete adaptation to ischemia (Elsasser et al., 1997). In man, false-positive or false-negative findings are frequent (Cabin et al., 1987). More recently, by transmural biopsy in patients at coronary bypass surgery, 16 of the 37 segments, which recovered function 2 to 3 months later, showed more wall thickness at low doses of dobutamine, higher thallium uptake, and less interstitial fibrosis, while coronary angiography findings could not predict functional recovery (Nagueh et al., 1999). Thus, the latter could be due to compensatory hypertrophy of myocardial cells around or within an asynergic zone.

At present, we can only speculate on the basis of the types of morphofunctional damage previously described admitting their reversable phase. Reperfusion induces hypercontraction. One may suggest that stunned myocardium persists in a contracted (maybe hypercontracted) state, only temporarily unable to relax. If hypercontraction without myofibrillar rhexis is a reversible agglomeration of contractile proteins, then Z-lines thickening should be included within the sliding theory of contraction as a reversible rolling up of myofibrils at the Z-line level. To the contrary, relaxation with temporarily abolished contraction should prevail in the hibernating myocardium.

Other Myocardial Cell Injuries

Other forms of myocardial cell injury may occur seemingly unrelated to the contraction–relaxation cycle. In general, degenerative changes, defined by various, often imprecise, descriptive terms such as cloudy swelling, hydropic or fatty or hyaline degeneration, vacuolization, and so forth, are part of a morphofunctional injury of the contractile cycle.

In subsequent chapters, myocardial lesions specific for diseases will be described. Here, we need to emphasize a few points. For instance, according to the Dallas criteria (Aretz, 1987), myocarditis is diagnosed when cytotoxic elements destroy myocardial cells. An infectious or toxic agent may trigger an inflammatory state that starts within the interstitium, may or may not produce a reactive exudation, and may kill myocytes. Particularly in viral myocarditis, cytotoxic T lymphocytes may destroy myocardial tissue, explaining congestive heart failure (see Chapter 5, section entitled "Myocarditis"). A diagnosis of myocarditis can be made by endomyocardial biopsy, even if diagnostic mismatches are frequent among expert cardiovascular pathologists (Shanes et al., 1987), suggesting a need for diagnostic quality control. However, the frequent association of contraction band necrosis and monocytes as seen in sudden and unexpected death (see below) can be misdiagnosed as focal lymphocytic myocarditis (Figure 15A), when, in reality, the finding represents an early repair by oligodendritic monocytes (Parravicini et al., 1991). This misinterpretation is avoided only by determining the monocyte phenotype immunohistochemically. Another observation concerns the finding of parasites crowding and destroying myocardial cells without an inflammatory response (Figure 21). In this condition, the term *myocarditis* seems inappropriate and should be replaced with the term *myocardial parasitosis*.

Cardiac asynergy may occur in any injury. However, a peculiar pattern exists in which asynergy is consequent not to myocardial cell injury but to an architectural disorganization of the myocardium. The latter loses the usual parallel alignment of its elements needed for pump function and assumes a star-like arrangement of the myocardial cells joined by short and hypertrophic myobridges plus an increased interstitial fibrosis. This *myofiber disarray* can be found focally in some areas of normal hearts and is more diffuse surrounding a scar or in malformed hearts and in some other diseases, for example, Friedreich's ataxia or hyperthyroidism (Goodwin, 1982). It is prominent in asymmetric hypertrophic cardiomyopathy (Teare, 1958) of apical or diffuse form (Ferrans et al., 1972; Maron et al., 1992). The focal disarray in normal hearts and around scar tissue may represent nodal junctions or centers of force where muscle bundles change their direction. When more extensive, it produces an asynergic myocardium working, or better hyperworking, without any useful function (see Chapter 5, section entitled "Cardiomyopathies of Unknown Nature").

Cellular Death

In a review, the concept of cellular death included two main patterns, oncosis or *ischemic* death characterized by cellular swelling (from the Greek term ογκος), vacuolization and blebbing ending in coagulation necrosis and karyolysis, and *apoptosis* by which is meant a genetically programmed (or physiological) death. The latter form of necrosis consists of early nuclear and cellular shrinkage, nuclear pycnosis (half-moon- or sickle-shaped nuclei) followed by karyorhexis, and breakup into a cluster of apoptotic bodies phagocyted by macrophages or neighboring cells ("cannibalism"). Oncotic or ischemic death massively involves many cells, while apoptosis kills here and there like leaves dropping (ptosis = fall) from (*apo*) a tree (Majno and Joris, 1995).

Such definitions of cell death may be valid for other tissue but not for the myocardium. In fact, myocardial oncosis or coagulative or ischemic or infarct necrosis never shows swelling or vacuolization or blebbing or coagulation. The same can be said for other conditions (myocarditis, storage disease, etc.); while, in contraction band necrosis and colliquative myocytolysis, coagulation and vacuolization are observed, respectively. On the other hand, we never observed changes peculiar to apoptosis (especially apoptotic bodies) in any type of myocardial necrosis.

In recent years, a crescendo of publications (Colucci, 1996) claim that apoptosis — suicide, execution, or murder (Martin, 1993) — detected by immunostaining of exposed nuclear molecular endings of DNA fragments, is present in many if not all cardiac diseases affecting, for example, smooth muscle cells and macrophages in the atherosclerotic plaque (Hand et al., 1995; Isner et al., 1995; Geng and Libby, 1995) after myocardial reperfusion (Gottlieb et al., 1994), in acute infarction (Bardales et al., 1996), in dilated cardiomyopathy and congestive heart failure (Katz, 1995; Narula et al., 1996; William, 1999), in arrhythmogenic right ventricular dysplasia (Mallat et al., 1996), in myocardial stretching (Cheng et al., 1995), in myocardial hibernation (Chen et al., 1997; Dispersyn et al., 1999; Lum et al., 1989), and in noradrenaline myotoxicity (Communal et al., 1998).

At present, the significance of these findings, in which cells disappear without any repair, is under discussion. A question is whether TUNEL (*in situ* nick end-labeling) is a sensitive and specific method for detecting apoptosis because rupture of the nuclear membrane with exposure of DNA endings may occur in different conditions, especially in contraction band necrosis. A recent contribution speaks of repair rather than apoptosis (Kanoh et al., 1999). Apoptosis was never demonstrated in dilated cardiomyopathy by electron microscopy (William, 1999), or was found in a range of 0.06 to 0.41% as "chromatin margination, condensation, clumping and discontinuity of nuclear membrane," but without mentioning apoptotic bodies (Guerra et al., 1999, p. 858), the latter being the pathognomonic sign of nuclear fragmentation, which was never found in our material. Therefore, TUNEL positivity must be revaluated (Jerome et al, 2000) and a consensus reached on "the pathology of myocardial cell death due to prolonged ischemia which includes coagulation necrosis or contraction band necrosis or both usually evolving through oncosis, but can result to a lesser degree from apoptosis" (Thygensen and Alpert, 2000, p. 964).

Cardiac Arrest

A cardiac arrest results from total loss of heart pump function with cessation of blood flow. When it occurs, this arrest is an obvious cause of death and can be primary when the mechanism(s) of the standstill is (are) intrinsic to the heart or secondary to various noncardiac diseases. Here it is important to discuss different types of cardiac arrest because their postmortem recognition may have relevance, particularly from a forensic point of view, in establishing the cause of death (this topic is also discussed in Chapter 10).

The heart may stop after *ventricular fibrillation* (Surawicz, 1985), generally preceded by a malignant arrhythmia; or in *asystole* (Milstein et al., 1989) as an end result of brady-cardia; or in *electromechanical dissociation*, that is, the loss of mechanical function despite a normal electrocardiogram (Fozzard, 1985). The latter condition is characterized by loss of pulse, blood pressure, heart sounds, and consciousness, and it is usually associated with pulmonary embolism, heart rupture with cardiac tamponade, and so forth. However, autopsy findings may be negative (Hackel and Reimer, 1993).

At present, the morphologic background for the different types of cardiac arrest is poorly defined, and apart from a few striking conditions (e.g., heart rupture plus tampon-ade), we cannot structurally diagnose the cause of a myocardial arrest. However, the different impairments of the contraction–relaxation cycle suggested a reconsideration of myocardial changes in this fatal, terminal event, both in patients and in normal subjects succumbing to accidental death.

People who die from carbon-monoxide intoxication provide an example of cardiac arrest in asystole. The myocardium appears relaxed without evidence of any other change, both in suicides using this gas, in those whom an adrenergic stress might be suspected, and in people poisoned accidentally. The finding is similar in rats exposed to a fatal dose of carbon monoxide. However, when preterminal animals were reoxygenated, foci of pathological crossbands and electrocardiographic signs of arrhythmia and ventricular fibrillation occurred and were prevented by a beta-blocking agent (Fineschi et al., 2000). These data suggest that severe hypoxia, even if protracted for hours, does not elicit any adrenergic response, rather it is triggered by reoxygenation. Therefore, focal necrosis, often described, in this intoxication is not caused by carbon monoxide, per se, but by reoxygen-ation that likely stimulates catecholamine release, a fact to be considered in the treatment of this poisoning. Reoxygenation following severe hypoxia does not reproduce a reflow necrosis after ischemia, because no hemorrhage was seen.

The other main type of cardiac arrest, predominant in sudden and unexpected coro-nary death (see below) is ventricular fibrillation resulting from arrhythmogenic conditions. From its electrocardiographic pattern, ventricular fibrillation was defined as "chaotic, random, asynchronous electrical activity of the ventricles due to repetitive re-entrant excitation and/or rapid focal discharge. Factors that enhance electrical synchrony facilitate, while factors that decrease electrical asynchrony hinder, the development of fibrillation" (Zipes, 1975, p. 3). The questions are whether an equivalent morphology of the electro-cardiographic pattern exists, what causes it, and how it evolves, whether simultaneously within the whole left ventricle or at a single focus.

In the old German literature, *fragmentation* of myocardial cells and *segmentation* or dissection at the intercalated discs were described. However, they were considered artifacts (Batsakis, 1968) and were ignored. Nevertheless, segmentation of hypercontracted myo-fibers was recognized (Hamperl, 1929) as a possible agonal event related to ventricular fibrillation (Stamer, 1907 quoted by Staemmler, 1961). In a large series, we studied patients who died of various causes and people who succumbed to accidental death, and the frequency, location, and extent of the following histological findings were observed (Baroldi, 2001) (see also Table 8 and Figure 22):

Table 8   Myofiber Breakup in Malignant Arrhythmia/Ventricular Fibrillation (Frequency, Location, and Extent)

Source	Total Cases	Frequency	Location LV	RV	IVS	LV + RV	LV + IVS	RV + IVS	LV + RV + IVS	Extent sites 1	2	3	4	5	6	7	8
Sudden/unexpected death							*Group 1[a]*										
Coronary heart disease	25	22	—	—	1	—	12	1	8	1	1	2	4	6	2	2	4
Chagas' disease	34	26	1	—	—	3	5	—	17	—	1	1	3	3	4	—	14
Intracranial hemorrhage	27	14	3	—	1	—	2	1	7	2	—	2	—	4	—	—	6
Transplanted heart	46	20	1	—	—	2	9	1	7	—	2	1	5	6	—	—	6
Acquired immunodeficiency syndrome	38	9	2	—	—	1	5	—	1	1	—	2	1	4	—	—	1
Congestive heart failure[b]																	
Coronary heart disease	63	—	—	—	—	—	—	—	—	—	—	—	—	—	—	—	—
Dilated cardiomyopathy	63	—	—	—	—	—	—	—	—	—	—	—	—	—	—	—	—
Chronic valvulopathy	18	—	—	—	—	—	—	—	—	—	—	—	—	—	—	—	—
Total cases	**314**	**91**	**7**	**—**	**2**	**6**	**33**	**3**	**40**	**4**	**4**	**8**	**13**	**23**	**6**	**2**	**31**
							Group 2										
Cocaine	26	11	11	—	—	—	—	—	—	—	—	—	—	—	—	—	—
Head trauma	45	29	29	—	—	—	—	—	—	—	—	—	—	—	—	—	—
Electrocution	21	19	19	—	—	—	—	—	—	—	—	—	—	—	—	—	—
Carbon monoxide	26	5	5	—	—	—	—	—	—	—	—	—	—	—	—	—	—
Total cases	**118**	**64**	**64**	**—**	**—**	**—**	**—**	**—**	**—**	**—**	**—**	**—**	**—**	**—**	**—**	**—**	**—**
Grand total	**432**	**155**	**155**	**—**	**2**	**6**	**33**	**3**	**40**	**4**	**4**	**8**	**13**	**23**	**6**	**2**	**31**

[a] Group 1: sampling anterior, lateral, posterior, left ventricle (LV), right ventricle (RV) and anterior, posterior interventricular-septum (IVS). Group 2: sampling of the anterior left ventricle.
[b] Hearts excised at cardiac transplantation.

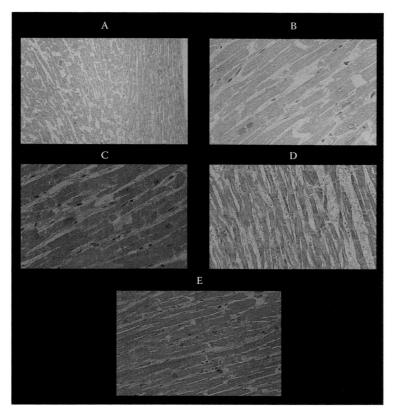

Figure 22 Myofiber breakup in ventricular fibrillation. (A) Myocardial bundle with segmentation (rupture of intercalated discs) in contracted myocells (left) associated with a bundle of relaxed ones (right). (B) Square nuclei expression of contraction. Stretching and detachment of intercalated discs between contracted myofiber elongation/separation of sarcomeres in myofibers connected with contracted ones (C, D). Myofiber breakup associated with a focus of contraction band necrosis (E).

1. Bundles of hypercontracted myocells alternated with bundles of hyperdistended myocardial cells.
2. Intercalated discs between hypercontracted elements were widened or stretched or segmented. The latter showed square nuclei instead of the usual ovoid or rectangular ones, as a consequence of hypercontraction.
3. Single or groups of hypercontracted myocardial cells were in line with hyperdistended ones with stretched sarcomeres or their detachment.
4. Typical pathological contraction bands and myocellular eosinophilia were absent.

These changes could be focal in a single region or diffuse in the whole myocardium and prevailed in the left ventricle. Not found in rigor mortis (Lowe et al., 1983; Vanderwee et al., 1981) nor in autolysis, the likelihood of them being an artifact related to sampling and histologic process is contradicted by their absence in 144 hearts excised at transplantation compared to their very high frequency in cases of sudden death. To exclude their occurrence secondary to ventricular fibrillation, the arrhythmia was induced by epicardial electrical stimulus or intracoronary infusion of KCl and was maintained for 30 min in ten anesthetized, open-chest dogs, without them developing. This result confirmed observations on calves on extracorporeal circulation with ventricular fibrillation lasting from one to several hours (Ghidoni et al., 1969).

These findings suggest a possible correlation between the observed myobreakup and electrocardiographic chaos. One notes that the many pathological changes proposed as a cause of reentry or focal discharge (focal fibrosis or necrosis, etc.) were not substantiated in our experiments or in human cases. Their presence was, in most instances, not associated with ventricular fibrillation.

What is the stimulus for arrhythmogenic ventricular fibrillation? An impairment of myocardial metabolism precedes ventricular fibrillation (Corday et al., 1977). In many experiments and in coronary heart disease patients, the ventricular fibrillation threshold was reduced or abolished by beta-blocking agents, suggesting an adrenergic role. By intravenous infusion of catecholamines in electrocardiographically monitored dogs, the only change was S-T segment depression and in no instance was a pattern of arrhythmia or ventricular fibrillation seen (Todd et al., 1985b). Only when noradrenaline was injected in one coronary artery were ventricular fibrillation plus typical catecholamine necrosis obtained (unpublished observation). This means that only regional or local catecholamine discharge could be responsible for ventricular fibrillation unbalancing the syncytial rhythm. The focal or diffuse myofibrillar breakup likely depends on the duration of the preceding malignant arrhythmia because when ventricular fibrillation starts, the cardiac pump ceases to eject blood with rapid abolition of contraction without time for further morphologic changes. In other words, a single focus of myofibrillar breakup may correspond to an instantaneous ventricular fibrillation, while an extensive one follows a relatively long-lasting malignant arrhythmia. In conclusion, segmentation and related findings seem to be a reliable histologic pattern for diagnosing cardiac arrest due to ventricular fibrillation.

Fibrillar Collagen Matrix

The following synthesis, on the adaptive changes of the collagen matrix, is derived from an exhaustive review (Weber, 1989) and more recent contributions (Ganote and Armstrong, 1993; Gunja-Smith et al., 1996; Stetler-Stevenson, 1996; Shirani et al., 2000; Jugdutt, 2003).

Normal myocardial protein synthesis and degradation can be altered in different conditions by several biochemical or traumatic factors. For instance, in contrast to volume-overload-related cardiac hypertrophy caused by anemia, or an arteriovenous fistula where no collagen changes are observed, pressure overload hypertrophy resulting from hypertension produces collagen modification. In the early stage, a more rapid noncollagenous protein synthesis is associated with a decline or degradation of collagenous proteins. At a late stage, when myocyte growth stabilizes, perimysial tendons, weave, and strands increased in number and dimension, leading to interstitial and perivascular fibrosis. In established hypertrophy, this *reactive fibrosis* is pericellular and distinct from *reparative fibrosis* subsequent to myocardial cell necrosis. Reactive or interstitial fibrosis is likely due to physical stress secondary to hypertension or is caused by local mitogens because hypertrophy may be limited to one ventricle. An adaptive change in the collagen matrix can be responsible for diastolic stiffness in the presence of interstitial fibrosis only, or of both diastolic and systolic stiffness whenever pericellular fibrosis is present.

In infarct necrosis, there is an early disruption of collagen matrix with proposed elongation of unprotected sarcomeres — a disruption likely induced by collagenase released from leukocytes or monocytes-fibroblasts. In reality, leukocytic infiltration is a late event, while early stretching of myocardial cells and fibrillar disruption are caused by intraventricular pressure (see above). A similar collagen matrix disruption was documented in stunned myocardium (Zhao et al., 1987), while abnormalities of the collagen matrix may be involved in inheritable diseases such as Ehlers-Danlos syndrome and Marfan syndrome. In dilated cardiomyopathy of unknown origin, the collagen tether is markedly reduced in its number of perimysial strands, thus explaining cardiac muscle fiber slippage, thinning of the cardiac wall by realignment of myocardial cells, and spherical dilation because intracavitary stress becomes equal in all directions. In contrast, an increase in collagen matrix size was observed in young patients with hypertrophic cardiomyopathy who died suddenly. For further discussion, see Chapter 5, "Congestive Heart Failure" section.

Extramural Coronary Arteries

Clinical and Postmortem Imaging of Coronary Arteries

Postmortem, the coronary arteries can be examined by different methods, as reported in Chapter 3. *In vivo, selective cineangiography,* meaning the injection of successive single coronary arteries, is the gold standard for the clinical diagnosis of coronary heart disease in terms of lumen reduction(s); or by changes such as "luminal irregularities, haziness with ill-defined margins, intraluminal luciencies, menstrum persistence" interpreted as the equivalent of thrombosis (Cowley et al., 1989, p. 109); or as coronary occlusion defined as a cut off of the vessel without visualization of its distal tract (De Wood et al., 1980). In addition to this routinely used method, others that are less frequently employed include the following:

Angioscopy permits direct vision, under very unnatural conditions, of the intimal surface only and distinction between white thrombosis in unstable angina and red thrombi in stable angina (Mizuno et al., 1992). A finding of a red thrombus in the infarct-related artery in 77% of acute infarct cases within 1 month (Van Belle et al., 1998) and the assumption that this method permits better definition of the plaque and its complications (Feyter et al., 1995) are untenable.

Intravascular ultrasound imaging may more precisely outline the shape of a stenosis ("sagittal, cylindrical or lumen cast," Roelandt et al., 1994) and provide some indication of plaque components (fibrosis, calcium) (Potkin et al., 1990; Tobis et al., 1991; Coy et al., 1992; Hodgson et al., 1993; Mints et al., 1995). However, the term *ultrasonic histology* is unrealistic.

Magnetic resonance imaging (Corti et al., 2002) and *near-infrared spectroscopy* (Wang et al., 2002a) are reported as promising tools for noninvasively detecting the age of arterial thrombosis or the chemical composition of a plaque, respectively.

Congenital Anomalies

In discussing the different patterns, histologic and *in vivo* imaging will be compared. Here what must be stressed is that all invasive techniques are performed in persons who are ill, with little possibility of studying the normal population — a possibility that the pathologist has.

The old distinction between minor or nonfunctional anomalies and major or dysfunctional anomalies is still acceptable with the admonition that some minor anomalies may, under certain circumstances, determine a dysfunction. An up-to-date revision of coronary anomalies as seen by cineangiography was published (Angelini et al., 1999). It is very useful in avoiding clinical pitfalls when diagnosing coronary disease.

Minor Anomalies of No Significance

Minor anomalies of no significance include the presence of multiple coronary ostia when, for example, a right accessory coronary artery exists or the left descending and circumflex branches originate directly from the aorta in the absence of the main left trunk.

Minor Anomalies That May Cause Myocardial Dysfunction

A rare finding is a single coronary ostium with a single coronary artery arising from the left aortic sinus. This artery may assume the distribution of the left coronary artery, the right coronary artery being a secondary main branch. Alternatively, the single coronary artery arises from the right aortic sinus and encircles the heart following the atrioventricular groove and ending as the anterior descending branch. These may be incidental postmortem findings, but dysfunction can result if the ostium or vessel is affected by disease.

Another anomaly is dislocation of the coronary ostia with a common origin of the two main coronary arteries from the same aortic sinus. When they arise from the anterior aortic sinus, the left coronary artery presents a sharp angulation with a slit-like opening and encircles the pulmonary artery. This anomaly may be associated with sudden death (see Chapter 5).

Major Anomalies Associated with Myocardial Dysfunction

Major anomalies are communications between the arterial system (high-pressure system) and right cardiac cavities, veins, or pulmonary artery (low-pressure system) with a left-to-right shunt. When the connection communicates with the left cardiac cavities, an aortic insufficiency-like condition exists. These anomalies elicit a variety of clinical symptoms and signs related to the number and location of the *fistulae* (Neufeld et al., 1961).

Perhaps the most peculiar anomaly is when coronary vessels originate from the pulmonary artery with the following patterns: both coronary arteries arise from the pulmonary artery, life being possible in the presence of pulmonary hypertension (Tow, 1931); an accessory coronary artery originates from the pulmonary artery and constitutes an autopsy finding or in life may determine a murmur similar to that of a patent ductus arteriosus; the right coronary artery arises from the vessel without or with minor dysfunction; the left coronary artery is a pulmonary vessel producing a left-to-right shunt with severe symptoms, cardiac insufficiency, and sudden death (Ogden, 1970). In this condition, enlarged collaterals drain into the pulmonary artery the aortic–coronary arterial blood flow.

Atherosclerosis

Atherosclerosis is an ancient disease that involves the arterial system and can be found in arterial remnants from Egyptian mummies (Ruffer, 1911). According to the injury theory (Ross, 1979), this disease is defined as "an excessive inflammatory-proliferative response to various forms of insult to the endothelium and smooth muscle of the artery wall. The earliest recognizable lesion is the so called 'fatty streak', an aggregation of lipid-rich macrophages and T lymphocytes within the innermost layer of the artery wall, the intima" (Ross, 1993, p. 4). The endothelial injury is not necessarily associated with endothelial denudation.

Physiologically, endothelial cells have many functions: they offer a nonthrombogenic surface; form a permeability barrier to the exchange of substances within the vessel wall; maintain vascular tone through nitric oxide, prostacyclin, and endothelin modulating vasodilation or constriction, respectively; form and secrete growth-regulatory molecules and cytokines; preserve basement membrane, collagen, and proteoglycans upon which they rest; provide a nonadherent surface for leukocytes; and have the ability to oxidize lipoproteins transported into the vessel wall. Changes in one or more of these functions due to one or several of many endothelial-noxious agents (hydrodynamic, immunologic, viral, adrenergic, hypercholesterolemic, etc.) may form the starting point of endothelial injury by trapping lipoproteins, particularly at branch points of arteries. Specific adhesive glycoproteins then appear on the endothelial surface with attachment of monocytes and T lymphocytes and their migration into the intima under the stimulus of growth regulatory molecules and chemoattractants released both by the altered endothelium, its adherent leukocytes, and possibly by underlying smooth muscle cells. The monocytes become macrophages that accumulate lipids ("foam cells") and together with T lymphocytes form a fatty streak. The sequence of events during progression of lesions in animals with induced hypercholesterolemia are proliferation of smooth muscle cells, macrophages, and lymphocytes; formation by smooth muscle cells of connective tissue matrix comprising elastic fiber proteins, collagen, and proteoglycans; and accumulation of lipid and mostly free and esterified cholesterol in the surrounding matrix and associated cells. This sequence is similar to that seen in human atherosclerotic coronary arteries in hearts excised at cardiac transplantation. Progression of atherosclerotic lesions is thus marked by the deposition of layers of smooth muscle cells and lipid-laden macrophages, the exposure of which at the site of retracted endothelial cells allows interaction with platelet and overlying mural thrombus formation. In this review article (Ross, 1993), all growth factors and chemoattractants, such as platelet derived, basic fibroblasts, heparin-binding epidermal, insulin-like, interleukin I, tumor necrosis, transforming b and colony-stimulating factors, monocyte chemotactic protein-I, and oxidized low-density lipoprotein are listed. The complexity of their interaction on smooth muscle cells, distinguished by a contractile phenotype responding to several vasoconstrictive or vasodilative substances and a synthetic phenotype capable of expressing genes for a number of growth-regulatory molecules and cytokines, and appropriate receptors and to synthesize an extracellular matrix are illustrated.

This reconstruction of the beginning and progression of the atherosclerotic plaque pertains to the experimental plaque obtained by feeding a hypercholesterolemic diet to animals. It corresponds in human pathology to a relatively small group of people who have lesions associated with familial or acquired hypercholesterolemia. Atherosclerotic changes observed in hearts excised at transplantation or in atherectomy or endarterectomy specimens are the end result of advanced plaques and cannot help define their early phase and progression. In reality, the need is to study all stages of the plaque in both normal subjects dying from accidental death and in patients with different patterns of coronary heart disease to establish its natural history in humans (Figure 23). One notes that each arterial tree has its own atherosclerotic natural history due to many differences in terms of hemodynamics, anatomical disposition, and wall structure. Extramural coronary arteries

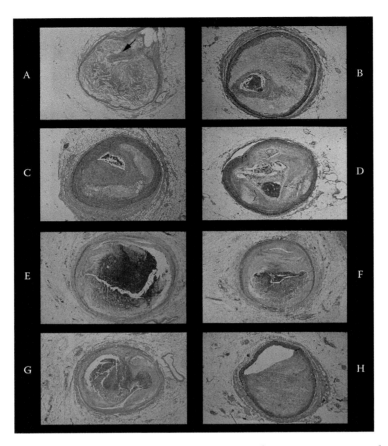

Figure 23 Coronary atherosclerosis. Different aspects of a severe, pinpoint lesion (arrow). Plaque with prevailing atheroma (A) or fibrosis (B). Plaque with pale, large zone of proteoglycan accumulation (C) or with small atheroma plus hemorrhage and proteoglycan associated with critical stenosis occluded by an acute thrombus (D). Sequence in the same plaque of rupture (E) followed by severe hemorrhagic atheroma with minimal, linear lumen (arrow) without occlusion (F). Occlusive thrombosis connected with hemorrhagic atheroma at the site of a critical stenosis (G). Semilunar stenosis (H) with a normal half wall and minor lumen reduction. The concept of vessel wall remodeling to compensate for plaque growth is not supported (very low frequency of this type of lesion versus severe concentric lumen reduction in the natural history of coronary heart disease).

are more prone to this disease in precocity, diffusion, and severity of lumen reduction than other muscular arteries of the same caliber. In a comparative study (Antoci et al., 1980) between 40 acute myocardial infarct patients and 41 acute brain infarct/hemorrhage cases, the frequency of severe atherosclerotic stenoses (70%) in different arteries ranged from 18 to 68% in the first group versus 2.5 to 6% in the second. The relatively high frequency (53%), and multiplicity (23%), of severe coronary stenoses in acute brain infarct/hemorrhage patients (all without cardiac disease) and the low frequency of severe stenoses in brain arteries of both groups is worth mentioning (Figure 24D).

The physiologic intimal thickening peculiar to extramural coronary arteries was described and discussed above. Erroneously interpreted as an early stage of atherosclerosis (Ehrich et al., 1931; Fangman and Hellwig, 1947; Moon, 1952) or the result of an inflammatory/immune process (Minkowski, 1947) or of platelet microthrombi (Likar et al.,

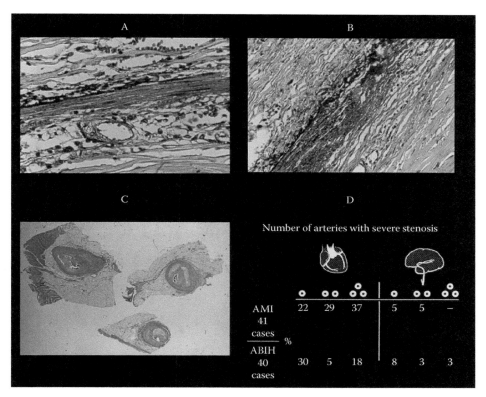

Figure 24 Alterations of the tunica media. In general, a corresponding thinning of the media (A) exists only at the level of an advanced plaque. Occasionally, focal destruction (B) with lympho-plasma-cellular reaction can be seen. However, a tunica media capable of spasm is always present. (C) Different patterns of damage along the course of the same stenotic plaque. For correct functional interpretation, it is necessary to consider all changes. Coronary atherosclerosis versus cerebral arteries: (D) percentage distribution of severe (>70%) stenosis in one or more vessels in 41 acute myocardial infarct cases and 40 cases of brain hemorrhage without heart disease. The coronary atherosclerosis in both groups is astonishingly more severe and diffuse than in cerebral arteries.

1969), it can be a predisposing factor for atherosclerosis related to hemodynamics and neural wall control. The few cases with an anomalous origin of one coronary artery from the pulmonary artery with severe atherosclerosis limited only to the coronary artery arising from the aorta provide a good example that emphasizes the importance of hemodymamics, as all other genetic and enviromental factors act in the same person (Kaunitz, 1947; Burch and De Pasquale, 1962; Blake et al., 1964).

We studied the frequency and quantified the extent of variables within plaques (Figure 23 and Figure 24; Table 9 and Table 10) in 3640 coronary histologic sections systematically sampled at the origin and at the site of maximal lumen reduction in eight selected tracts of extramural coronary arteries from 100 cases of acute myocardial infarction, 50 cases of chronic coronary heart disease, 208 cases of coronary sudden and unexpected death within 10 min from the onset of the terminal episode, and 97 cases of accidental death in individuals without a history of disease and no pathological findings at autopsy (Baroldi et al., 1988).

Table 9   Percentage Distribution of Morphological Variables and "Fibrous Plaque" in Different Groups in Relation to Lumen Reduction and Intimal Thickness

Group (Sections)	Morphological Variables									
	Absent	FP	P	AT	IV	HR	CA	ILI	ALI	ILI + ALI
No lumen reduction										
AMI (309)	96	—	2	1	2	1	0.6	1	1	2
CC (225)	99	—	0.4	—	—	—	0.4	0.9	—	0.9
SUD (580)	99	—	0.4	—	0.4	—	0.3	0.4	0.9	0.4
AD (405)	100	—	—	—	—	—	—	—	—	—
Total (1.519)	99	—	0.5	0.3	0.4	0.2	0.3	0.5	0.5	0.9
Lumen reduction ≤69%										
AMI (264)	9	17	72	56	43	16	50	55	46	67
CC (77)	23	43	40	43	40	9	43	51	40	57
SUD (346)	23	36	60	37	51	3	36	36	29	42
AD (307)	28	37	59	30	41	1	27	28	20	35
Lumen reduction >70%										
AMI (227)	0.4	11	60	78	72	43	74	74	65	85
CC (98)	4	14	44	73	68	32	63	76	67	87
SUD (413)	2	13	74	74	84	20	59	70	60	76
AD (64)	3	8	80	75	72	8	44	64	42	72
Grand total (2.121)	14	25	63	53	58	13	47	50	42	59
Intimal thickness (µm) ≤599										
AMI (43)	26	39	58	21	28	5	21	39	21	44
CC (3)	33	67	33	33	67	—	—	67	33	67
SUD (105)	64	76	19	9	12	—	13	11	6	13
AD (112)	66	71	28	6	12	—	11	6	6	9
Intimal thickness (µm) >600										
AMI (448)	3	12	67	71	59	31	65	66	58	78
CC (172)	12	26	42	60	56	22	55	65	56	74
SUD (466)	4	16	75	61	75	12	52	55	48	65
AD (259)	5	15	77	52	61	4	39	46	32	55

Note: Abbreviations: AD, accidental death in healthy subjects; ALI, adventitial lymphocytic infiltration; AMI, acute myocardial infarction; AT, atheroma; P, proteoglycans; CA, calcification; CC, chronic coronary heart disease; FP, fibrous plaque without atheroma or basophilia; HR, intimal hemorrhage; ILI, intimal lymphocytic infiltration; IV, intimal vascularization; SUD, sudden and unexpected death.

The more important findings were as follows:

1. A significant prevalence of atheroma, hemorrhage, calcification, and lympho-plasma-cellular infiltrates in a coronary group independent of the degree of lumen reduction and intimal thickness.

2. Intimal hemorrhage was the least frequent variable, being found in concentric, severe stenoses in the infarct-related coronary artery or branch.

3. Demonstration by serial sections of a dramatically increased vascularization around and within atherosclerotic plaques with giant capillary-like vessels communicating with proximal and distal secondary branches and forming an adventitial anastomotic network connecting with newly formed arterioles and angiomatous plexuses within the intima, in turn draining into the residual lumen. This finding was

Table 10   Main Significant Variations of Coronary Atherosclerotic Plaque Variables in Different Groups

	AMI	CC	SUD	SUDF	AD
All stenoses	+	−	+	ns	−
Severe stenosis (70%)	+	+	ns	ns	−
Concentric plaque	+	ns	−	ns	ns
Semilunar plaque	−	ns	+	ns	ns
Short stenoses (3 mm)	+	ns	ns	ns	+
Long stenoses (30 mm)	−	ns	ns	ns	ns
Intimal thickness 2000 μm	+	+	ns	ns	−
Intimal thickness 299 μm	−	−	−	−	+
Medial thickness 200 μm	ns	+	ns	ns	ns
Medial thickness 99 μm	+	ns	+	ns	ns
Atheroma/advential-intimal lymphocyticinfiltrate/calcification	+	+	ns	ns	−
Same variables in most stenoses in single case	+	+	ns	ns	−
Intimal hemorrhage	+	ns	−	−	−
Proteoglycans	+	−	ns	+	+
Intimal vascularization	−	−	+	+	ns
Fibrosis plaque	−	−	+	−	+
Thrombosis					
Acute occlusive	+	ns	ns	ns	−
Acute mural	+	ns	ns	ns	−
Previous occlusive	ns	+	ns	ns	−
Previous mural	ns	ns	ns	ns	−

Note: AD, accidental death in normal subjects; AMI, acute myocardial infarction; CC, chronic coronary heart disease; SUD, sudden and unexpected coronary death without myocardial fibrosis; and SUDF, sudden and unexpected coronary death with >10% myocardial fibrosis. + in excess; − in deficit; ns in expected range.

confirmed by plastic material injection (Baroldi and Scomazzoni, 1967; Zamir and Silver, 1985; see Figure 25).

4. Plaque calcification did not correspond to lumen reduction and is not synonymous with severe stenosis — a conclusion confirmed clinically (Sangiorgi et al., 1998). Coronary artery calcification does not predict the outcome in high-risk patients (Detrano et al., 1999).

5. The tunica media showed a reduced thickness, with occasional focal destruction only at the level of a severe plaque (Figure 24).

6. Intimal and adventitial inflammatory infiltrates were present in the plaques of all ischemic patients and significantly less were seen in normal subjects with similar lumen reduction and intimal thickness (Table 11). This inflammatory process, described long ago (Morgan, 1956; Baroldi, 1965; Kochi et al., 1985; Velican et al., 1989), was considered a minor complication related to plaque size (Schwartz et al., 1962) and was interpreted as an autoimmune process (Parums et al., 1981; van der Wal et al., 1989). In our experience, it starts after proteoglycan accumulation and is already present in small atherosclerotic plaques with minor (<50%) lumen reduction or low (<600 μm) intimal thickness (Table 12). In ischemic patients, the change was present in all or most plaques in the arterial coronary tree. In contrast, it was absent in most normal subjects or was found in one plaque only. This inflammatory reaction correlated with the presence of an acute luminal thrombus, infarct necrosis, and short severe stenoses mainly found in acute infarct cases. No correlation was observed with myocardial fibrosis, previous thrombi, or heart weight. A peculiar tropism of

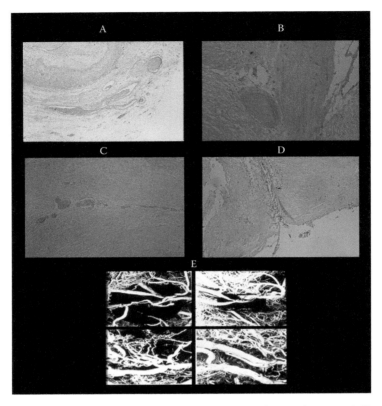

Figure 25 Vascularization of a coronary atherosclerotic plaque showing different aspects of neovascularization. By serial sections of postmortem injected plaques, giant advential capillary-like vessels (A) are connected with secondary branches proximal and distal to the plaque and with new arterioles (B) with a well-developed tunica media (indication of functioning blood flow), within the thickened, atherosclerotic intima, in turn, joined through angiomatous plexuses (C) to the residual lumen (D). (E) Plastic casts of plaques with different aspects of vascularization.

Table 11 Lympho-Plasma-Cellular Inflammation (LPI) in Coronary Atherosclerotic Plaque in Different Groups of Coronary Heart Disease and Normal Controls

	Number of Cases	LPI (%)	Number of Stenoses	LPI (%)	Mild	Moderate	Severe
Acute infarct	100	100	491	75	34	21	20
Chronic coronary heart disease	50	88	175	74	32	13	29
Sudden/unexpected death	208	83	1084	55	29	16	10
Healthy controls	97	64	371	41	22	14	5

Table 12 Percentage Distribution of Lympho-Plasma-Cellular Inflammation (LPI), Proteoglycan Accumulation (PA), and Atheroma (AT)

Intimal Thickness (μm)	LPI	PA	AT	Lumen Reduction (%)	LPI	PA	AT
300	2	3	2	<50	32	52	27
600	28	50	16	50–69	62	70	54
1000	54	69	43	70–79	74	71	75
2000	73	75	69	80–89	84	71	77
>2000	79	57	76	>90	82	58	74

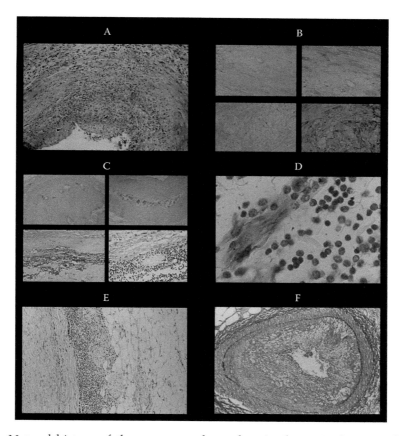

Figure 26 Natural history of the coronary atherosclerotic plaque in the general population, including most coronary heart disease patients. The starting point is a nodular hyperplasia of smooth muscle cells and elastic tissue with progressive fibrous replacement. No other changes, such as subendothelial lipoprotein-cholesterol storage, inflammatory process of any type, platelet aggregation, or fibrin-platelet thrombi, are found (A). Proteoglycan accumulation in the deep intima between tunica media and the fibrous cap is the second step (B). In this proteoglycan pool, lipoprotein/cholesterol clefts, in macrophages (foam cells), and Ca++ salts appear. Vascularization of the plaque and hemorrhage (C) follow. In the stage of proteoglycan accumulation, lympho-plasma-cellular infiltrates occur in the adventitia and intima (C) with specific localization, around adventitial nerves closed to the tunica media (medial neuritis) (D, E). This natural history is totally different from that obtained experimentally by a hypercholesterol diet in animals free of spontaneous atherosclerosis or in the small group of patients with familial hypercholesterolemia (F), in which transendothelial lipoprotein insudation is the starting point.

plasma cells and T lymphocytes for a nervous structure adjacent to the tunica media only at the plaque site was noted and defined as *medial neuritis* (Figure 26).

7. Platelet aggregates, fibrin-platelet mural or subintimal thrombi, subendothelial fatty streaks, lipoprotein/cholesterol infiltration, and macrophages loaded with the latter (foam cells) were not seen in 1519 coronary arterial sections with physiological thickening and a normal lumen and in 1319 sections with less than 70% lumen diameter reduction and in 743 sections with intimal thickening less than 1000 μm. Occasional lipoprotein/cholesterol subintimal deposits were seen in a few severe (70%) stenoses in ischemic patients.

This study permitted a reconstruction of the natural history of the coronary atherosclerotic plaque found in the general population (Figure 26). It starts with a nodular hyperplasia of smooth muscle cells with subsequent elastic hyperplasia ending in a fibrous nodule or early plaque, found in healthy subjects less than 20 years of age (Angelini et al., 1990). The second stage is a basophilic proteoglycan accumulation deep to the plaque's fibrous cap, close to the media, with the subsequent appearance of adventitial perinerve inflammatory infiltrates. Then, within the deep proteoglycan pool, lipoprotein/cholesterol plus foam cells or calcium salts become obvious in keeping with the chemical affinity of glycosaminoglycans for lipoproteins and calcium salts (Wight et al., 1983). A recurrence of these main changes — nodular myocell hyperplasia, fibrosis, deep proteoglycan accumulation, lipoprotein or calcium salts storage plus macrophagic reaction — explains the tridimensional growth of a plaque, that is, its radial, circumferential, and longitudinal expansion. The need is, as indicated above, to distinguish a *myohyperplastic atherosclerotic coronary plaque* (Table 13) present in the majority of the general population and a *hypercholesterolemic plaque* (Figure 26F) induced experimentally by hypercholesterolemic diet and found in limited conditions of human pathology. In our opinion, a plaque can be defined as "complicated" when it presents perimedial lympho-plasma-cellular neuritis, intimal hemorrhage, and thrombosis with or without rupture. A vessel lumen filled by atheromatous-like material can represent the degenerative changes in a long-lasting coagulum rather than be due to atheromatous plaque rupture.

Regression of atherosclerotic plaques in terms of a reduction or disappearance of luminal stenosis was rarely reported in angiographic studies (Bemis et al., 1973; Gensini and Kelly, 1972; Laks et al., 1979; Rafflenbeul et al., 1979; Haft and Al-Zarka, 1993) and is generally referred to as a recanalization of a thrombus, lysis of an embolus (O'Really and Spellberg, 1974), or resolution of spasm. A regression of angiographic stenoses associated with clinical improvement was obtained by enforcing a lipid-lowering diet in patients with a high level of apolipoprotein B (Brown et al., 1990). However, clinical benefits by

Table 13   Natural History of Human Coronary Myohyperplastic Atherosclerotic Plaque

Intimal Thickness (μm)				Lumen Diameter Reduction (%)
≤300	Nodular smooth muscle cell hyperplasia	→	Elastic hyperplasia ↓ Fibrosis	≤50
600	Proteoglycan pool in deep intima ↓ Lipoprotein-cholesterol Ca^{++} salts	→	Lymph-plasma-cell Inflammation Perimedial nerves Advential/intima	50–69
≥1000	Plaque tridimensional growth Myohyperplasia Fibrosis Proteoglycan Atheroma Ca^{++} salts Vascularization		Complications Hemorrhage Laminar/occlusive thrombosis with or without fissuration/rupture	≥70

statin therapy were not paralleled by angiographic changes (Vaughan et al., 1999). In the international nifedipine trial on antiatherosclerotic therapy, regression (20% of diameter stenosis) was observed in only 4% of 1063 coronary segments (Jost et al., 1993). Thus, while hypercholesterolemic plaque may regress if blood cholesterol is lowered, as happens in animal models (Wissler, 1978), it is unlikely that a myohyperplastic plaque would regress with this type of therapy.

At present, the concept of an active plaque is related to the interaction of several substances (metalloproteinase lysing the fibrous cap, etc.) released by different elements (macrophages, endothelial cells, smooth muscle cells, platelets, etc.) leading to plaque fissure or rupture plus thrombosis and embolization (Libby, 1995). In reality, the activity of a plaque is a more complex phenomenon in which other ignored (Heistad, 2003), but still present, factors must be considered (see below).

In this context, mention should be made of several clinical studies supporting an inflammatory process underlying coronary heart disease manifested by C-reactive and serum amyloid A proteins as markers of a hypersensitive inflammatory system predictive of higher mortality (Morrow et al., 1998; Liuzzo et al., 1998; Vakeva et al., 1998) and a systemic monocytic or granulocytic activation (Neri Serneri et al., 1992; Mazzone et al., 1993). These observations and the association with infectious agents revitalize a very old concept that an inflammatory process causes atherosclerosis ("chronic endoarteritis deformans"; Virchow, 1856). Recently, a direct immunofluorescence test for serum antibodies to *Chlamydia pneumoniae* was reactive in 86% of cases with severe atherosclerosis and in 6% with mild coronary atherosclerosis, whereas immunoperoxidase staining was, respectively, reactive in 80% and 38%. Elevated IgG and IgA levels against this microbiologic agent did not differ according to the degree and extent of coronary atherosclerosis (Ericson et al., 2000).

In our study of the early phase of myohyperplastic plaques, no inflammatory markers were detected, while medial neuritis was a later event at the time of proteoglycan accumulation. Therefore, the only inflammatory process, advential-medial-intimal lympho-plasmacellular phlogosis (in our opinion, macrophagic reaction is a repair and not an inflammatory process), is a secondary phenomenon of unknown nature (whether it be autoimmune or infectious), and its relationship with an inflammatory clinical component deserves further study (see Chapter 10 for further discussion).

Nonatherosclerotic Extramural Coronary Arteriopathies

Obliterative Intimal Thickening

Among this group, *allograft arteriopathy* is characterized by the most impressive change. The histologic pattern is seen in both human (Thomson, 1969; Cooley et al., 1969; Bieber et al., 1970; Smith et al., 1987b; Billingham, 1988; Rose et al., 1991) and experimentally (Kosek et al., 1969) transplanted hearts as a typical example of a subocclusive intimal proliferation. The medial and internal elastic membranes are intact, and the circumferential intimal thickening is a result of smooth muscle cell proliferation, increased ground substance, and minor interstitial fibrosis without elastic hyperplasia (Figure 27A). This lesion and those described below are often erroneously defined as "accelerated atherosclerosis" (Bulkley and Hutchins, 1977) despite the absence of typical variables and the natural sequence proper to the atherosclerotic process. In long-lasting cases (>1 yr), a superimposed atherosclerotic process can develop deep in the thickened intima. This arteriopathy,

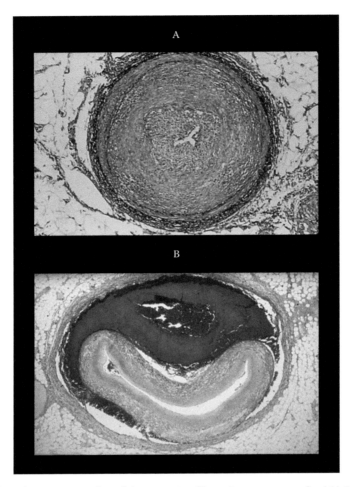

Figure 27   Subocclusive intimal proliferation in allograft coronaropathy (A). Note the absence of any atherosclerotic variable. Atherosclerosis may be superimposed in transplanted hearts with a long survival. (B) Dissecting aneurysm of an apparently normal coronary artery, with occlusion of the lumen. (Courtesy of Prof. G. Thiene.)

often defined improperly as chronic rejection, involves extramural coronary arteries and branches of any size. In 27 consecutive cases with a survival greater than 1 month, its frequency was 15%, while typical atherosclerosis, in the absence of allograft coronaropathic changes, was seen in 12 cases (44%). In no instance did this lesion occur in intramural vessels (Baroldi et al., 2003). However, it is still unclear if it is caused by an immunological process, which should occur in all cardiac vessels, or if other still undetermined factors, among which an adrenergic response to denervation, should be considered.

The same coronary artery obliterative intimal thickening may be observed in many other conditions, for example, coarctation of the aorta (Vlodaver and Neufeld, 1968), aortocoronary saphenous vein graft (Johnson et al., 1970; Marti et al., 1971; Vlodaver and Edwards, 1971; Brody et al., 1972; Kern et al., 1972; Virmani et al., 1991), tuberculosis (Gouley et al., 1933), rheumatic fever (Gross et al., 1935), systemic lupus erythematosus (Bonfiglio et al., 1972), experimental hypertension (Spiro et al., 1965; Oka and Angrist, 1967; Still, 1968; Esterly et al., 1968; Cohn et al., 1970; Constantinides, 1970; Huttner et al.,

1970; Wolinsky, 1972), flow volume variations (Rodbard, 1956), and trauma (Hassler, 1970).

Coronary Ostial Stenosis or Occlusion

Nonatherosclerotic coronary ostial stenosis or occlusion may occur in degenerative and inflammatory reparative processes and affect the ascending aorta, such as in syphilis, Takayasu's syndrome, spondylitis, aortic dissecting aneurysm, trauma (see below), adhesion of aortic semilunar valve cusp, embolism, or papillary fibroelastoma of an aortic cusp (see Chapter 6).

Infantile Arterial Calcification

The cause and pathogenesis of this rare disease are still unknown. It is characterized by early degeneration and progressive calcification of the internal elastic membrane, with atrophy of the media and intimal proliferation with foreign-body giant cells when calcification is massive. This systemic disease is particularly prominent in the extramural coronary arteries with severe lumen reduction and is occasionally associated with myocardial infarct (Brown and Richter, 1941; Stryker, 1946; Sladden, 1952). This pattern, not to be confused with Mönckeberg's sclerosis in which calcification is limited to the media, is similar to that observed in cases with pseudoxanthoma elasticum (Eddy and Farber, 1962) and in other, not well-defined cases with obstructive intimal proliferation, particularly of the extramural coronary arteries with, or without, calcification and myocardial infarction (Jokl and Greenstein, 1944; Rosenberg, 1973). This raises the possibility of a spectrum of different aspects of the same entity (Witzleben, 1970). The relationship with other diseases, such as Hurler's syndrome (gargoylism) and Marfan syndrome in which obliterative intimal thickening of the extramural coronary arteries is present (Lindsay, 1950; Schiebler et al., 1962; Vlodaver et al., 1972), merits further investigation (see Chapter 7).

Infectious–Immune Processes

Obliterative intimal thickening occurring as a common reaction in many pathologic conditions was discussed in a preceding paragraph. Here, other distinctive morphologic features seen in specific diseases are summarized. In *tuberculosis*, coronary arteries may show typical granuloma of the wall, particularly when this disease involves pericardium and myocardium. *Polyarteritis nodosa* affects the extramural coronary arteries in 38% of cases (Holsinger et al., 1962) with symptoms and signs of coronary heart disease associated with severe lumen obstruction. *Giant cell arteritis* is rarely observed in coronary arteries — one of 16 cases with temporal arteritis (Harrison 1948) — and shows the typical pattern of giant cells associated with degeneration of the internal elastic membrane, medial necrosis, mononuclear infiltrates, and intimal proliferation (Kimmelstiel et al., 1952). Less specific changes occur in extramural coronary arteries, with minor intimal thickness in rheumatic fever (Karsner and Bayless, 1934), in lupus erythematosus (Bonfiglio et al., 1972) and in Buerger's disease, the latter being a controversial entity (Wessler et al., 1960).

Coronary Aneurysm

A distinction must be made between vascular ectasia and aneurysm, terms that are often used synonymously, particularly by angiographists. By ectasia we mean a dilatation of the whole vessel, for instance, as a result of aging, while an aneurysm is a local fusiform or saccular dilatation of the wall due to a congenital or acquired defect of wall structure. Among the latter, any inflammatory-necrotizing process, such as polyarteritis or panarteritis nodosa (Sinclair and Nitsch, 1949) or Kawasaki disease, considered the juvenile form of the latter, may cause an aneurysm with risk of rupture. Atherosclerosis, septic emboli, as well as any arteritis may produce an aneurysm of the extramural coronary arteries. They were detected by cineangiography in 25% of 1660 patients (right coronary artery [RCA] 53%, left anterior descending [LAD] 25%, and left circumflex [LCX] 22%; Aintablian et al., 1978). A total of 127 coronary aneurysms were reported in a review of 89 autopsy cases involving 66 men and 23 women. They were congenital in 17%, atherosclerotic in 52%, mycotic in 11%, syphilitic in 4%, dissecting (see below) in 11%, and undetermined in 5%. Thirty-five were located in the left main trunk, 15 in the LAD, 15 in the LCX, 1 in the left posterior descending branch, and 61 in the RCA. The main complications were rupture with hemorrhage, thrombosis with embolism, and compression of adjacent structures (Daoud et al., 1963). In relation to the frequency and severity of coronary atherosclerosis, one notes how rare coronary atherosclerotic aneurysms are. In our experience, in hundreds of cases with and without aortocoronary injection by plastic and radiopaque material, we never observed a spontaneous rupture of an atherosclerotic aneurysm producing pericardial hemorrhage.

Coronary Dissecting Aneurysm

We noted that an aortic dissecting aneurysm may involve the ostium or extend along a coronary artery. Such a dissection may also be a complication of Marfan syndrome, hypertension, Takayasu's syndrome, or other aortopathies. In 4 of 20 cases, the dissection extended into the left main coronary trunk — in one it also extended into the RCA — with an increase of proteoglycans (cystic necrosis) in the dissected coronary vessels (Virmani et al., 1984).

In rare instances, mainly in women, often in the peripartum period, coronary dissection is a primary and isolated event. Among 25 cases (21 women, 8 peripartum; 4 men), 18 died suddenly and unexpectedly and, at autopsy, 6 had an acute myocardial infarction that had preceded death by days. Most dissections occurred in the LAD, likely through a break of the intima (Claudon et al., 1972). The dissection involved the outer tunica media and produced an acute occlusion by lifting the dissected wall into the lumen (Figure 27B). The etiopathogenesis is unknown. Nonspecific histological changes were focal granulation tissue, scar, degeneration of internal elastic lamina, and more or less extensive infiltration of eosinophils, lymphocytes, plasma cells, and histiocytes, but rarely polymorphonuclear leukocytes (Silver, 1968; Claudon et al., 1972). These findings associated with a thrombus in a pregnant 23-year-old woman who died suddenly after 9 h of nausea and chest pain and without any other pathologic findings (Ahronheim, 1977) suggest that the primum movens of this entity is a periarteritis frequently observed in these cases, the dissection being due to the lytic action of substances such as lysosomes, phospholipase, and so forth, released by eosinophils (Virmani et al., 1984).

Coronary Embolism

Today, a coronary artery embolism is a rare event because of the low frequency of valvular endocarditis in Western countries. In a previous study of 74 cases, the source of the embolus was an endocarditic valve (64%), ventricular thrombosis (11%), luetic aortitis (5%), ulcerated atherosclerotic plaque of the aorta (4%), paradoxic emboli of various types (4%), proximal aortocoronary thrombosis (6%), pulmonary thrombosis, caseous tuberculous material, neoplastic tissue, calcific valve fragment (5%), and undetermined (1%). The left main trunk and first tract of the LAD were most frequently involved (68%). The acute coronary occlusion resulted in sudden death in 60% of these cases (Wenger and Bauer, 1958). Embolism may also occur after cardiovascular investigative or therapeutic procedures (see below).

Trauma

In another section, cases with a thoracic wound and associated laceration of a coronary artery that was repaired surgically by vessel occlusion and not followed by a myocardial infarct were mentioned. Nonpenetrating trauma of the chest was considered a possible cause of myocardial infarction (Cheitlin et al., 1975; Jones, 1970; Vlay et al., 1980), even in the absence of coronary occlusion on late coronary cineangiography (Candelle et al., 1979; Oren et al., 1976). Experimentally, a thrombosis occurs only when the intima breaks (Moritz and Atkins, 1938). Two unusual examples are worth mentioning. One reported the formation of a coronary fistula to the right ventricle after trauma by a bullet (Alter et al., 1977). The other reported a coronary occlusion in a football player caused by a calcific embolus from an atherosclerotic plaque after thoracic contusion (Cheitlin et al., 1975) (see Chapter 12).

Iatrogenic Coronary Artery Damage

An isolated coronary artery dissection may follow catheterization or angioplasty or may develop at the site of insertion of a coronary bypass graft. A localized and controlled dissection is expected following endoarterectomy. Fatal coronary artery dissection is an extremely rare event (0.02 to 0.05%) during coronary cineangiography (Virmani et al., 1984) and may cause sudden death. The reestablishment of a normal lumen by angioplasty implies fragmentation and dissection of the atherosclerotic wall. Nevertheless, as proved by thousands of cases, this procedure has a mortality rate of 0.9% with minor early complications in 14% (Dorros et al., 1984). The fissured intima may heal regularly, but evidence of a healed dissection tract is often obvious.

Obliterative intimal thickening followed high-pressure coronary perfusion during surgical replacement of an aortic valve (Silver et al., 1969; Brymer, 1974; Molina, 1993) and occlusion of a coronary artery by erroneous ligation may occur during surgical valve replacement.

Another suspected iatrogenic cause of coronary heart disease in young women is the use of oral contraceptives, particularly when associated with hyperlipidemia, hypertension, and smoking (Oliver, 1970). Of interest is a case of a castrated 37-year-old woman without any risk factor who received substitute estrogen therapy. She had two myocardial infarcts, the last being fatal. At postmortem, there was a severe stenosis (> 70%) of the RCA with circumferential intimal thickening and small laminar thrombi, without evidence of an atherosclerotic process; the residual lumen was occluded by a fresh thrombus. A similar pattern was seen in the LAD. Its stenosed lumen was occluded by an organized, vascularized thrombus. This case suggested that progressive laminar thrombosis with organization of older thromboses was caused by estrogen therapy (O'Really and Spellberg, 1974). Finally, chest radiation for mediastinal malignant tumors may induce adventitial and medial coronary artery fibrosis with unusual fibroblasts and obliterative intimal thickening (Applefeld and Wiernik, 1983).

Intramural Arterial Vessels

Atherosclerosis

Among the thousands of hearts we studied, including those with severe extramural atherosclerosis, not one had changes caused by atherosclerosis affecting intramural vessels. Only in type 2 or type 4 hyperbetalipoproteinemia will such a lesion be observed (Vlodaver et al., 1972).

Obliterative Intimal Thickening

An obliterative intimal thickening of arterioles is frequently observed at the periphery of a myocardial infarct in the early phase of healing (Figure 13E). The cause of this secondary change is not unknown. A possible role for adrenergic stimulation, which acts at the periphery of an infarct, could be worth investigation. On the other hand, fibrous intimal thickening and medial hyperplasia are normally observed around and within myocardial scars, and disarray may have an origin associated with blood flow hindrance in nonfunctioning myocardium. A similar thickening is frequently seen, even in normal subjects, around the annulus fibrosus and in the membranous septum, interventricular septum, conduction system, and papillary muscles — changes likely related to the peculiar functional anatomy in these cardiac areas (Boucek et al., 1965).

Perivascular Fibrosis

This is the typical change seen in rheumatic heart disease, often associated with a classic nodular granuloma. On the other hand, perivascular fibrosis is a common finding in the myocardium following repair. In cardiac hypertrophy, it is a part of myocardial adaptation to increased function (compensatory hypertrophy) or to a dysfunctional disorder (pathologic hypertrophy).

Amyloidosis

Rare in extramural coronary arteries, amyloidosis is more often seen infiltrating the wall of intramural vessels and the interstitium. The vascular deposit does not determine any other anatomical change and seems innocuous in terms of function. Some nodular hyaline subendothelial deposits are often present in the elderly. They are not and must be distinguished from amyloidosis.

Fibrin-Platelet Thrombi or Emboli

Occlusive vascular changes or spasms are often invoked to explain acute coronary syndromes in the absence of cineangiographic coronary obstruction (Eliot and Bratt, 1969; Oliva et al., 1973; Maseri et al., 1978a; Sasse et al., 1975; Rosenblatt and Selzer, 1977). In postmortem studies, fibrin-platelet thrombi and emboli, presumably derived from mural thrombi or atheromatous material, were seen in a relatively small number of vessels in acute coronary syndromes and were considered the cause of focal necrosis, myocytolysis, microinfarcts related to sudden death (Jorgensen et al., 1968; Mustard and Packham, 1969; Haerem, 1972; Frink et al., 1978; El-Maraghi and Genton, 1980; Falk, 1985; Davies et al., 1986).

Thrombotic microangiopathy of intramural arterial vessels is an impressive morphologic change observed in the myocardium of patients dying from *thrombotic thrombocytopenic purpura* (TTP) or Moschcowitz's disease (Moschcowitz, 1924). Of 39 cases from

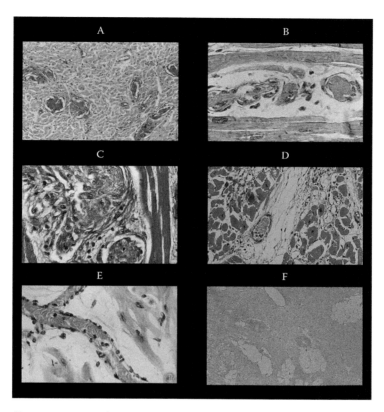

Figure 28 Different aspects of severe intramural arteriolar obstructions without myocardial damage in cases of thrombotic thrombocytopenic purpura (A, B) in the absence of fibrosis of the pseudothrombotic material (C) and numerous occlusive platelet aggregates in normal arterioles (D). Despite severe hypoxia, hemorrhage, hemolytic anemia, and this obstructive microangiopathy, no one case had symptoms and clinical or histologic signs of coronary heart disease. Similarly, despite the entanglement of sickle red cells (E), no coronary heart disease occurs in sickle cell anemia. (F) Thickening of the media of intramural arterioles secondary to myocardial dysfunction, for example, scar formation.

the files of the Armed Forces Institute of Pathology, Washington, D.C., studied personally, 90% died after a survival time ranging from 4 to 300 days (average 37). The pathognomonic microangiopathy, often defined as hyaline thrombosis or platelet-fibrin thrombosis, was present in 100% of hearts, adrenals, pancreas, and lymph nodes; 93% of posterior hypophyses; 85% of brains, livers, kidneys, and gastrointestinal tract; 75% of thyroids, spleens, bone marrow; and 13% of lungs. This arteriolar microangiopathy particularly diffuse in the myocardium was formed by an intense PAS-positive, diastase-resistant material located in the subendothelium. It bulged and suboccluded the lumen without any evidence of thrombus formation (Figure 28). Occasionally, the involved vessel assumed a glomerular-like aspect. Healing or fibrosis was never observed, even in serial sections of affected vessels. A glomerular-like aspect suggesting a recanalization of a thrombus is likely due to subendothelial PAS material deposition at the bi- or trifurcation of a vessel. Occasionally, fibrin was demonstrated on the endothelial surface. No infiltration of any type of inflammatory cell or medial necrosis or capillary involvement was seen. Giant capillaries interpreted as aneurysmal dilatations (Orbison, 1952) by serial section resulted from collaterals joining two arterioles. In normal arterioles, occlusive platelet aggregates without fibrin were frequently observed. The tissue damage in involved organs could not explain symptoms and signs, and no one had an acute myocardial infarction or angina (Baroldi and Manion, 1967). From the files of the same institute, the author also examined the myocardium in the following diseases: idiopathic thrombocytopenic purpura (8 cases, 22 surgically excised spleens), necrotizing angiitis (9 cases), Schönlein-Henoch purpura (6 cases), polyarteritis nodosa (23 cases), rheumatoid arthritis (25 cases), acute rheumatic cardiopathy (25 cases), lupus erythematosus disseminatus (27 cases), atypical endocarditis (18 cases), bacterial endocarditis (22 cases), acute (20 cases) and chronic leukemia (10 cases), acute necrotizing hepatitis (10 cases), liver cirrhosis (25 cases), diabetic glomerulitis (20 cases), carcinoma of various organs (22 cases), amyloidosis (20 cases), hemangioma-thrombocytopenia Kasabach-Merritt syndrome, 1949; (1 case, 1 biopsy) cavernous hemangioma (10 of the liver, 7 spleen, 8 pancreas), sickle cell anemia (50 cases), prolonged shock (50 cases), malignant nephrosclerosis (26 cases), afibrinogenemia in abruptio placentae (7 cases). In none of the cases was a microangiopathy like that found in TTP, but in a few vessels in cases of Kasabach-Merritt syndrome it was observed. A review of 220 TTP cases reported in the literature confirmed my conclusion that in all conditions with morphologic involvement of intramural vessels frequently associated with severe hypoxia, no symptoms or morphologic changes of coronary heart disease occur.

Red Blood Cell Plugging

Disseminated intravascular coagulation (Hardaway et al., 1961; McKay, 1965) was proposed to explain ischemic conditions. In sickle cell anemia, the plugging of sickled erythrocytes can be demonstrated *in vivo* (Knisely, 1961; Levine and Welles, 1964) by examination of the retina and histologically in the myocardium (Figure 28) by a diffuse sickle cell aggregation in arterioles and capillaries without evidence of myocardial damage (Baroldi, 1969). This is another example that shows a lack of a relationship between microcirculatory alteration and ischemia.

Medial Hyperplasia Obliterans

This change already described in Chapter 2 is not related to age, gender, heart weight, myocardial fibrosis, or any other variable (Table 14). It is different from the medial hyperplasia found within myocardial scar tissue (Figure 28) and in hypertrophic cardiomyopathy,

Table 14   Medial Hyperplasia Obliterans (MHO) in Relation to Age and Gender in 208 Coronary Sudden/Unexpected Death (SD) Cases and 97 Otherwise Healthy Subjects of Accidental Death (AD)

		SD				AD		
Decade	Total Cases	M	W	(%)[a]	Total Cases	M	W	(%)[a]
40	35	13	2	15 (43)	10	6	1	7 (70)
40–49	33	16	2	18 (54)	13	7	2	9 (69)
50–59	68	33	2	35 (51)	28	24	—	24 (86)
60–69	45	21	3	24 (53)	23	14	3	18 (78)
70	27	11	6	17 (63)	23	17	1	18 (78)
Total	208	94	15	109 (52)	97	69	7	76 (78)

Note: M, men; W, women.

[a] Percentage of total cases per decade.

where longitudinal bundles are not so frequent. Both of these medial changes, often interpreted as the cause of myocardial damage, are, rather, related to particular myocardial function (papillary muscles) or dysfunction (scar, disarray) (Baroldi and Silver, 1995). This must be considered when an endomyocardial biopsy is examined.

Cardiac Lymphatic Vessels

It is not known if cardiac lymphatic vessels develop adaptive changes or whether such changes might influence cardiac pathology. It seems reasonable that any mediastinal lesion that blocks lymph flow into the thoracic duct could affect the normal status of the cardiac interstitium. In an extensive review of experimental obstruction of cardiac lymph flow, Miller (1982) found that acute (EKG changes, impairment of ventricular function, ischemic myocardial necrosis, etc.) and chronic (endocardial fibroelastosis, larger infarct size, alteration of healing process, etc.) findings were inconstant and possibly secondary to the experimental technique. Similarly, there is a long list of human conditions (rheumatic fever with lymphangitis forming Aschoff's bodies, radiation damage, endocardial fibroelastosis, etc.) in which lymphostasis could have a pathologic role. However, in these conditions, it is difficult to distinguish primary from secondary events. At present, no objective proof of any cause–effect relationship exists, and the claim that endocardial fibroelastosis and other intramyocardial changes are related to lymphostasis (Miller, 1982) lacks solid documentation. Only in transplanted hearts where lymphatic drainage is severed surgically do we have a model to study, at least in the acute phase because lymph recanalization occurs rapidly (Miller, 1982). Possibly, the massive myocardial interstitial exudation seen in fatal acute rejection is aggravated by an abolished lymph drainage (see below). The supposition that lymphostasis causes accelerated atherosclerosis by impairing intimal repair (Mehmet et al., 1987) is not confirmed.

Cardiac Nerves

The intrinsic cardiac nervous system includes afferent neurons, local circuit neurons (interconnecting neurons), and both sympathetic and parasympathetic efferent postganglionic neurons — a very complex interactive regulatory system that participates in many cardiopathies (Armour, 1999), particularly sudden death.

At present, morphologic knowledge of reversible and irreversible alterations affecting intramyocardial nerves is very limited (Paessens and Borchard, 1980; Marron et al., 1995). Indirectly, using iodine-123-metaiodobenzylguanidine (MIBG) scintigraphy or other methods, a reduction or denervation of adrenergic nerves and their terminal endings was observed in experimental and human infarction (Barber et al., 1983; Stanton et al., 1989; Dae et al., 1991; Spinnler et al., 1993; Kramer et al., 1997); adjacent noninfarcted regions (Matsunari et al., 2000); ischemic heart disease (Calkins et al., 1993); absence of coronary disease (Mitrani et al., 1993); relation to malignant arrhythmia (Schaeffers et al., 1998); vasospastic angina (Kaski et al., 1986; Sakata et al., 1997; Takano et al., 1997); arrhythmogenic right ventricular cardiomyopathy (Wichter et al., 1994, 2000); dilated cardiomyopathy of unknown origin (Schofer et al., 1988; Ungerer et al., 1998); congestive heart failure (Bean et al., 1994; Kaye et al., 1994; Nakata et al., 1998); Chagas' disease (Emdin et al., 1992); and inherited ventricular arrhythmia (Dae et al., 1997). The magnitude and dispersion of local repolarization responses are related to the severity of denervation and neural versus humoral stimulation (Yoshioka et al., 2000). Furthermore, in a model of global denervation (i.e., the orthotopically transplanted heart), the process of reinnervation is still unclear (Wilson et al., 1993; Halpert et al., 1996) and late (De Marco et al., 1995), requiring 15 years (Bengel et al., 1999). Denervation limited to one myocardial zone may have a more rapid reinnervation. However, at present, the nature of dysfunction and recovered function of viable myocardium (Udelson et al., 1994) is a matter of speculation, and denervation is one possibility. Experimentally, stunned myocardium following reperfusion after ischemia can recover with catecholamine infusion, if not by stellate ganglia stimulation, a fact that raises the suspicion of denervation (Ciuffo et al., 1985). This poses the question of whether denervation is caused by ischemia, or an inflammatory interstitial process (e.g., myocarditis), or tensile mechanical forces (stretching or hypercontraction of myocardial cells), with different conditions having different causes. One fact is important: no matter its nature, denervation can trigger myocardial supersensitivity to catecholamines, resulting in malignant arrhythmias (Donald, 1974; Bevilacqua et al., 1986; Inoue and Zipes, 1987; Rundqvist et al., 1997) associated with pathological contraction bands (Baroldi et al., 2001a). Autonomic neuropathy in diabetic patients increases the sympathetic tone, and sudden death is frequent in this condition (Jacoby and Nesto, 1992; Di Carli et al., 1999).

Specific Heart Diseases and Sudden Death

5

GIORGIO BAROLDI AND VITTORIO FINESCHI

Contents

Heart disease and heart failure may result from congenital anomalies or acquired conditions such as coronary artery disease, hypertension, valvular heart disease, myocarditis, and so forth. Then, there is a group of diseases, the cardiomyopathies, where the cardiac defects originate within the organ and may alter its anatomy if definitely altering its function adversely. These and other pathological conditions will be discussed in subsequent sections.

Coronary Heart Disease

Ischemic heart disease or *coronary heart disease* is the leading cause of morbidity and death in Western countries (Kannel and Thom, 1994) and in any affluent developing society. Its 40% frequency, as cause of death, is double that of malignant tumors. More important, and in contrast to other diseases, it affects people at the time of their maximal experience and contribution to society with an incalculable adverse effect.

The present concept in defining this disease is as follows:

> Evidence from serial coronary arteriography and that obtained after reperfusion by thrombolysis, at operation during acute coronary syndromes and from postmortem arteriography have also confirmed the importance of plaque disruption and thrombosis. Indeed, these acute or subacute changes in coronary arterial anatomy appear to be the most frequent cause of all the acute coronary syndromes including unstable angina, myocardial infarction and ischemic sudden death. If we accept the premise that all three acute coronary syndromes may evolve from acute plaque disruption followed by thrombosis or spasm or both, we can construct an unifying theory. (Gorlin et al., 1986)

This unifying theory was suggested by (1) angiographic demonstration of coronary occlusion in 87% of patients with Q wave myocardial infarct (De Wood et al., 1980, 1986); (2) recovery of a layered thrombus proximal to a stenosis in most patients undergoing emergency bypass surgery for an acute infarct (De Wood et al., 1980); (3) by fissuring or rupture of an atherosclerotic plaque with thrombus formation or embolization causing an infarct or microinfarcts, respectively (Davies and Thomas, 1984; Davies et al., 1986; Falk, 1985); and (4) progression of the stenotic obstruction by silent mural thrombi (Davies, 1990). The theory linking coronary artery changes and acute coronary syndromes is as follows:

1. Atherosclerotic coronary artery obstruction results in chronic ischemia and angina pectoris.
2. Coronary artery thrombotic occlusion following plaque rupture determines a blockage of the nutrient flow with infarct necrosis in the myocardium related to the occluded artery and sudden death.
3. Intramyocardial embolization from a ruptured plaque causes microinfarcts and sudden death.

This theory does not provide for the following:

1. Recognition of different forms of myocardial injury present in coronary heart disease
2. Existence of compensatory collateral flow

Unfortunately, clinical imaging of coronary arteries (see above) is not ideal. In addition, all *in vivo* images and postmortem findings pertain to patients affected by the disease. At present, we cannot visualize the beginning of an acute event, and no experimental model exists. Therefore, we are dealing with hypotheses. To be sustained, they must be constantly tested. Our studies contradict points in the current theory.

Chronic Ischemia and Atherosclerotic Coronary Artery Obstruction

The majority of ischemic cases have severe atherosclerosis of the extramural coronary arteries and their branches when an acute event occurs for the first time in an apparently healthy subject (first episode), or one develops in a patient with previous disease manifestations (second episode or chronic). This means that severely obstructive coronary atherosclerosis, even affecting several vessels, preexists long before the first symptom and without any evidence of ischemia despite an often stressful life. The second observation is that the same conditions exist in noncardiac patients and in otherwise healthy people succumbing to accidental death (Table 15). One is forced to conclude that collateral vessels were able to compensate for any blood flow reduction induced by severe coronary stenoses. No proof exists that a failure of collateral flow initiates symptoms, keeping in mind that the endothelial capillary-like structure (with an absence of a smooth muscle cell coat) of collaterals does not support the likelihood of spasm of these vessels. The increased incidence of coronary atherosclerosis in chronic patients will be discussed subsequently. Here it is

Table 15   Percentage Distribution of Maximal Lumen Reduction and Number of Atherosclerotic Coronary Arteries with Severe (>70%) Lumen/Diameter Stenosis

Total Cases	AMI 1st (145)	AMIC (55)	SD 1st (133)	SDC (75)	NCA (100)	AD (97)
Stenosis						
0	2	—	8	—	7	8
<50	2	2	13	—	10	21
50–69	7	—	13	7	17	32
70	21	14	16	11	11	19
80	31	20	29	18	24	14
90	37	64	20	64	31	6
Vessel >70%						
1	42	29	30	17	26	23
2	34	40	25	35	18	13
3	13	29	10	41	22	3

Note: AMI 1st, first episode of acute myocardial infarction; AMIC, in chronic patients; SD 1st, sudden and unexpected coronary death as first episode; SDC, in chronic patients; NCA, noncardiac patients who died from noncardiac causes; AD, otherwise healthy subjects who succumbed to accidental death.

Table 16   Distribution of Maximal Lumen-Diameter Reduction and Number of Vessels with Severe Stenosis in Relation to Age in 200 Acute Myocardial Infarct (AMI) 208 Sudden and Unexpected Coronary Death (SD) and 97 Normal Cases Who Succumbed to Accidental Death (AD)

Age (Years)	Total Cases	Lumen Diameter Reduction (%)					
		0	≤69	≥70	1	2	3 Vessels
≤39							
AMI	4	1	—	3	2	—	1
SD	35	10	8	17	7	6	4
AD	10	2	7	1	—	1	—
40–69							
AMI	133	2	11	120	46	50	24
SD	146	—	28	118	37	45	36
AD	64	6	33	25	14	9	2
≥70							
AMI	63	—	3	60	29	21	10
SD	27	—	5	22	9	9	4
AD	23	—	11	12	8	3	1

Table 17   Percentage Distribution of Infarct Size in 200 Acute Fatal Cases First-Episode (AMI 1st) and in Chronic Patients

AMI		Infarct Size (%)					
		≤10	11–20	21–30	31–40	41–50	>50
First	145 (100%)	22	21	22	16	10	9
Chronic	55 (100%)	51	13	22	5	7	2
Total	200 (100%)	30	19	22	13	9	7

important to stress that the frequency of severe, multiple coronary artery stenoses increased progressively from 90 to 95% in acute myocardial infarction and from 39 to 51% in controls, while in sudden and unexpected death, it remains constant when groups were subdivided into those less than 39 years old, between 40 and 69 years of age, and those older than 70 years (Table 16). The higher values in the oldest people were not statistically significant, indicating that a considerable progression of coronary atherosclerosis does not occur in old age. However, the increasing life expectancy and prolonged survival of patients will change these figures, while more exhaustive studies in healthy elderly subjects are needed. We note that collaterals are present at any age.

Another point to be stressed is the lack of relationship between death and infarct size, that is, the percent of the left ventricular mass inclusive of the interventricular septum, and infarct size and vascular territory of the related artery or main branch. In Table 17, the percentage distribution of infarct size measured in 200 fatal acute infarcts is reported. In 43% of first-episode cases and in 64% of second episode cases, the size was less than 20%. In other words, the infarct involved only a small part of the myocardial mass and the whole vascular territory of the related artery, while the largest infarcts invaded vascular territories of other nonoccluded arteries.

Table 18 Percentage Distribution of Acute Infarct Size in Relation to Grade and Number of Coronary Arterial Stenoses

Infarct Size (%)		Lumen Reduction (%)				
		<69	≥70	1 Vessel	2 Vessel	≥3 Vessels
≤20	97	7	93	40	38	14
>20	103	10	90	37	33	20
Total	200	8	91	38	35	17

Note: p = ns.

Table 19 Frequency of Previous Coronary ≥90% Lumen-Diameter Reduction without Extensive (≥10% Left Ventricular Mass) Myocardial Fibrosis (EF)

Source	Number of Cases	≥90% Stenosis (%)	1 Vessel	2 Vessels	≥3 Vessels
AMI	200	89 (100)	68 (100)	16 (100)	5 (100)
No EF	145	54 (61)	45 (66)	6 (37)	3 (60)
SD	208	75 (100)	51 (100)	22 (100)	2 (100)
No EF	133	27 (36)	19 (37)	7 (32)	1 (50)
NCA	100	40 (100)	27 (100)	11 (100)	2 (100)
No EF	81	31 (77)	21 (78)	9 (82)	1 (50)
AD	97	6 (100)	3 (100)	2 (100)	1 (100)
No EF	92	6 (100)	3 (100)	2 (100)	1 (100)

Note: AMI 1st, first episode of acute myocardial infarction; SD 1st, sudden and unexpected coronary death as first episode; NCA, noncardiac patients who died from noncardiac causes; AD, otherwise healthy subjects who succumbed to accidental death.

The adequate compensatory function of coronary collaterals is also confirmed by a lack of a relationship between infarct size and the number of severe stenoses (Table 18). This is a contradictory fact, because more stenoses should imply a greater flow reduction with consequent larger infarct, and there is an absence of extensive myocardial fibrosis in most cases with previous severe (90%) coronary artery stenosis (Table 19).

Finally, adequate collateral function was proven experimentally in dogs by occlusion of a previously controlled stenosis maintained for 5 to 7 days. That produced a dramatic increase in the diameter of collaterals that totally protected from the occlusion, as shown by the absence of infarct, dysfunction, or malignant arrhythmia (Khouri et al., 1968). All these observations support the concept that coronary collaterals can compensate for an acute occlusion of a severe preexisting stenosis as is found in the majority of ischemic cases. In other words, in chronic obstructive coronary disease, any acute coronary syndrome occurs in the presence of an already enlarged collateral system, as shown by postmortem coronary plastic casts (see below). However, the question remains open for the relatively rare cases of acute embolism or dissection of normal coronary arteries — conditions similar to the experimental acute occlusion of a normal coronary artery. Should this imply an insufficiency of normal collaterals? A few human cases were reported where a lacerated normal main coronary artery following a chest wound was surgically occluded acutely without a subsequent infarct (Pagenstecher, 1901; Bradbury, 1942; Zerbini, 1943; Bean, 1944; Carleton and Boyd, 1958; Parmley et al., 1958).

Collaterals and Coronary Atherosclerosis

The association between coronary atherosclerosis and coronary heart disease is incontrovertible but, per se, does not prove a cause–effect relationship between lumen reduction and ischemia as is generally believed. That assumption is based on ignorance of the compensatory function that native collaterals supply. The latter has led to the suggestion of gene therapy to promote new vessel formation, including collaterals, to restore nutrient blood flow (Unger et al., 1990). We reported that postmortem tridimensional plastic casts of the normal coronary arteries show innumerable anastomoses joining at any level the arterial intramural branches, forming an extensive network (Figure 6). These normal collaterals enlarge in cardiac hypertrophy and in chronic hypoxic states (e.g., anemia) in the presence of normal coronary arteries (Figure 9D) and overall when coronary lumen obstruction occurs. Enlargement is maximal when severe coronary atherosclerotic lesions are multiple (Figure 29). Because coronary atherosclerosis develops slowly, there is plenty

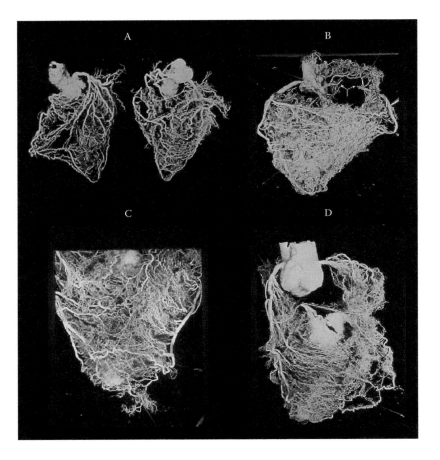

Figure 29 Collateral enlargement in topographical relation (satellite) with severe stenosis or occlusion. (A) Double occlusion of LAD (anterior view) and occlusion of RCA (posterior view) apparently compensated by enlarged collaterals in a non cardiac patient dead from brain hemorrhage. (B) Similar condition in cases with RCA occlusion (arrow) without corresponding myocardial infarct with numerous homo and intercoronary collaterals of the anterior wall (C), and (D) septum.

of time for collateral enlargement (Gregg, 1974; Gregg and Patterson, 1980). Experimentally, the latter develop within a few days (Khouri et al., 1968), disappearing when a stenosis is removed and reappearing as soon as the latter is reestablished (Khouri et al., 1971). Therefore, even at a first episode of the disease, we deal with an enlarged and already well-functioning collateral system; the image obtained by plastic casts provides a unique way to study the natural history of human anastomoses by establishing frequency and enlargement in different conditions (Table 20). This enlargement is likely stimulated by hypertrophy/hypoxia and pressure gradients when related to a stenosis rather than ischemia (Chilian et al., 1990). In the latter instance, the enlargement should be diffuse to the whole ischemic zone, which is not the case (Figure 29F).

The limited view of the collateral system obtained by coronary cineangiography is the main reason that angiographers underestimate it (Bodenheimer et al., 1977; Cohn et al., 1980; Vanoverschelde et al., 1993; Helfant et al., 1970), with the misleading belief that it is ineffectual. This assumption is based on the present, very poor cineangiographic resolution of intramural vessels, including anastomotic channels, and the use of selective single coronary artery injection with radiopaque blood flow competing with the simultaneous nonradiopaque flow from other coronary arteries. In no instance of coronary occlusion demonstrated by plastic casts did we have a vessel cutoff, as is seen by cineangiography; the whole artery distal to an occlusion was filled via collaterals. In reality, a severe coronary obstruction may show many relatively small, angiographically invisibile satellite anastomoses or a relatively few highly enlarged ones visible by angiogram (Figure 30). The different patterns of flow redistribution likely depend on flow changes affected by many factors, such as the development of new critical stenoses in other vessels and the disappearance of

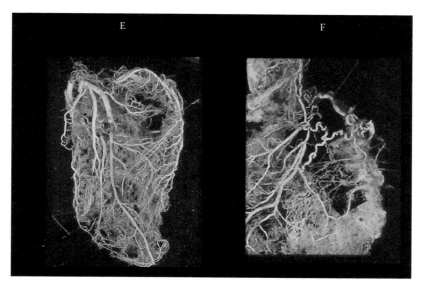

Figure 29 (continued). Occlusion of LAD without evidence of other stenotic changes of the coronary arteries in a 39-year-old woman with rheumatic heart disease and mitral insufficiency (E). In this case, arteritis was documented histologically by sampling before corrosion. An acute infarct (avascular area at the apex, arrow) was present. (F) A single, high enlarged collateral from LCX, supplying the distal tract of an occluded LAD. Note numerous normal anastomoses. This indicates that ischemia is not the cause (no diffuse enlargement of all collaterals in the whole ischemic area) but rather pressure gradient induces selective compensatory routes.

Table 20 Maximal and Average Coronary Collateral Diameter and Anastomotic Index in Various Conditions

Coronary Arteries	Collateral Diameter (μm)		Anastomotic Index
	Maximal	Average	
Normal			
+ Normal myocardium	280 (180–350)	200 (150–200)	4.7 (3.4–6.2)
Atrophy	240 (150–390)	170 (100–250)	3.7 (2.5–6.4)
Hypertrophy	350 (200–500)	221 (130–350)	7.4 (4.5–14.0)
Chronic hypoxia	395 (299–395)	249 (180–400)	12.4 (8.9–19.3)
Mild atherosclerotic stenosis (69%)	320 (250–425)	209 (165–225)	5.7 (4.8–7.1)
Severe (70–90%)	345 (260–400)	170 (180–200)	7.0 (6.6–8.0)
Subocclusion (90%)	685 (350–1250)	347 (200–450)	14.3 (7.5–23.0)
Multiple	786 (280–2000)	467 (100–600)	20.5 (4.8–38.0)

Note: Association of these conditions increments collateral size. For instance, for atrophy + hypoxia, the figures are 315 (260–400), 218 (180–280), 6.4 (4.5–12.4). To compare different conditions, an anastomotic index was adopted (Baroldi and Scomazzoni, 1967): Anastomotic index = Maximum diameter + (average diameter × frequency)/100, where average diameter of all anastomoses grater than 100 μm and frequency of the latter refer to normal adult hearts considered equal to 1.

collaterals in asynergic and infarcted myocardium (avascular area; Figure 13). In the latter situation, because the pressure gradient induced by the preexisting critical stenosis is maintained, the surviving collaterals will enlarge to carry the same amount of flow. In an infarct scar, giant, capillary-like, anastomotic channels are frequently seen histologically (Figure 30) — a finding that explains why only in scars is it possible to see collaterals by a technique that shows only very large vessels (Spain et al., 1963).

A last comment relates to angiogenesis (Unger et al., 1990) within the myocardium. There is no morphologic evidence of it happening. Plastic casts of hypertrophic hearts show an increase in vessel size without any histologic pattern of endothelial budding. For example, new vessel formation occurs when there is repair through granulation tissue on a background of fibrin components. In Figure 30H, a recent infarct associated with an endocardial thrombus in a heart injected postmortem with intracoronary radiopaque material shows a rich neovascularization of the thrombus in contrast to the avascularity of the infarcted myocardium that repairs by collagenization of empty sarcolemmal tubes and not by granulation tissue. This is an example that shows that when neoangiogenesis is present, it is visible. So far, only when there is a wound in the myocardium followed by granulation tissue repair do we see new vascularization, as, for instance, in the now abandoned Vineberg operation, in which some anastomoses were created at the site where an internal mammary artery was implanted in ischemic myocardium. However, lysamine-dye injected into the implanted artery remained in the limited area where the implant was done, with its expansion occurring only after clamping of the related coronary artery beyond its occlusion (Mantini et al., 1968). This means that newly formed collaterals at the wound site competed poorly with preexisting ones. At present, we prefer to speak of angiohyperplasia, due to enhanced flow (Flynn et al., 1993) and growth factors (D'Amore and Thompson, 1987; Rajanayagam et al., 2000), that is, enlargement of a vessel by hyperplasia of its wall components rather than angiogenesis in hypertrophy/hypoxia and coronary heart disease, explaining the relationship, detected in patients with one to three vessels diseased at angioplasty, between collateral flow and intracoronary growth factor concentration (Fleisch et al., 1999). The formation of an angiomatous vascularization by genes

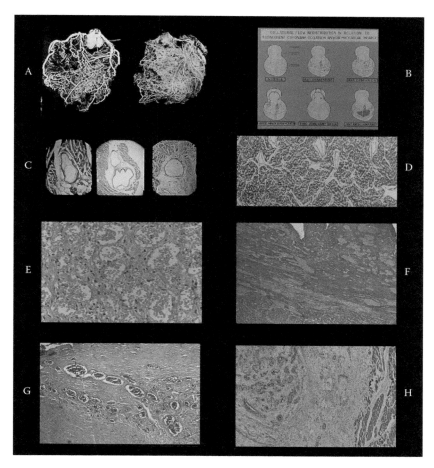

Figure 30   Different aspects of collateral compensation in the presence of the same occlusive pattern of LAD. (A) Relatively few very enlarged collaterals and numerous relatively small collaterals. This divergency may be due to progressive atherosclerotic obstruction of other main vessels or loss of the intramural vasculature, including collaterals, following an infarct. Chart (B) shows all the possibilities of flow redistribution. The histology of the enlarged anastomoses corresponds to a capillar-like wall, even in the rare extramural collaterals with rudimentary focal tunica media (C). (D) Enlarged collaterals in a case of anomalous origin of LAD from the pulmonary artery and (E–G) different aspects of giant capillaries (or plexus) in various stages of an acute/old infarction. The absence of new vessel formation is well documented in recent infarcts associated with endocardial thrombus (G). In the latter numerous new vessels form in the granulation tissue repair of the thrombus in contrast to their absence in infarct (H).

implanted in the subendocardium of an infarcted myocardium in rats (Schwartz et al., 2000a), as shown by histology, seems more a vascularization of an intracavitary thrombus than neovascularization of infarcted myocardium (see Figure 30H). Neoangiogenesis following gene injection within the myocardium has to be proven.

Plastic casts offer an explanation for many facts previously listed in the natural history of coronary heart disease (Table 21 and Table 22), supporting an adequate compensatory function for collaterals also confirmed by some cineangiographic studies that demonstrated the prevention of a postinfarct aneurysm: reduction of infarct size, improved myocardial function (Rigo et al., 1979), and the persistence of myocardial viability (Sabia et al., 1992)

Table 21   Frequency Distribution of Coronary Atherosclerotic Stenosing Plaques in 97 Adult "Normal" Subjects (88 Men, 9 Women) Succumbing to Accidental Death, without History and Postmortem Finding of Disease

Decades	Number of Cases	Maximal Lumen/Diameter Reduction (%)						Stenosis versus Number of Vessels					
								Any Type			Severe (≥ 70%)		
		0	<50	50–69	70–79	80–89	>90	1	2	≥3	1	2	≥3
<39	10	2	3	4	1	—	—	3	3	2	—	1	—
40–49	13	2	6	2	2	1	—	1	5	5	2	1	—
50–59	28	4	7	7	5	4	1	3	6	15	6	3	1
60–69	23	—	2	9	5	4	3	5	3	15	7	4	1
>70	23	—	2	9	6	4	2	—	4	19	7	4	1
Total	97	8	20	31	19	13	6	12	21	56	22	13	3

Table 22   Main Facts Supporting the Adequacy of Collateral Function

	Coronary Atherosclerosis (%)	
Absence of Previous Infarct	All Grades	≥70 and Multiple
Normal subjects	92	39
Noncardiac patients	95	66
Acute myocardial infarction, first episode	98	89
Sudden death, first episode	92	65

Note: Infarct size is not related to the number of severe stenoses and extent of vascular territory of the obstructed artery. Hypokinesis *in vivo* extends in normally perfused myocardium (Ahrens et al., 1993). Experimental occlusion of a critical stenosis without ischemic signs, plus demonstration of increased collateral flow (Khouri et al., 1968).

in the presence of collaterals. Not only that, but also the occlusion of a critical stenosis by an angioplastic balloon inflation is followed within 90 sec by retrograde anastomotic filling (Rentrop et al., 1988). Repetitive angioplastic occlusion of a stenosis (Deutsch et al., 1990; Cribier et al., 1992; Sakata et al., 1997c; Billinger et al., 1999; Sand et al., 2000) or of a normal coronary artery (Yamamoto et al., 1984) induces a progressive myocardial adaptation to ischemia through collateral recruitment. Electrocardiogram (ECG) changes and chest pain at first occlusion regress rapidly when repetitive occlusions are performed. This means that functioning collaterals exist before the angioplastic occlusion. The latter is a highly traumatic procedure that may trigger coronary spasm or regional myocardial asynergy by medial wall nerve stretching with abolition of any flow, including a collateral one. In turn, any measurement made distal to a stenosis by a catheter equipped to measure pressure, flow, etc. is unreliable. On this subject, a comment, about catheter size and coronary residual lumen is appropriate. The frequency of maximal lumen reduction calculated for a total of 408 cases (200 AMI, 208 SD; Table 16) was less than 69% in 68 cases (17%), 70% in 67 (16%), 80% in 109 (27%), and 90% in 164 (40%). In Table 23, the normal diameters of the main coronary arteries are compared with those resulting after different degrees of stenosis. Because the diameter of a catether is approximately 1500 μm, the question is how and where the catheter crosses a stenosis greater then 70%, that is, <1080 μm in diameter, the latter ranging from 720 to 300 μm in 67% of these cases. It seems proper to state that in most instances the plaque is dissected (Figure 23) with consequent, unvaluable functional and morphologic complications — a criticism valid for any type of coronary catheterization (intravascular ultrasound, angioscopy, etc.).

Table 23   Lumen Reduction in 408 Cases of Coronary Stenosis (μm)

Lumen μm	LAD μm	LCX μm	RCA μm
Normal (medium)[a]	3600	3000	3200
Stenosis %			
70	1080	900	960
80	720	600	640
90	360	300	320

[a]   Calculated on plastic coronary casts of vessels in dilatation. The diameter of a catheter is approximately 1500 μm.

Finally, among many others, an impressive example of functioning collaterals is offered by two cases where marathon runners (Noakes et al., 1979) died suddenly, one unexpected and the other expected. The first was a 44-year-old man without a history of coronary heart disease. During a race, after 19 km, without distress, he stopped to adjust a loose shoelace and died instantaneously. Autopsy findings were a normal heart weight, extensive monofocal anteroseptal scar (previous silent infarct), atherosclerotic severe stenosis of left anterior descending branch with occlusive organized and recanalized thrombus, atherosclerotic severe stenosis of the left circumflex branch, and a normal conduction system. The other marathon runner was a 41-year-old man who suffered an acute myocardial infarct 2 years before the last episode, with angiographic occlusion of the left circumflex branch and 50% narrowing of his proximal right coronary artery. Despite the infarct, he persisted in his athletic activity for 2 years. After two episodes of unstable angina, angiography confirmed the occlusion of the left circumflex, total occlusion of the right coronary artery, and 80% stenosis of the left anterior descending branch. Listed for bypass surgery, he developed severe chest pain. An ECG indicated a myocardial infarct, and despite therapy, he died within 30 min. At autopsy, a heart of normal weight showed atherosclerotic stenosis of the three main vessels, with organized occlusive thrombus of the left circumflex and right coronary artery and a fresh thrombus of the stenosed left descending branch, an anterior healed infarct, and no evidence of an acute one. Both cases show that heavy physical activity is possible for a long period despite severe, multiple coronary stenoses, in silent and nonsilent chronic coronary heart disease. We had the opportunity to examine the histologic slides, and both hearts had many foci of contraction band necrosis.

Coronary Occlusion

Extramural Coronary Arteries

Apart from equivocal histologic signs of medial smooth muscle cell contraction (Factor, 1985; James and Riddick, 1990), a pathologist has no way to document coronary artery spasm. He can see a severe stenosis caused by an atherosclerotic plaque or an occlusion by a thrombus, the age of which may range from early platelet-fibrin composition to organized, often vascularized tissue. Other acute occlusions easily recognizable are the rare dissecting aneurysm, embolism, and thrombotic arteritis, all not pertinent to the natural history of coronary heart disease. In the latter, when present, the thrombus, with or without plaque fissuring or rupture, is located mainly in an area of critical luminal stenosis, as shown by all pathologic studies and *in vivo* after fibrinolytic recanalization (Cowley et al.,

1981; Ganz et al., 1981; Mathey et al., 1981; Reduto, 1981). A fact (Chandler et al., 1974; Baroldi and Silver, 1995; Fishbein and Siegel, 1996) which contradicts the illusion of some authors that a small plaque may rupture causing a thrombotic occlusion followed by an infarct or sudden death (Little et al., 1988; Hackett et al., 1989; Webster et al., 1990). This assumption is

> In many patients (78%) who subsequently developed myocardial infarction, prior angiography revealed lesions that were <50% occlusive in the infarct-related artery. Although the degree of narrowing in these arteries just before the onset of infarction was unknown, it was assumed that a more significant narrowing in the infarct-related artery had not slowly developed before the acute event. We suspect that this may have occurred because progression of coronary artery disease at restudy was uncommon in noninfarcted-related lesions. Therefore, disruption of a mild or moderate atherosclerotic plaque with resultant thrombosis and total or subtotal occlusion probably explained the myocardial infarction. In patients with a previously normal appearing infarct artery, we assume that some degree of diffuse coronary disease was indeed present, but was not detectable by these angiographic techniques. (Ambrose et al., 1988)

This assumption was never demonstrated by postmortem studies and ignores progression of a plaque by increased wall stresses related to increased peripheral resistance following regional myocardial asynergy. The latter precedes, by a long period, an infarct or sudden death (see below the monitored case at the time of infarction).

Any discussion of the frequency of occurrence of an occlusive thrombus is meaningless. What is essential is to determine whether it stops the nutrient flow to the dependent myocardium. A positive answer would permit correlation and allow a cause–effect relationship. Unfortunately, this absolute correlation is not possible. For example, an occlusive thrombus may not be associated with a myocardial infarction either at postmortem or by angiography (Ambrose et al., 1985a,b). Also, there may be discrepancy between the age of a thrombus and that of an infarct. The *coeval* relationship, according to different histological evolutive stages of a thrombus in coronary arteries (Irniger, 1963), between thrombus and infarct was investigated (Table 24) with a possible coeval correlation in only 48 of 98 acute myocardial infarcts associated with an occlusive thrombus out of a total of 207 acute myocardial infarct cases (Baroldi, 1965). The presence of a coronary artery thrombus does not automatically permit a pathologist an absolute diagnosis of a myocardial infarct (or of coronary sudden death), particularly in cases dying before histological changes of an infarct become apparent.

Table 24   Correlation between Age of Occlusive Thrombus and Myocardial Infarction

Thrombus	Time	Infarct Necrosis
Early thrombus	6–8 hours	Normal myocardium or polymorphonuclear infiltration
Endothelial proliferation/capillary growth	7 days	Initial macrophagic reaction at periphery
Early organization/collagen fibril deposition	15 days	Diffuse macrophages alveolar appearance
Early to late organization	>20 days	Recent to previous fibrosis

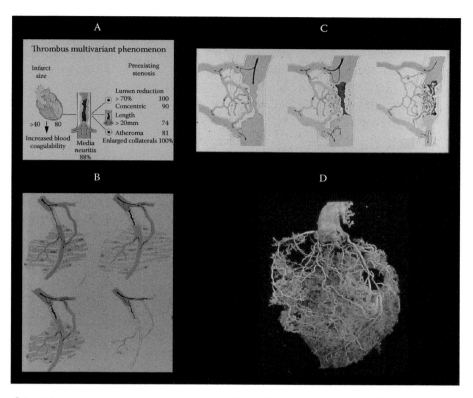

Figure 31 The coronary thrombus is a multivariant phenomenon (A), including medial neuritis. Its location in severe (70) stenosis associated with other factors (retrograde collateral flow, reduced fibrinolytic activity, etc., see text) justifies the concept that it is a secondary phenomenon. Any time there is an increased peripheral resistance (B) (spasm, intramural extravascular compression following infarction, etc.), stasis in the related main vessel and in collaterals both outside and within the plaque is expected with hemorrhage, plaque rupture, and thrombosis (C). On the other hand, it is difficult to accept that acute occlusion of a pinpoint lumen bypassed by preexisting functioning collaterals (D) may result in infarct necrosis. Even experimentally, occlusion of a severe chronic (7 days) stenosis does not produce any ischemic dysfunction.

On the contrary, the factors correlated with the thrombus (Figure 31) and preexisting conditions at the plaque level support the possibility of secondary thrombus formation. Within the residual lumen, a peculiar hemodynamic exists because proximal blood flow reduction is counterbalanced by distal retrograde collateral flow with hindrance in the irregular residual lumen in which, angiographically, a systolic-to and diastolic-fro movement can be observed (Maseri, personal communication). This flow impairment is associated with reduced fibrinolytic activity of the atherosclerotic plaque (Myasnikov et al., 1961); possible concomitant spasm of the atherosclerotic/stenotic infarct-related artery (in the few instances in which a cineangiographic spasm was seen before death, it occurred in a severely obstructed vessel; Maseri et al., 1978; Roberts et al., 1982a); increased coagulability any time there is tissue necrosis; and mechanical action, especially upon exertion, of the contracting myocardium on the atherosclerotic-rigid coronary wall (Black et al., 1965). All these factors favor thrombus formation as well as hemorrhage (mainly seen in the infarct-related artery) and plaque rupture, especially after blockage of flow secondary to an infarct (Figure 31); plaque hemorrhage, fissuring, and rupture being found particularly after exertion (Burke et al., 1997a, 1999b). This hypothesis is further supported by a significant lower angiographic recanalization (28%) observed in the presence of collaterals rather than in their absence (55%; Araie et al., 1990) and by the occlusion of a stenotic artery after surgical bypass (Aldridge and Trimble, 1971; Griffith et al., 1973), the latter being the equivalent of a collateral bypass. In turn, a critical stenosis may apparently be aggravated by mural thrombi secondary to local myocardial asynergy inducing extramural coronary stasis without effective nutrient blood flow reduction.

In a recent revisitation and revision of the role of thrombi in acute coronary syndromes by an angiographer, their causal role was reinforced because of their angiographic presence (Rentrop, 2000) without a correct definition of what a thrombus is and the risk of confusing it with a coagulum. The structural differences between the thrombus and coagulum are too often forgotten (Boyd, 1961) and need to be recalled. The first starts in flowing blood on an eroded endothelium, eliciting a platelet plug and fibrin deposition with a layered growth (Zahn's lines), ending in granulation tissue and healing. The thrombus at its initiation is firmly attached to the vessel wall and is not easily removed to leave fragments *in situ*. On the contrary, a coagulum, erroneously defined as a red thrombus, maintains the usual blood composition (i.e., prevailing red cells plus leukocytes and platelets and a fine network of fibrin), is not attached to the endothelium, and can be easily removed. The coagulum is secondary to stasis found in the cul de sac proximal and distal to a severe stenosis in an infarct-related artery and is visible by angioscopy (van Belle et al., 1998). Therefore, it cannot hinder blood flow as some believe. This needs to be considered when one speaks of the frequency and extent of a thrombus in material obtained by directional atherectomy, a highly traumatic procedure. Using it, 35% of unstable versus 17% stable angina patients (Arbustini et al., 1995) and 44% of patients with unstable angina and recent (2-week-old) myocardial infarct had a large, fresh "thrombus" formed by layered erythrocytes, platelet aggregates, and fibrin (Rosenschein et al., 1994). Any lytic agent or the reestablishment of blood flow would easily dissolve a coagulum with the angiographic impression of embolization. The fate of the coagulum is not known. In ischemic patients who had surgical bypass long before heart transplantation, the excised hearts showed venous grafts with mild nonatherosclerotic wall thickening but a lumen completely filled by pultaceous, yellow material that was easily squeezed from it like toothpaste. Histologically, it had the appearance of atheromatous material (Figure 32). At first glance in a single section, this could give the false impression of rupture of a nonexistent giant atheromatous plaque. This could be the fate of a coagulum in a cul de sac. This finding must be keep in mind, because this transformation into atheromatous-like material was proposed as a mechanism for atheroma formation within a plaque following microhemorrhage (Morgan, 1956) and was confirmed more recently (Kolodgie et al., 2003).

The claim that a thrombus may lyse in the agonal period or at postmortem, explaining an infarct in the absence of an occlusive thrombus, is contradicted by experimental thrombosis that remains unchanged for 17 days, retraction of the thrombus being a late, limited phenomenon (Weisse et al., 1969). Furthermore, the question of therapeutical lysis of a thrombus remains an open question. Are we lysing a coagulum at the site of a pseudo-occlusion? Spontaneous reflow in an occluded coronary artery was observed within 3 to 14 h from the beginning of an infarct (Rentrop et al., 1989).

Some other points deserve revision. An argument against rupture or fissuring of a coronary artery plaque is the very low frequency of cholesterol-emboli found within the

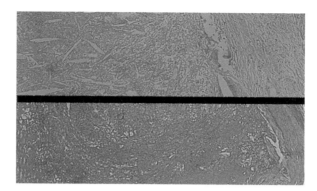

Figure 32 The possible fate of a coagulum — transformation in an atheromatous-like material seen in a venous bypass without atherosclerosis. This is a pattern not to be confused with rupture of an extremely large atheromatous plaque.

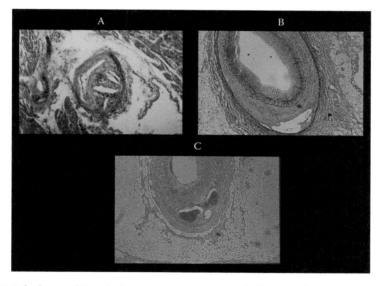

Figure 33 A "cholesterol" embolus in an intramyocardial arteriole (A) was a unique finding in hundreds of systematically examined hearts. It was found in a case of sudden and unexpected coronary death without related myocardial damage. Despite rupture of innumerable plaques by angioplasty, intramural cholesterol emboli are rare. Angioplasty means fracture of the atherosclerotic wall with possible formation of a dissecting aneurysm (B, C).

myocardium in contrast to those observed in kidneys, spleen, brain, etc. mainly related to an atherosclerotic aorta. In more than a thousand hearts that had not undergone angioplasty that were studied systematically with an average of 16 total wall samples from different cardiac regions, we found only a single cholesterol-embolus in a small, normal, intramural arteriole from a sudden and unexpected death case without related myocardial changes (Figure 33). One notes that thousands, if not millions, of angioplastic procedures are performed worldwide with minimal complications, despite the fact that reopening a stenosis by this method produces an iatrogenic plaque rupture (Waller, 1991; Saber et al., 1993) with distal embolization of atherosclerotic or thrombotic material in 81% of 32 patients examined, with a mean number of 3.9 emboli when a rupture of the plaque was present (14 patients). Emboli were thrombotic in 49%, atheromatous in 29%, and mixed in 17%. It is interesting to note that of 39 controls, 12 normal, 13 with coronary atherosclerosis but no myocardial infarct, and 14 with an acute myocardial infarct, no emboli were found, even in the presence of plaque rupture. Experimental coronary wall rupture by angioplasty does not elicit any superimposed thrombus (Gravanis et al., 1993). Furthermore, angiographically dissected plaques had no different incidence of restenosis and clinical complications than nondissected ones (Hermans et al., 1992), with the observation that the absence of a true dissecting aneurysm (see above) in coronary heart disease, in contrast to those occasionally seen after angioplasty (Figure 33), confirms the lack of a significant role of plaque rupture distal embolization within the myocardium; rupture is often artifactual.

On this subject, a sign of intimal fissuring was accepted any time radiopaque material was seen deep to the endothelium, with the supposition that it penetrated through an intimal break (Davies and Thomas, 1984; Davies et al., 1976). This viewpoint totally ignores plaque satellite collaterals joining the adventitial and intimal vascular network and diffusion of the menstruum following rupture of such intimal vessels.

Considering the mechanism of myocardial infarction formation, if plaque rupture and thrombosis are excluded as causes, spasm of resistance intramural vessels linked with endothelial-derived vasoactive substances (Sambuceti et al., 1997; Marzilli et al., 2000) is an alternate hypothesis. Another was deduced by the demonstration of an avascular area in early acute infarcts (Figure 13). This suggested a possibile extravascular compression by myocardial cells, stretched, as is shown also histologically, by intraventricular pressure (Baroldi, 1965). This theoretic proposal is supported by the case reported below and by the fact that hypokinesis of the myocardium precedes chest pain and ECG ischemia (Nesto and Kowalchuk, 1987), confirmed in ambulatory patients who died suddenly (Taki et al., 1987). In an attempt to demonstrate this ischemic role of dyskinesis-akinesis, I examined from the files of the Armed Force Institute of Pathology, Washington, D.C., 18 cases with severe constrictive pericarditis associated with markedly obstructive coronary atherosclerosis. The thesis was that constrictive pericarditis might control bulging with a minor degree of resulting myocardial necrosis. In not one of these cases were myocardial changes related to coronary disease found. Recently, the experimental prevention of cardiac wall bulging in sheep after acute coronary occlusion showed the absence of a huge scar or aneurysm, the shape of the left ventricular chamber being similar to controls (Kelley et al., 1999). This bulging or akinetic effect of intramural blood flow blockage means, in turn, that therapeutic agents could not penetrate the necrotic tissue within 1 h of its onset. On this subject, it is pertinent to mention that a similar cardiac support device (i.e., a thin polyester mesh that does not generate diastolic constriction) protected against intracoronary microembolization (Saavedra et al., 2002).

In conclusion, from a practical standpoint, the presence of an occlusive coronary thrombus and other plaque activities, per se, does not allow us to diagnose an acute coronary syndrome.

Small Vessel Disease

Changes affecting small coronary arteries are included under the heading of *pathology of small coronary arteries* (James, 1967) or *small vessel disease*. This means that it is a disease of the intramural vessels and not of small extramural arteries.

The concept that platelet aggregation might induce ischemia was derived from the experimental infusion of adenosine diphosphate with transient circulatory collapse, ECG signs of myocardial ischemia, possible ventricular fibrillation, thrombocytopenia by platelet sequestration in microvessels, and myocardial infarcts (Jorgensen et al., 1967, 1970). The conclusion was that any factor, for example, tissue necrosis or hemolytic anemia, capable of producing adenosine diphosphate could cause an acute ischemic event following platelet sequestration in small vessels (Mustard and Packham, 1969). One notes that platelet aggregation or sequestration not associated with fibrin is a phenomenon distinct from thrombosis. However, this sequestration occurs in human pathology. For example, thrombotic thrombocytopenic purpura (TTP) is a mixture of ischemic and hypoxic factors: extremely severe hemolytic anemia, diffuse myocardial obstructive angiopathy of small intramural arterioles, thrombocytopenia with platelet sequestration in normal small arterioles, hemorrhage, neurologic disorders, and so forth. Yet, not one personally studied case or those reviewed from the literature had a cardiac infarct, died suddenly, or had angina pectoris. Therefore, TTP can be considered an experimental human condition to test many supposed ischemic events, including the unproven hypothesis that platelet aggregates may release vasoactive factors inducing spasm of small arterioles and then quickly disaggregate and, therefore, will not be found at autopsy.

Similarly, in sickle cell anemia, despite small vessels plugged by sickled erythrocytes demonstrated also *in vivo* (Knisely, 1961; Levine and Welles, 1964), no history or histologic signs of ischemic heart disease were detected (Baroldi, 1969), and no evidence of ECG changes or of MB isoenzymes of creatine phosphokinase were seen during sickle cell crises (Val Mejias et al., 1974). This is a fact that speaks against disseminated intravascular coagulation as a mechanism of myocardial ischemia. On the other hand, no platelet sequestration or fibrin-platelet thrombi were observed in cases of fatal infarct necrosis (Figure 34) in the absence of, or with, minimal coronary lumen reduction (Eliot et al., 1974; Fineschi et al., 2001b). In 208 sudden and unexpected death cases and in 97 controls, platelet aggregates in normal vessels were seen in both groups with a similar low incidence (Table 25), with this having a significant impact on the individual's survival time. Platelet aggregates without fibrin were significantly more frequent, especially in the control group

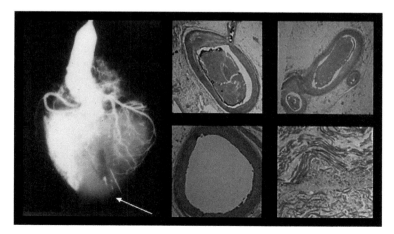

Figure 34 A case of transmural myocardial infarction of the left ventricle (>30% size) with wall rupture and tamponade without coronary occlusion or stenosis or small vessel disease. This is a postmortem injection showing a normal coronary artery and an inferior avascular area (arrow) with normal coronary lumen (confirmed histologically) and massive myocardial infarct necrosis with hemorrhage consequent to wall rupture.

Table 25   Frequency of Platelet Aggregates and Number of Occluded Intramural Arterial Vessels in Sudden and Unexpected Death (SD) and Healthy Controls (AD)

Total Cases	SD	(%)	AD	(%)
Arterial platelet aggregates	208	(100)	97	(100)
Absent	61	(29)	23	(24)
Present in	147	(71)	74	(76)
5 vessels	77	(37)	35	(36)
5–10 vessels	37	(18)	14	(14)
11–20 vessels	22	(10)	11	(11)
21–30 vessels	7	(3)	5	(5)
31–60 vessels	4	(2)	9	(9)
Total sections	3328	(100)	1552	(100)
Absent	2793	(84)	1273	(82)
Present in	535	(16)	279	(18)
1 vessel	269	(8)	124	(8)
2 vessels	134	(4)	62	(4)
3 vessels	61	(2)	40	(3)
4 vessels	26	(1)	21	(1)
5–15 vessels	45	(1)	32	(2)

($p < 0.05$ for trend), with a longer survival time and in the presence of blood either in arterial or venous vessels (Table 26). We interpreted these findings as a terminal layering effect of blood elements following terminal blood flow reduction (separation of blood elements of different weights; Figure 35). This finding should be considered when this flow reduction occurs in proximal and distal cul-de-sacs at the level of an occluded atherosclerotic plaque.

In conclusion, intramural platelet aggregates or fibrin-platelet thrombi seem to play a small role in the natural history of any acute coronary syndrome. The assumption that "without alternative explanations the differential diagnosis leaves embolization with microvascular obstruction as the leading suspect" (Topol and Yadav, 2000) is not supported. The endothelial and platelet interrelationship is complex and is currently being investigated. Some diseases, such as atherosclerosis, diabetes mellitus, uremia, hypertension, hypercholesterolemia, eclampsia, and others, may impair the release of nitric oxide/endothelium-related relaxing factor (vasodilatation) and platelet aggregation factor (anticoagulation-fibrinolysis) from one site and prompt endothelial vasoconstrictive and platelet aggregating factors from another (Ware and Heistad, 1993; Anderson, 1999). The local trigger for an endothelial reaction limited to a minimal part of the immense endothelial surface of human body, including lymphatics, remains undetermined. On this subject, one should remember the following:

1. Platelet aggregation, both histologically and ultrastructurally, has not been demonstrated during experimental infusion of catecholamines (Todd et al., 1985a,b) or by Cr-labeled platelets (Moschos et al., 1978).
2. The postulate that the plugging/adhesiveness of polymorphonuclear leukocytes (Mazzone et al., 1993; Entman and Ballantyne, 1993) releases vasoactive factors as a pathogenic mechanism for ischemia has no pathologic evidence. In coronary

Table 26 Frequency of Arterial (AP) and Venous (VP) Platelet Aggregates in Relation to Intramural Blood Stasis (Total Sections), and Platelet Aggregates (PAs) versus Survival (Total Cases) in Sudden and Unexpected Coronary Death (SD) and Healthy Controls (AD)

Stasis	SD (%)		AD (%)	
No demonstrable stasis	1005	(100)	629	(100)
+ AP	60	(6)	48	(8)
+ VP	29	(3)	5	(1)
Arterial + venous stasis	1512	(100)	620	(100)
+ AP	454	(30)	217	(35)
+ VP	418	(28)	210	(34)
Venous stasis alone	811	(100)	303	(100)
+ AP	21	(3)	14	(5)
+ VP	283	(35)	107	(35)
Total sections	3328	(100)	1552	(100)
+ AP	535	(16)	279	(18)
+ VP	730	(22)	322	(21)

Survival (min)	SD (%)		AD (%)	
<10	151	(100)	75	(100)
No PA	51	(34)	22	(29)
+ PA	100	(66)	53	(71)
<5	52	(34)	30	(40)
5–10	26	(17)	11	(15)
>10 vessels	22	(14)	12	(16)
≥10	57	(100)	22	(100)
No PA	10	(17)	1	(5)
+ PA	47	(82)	21	(95)
<5	25	(44)	5	(23)
5–10	11	(19)	3	(14)
>10 vessels	11	(19)	13	(59)

heart disease, neutrophils appear in the myocardium only 6 to 8 h after infarct necrosis, without any documented complication. Infarct size reduction, experimentally studied by abolishing neutrophil cytotoxicity (Amsterdam et al., 1993), may have a different explanation (see above).

3. Any therapeutic benefits may be due to mechanisms other than those believed. For instance, in isolated rat hearts during global ischemia, and therefore independent of platelets, streptokinase improves contraction after reperfusion (Fung and Rabkin, 1984). Furthermore, prevention and control of the often increased coagulability by fibrinolytic agents in any acute event is always beneficial, even if thrombolytic therapy is ineffective in unstable angina (Neri Serneri et al., 1992).

Functional Occlusion—Coronary Spasm

Long ago, Leary (1935) proposed vascular spasm as a possible mechanism of occlusion causing an acute coronary syndrome. At present, spasm is apparently demonstrated based on the cineangiographic imaging of an occlusion of an extramural artery that reopens spontaneously or following administration of a vasodilator, in cases of both angina pectoris

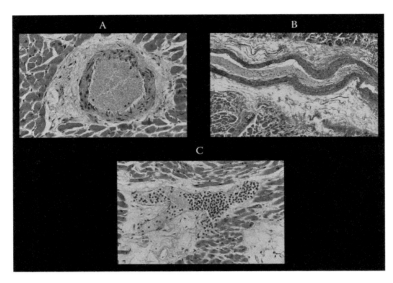

Figure 35 Platelet aggregates occluding a normal small intramyocardial artery in a normal subject (A). There is a layering effect during progressive terminal stasis, with separation of red cells and platelets (B) and leukocytes (C).

(Dhurandhar et al., 1972; Oliva et al., 1973) and myocardial infarction (Cheng et al., 1972; Oliva and Breckinridge, 1977; Maseri et al., 1978a; Vincent et al., 1983); and by imaging an intramural blood flow reduction (Sambuceti et al., 1997; Marzilli et al., 2000). Already we question if an angiographic coronary occlusion is a pseudo-occlusion due to stasis caused by increased peripheral resistance (see below). The possibility that isolated coronary spasm associated with an autonomic disorder (Lanza et al., 1996) can determine an arrhythmogenic cardiac arrest is supported by a positive ergonovine test in resuscitated people with a normal angiogram (Chevalier et al., 1998), even if angiography may fail to depict coronary artery narrowing (Dietz et al., 1992). A normal angiogram cannot exclude the presence of small neuroactive plaques, as demonstrated by intravascular ultrasonography at the site of focal spasm induced by ergonovine maleate (Yamagashi et al., 1994), keeping in mind that 37% of unstable angina patients are hypersensitive to spasmogenic stimuli compared to 4% with stable angina (Bertrand et al., 1982). It remains to be proven if coronary artery spasm occurs, whether it is a primary or secondary event, how long it lasts, and if it involves the whole vessel or is limited to a segment. Nevertheless, the spasm hypothesis of extramural and intramural (Hellstrom, 1982) vessels remains a field open to further investigation, particularly in relation to sudden death. Experimental coronary occlusion immediately induces cyanosis and myocardial paralysis with bulging of the ventricular wall and is often associated with a malignant arrhythmia ending in cardiac arrest due to ventricular fibrillation. Theoretically, a spasm should last less than 20 min because most sudden and unexpected death cases do not develop an infarct and do not show reflow necrosis.

Finally, flow reduction was proposed when collaterals steal flow from the territory of a parent vessel ("steal syndrome," Leachman et al., 1972). This mechanism is difficult to assess for the myocardial anastomotic network when occlusion of the parent vessel occurs ("infarct at distance," Blumgart et al., 1940), or when perivascular fibrosis exists to limit compensatory vasodilation (Reagan et al., 1975). The absence of ischemic heart disease in rheumatic hearts with extensive perivascular fibrosis contradicts this hypothesis. Finally, the so-called *no-reflow phenomenon*, which consists of interstitial exudation or hemorrhage associated with parenchymal cell swelling because of vascular compression, was described in the brain, kidney, and heart (Sommers et al., 1972; Majno et al., 1967; Kloner et al., 1974) but is not found in the natural history of coronary heart disease in man.

Relative Coronary Insufficiency

In the absence of vessel damage, relative coronary insufficiency is often proposed to cause disease and death any time there is a discrepancy between nutrient flow and myocardial demand. Cardiac hypertrophy is a typical example. Fibrotic foci often found in hypertrophic myocardium were related to coronary insufficiency (Buchner, 1950), and myocardial failure was thought to depend upon the latter (Linzbach, 1947). However, a hypertrophied heart may work for many years without concomitant ischemia, while patients with an acute myocardial infarction or sudden death first episode have a heart weight of 500 g in 39% and 43% of cases, respectively, without any relationship between heart weight and infarct size (Table 27). These data show that relative coronary insufficiency due to hypertrophy is a questionable hypothesis. Cor pulmonale provides a good example, because the right ventricular mass may weigh as much as the left one. In no instance, even with severe atherosclerotic obstruction of the right coronary artery, was an

Table 27 Maximal Lumen Reduction and Number of Vessels with Severe Stenosis (≥70%) in Relation to Heart Weight in Different Groups

Heart Weight (g)	Lumen Reduction (%)		Stenosis ≥70% in			
	<69	≥70	Total	1 Vessel	2 Vessels	3 Vessels
AMI 1st						
<500	6	37	43	16	17	4
500	4	23	27	13	6	4
Total	10	60	70	29	23	8
AMIC						
<500	—	14	14	3	7	4
500	1	15	16	6	6	3
Total	1	29	30	9	13	7
SD 1st						
<500	24	52	76	29	21	2
500	22	35	57	11	13	11
Total	46	87	133	40	34	13
SDC						
<500	2	16	18	3	6	7
500	3	54	57	10	20	24
Total	5	70	75	13	26	31
AD						
<500	52	35	87	21	11	3
500	7	3	10	1	2	—
Total	59	38	97	22	13	3

Note: In acute myocardial infarction (AMI), the relationship between heart weight and lumen reduction was calculated in 100 cases. Abbreviations: AMI 1st, first episode of acute myocardial infarction; AMIC, in chronic patients; SD 1st, sudden and unexpected coronary death as first episode; SDC, in chronic patients; AD, otherwise healthy subjects who succumbed to accidental death.

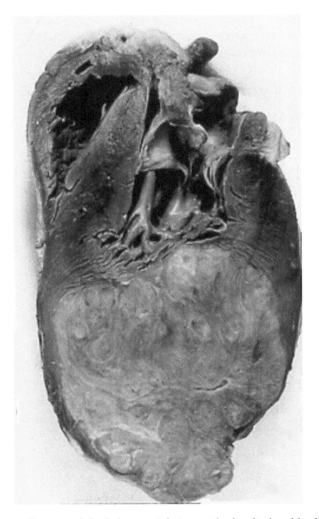

Figure 36 Enormous fibroma of the left ventricle in a girl who died suddenly and unexpectedly after a swimming race. This tumor — a previous lesion — did not induce symptoms and signs, despite training and racing. (Courtesy of Dr. Heat.)

isolated right ventricular infarct found (Baroldi, 1971). It seems likely that coronary vasculature is able to cope with hypertrophy for a long time (Figure 36).

The distinction between *concentric* or *tonogenic* or *compensatory* and *eccentric* or *myogenic* or *failing* hypertrophy at a critical heart weight 500 g (Linzbach, 1947) has a practical comparative value, keeping in mind that congestive failure may occur without eccentric dilatation and with a normal heart weight or one <500 g.

In this review on myocardial mass changes versus function, atrophy of the heart is worth mentioning. Its lack of relationship with heart failure (Hellerstein and Santiago-Stevenson, 1959) questions if atrophy is a pathologic entity or more likely an adaptation to reduced work in relation to a diminished body mass, without any diagnostic meaning.

Another condition in which a relative coronary insufficiency should be present is cardiogenic shock. In patients who died from this condition with a survival longer than 8 h, ischemic necrosis was not observed histologically, despite a period of time sufficient for it to develop (Baroldi, 1969). Infarcts that are mainly subendocardial following accidental shock with resuscitation are examples of reflow necrosis (see above).

This indicates a need to revise the previously quoted unifying theory that proposed a unique etiopathogenesis for all three acute coronary syndromes, namely, acute myocardial infarct, unstable angina, and sudden death. Each pattern requires a better definition, keeping in mind that each acute syndrome is a clinicopathologic entity that may or may not transform into another.

Acute Myocardial Infarction

From the first reports (Hammer, 1878; Herrick, 1912, 1919) of acute myocardial infarcts (AMIs), an occlusive coronary thrombus was considered the cause. In the pathological literature, such an occlusive thrombus was found in 35 to 85% of cases (Chandler et al., 1974; Baroldi, 2001; Fineschi and Baroldi, 2004) and was mainly associated with plaque rupture (Chapman, 1974) and large infarcts (Freifeld et al., 1983). Its frequency varied in first-episode or chronic cases (Table 28) and directly correlated with a severe degree of preexisting lumen reduction due to atherosclerosis; the length of the atherosclerotic plaque; its type, that is, concentric, atheromatous, and so forth; the presence of lympho-plasma-cellular infiltration surrounding nerves of the tunica media; the infarct size; and survival (Table 29). In other words, a thrombus, when present, is found in 98% of AMI cases at the level of a severe (>70%) stenosis, in general, longer than 5 mm, and is related to other variables (Fineschi and Baroldi, 2004). The latter could explain the variation in thrombus frequency reported in the pathology literature. If only large transmural infarcts with a related coronary artery severely stenosed (90%) by concentric-atheromatous-inflamed plaque are selected, the likelihood of finding an occlusive thrombus is 100%. However, this does not mean that the latter caused the infarct; it may merely represent an epiphenomenon related to changes within the plaque or elsewhere (Baroldi et al., 2001; Fineschi and Baroldi, 2004). On this point, Wilson's (1952) statement is appropriate: "If one doubts the necessity of controls reflects on the statement: it has been conclusively demonstrated by hundreds of

Table 28   Frequency of Acute Occlusive and Mural Thrombi in Acute Infarcts and Sudden Coronary Deaths in Relation to Extensive Myocardial Fibrosis as a Sign of Chronic Disease

Source	Total (%)	Occlusive	Mural
AMI 1st	145 (100)	60 (41)	29 (20)
AMIC	55 (100)	22 (40)	7 (13)
Total	200 (100)	82 (41)	36 (18)
SD 1st	133 (100)	11 (8)	14 (10)
SDC	75 (100)	21 (28)	8 (11)
Total	208 (100)	32 (15)	22 (10)

Note: In the healthy control group, only one mural thrombus was found. Abbreviations: AMI 1st, first episode of acute myocardial infarction; AMIC, in chronic patients; SD 1st, sudden and unexpected coronary death as first episode; SDC, in chronic patients; NCA, noncardiac patients who died from noncardiac causes; AD, otherwise healthy subjects who succumbed to accidental death.

Table 29   Acute Coronary Thrombus in 200 Acute Myocardial Infarcts (AMIs) and 208 Sudden and Unexpected Coronary Deaths (SUDs) in Relation to Atherosclerotic Plaque Variables, Infarct Size, and Survival (Percentage Distribution)

	AMI		SUD	
	Occlusive	Mural	Occlusive	Mural
Thrombus				
Total	41	18	15	11
ATS plaque stenosis (%)				
69	7	14	—	9
70–79	33	36	16	20
80–89	35	19	47	45
>90	24	31	38	32
Length (mm)				
<5	6	19	6	9
5–20	38	39	19	27
>20	56	42	75	64
Type				
Concentric	100	100	94	91
Atheromatous	84	81	75	82
Lymph-plasma inflammation	92	82	92	79
Infarct size (%)				
10	20	17	—	—
11–20	32	19	—	—
21–30	48	33	—	—
31–40	44	19	—	—
41–50	78	8	—	—
>50	86	3	—	—
Survival — days				
2	29	—	—	—
3–10	51	—	—	—
11–30	45	—	—	—
Survival — minutes				
<10	—	—	12	—
10–60	—	—	23	—
61–180	—	—	30	—

experiments that the beating of tom-toms will restore the sun after an eclipse." Then, how do we interpret the cineangiographic finding of a "coronary occlusion" — found in 87% of acute infarct cases examined within the first 4 h of onset of symptoms (De Wood et al., 1980, 1986) or in 70% within the first 6 h (Rentrop, 2000)? Does this angiographic occlusion precede or follow the infarct? What we need is to monitor what happens before, at the onset, and after the formation of an infarct or any other acute coronary syndrome.

At present, in the literature, there is only one case where it was possible to demonstrate this sequence (Baroldi et al., 1990b), in a 48-year-old man with unstable angina of 4 months duration. At cineangiography, a severe stenosis of the LAD and a stenosis of the first part of the RCA were seen. Radioventriculography showed an extensive hypokinetic zone of the anterior left ventricle. Suddenly, during cineangiography, ischemic ECG changes developed without any symptoms or any other signs, including angiographic changes at the level of stenosis. Several left anterior angiograms remain unchanged for 20 min when a progressive disappearance ending in a cutoff of this vessel was seen, again without other

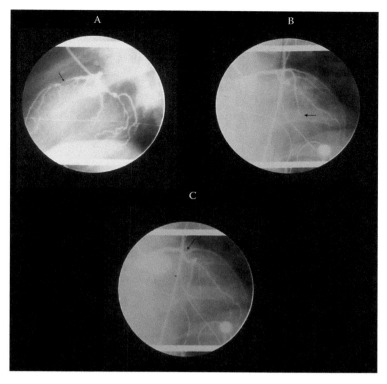

Figure 37   Cineangiographic monitoring in a patient with nonocclusive left anterior descending (LAD) stenosis (A) who developed an extensive infarct without angiographic occlusion. The subsequent imaging of occlusion began distally (B) and ascended to the origin (C) of the vessel (arrow), indicating that the angiographic pseudo-occlusion was due to stasis for increased peripheral resistance and not to thrombosis, not shown morphologically (see text).

symptoms and signs. In rapid sequence, intracoronary vasodilator, fibrinolytic agent, angioplasty with establishment of a normal lumen, and bypass surgery were performed without flow restoration. Only on a few occasions did a brief period of reflow occur showing that the angiographic occlusion of the LAD started distally and ascended progressively to the vessel's origin — and not at the plaque level — where flow in the left main trunk and LCX branch was maintained (Figure 37). No further angiographic changes of the vessel wall were seen. Severe chest pain started 90 min after the first ischemic ECG change and coincided with a successful angioplasty that reestablished a normal lumen. At surgery, a coagulum — not a thrombus — was easily removed, and inspection showed a normal lumen. Nevertheless, reflow, as shown by a flowmeter, was lacking. The patient recovered from a large anterolateral infarction, but because of progressive congestive heart failure, the patient underwent heart transplantation 12 months later, giving an opportunity to examine the heart.

Morphologic findings confirmed the lack of thrombosis within the LAD, which, however, showed a severe atherosclerotic lumen reduction along its whole course. No emboli or other changes were seen in intramyocardial vessels. The only myocardial finding was an extensive anterolateral scar. The RCA had an organized occlusive thrombus within a severe atherosclerotic stenosis, without evidence of damage in the subtending myocardium. The conclusions were as follows:

1. The pathogenesis of this infarct was an increased peripheral resistance blocking intramural flow by aggravating the previous hypokinetic zone and extravascular compression by this hyperdistended hypokinetic myocardium through intraventricular pressure with extramural coronary stasis — not a primary occlusion of the related artery.
2. There was an absence of occlusive or mural thrombi in the related artery without rupture of a plaque or intramural embolization.
3. There was a progressive, rapid (12 months), subocclusive intimal sclerosis of a previously documented normal lumen both following angioplasty and at bypass surgery.
4. There were a lack of ischemic clinical and pathological injuries related to congestive heart failure.
5. There was organized thrombotic occlusion of the RCA without evidence of damage in the related myocardium.
6. There was an unreliable time-to-treatment interval in clinical studies (Schomig et al., 2003). A period of 90 min between asymptomatic first electrocardiographic ischemic changes and chest pain was observed.

This unique case monitored before, at the onset of, and following an infarct by up-to-date diagnostic and therapeutical approaches, poses a basic question: how many of the 87% occlusions reported by cineangiography within hours of the onset of an infarct (De Wood et al., 1980, 1986) are pseudo-occlusions and a consequence rather than a cause of the infarct? One case is just that, but it is invaluable in showing the natural history of an event.

Another question relates to blockage of intramural blood flow following temporary or permanent myocardial asynergy, with an increase in wall stress pressures in the related extramural artery or branches and enhanced atherosclerotic progression as shown in this case. If, as recognized today, regional hypokinesis is the first ischemic change, and it may last for a long period, one may expect that increased pressure in related arteries will accelerate plaque progression, thereby explaining the higher frequency of severe stenoses in chronic patients (Table 15).

Finally, the association of infarct necrosis with other forms of myocellular morphofunctional death discussed above is an important factor in explaining complications and cause of death. Contraction band necrosis is always extensive when in continuity with infarct necrosis and focal, often confluent, in noninfarcted myocardium. It is specifically stained by hematoxylin-basic fuchsin-picric stain (Lie et al., 1971), while colliquative myocytolysis (Saram, 1957) is observed in 38% of acute infarcts in the subendocardial myocardial layer and surrounding vessels, where myocardium is preserved from infarct necrosis (Fineschi and Baroldi, 2004). Then, their role in the following complications of an acute infarct must be considered.

Complication of Myocardial Infarction

Rhythm Irregularities

From the experimental infarct model induced by coronary occlusion, we know that an infarct is completely developed within 1 h and is often associated with ventricular fibrillation. Similarly, in humans, the first complication is *ventricular fibrillation* leading to cardiac arrest. It may occur at any time during the repair process, being more frequent within the first 24 h. *Asystole,* as an expression of acute heart failure, is the other main fatal complication.

Cardiogenic Shock

This relatively frequent complication in the past was reduced in occurrence during the last decades. In a series of 100 AMIs studied between 1971 and 1974 (Baroldi et al., 1974b) and in another one that occurred between 1976 and 1980 (Silver et al., 1980), distribution of infarct size was similar in cases with and without this complication. This finding contradicts the belief that cardiogenic shock was related to an infarct size greater than 40% (Page et al., 1971). The reason for the dramatic reduction of cardiogenic shock is obviously related to current therapy, even if the specific therapeutic agents are unknown. This acute failure is also responsible for pulmonary hypertension, deep vein thrombosis, right ventricular dysfunction, and so forth.

Fibrinous Pericarditis

A pericardial inflammatory process was reported in 7 to 16% of clinical cases (Kahn, 1975) and 13 to 45% of those examined at autopsy (Toole and Silverman, 1975). Rarely, the healing of this inflammatory process, linked with transmural infarction, leads to a constrictive pericarditis. Fibrinous pericarditis must be distinguished from Dressler's syndrome, an autoimmune process that starts 10 days to 2 yr after infarction (Kennedy and Das, 1976).

Cardiac Rupture

In our series of 200 AMIs, 27 died of cardiac tamponade (Table 30) following rupture of the free left ventricular wall associated with a transmural infarct. In two other instances, there was perforation of the interventricular septum, and in five other instances, there was rupture of left anterior (two cases) or posterior (three cases) papillary muscles. Ninty-three percent of the ruptures occurred in first-episode AMI. No relationship was observed between occlusive coronary thrombus, number of severe coronary stenoses, or heart weight.

The rupture starts with an endocardial tear near the margin of an infarct with a subsequent serpiginous hemorrhagic dissection through necrotic wall driven by the systolic blood pressure. The morphologic end result depends upon the speed, direction, and extent of dissection through the ventricular wall and whether its ventricular ostium gapes open. An extension through the ventricular wall with rupture into the pericardial cavity produces a 250 to 300 ml hemopericardium and acute cardiac tamponade. The latter is easily

Table 30   Heart Rupture in 200 Consecutive Acute Myocardial Infarct First-Episode Patients (AMI 1st) and in Chronic Patients (AMIC)

Source	Number	Infarct Size (%)					Lumen Reduction (%)					Occlusive Thrombus
		<10	20	30	40	>50	≤69	≥70	1 Vessel	2 Vessels	3Vessels	
AMI 1st	145	32	30	32	24	27	16	129	61	49	19	60
+ Rupture	31	2	10	11	3	5	3	28	15	9	4	15
AMIC	55	28	7	10	3	5	1	54	16	22	16	22
+ Rupture	3	1	—	2	—	—	—	3	2	1	—	2

Note: Abbreviations: AMI 1st, first episode of acute myocardial infarction; AMIC, in chronic patients.

recognized clinically by both electromechanical dissociation and by echocardiography. Pathologically, there is a bulging of a bluish pericardium. The rupture may be *rapid* within 24 h of infarct onset or *slow* within 3 to 10 days (Silver et al., 1993) and may be due to destruction of the cardiac skeleton, massive polymorphonuclear infiltration with dissolution of necrotic myocardium by proteolytic enzymes, or exertion or hypertension before or after infarction. The high frequency of rupture of left ventricular acute infarcts (73%) in mental patients (Jetter and White, 1944) raises the possibility of a heart–brain relationship with adrenergic stress. Rupture of the right ventricle and its papillary muscles and atria are rare (Vlodaver and Edwards, 1977). Also, it is rare to see a healed dissection.

Aneurysm

We may distinguish an early or *acute* aneurysm determined by the paradoxical bulging of a myocardium in irreversible relaxation with early thinning of the infarcted wall. Also, because the healing phase, especially of a large, transmural infarction, occurs in the presence of a maintained intraventricular pressure, an aneurysm may also develop over a protracted period. The latter, particularly seen after anterior transmural infarction, shows a thin wall with possible mural intracavity thrombi, calcium deposition, and an evolving scar that may contain viable, stretched myocardial cells (Figure 38A). Furthermore, lipomatous metaplasia (see below) can be found in 55% of surgically excised specimens. In this context, the marked decrease of tissue norepinephrine (denervation) with an increase of β1-adrenoreceptors is a possible cause of recurrent sustained ventricular tachycardia with fatal malignant arrhythmia, as proposed by Bevilacqua et al. (1986). The other main complications of a ventricular aneurysm are congestive heart failure, a result of many concurrent factors (see below); embolization from endocardial mural thrombi; and mitral insufficiency.

A rare event is a *dissecting aneurysm* of the infarcted cardiac wall. In a personal case, the left ventricle was formed by two equally sized chambers communicating through a 2 cm hole, the external one being totally filled by a thrombus. Postmortem injection showed a normal distribution of the coronary arteries on the surface of the whole left ventricle. This excluded a *pseudoaneurysm*, that is, a hematoma limited by the pericardium after free-wall rupture (Scomazzoni et al., 1957).

Denervation

The involvement of cardiac nerves in different pathologies, resulting in zonal or regional denervation, was discussed. The denervation around an acute infarct (Barber et al., 1983; Stanton et al., 1989; Dae et al., 1991; Spinnler et al., 1993; Kramer et al., 1997) is thought to be consequent to the destruction of axons crossing the necrotic area. Therefore, myocardial denervation around an acute infarct can be defined as a complication, because a denervated myocardium becomes hypersensitive to catecholamine, triggering malignant, fatal arrhythmias (Inoue et al., 1987).

Papillary Muscle Dysfunction

This physiopathological impairment can be due to the acute infarct, scarring, or postinfarct atrophy and the dyssynchronous arrival of nerve impulses to contract the papillary muscle (denervation?), all leading to *mitral valve regurgitation* that may be aggravated by congestive heart failure with separation of papillary muscles, or to *tricuspid valve incompetence*.

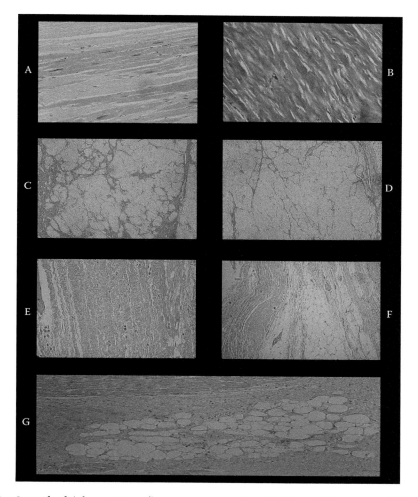

Figure 38 Stretched (elongation of) sarcomeres and nuclei in viable myocardial cells in an aneurysm of the left ventricular wall with dense fibrosis (A). Dense and compact fibrosis is the end result of a repair process of an infarct. The collagen fibers are straight and close together (B). In contrast, in hearts with congestive heart failure, the myocardial fibrosis is very mild (C, D) showing a corkscrew aspect of collagen fibers (E). This reflects an adaptation of collagenogenesis to the contraction cycle without any capability to reduce or stop the latter. Furthermore, fibrous tissue may by metaplasia become adipose tissue that substitutes for a large fibrous area (F, G). This is a fact to keep in mind when quantifying the size of a scar or measuring myocardial viability by the nuclear method.

Endocardial Thrombosis and Thromboembolism

Rarely, thromboembolism is the presenting clinical feature of a myocardial infarct, or it may occur during its course. The complication may develop even in the absence of an aneurysm. Mural thrombi can be detected *in vivo* by several methods, echocardiography being the best with 90% specificity and sensitivity. Atrial thrombi are demonstrated by the transesophageal approach. Mainly found on the left anterior ventricular wall (39%), mural thrombi are rare with inferior infarcts (1%) (Dantzig et al., 1996). In a clinical study, an endocardial thrombus was found in 31% of 124 patients with an anterior infarct imme-diately before discharge from the hospital, with none in 74 patients with an inferior infarct. No relation was found with thrombolytic therapy. A low ejection fraction (< 35%) and apical dyskinesia were correlated with the intracavitary thrombus, and follow-up showed its disappearance in 48% of cases. Systemic thromboembolism occurred in six patients (Keren et al., 1990). The healing of endocardial thrombosis results in a fibrous thickening of the endocardium with obliteration of trabeculae (see "Pathogenetic Factors" discussed in the "Congestive Heart Failure" section below for more information on endocardial thickening due to myoelastofibrosis).

Unstable Angina

Unstable angina pectoris was linked to coronary stenosis associated with mural thrombi and platelet/thromboemboli to the myocardium (Davies and Thomas, 1984).

Different angiographic criteria such as luminal irregularities, a haziness with ill-defined margins, a smudged appearance, inhomogeneous opacification within the lumen, and changes suggestive of ulceration or plaque rupture were used as evidence of a thrombus associated with a complex or acute active plaque seen in 58% of 69 patients with unstable angina in contrast to only 5% of 20 patients with stable angina (Cowley et al., 1989) and in 57% of 37 patients with prolonged angina at rest, who were improved by fibrinolytic therapy with a marked reopening of their narrowed coronary artery segments despite a recurrence of symptoms and signs in 71% (Gotoh et al., 1988).

In another study, angiographic stenoses were classified as *concentric* if they presented a symmetric and smooth narrowing; *eccentric* if an asymmetric narrowing, with *type I* having a smooth border and broad neck and *type II* a convex intraluminal obstruction with a narrow base due to one or more overhanging edges or extremely irregular or scalloped borders or both and multiple irregularities with three or more serial and closely spaced severe obstructions or severe diffuse irregularity. Authors considered type II eccen-tric stenoses either a disrupted atherosclerotic plaque or a partially occlusive or lysed thrombus (coronary plaque in evolution, precursor of impending infarction). They were found in 71% of patients with unstable angina, 16% with stable angina, and 66% with acute (within 12 h of symptoms onset) or recent (1 to 2 weeks) or healing (2 to 10 weeks) myocardial infarctions (Ambrose et al., 1985a,b).

A simpler subdivision was done in a postmortem coronary angiographic/histologic study of 73 stenoses ranging from 50 to 99% of luminal diameter from 39 fatal infarcts or following bypass surgery. Stenoses were classified as those with a smooth border, an hourglass configuration, or no intraluminal lucencies and those with an irregular border and intraluminal lucencies. Of 35 angiographic stenoses of the first type, 11% presented a complicated pattern (i.e., showed rupture, hemorrhage, mural thrombus, or recanalized thrombus), with most being uncomplicated (i.e., fatty or fibrous plaque with intact intimal surface and no thrombus). In contrast, of 38 of the second type of stenosis, 79% were complicated (Levin and Fallon, 1982). These findings support the previous angiographic classification. However, angiography fails to visualize all severe coronary lesions even after many years (Haft and Al-Zarka, 1993), and clinically stable angina is not necessarily associated with a histologically stable plaque (van der Wal et al., 1996). Furthermore, angiographic evidence of thrombi may only image a vascularization of the atherosclerotic plaque demonstrated by serial section as previously described. An unstable plaque (Heistad,

2003) more often seen in unstable angina associated with an acute infarction is a secondary phenomenon due to intramyocardial blockage of flow, with hindrance of the latter at plaque level.

In conclusion, unstable or stable angina can be observed clinically. A pathologist can examine only cases of angina associated with a fatal AMI or SD. Findings will include those pertinent to these acute syndromes.

Sudden Coronary Death

Sudden death may occur in the course of any disease or condition. In history, several examples are reported, the best known being Phidippides, a soldier, who died suddenly in 490 B.C. after a 22 ml and 1.7 yd run from Marathon to Athens to herald a Greek victory against the Persians. However, only twice in history is an epidemic of sudden death referred to. One occurred in 1705 in Rome. Its virulence induced Pope Clement XI to order an inquisition by his *archiatra* or Chief of Medicine, Giovanni Maria Lancisi in 1707. The result was the first publication (*De Subitaneis Mortibus*; Lancisi, 1745) based on autopsy findings that suggested that sudden death is frequently of cardiac origin. The second epidemic is the one we are witnessing and is a major cause of death in technological–consumeristic societies. Again, most of these cases are linked to cardiovascular pathologic changes, as, for instance, was shown in an Italian study (Baroldi et al., 1979) of 208 sudden and unexpected deaths (SD) selected from 765 out-of-hospital sudden deaths, 606 of whom were diagnosed by the coroner as suffering "coronary disease" due to the presence of coronary atherosclerosis as the main pathology. In 72 cases, there was a brain infarct/hemorrhage; 27 had aortic rupture after dissection; 2 had rheumatic heart disease; 2 had congenital heart disease; and 56 had noncardiac vascular conditions, such as rupture of esophageal varices in liver cirrhosis, pulmonary infarct or pneumonia, and so forth.

Sudden death is currently a major topic in health care, with researchers striving to determine its cause. Forensic pathologists, by law and like Lancisi, have a major responsibility. The first need is to put forth a clear-cut definition of the entity, and the second is to delineate a more clear-cut classification of its various categories.

Definition of Sudden Death

Lancisi formulated an exhaustive list of different types of death:

> Indeed this absolutely complete cessation of animal movements and this departure of the soul from the body, even though it happens at all times more swiftly than thought itself, is nervertheless divided for the sake of common parlance and for greater clarity of teaching, into natural, untimely and violent death, and those again individually into slow and sudden death, into those that are foreseen and forefelt and finally into such as are unforeseen, imperceptible and unexpected. (Lancisi, 1745, translated by White and Boursey, 1971)

Little can be added. In line with Lancisi's teaching, we can distinguish the following:

1. *Physiologic or genomic death*: This is death programmed by the genome of each living organism. At present, it is difficult to discriminate this form of death because of overlying concurrent chronic diseases, some of which may have a genetic basis.

2. *Natural or pathological death*: Here, diseases or deficiencies or malnutrition have followed their natural history in killing the individual. A pathological death can be *sudden* and either *forefelt/foreseen* or *unforeseen/unexpected*.

3. *Violent death*: This is a consequence of trauma, poisoning, and so forth, and may be the result of accident, suicide, or manslaughter/homicide.

4. *Unexplained death*: This term is used when careful investigation produces findings insufficient to explain death. This situation is more frequent than one might believe when findings are reviewed critically and quantified.

The definition of sudden death that we adopt is "a rapid (without any specific chronologic limit) and unexpected or unforeseen — both subjectively and objectively — death which occurs without any clinical evaluation and in apparently healthy people in normal activity (*primary* or *unexpected sudden death*) or in patients in an apparently benign phase during the course of a disease (*secondary* or *expected sudden death*)" (Baroldi and Silver, 1995; Baroldi et al., 2001; Fineschi and Baroldi, 2004).

This definition is perfectly tailored for coronary heart disease, with the awareness that etiopathogenetic uncertainty makes any definition a useful working one. However, as a corollary to the definition, there is a need to clearly define the criteria for the selection of sudden and unexpected coronary cases, that is, cases involving apparently healthy subjects living their normal lifestyles, without a history of any disease correlated to sudden death, not under medical care, who died out-of-hospital without or with recorded terminal therapy and resuscitation attempts in the presence of witnesses, and with reliable information on the clinical history from family members and friends (sudden and unexpected coronary death) or in patients with known CHD (coronary heart disease; sudden and expected coronary death). In both conditions, the gross and histologic documentation of extensive monofocal fibrosis (10% of the myocardial mass) is a reliable marker of a previous infarct that may have been silent (see below) and, therefore, not attested to in the clinical history. This permits a further distinction between first-episode and second-episode or chronic cases.

Another point concerns the difficulty of comparing data in the literature because of divergent definitions and selective criteria. Frequently, cases are selected when death occurs within 1 to 24 h, an unacceptably long survival to discriminate causal factors from agonal events and changes due to cardiopulmonary resuscitation. Even at 1 h, the findings are a miscellanea of the latter, and survival has to be ascertained just to discriminate the physiopathology of cardiac arrest from other concurrent factors. A last preconceived criterion is to consider coronary death as occurring only in those cases with more than 50 or 75% of coronary lumen reduction (Roberts and Jones, 1979; Farb et al., 1995), because ischemic heart disease may occur in the presence of mild coronary stenosis as documented in many reports (Baroldi and Silver, 1995; Fineschi and Baroldi, 2004).

Pathological Findings in Sudden and Unexpected Coronary Death

We reported the pathological findings in sudden coronary death in association with those found in acute myocardial infarct. There is a similar trend in these two patterns of acute CHD with respect to controls, despite some strong divergencies: in sudden coronary death we found a higher frequency in men (M/W = 7) than in AMIs (M/W = 2), with the still controversial protective action of estrogen in women (Schulman et al., 2002); a higher frequency of cases with no, or minimal (50%), lumen reduction (14% vs. 3%); a significantly reduced frequency of hemorrhage within the atherosclerotic plaque (Table 9 and Table 10); a significant low frequency of both acute and mural thrombi (Table 28); and a lack of relationship with heart weight >500 g (Table 27) and with activity and symptoms preceeding death (Table 31). In 35 of 208 cases of sudden and unexpected death, an infarct was histologically documented associated with an occlusive thrombus in 11, while contraction band necrosis was present in all of these 35 cases and as a unique acute lesion in 72% of the other 173 (Table 32).

In an extensive review of the literature (Baroldi and Silver, 1995), despite difficulty in comparing material differently defined and selected, most data were in agreement with our findings. In only two contributions on sudden death in unstable angina (Davies and Thomas, 1984; Falk, 1985) was the frequency of occlusive thrombi similar to that found in acute myocardial infarction. This discrepancy is easily understood, because most of the reported cases were histologically documented myocardial infarct patients who died within 6 h, and these cases were really beyond the definition of a sudden death. They should be regarded as infarct cases, with patients who died rapidly or were expected sudden death cases.

The misleading interpretation offered by inclusion of these cases raises a need to discriminate between acute myocardial infarct and sudden and unexpected coronary death because they are two distinct entities with their own morphologic and pathophysiologic patterns. In fact, most patients (81%) resuscitated by defibrillation did not show an infarct (Cobb et al., 1975, 1980; Goldstein et al., 1981), and of patients who suddenly died while wearing a Holter monitor, 83% had a cardiac arrest due to ventricular fibrillation without evidence of a preceding infarct (Bayés de Luna et al., 1989, 1990). Therefore, acute myocardial infarcts, even if a patient dies rapidly of ventricular fibrillation, are not synonymous with sudden coronary death.

On this subject, sudden and unexpected death cases without critical coronary atherosclerosis merit particular attention. In 28 of our cases — all first-episode cases — with a normal lumen or a stenosis less than 50% and no coronary thrombus, the frequency of contraction band necrosis was 78% with an extent similar to cases with severe atherosclerotic obstruction. Contraction band necrosis was present in cases of sudden and unexpected death both in a patient with acquired immunodeficiency syndrome (AIDS) (Baroldi et al., 1993) and in seropositive subjects with silent Chagas' disease with extensive lymphocytic myocarditis, all with normal coronary arteries (Baroldi et al., 1997b) (see "Myocarditis" section below).

Diagnostic Criteria

Due to the frequent absence of objective data in diagnosing CHD postmortem, the following can be proposed, providing that clinical features are compatible:

1. Coronary atherosclerosis of any degree and number of stenoses without plaque complications and associated with a normal myocardium does not allow for a diagnosis of CHD.
2. Uncomplicated plaques plus extensive myocardial fibrosis without a history of other damaging diseases allows for a diagnosis of chronic but silent or inactive CHD but not of sudden coronary death.
3. With uncomplicated or complicated plaques and contraction band necrosis in different evolutive stages and to an extent greater than 9 ± 6 foci and 102 ± 143

Table 31 Survival, Activity, and Symptoms before Sudden Death in Relation to Coronary and Myocardial Damage and Heart Weight

| | Total Cases | Lumen Reduction | | | | Acute Occlusion Thrombus | Acute Infarct | Myocytolysis | | Heart Weight ≥500 g |
		≤69%	≥70% in 1 Vessel	2 Vessels	≥3 Vessels			Coagulative	Fibrosis	
Survival										
<10 min	151	41	39	41	30	18	25	110	123	68
10–60 min	47	9	11	17	10	11	8	33	37	18
>60 min	10	1	3	2	4	3	2	6	10	1
Total	208	51	53	60	44	32	35	149	170	87
Activity										
Working	52	10	22	11	9	5	6	37	40	22
Walking	44	7	11	13	13	5	7	37	40	17
Driving	13	—	1	9	3	1	2	9	12	5
Sleeping	22	9	5	5	3	4	5	13	20	9
Resting	72	23	12	21	16	17	14	51	56	33
Unknown	5	2	2	1	—	—	1	2	2	1
Total	208	51	53	60	44	32	35	149	170	87
Symptoms										
Angina	48	4	20	12	12	11	9	32	35	16
Dyspnea	29	12	6	8	3	3	6	24	25	14
Paresthesia	3	—	1	2	—	2	1	3	3	—
Vertigo	4	—	1	3	—	—	1	4	4	2
Unknown	124	35	25	35	29	16	18	86	103	55
Total	208	51	53	60	44	32	35	149	170	87

Table 32 Type of Acute Myocardial Necrosis and Fibrosis, Lumen Reduction, and Acute Occlusive Thrombus in Coronary Sudden and Unexpected Death

Myocardial Damage	Stenosis (%)			≥70% in			Acute Occlusive Thrombosis
	<69	≥70	Total	1 Vessel	2 Vessel	3vessels	
Infarct size							
10	—	11	11	3	4	4	1
11–30	—	16	16	4	8	4	6
>30	2	6	8	5	1	—	4
Total	2	33	35	12	13	8	11
Coagulative myocytolysis							
Minimal	23	65	88	16	25	24	7
Moderate	6	32	38	8	14	10	9
Extensive	5	18	23	8	9	1	9
Total	34	115	149	32	48	35	25
Early	27	68	95	21	25	22	9
Alveolar	1	27	28	7	9	11	9
Scarring	6	20	26	4	14	2	7
Total	34	115	149	32	48	35	25
Colliquative myocytolysis							
Minimal	—	16	16	1	5	10	2
Fibrosis							
No or minimal	39	64	103	32	24	8	15
Moderate	4	44	48	13	15	16	4
Extensive	1	56	57	12	15	29	5
Total	44	164	208	57	54	53	24

necrotic myocytes $\times$ 100 mm^2, the diagnosis of sudden coronary death is pertinent. It must be stressed that unquantified foci and myocardial cells showing only early contraction band necrosis, per se, do not allow for a diagnosis of early myocardial infarction to be made, as proposed by Holpster et al. (1996) and Virmani et al. (1996).

4. Complicated plaques with a normal myocardium or myocardial fibrosis are compatible with a probable diagnosis of CHD/sudden death.

5. With intimal hemorrhage plus or minus a mural or occlusive thrombus in one artery with a dependant normal or fibrotic myocardium, the diagnosis of an early myocardial infarction of less than 1 h is likely; with a stretched myocardium plus peripheral contraction band necrosis, the diagnosis of early acute infarct with rapid death is legitimate.

6. For acute or recent myocardial infarction with or without complicated and severe plaques, the diagnosis is rapid death following an infarct.

7. The myobreakup of myocardial cells allows for a diagnosis of cardiac arrest by ventricular fibrillation and nothing else. When associated with previous findings, it may indirectly confirm a diagnosis of sudden coronary death.

It must be stressed again that by "complicated" or "active" plaque, we mean a plaque showing medial neuritis, intimal hemorrhage, or thrombosis with or without rupture. Keep in mind that a large lumen occluded by atheromatous material may be due to degeneration of a coagulum (Figure 32) rather than rupture of an atheromatous plaque. Serial sections are needed for a correct diagnosis to be made.

Silent Acute Coronary Heart Disease

In the clinical setting, permanent akinetic areas of myocardium and ischemic signs can be observed without a history of a previous myocardial infarct (Cohn, 1989). Painless myocardial infarction seems prevalent in diabetics with autonomic neuropathy (Faerman et al., 1977). A pathologist recognizes that a silent cardiac infarct has occurred among sudden and unexpected death cases when patients who died within a few minutes without evidence of chest pain or other ischemic symptoms or signs have a histologic myocardial infarction hours to days old (Table 33). Recently, we observed a 7- to 10-day-old transmural 30% infarct associated with widespread contraction band necrosis and normal coronary arteries including intramural vessels in a 28-year-old bodybuilder who used steroids and died suddenly during strenuous exercise in a gymnasium (Fineschi et al., 2001b). This is really a case of silent myocardial infarction in the absence of coronary heart disease. His clinical

Table 33   Silent Myocardial Infarct versus Clinical Infarct and Sudden Death without Histologic Infarct

Morphologic Variables	Silent Infarct (35 Cases)[a]	Clinical Infarct (200 Cases)	SD with No Infarct (173 Cases)
Coronary stenosis			
<69	3	17	48
≥70	32	183	125
≥70 in			
1 vessel	12	77	41
2 vessel	12	71	48
3 or more vessels	8	35	36
Occlusive thrombus			
Acute	11	83	21
Previous	2	30	12
Acute infarct			
Size (%)			
<20	20	97	—
20	8	103	—
Age (days)			
<2	19	70	—
2–10	4	74	—
11–30	12	56	—
Location			
Subendocardial	6	26	—
Internal half	18	52	—
Transmural	11	122	—
Myocardial fibrosis			
Absent/microfocal	31	145	148
Extensive	4	55	25
Heart weight			
<500 g	23	96	98
500	12	104	75

[a] Apparently normal people dying suddenly and unexpectedly without symptoms of an acute infarct.

history was negative. However, a week previously, he had a short episode of chest pain interpreted as skeletal muscle pain due to physical effort.

Chronic Coronary Heart Disease

Often defined as "ischemic cardiomyopathy," this condition will become the most frequent pattern of CHD because of the increasing mean age of the population and the therapeutic prevention of death in acute coronary syndromes, particularly following acute myocardial infarction. Chronic coronary heart disease mainly results in congestive heart failure (see below).

Myocarditis

Myocarditis is a *primary* or *secondary* inflammatory process that involves the myocardium. By primary or secondary, we mean, respectively, a process that begins within the myocardium or in other organs or systems. A further distinction defines myocarditis of *known* or *unknown* causes, avoiding the ambiguous term "idiopathic." A third distinction relates the site of action of noxious agents either in the myocardial interstitium or within the myocardial cell.

The aim of this section is to review the main pathological findings in relation to myocardial dysfunction and cardiac arrest. Accordingly, a quantification of the process in terms of its extent versus the total ventricular mass is essential. In fact, any inflammation may range from a small focus to a diffuse and massive exudation capable, per se, of explaining an acute cardiac insufficiency, as for instance in acute rejection (see section entitled "Orthotopic Transplanted Human Heart" below). The method of heart examination we use (see above) is adequate to correctly measure the phenomenon in relation to the histologic area in square millimeters of all cardiac regions. This approach implies a need to review the "Dallas criteria" (Aretz, 1987) adopted for diagnosis in endomyocardial biopsies, that is, an inflammatory cytotoxic infiltrate determining a myocardial cell necrosis. It is obvious that an endomyocardial biopsy cannot provide the real extent of an inflammatory process in the myocardium.

Myocardial Biopsy

Myocardial biopsy includes transmural and endomyocardial sampling done *in vivo*, specimens removed at surgery, and the whole heart excised at transplantation. The left ventricular transmural biopsy (Shirey et al., 1972) was rapidly abandoned because of complications following the insertion of a needle into the beating cardiac wall. The endomyocardial biopsy obtained from the apical-septal region of the right ventricle via the jugular vein or from that area of the left ventricle by a biotome inserted into a catheter passed via a femoral artery has several limitations. They include the following:

1. Tissue samples are small, subendocardial (Figure 39), and removed from a specific zone. This can produce false-negative results.
2. Discordance exists among expert cardiovascular pathologists in diagnosing elementary parameters (Shanes et al., 1987) as a possible source of false-positive findings.
3. The presence of an artifactual contraction band-cutting edge lesion (see above) (Todd et al., 1985a,b) (Figure 17) must be distinguished from other types of myocardial damage, especially from catecholamine necrosis.

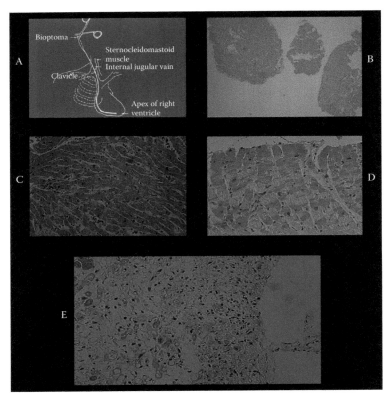

Figure 39　Endomyocardial biopsy. (A) Scheme showing bioptome introduced in the right ventricle via the jugular vein. (B) Samples obtained by this method. (C) Artifact contraction bands are sometimes difficult to distinguish from contraction band necrosis (CBN) (D). Alveolar CBN (E). (Part A: Courtesy of Dr. M.D. Silver.)

4. A misinterpretation of natural monocyte infiltrates can occur.
5. Obliterative medial hyperplasia (Baroldi and Thiene, 1998; Figure 8) may be found in columnae carneae and in papillary muscles in normal hearts of healthy people and may be erroneously interpreted as small vessel disease.

Morphologic Types

In general, the different patterns of myocarditis (Winters and McManus, 2001) are distinguished according to the type of inflammatory cell reaction elicited by the causal agent (Table 34), even if, all too often, the etiologic factor can be demonstrated neither histologically nor by culture. AIDS and iatrogenic (transplanted heart) immunodeficiency syndrome are conditions in which opportunistic infections occur frequently, and microorganisms can be seen. However, when the latter are not visible, a specific chemotactic cellular response may indirectly suggest etiology.

Polymorphonuclear or Neutrophilic

This interstitial inflammatory process is characterized by a marked exudative infiltration of neutrophil leukocytes with or without abscess formation (Figure 40). Often it coincides with a septicemia caused by bacterial cocci and mycotic agents. The main complications

Table 34   Myocarditis — Classification in Relation to Type of Inflammatory Reaction and Causal Agents

Myocarditis	Cause	Observations
Polymorphonuclear leukocytes *Polymorphonuclear myocarditis*	Staphy-strepto-meningococci	Focale or diffuse with abscess formation
	Actinomyces, aspergillus, candida	Septicemia
Lymphocytes *Lymphocytic myocarditis*	Virus	Focal or diffuse

	Corynebacterium difteria,
	  Trypanosoma cruzi, Borrellia,
	  Toxoplasma arsenicum,
	  chloroquine euretine,
	  phenothiazine (autoimmunity)
		Dilated cardiomyopathy, Lyme disease, Chagas' heart disease, transplant rejection

Eosinophils *Eosinophilic myocarditis*	Hypersensitivity Helminthiasis	
Granuloma *Granulomatous myocarditis*	Tuberculosis	Plurifocal-diffuse
	Rheumathic fever, lupus erythematosus, sarcoidosis, fungal infection, foreign bodies	
Giants cells *Giant cell myocarditis*	?	Associated with lymphocytes, plasma cells, eosinophils

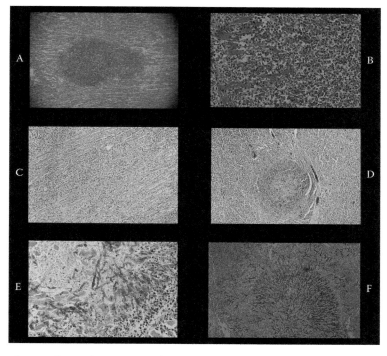

Figure 40   Polymorphonuclear myocarditis. Focal myocardial abscess at low (A) and higher (B) enlargement due to unknown agent. Aspergillus fumigatus (C, D) as a cause of focal (C) or diffuse (D, E, F) myocarditis.

linked with sudden death are rupture of a ventricular wall or papillary muscle and involvement of cardiac valves or the conduction system. In some cases, the neutrophilic infiltration may mimic a myocardial infarct (see below).

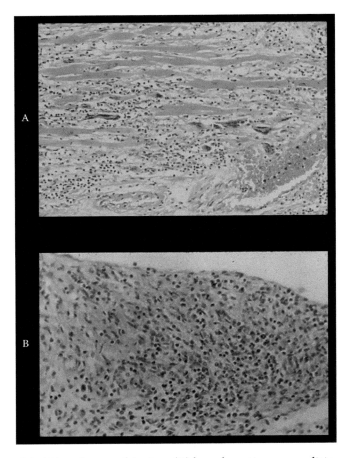

Figure 41   Idiopathic (A) and coxsackievirus (B) lymphocytic myocarditis.

Lymphocytic

Several noxae may induce a lymphocytic myocarditis. In particular, many viruses, such as enteroviruses, (coxsackie viruses), adenoviruses, cytomegalovirus, and human immuno-deficiency virus (HIV) seem responsible for this inflammatory process, which is possibly related to the genesis of dilated cardiomyopathy (see below).

The histologic diagnosis of lymphocytic myocarditis (Figure 41) is one of the more difficult tasks for a pathologist. In fact, lymphocytic infiltrates are present in the myocardial interstitium of healthy subjects (see below). We previously mentioned the need for phe-notyping the intramyocardial monocytes by monoclonal antibodies (Parravicini et al., 1991). Furthermore, viral antigens or frozen sections as well as immunofluorescence, poly-merase reaction, genoma identification, etc. (Winters and McManus, 2001; see Chapter 7), may help confirm the presence of a virus. Nevertheless, the latter is not enough to prove a cause–effect relationship.

Three conditions characterized by lymphocytic myocarditis that may help in our understanding are now discussed:

> *Acquired immunodeficiency syndrome (AIDS)*: Among 26 cases, with patients who died in a relatively short time (8 ± 3 months) when therapy did not prolong life (Baroldi et al., 1988), the main cardiac findings were Kaposi's sarcoma in 2 (Figure 42),

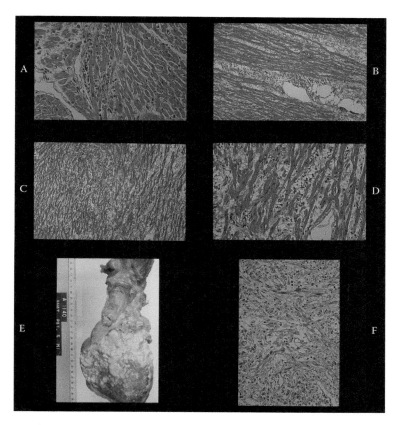

Figure 42 Lymphocytic myocarditis in AIDS patients. Focal with myocardial cell necrosis (A, B) and diffuse in a patient dying suddenly and unexpectedly (C, D), in whom the diagnosis of AIDS was confirmed at autopsy (HIV positivity, diffuse lymphoadenopathy, Kaposi's sarcoma in a young, drug abuser woman). Kaposi's sarcoma is frequent in AIDS patients, occasionally grossly visible (E), and is not to be confused with a myocarditis (F).

microfocal myocardial abscesses in 1, a microfocal infarct in 2 without coronary disease, foci of catecholamine necrosis in 13, lymphocytic myocarditis in 9, and lymphocytic infiltrates without myocardial necrosis in 7 (Figure 42). In contrast to other reports, not one had dilated heart cavities or died of congestive heart failure. An interesting case not included in the previous group concerned a 25-year-old woman drug abuser who died suddenly and unexpectedly from ventricular fibrillation documented on her arrival at the emergency ward complaining of vague symptoms. A serum test was positive for HIV, and AIDS was confirmed at autopsy (a diffuse lymphoadenopathy, splenomegaly, interstitial pneumonia, Kaposi's sarcoma). The cardiac findings were diffuse microfoci of TCD3$^+$ lymphocytes without myocardial cell necrosis, that is, lymphocytic myocarditis in the absence of associated myocyte necrosis, diffuse microfoci of catecholamine necrosis without cytotoxic elements, and myocardial cell segmentation as a sign of the ECG-documented ventricular fibrillation. The number of lymphocytic infiltrates × 100 mm^2 in this case was 9.5 versus 1 ± 2 in other AIDS cases. The figures for the number of foci and myocells with catecholamine necrosis were 6 and 114 versus 3 ± 1 and 10 ± 21, respectively.

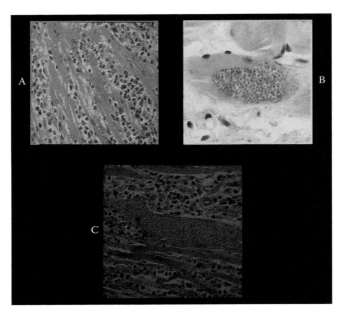

Figure 43 Lymphocytic myocarditis in acute Chagas' disease (A) showing amastigote forms of *Trypanosoma cruzi* within myocardial cells (B, C).

Chagas' disease: This disease (Figure 43), defined as a "denervation syndrome" due to *Trypanosoma cruzi*, is a medicosocial problem in South America as well an extremely interesting model in human cardiology. Often silent, it begins as an acute myocarditis or meningoencephalitis with a 10% mortality. The main heart finding is a severe and diffuse lympho-plasma-cellular inflammation of the myocardium. Most patients survive, but after an asymptomatic period of 10 to 20 years, a second clinical phase is characterized by mega esophagus-stomach-colon or dilated cardiopathy with a typical apical (47%) or left posterior (27%) aneurysm associated with congestive heart failure. Sudden death is reported in 15 to 30% of these cases. This high frequency suggested a specific protocol (Baroldi et al., 1997b) to study apparently healthy people who died suddenly and unexpectedly, out-of-hospital, without history of any disease and without resuscitation attempts, all with a positive serum test for Chagas' disease. In all 34 studied subjects, the unique findings were numerous and often extensive lymphocytic infiltrates (Figure 43), occasionally with some giant cells, associated with myocardial cell necrosis (91%); catecholamine necrosis without (50%) or with macrophagic monocytes (50%), most showing later stages of this form of necrosis (alveolar, healing, healed) (Figure 44); extensive (>20%) myocardial fibrosis in 21%; and apical aneurysm (47%) formed by a thin nonfibrotic myocellular wall (Figure 45). The lymphocytic infiltrates were significantly more frequent and extensive than in controls ($p < 0.001$), and there was a significant higher frequency ($p < 0.0001$) of lymphocytic infiltrates without myocardial necrosis. When present, the latter showed contraction bands typical for catecholamine necrosis.

In hamsters infected with *Trypanosoma cruzi*, the marked lymphocytic myocarditis seen soon after infection did not correspond to the extensive fibrosis observed

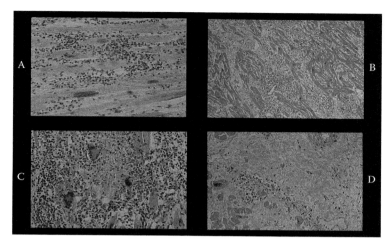

Figure 44 Lymphocytic myocarditis in subjects apparently healthy who died suddenly and unexpectedly; serum positive for Chagas' disease at autopsy. Diffuse lymphocytic infiltration without myocardial cell necrosis (A), even with marked interstitial exudation (B). Occasionally, giant cells are present (C). Myocardial fibrosis (D) is not frequent and never extensive. Exceptionally, *Tryponosoma cruzi* is found in a myocardial cell.

in infected animals, sacrificed after 2 months (Figure 46). These human and experimental data confirm our impression that lymphocytic myocarditis is not the cause of extensive fibrosis, rather, when present, it is the consequence of catecholamine necrosis.

Dilated cardiomyopathy of unknown cause: This condition will be discussed again in the sections below entitled "Cardiomyopathies of Unknown Nature" and "Congestive Heart Failure" and in Chapter 7. Its mention here is justified by the current hypothesis that we are dealing with a process — similar to Chagas' disease — that starts as viral lymphocytic myocarditis and, after a long asymptomatic latency, possibly through an autoimmune mechanism, ends in fibrotic dilated cardiomyopathy. This is a hypothesis mainly derived from experimental models (Kawai and Abelman, 1987; Baroldi et al., 1990a; Camerini et al., 1998) and supported by increasing knowledge on the relationship between host and virus in respect to the action on the molecular process of their genetic determinants and environmental factors (Winters and McManus, 2001). Here, the need is to establish whether or not lymphocytic myocarditis may explain cardiac insufficiency by killing myocytes substituted by repair fibrosis. Our findings in 63 hearts excised at transplantation (Table 6, Table 35 through Table 38) excluded this pathogenetic mechanism (see section below entitled "Congestive Heart Failure").

Eosinophilic

Eosinophilic leukocytes are the distinctive inflammatory elements that prevail in the myocardial interstitium and around vessels in this pattern of myocarditis. Even in this condition a disproportion exists between the severity of the interstitial exudate and myocellular necrosis, including the acute necrotizing eosinophilic myocarditis often associated with sudden death (see Chapter 7). Many substances — more than 30 (Lewis and Silver, 2001) — are responsible for this hypersensitivity reaction, mainly due to penicillin, methyldopa,

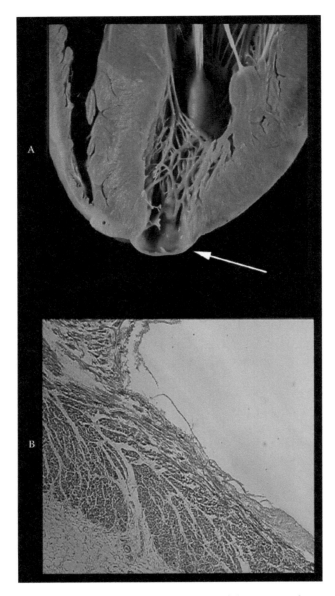

Figure 45   Typical aneurysm of the cardiac (left ventricle) apex in Chagas' disease. (A) Gross and (B) histologic aspects of its thin wall formed by normal myocardial cells, without evidence of fibrosis and infiltrates.

and sulfonamides, as well as some parasites such as *Taenia solium, Echinococcus granulosus, Schistosoma, Trichinella spiralis,* etc. (Winters and McManus, 2001).

Granulomatous

This inflammatory reaction assumes a granulomatous aspect with (in tuberculosis) or without (in sarcoidosis) a central caseous necrotic core. The reactive cells are giant cells,

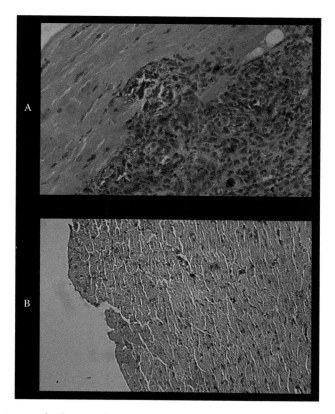

Figure 46 Experimental Chagas' disease in hamster infected with *Trypanosoma cruzi*. Acute myocarditis (A) and absence of myocardial damage including myocardial fibrosis in animal sacrificed at 60 days after acute phase (B).

lymphocytes, fibroblasts, and Anitschow macrophages (Figure 47). The granuloma results in fibrous tissue with the peculiar perivascular fibrosis seen in chronic rheumatic hearts.

Giant Cell

This rapidly fatal condition of unknown cause is rare. Giant cells, lymphocytes, plasma cells, and eosinophils form the infiltrate and are often located around necrotic myocells (Figure 48). It is still unclear as to the myogenic or macrophagic (or both) nature of the giant cell and its relationship with several conditions, such as rheumatic fever, thymoma associated or not associated with myasthenia gravis, lymphoma, systemic lupus erythematosus, dermatomyositis, and so forth (Winters and McManus, 2001).

Parasitosis

Opportunistic infections in acquired or iatrogenic immunodeficiency syndromes frequently show an intramyocardial cell invasion by several agents (for instance, toxoplasma, sarcosporidium, cryptococus; Baroldi et al., 1988) without inflammatory reaction (Figure 49). This condition cannot be defined as a myocarditis, and its functional damage is difficult to validate histologically.

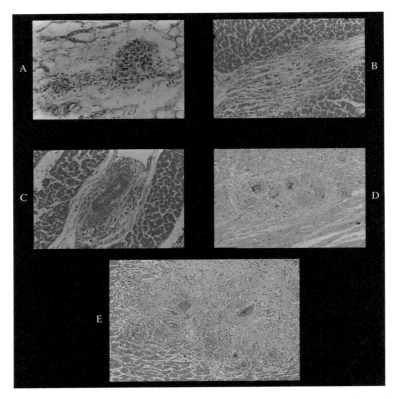

Figure 47   Granulomatous myocarditis. Rheumatic granuloma or Aschoff's nodes in perivascular interstitium, resulting in perivascular fibrosis (A, B, C). (D) Sarcoidosis was the only finding in a young, apparently healthy man who died suddenly and unexpectedly (E).

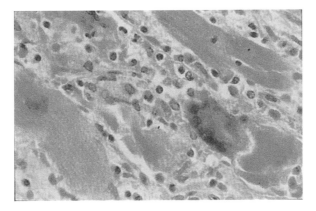

Figure 48   Giant cell myocarditis associated with contraction band necrosis (CBN) in a 32-year-old man, the death of whom was sudden and unexpected.

Pseudomyocarditis

Within the normal myocardium, oligodendritic monocytes, easily confused with lymphocytes, exist (Parravicini et al., 1991). They are activated as macrophages in the early repair of catecholamine necrosis (Figure 15), giving a false impression of a myocarditis, that becomes extensive around an acute infarct where large areas are involved (Figure 50) as

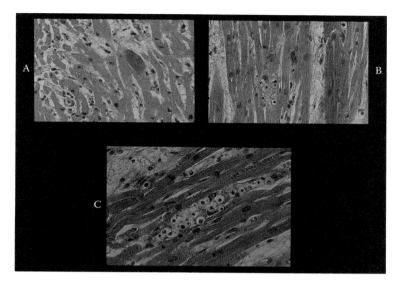

Figure 49 Noninflammatory parasitosis. Intramyocardial cell toxoplasmas (A) and crypto-cocci (B, C) without any type of cellular or exudative interstitial reaction. In these cases, the definition of parasitosis seems more correct than the definition of myocarditis. Dysfunction should be related to the number of myocardial cells invaded.

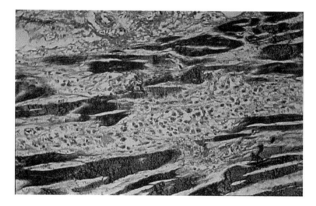

Figure 50 Acute human myocardial infarct 7 days old with extensive monocyte infiltration in repairing the contraction band necrosis (CBN) at the periphery of the infarct — a finding not to be confused with a myocarditis.

well as in other noncoronary conditions (Figure 51). The association with typical patho-logical hypercontracted bands may discriminate this pseudomyocarditis in relation to the Dallas criteria. One notes that in the first report (Szakacs and Cannon, 1958), the cate-cholamine lesion was misinterpreted as a myocarditis. Furthermore, a definitive distinction between cytotoxic T lymphocytes and dendritic monocytes can be achieved immunohis-tochemically by monoclonal antibodies (Parravicini et al., 1991).

Another pattern of pseudomyocarditis is the polymorphonuclear leukocytic infiltra-tion of the infarcted myocardium 6 to 8 h from its onset (Figure 52). Occasionally called inflammatory phenomenon, in reality, it is a secondary response to some chemotactic trigger, the nature and meaning of which are still undetermined.

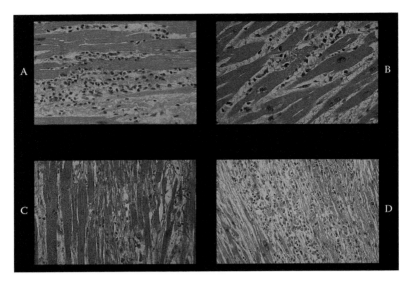

Figure 51   Lymphocytic pseudomyocarditis. The early repair of contraction band necrosis (CBN) is formed by macrophagic monocytes. This histologic pattern may be erroneously interpreted as lymphocytic myocarditis (A–D). Pathologic contraction bands permit a differential diagnosis, including the typical alveolar pattern full of lymphocytes. Immunohistochemically, the cellular phenotype can be determined by distinguishing dendritic monocytes from T cytotoxic lymphocytes.

Myocarditis and Sudden Death

Many years ago, a few small lymphocytic infiltrates were diagnosed as a myocarditis responsible for sudden death in three cases (Helwig and Wilhelmy, 1939). As already discussed, a few foci of lymphocytes, per se, do not entitle a pathologist to relate them causually with sudden death. In our experience (AIDS, Chagas', drug hypersensitivity), sudden death, even unexpected, was associated with extensive (Table 35) lymphocytic myocarditis. However, the cause of the sudden cardiac arrest in most cases with myocarditis of any type is still a matter of speculation. The frequent coexistence in our cases of myocarditis and catecholamine necrosis suggests that several mechanisms could be involved, including destruction (or irritation) of intramyocardial nerves that alters sympathetic control and induces arrhythmias (Baroldi et al., 1997a). In conclusion, no causal link exists between sudden death and the limited foci of a few inflammatory elements. In the rare instances of diffuse myocarditis, the cardiac arrest may be due to myocardial insufficiency consequent to massive interstitial exudation (e.g., acute rejection of a transplanted heart), or other mechanisms involving adrenergic nerves resulting in ventricular fibrillation (e.g., sudden death in Chagas' disease in which diffuse lymphocytic myocarditis and catecholamine necrosis coexist) (Baroldi et al., 1997b).

Cardiomyopathies of Unknown Nature

Cardiomyopathies are fully discussed in Chapter 7. Here, the aim is to comment on some specific points, starting with the meaning of the term "cardiomyopathy" and presenting

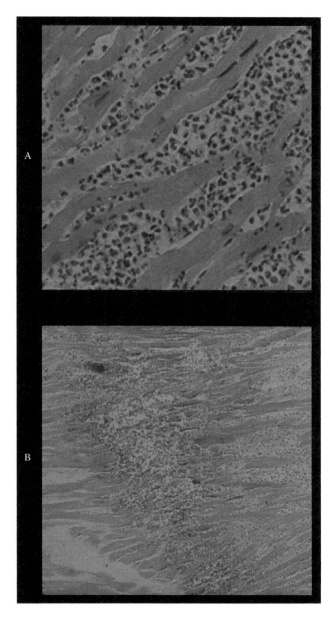

Figure 52 Polymorphonuclear (PMN) pseudomyocarditis. In infarct necrosis, the centripetal and marked infiltration of PMN starts after 6 to 8 h (A) and disappears in a few days without evidence of further changes of necrotic myocardial cells. In large infarcts, the blockage of this infiltration along a line may assume the aspect of an abscess (B). The nature and meaning of this transitory infiltration is not known. Myocardial infarction cannot be defined as a myo-carditic process.

the main pathologic findings in adults and referring the reader to clinicopathologic cor-relation, as discussed by Gallo and d'Amati (2001).

The German term "idiopatische Herz-muskelkrankungen" introduced more than a century ago (Krehl, 1891) and its English equivalent "cardiomyopathy" introduced almost

Table 35   Number of Intramyocardial Lymphocytic Foci × 100 mm² (SE) of the Total Histologic Area Examined in Each Case

Disease	Number of Cases	Intramyocardial Lymphocytic Infiltrates		
		Absent	Foci[a]	Mean Elements × Focus
AIDS	38	19	1.5 ± 1	<20
SD AIDS	1	—	10	10–50
SD Chagas	34	—	13 ± 2	>50
CHF Chagas	9	—	14 ± 1	20–50
SD coronary	25	6	2 ± 0.4	<20
Dilated cardiomyopathy[b]	63[b]	9	0.2 ± 0.3	<20
Chronic coronary disease[b]	63[b]	8	0.4 ± 1	<20
Valvulopathy[b]	18[b]	3	0.3 ± 0.4	<20
Intracranial hemorrhage	27	17	2 ± 2	<20
Transplanted hearts	46	5	10 ± 2	<20
Head trauma in healthy individuals	45	29	5 ± 3	<20
Survival minutes	26	18	1 ± 0.2	<20
Survival 1 h	19	11	9 ± 5	<20

Note: AIDS, acquired immunodeficiency syndrome; SD, sudden/unexpected death; CHF, congestive heart failure.

[a] Each focus formed by ten or more elements.
[b] Hearts excised at transplantation.

70 years later (Bridgen, 1957) were used to indicate a primal, noncoronary disease of the myocardium. In the 1984 World Health Organization (WHO) report, cardiomyopathies were distinguished as *heart muscle diseases of unknown cause* and *specific heart muscle diseases of known cause or associated with disorders of other systems.* Finally, a 1996 WHO report defined most cardiac diseases as cardiomyopathies. This is an unsatisfactory classification, because it uses ambiguous definitions including ischemic cardiomyopathy, which "presents as dilated cardiomyopathy with impaired contractile performance not explained by the extent of coronary artery disease or ischemic damage." If so, how and why is this condition considered ischemic?

In medical science, most classifications are temporary groupings forming "a bridge between ignorance and knowledge" (Goodwin quoted in the 1996 WHO report). Any classification should not use preconceived terms and definitions. For this reason and despite the increasing evidence of a possible causal relation with genetic factors, I prefer to be conservative in maintaining the 1984 distinction between cardiomyopathies of unknown cause and specific heart muscle diseases with the following considerations:

1. "Myocardiopathy" seems to be a more precise term for a primal disease of the myocardium.
2. The myocardium is a complex structure that includes vessels, nerves, interstitium, and collagen matrix, any one of which could be primarily involved, for example, interstitium with secondary damage of myocytes as in many inflammatory processes or a comprised primary nerve as in denervation.
3. The "known" causes of myocardial disease are often only hypothetical.

4. Dilated cardiomyopathy, associated with so many cardiac and systemic disorders (Silver et al., 1991), does not mean that we know the cause of this association.

5. Cardiomyopathies of unknown causes (i.e., dilated, hypertrophic, restrictive, or arrhythmogenic) were defined as such despite some being recognized as familial or genetic diseases. The increasing identification of causal genetic defects does not explain how and when they act or the overall pathogenetic mechanisms involved with a possible interaction between genetic and phenotypic factors. This is the main reason why I used the heading "unknown nature" to emphasize that pathogenesis is still obscure.

All of these considerations emphasize how difficult it is for a forensic pathologist, generally without a clinical history, to assure a correct diagnosis.

Dilated

This condition is characterized clinically by progressive congestive heart failure with the dilemma of whether its increasing incidence is a fact or only a consequence of a combination of increasing awareness and more advanced diagnostic techniques. For example, in Malmo (Sweden), its frequency was retrospectively calculated as 5 to 10 patients/100,000 population/years; in China, among 66,000 factory workers, government employees, and teachers, 52 subjects presented a clinical pattern of dilated cardiomyopathy and 4 had hypertrophic cardiomyopathy; at Chandigarh (India), a town with an autopsy rate of 90%, 28 cases of dilated cardiomyopathy (2.8%), 9 of restrictive cardiomyopathy (0.9%), and 1 of hypertrophic cardiomyopathy (0.1%) were diagnosed per year (WHO report 1984). In the Mayo Clinic geographic area (Rochester, NY), the corresponding incidence was 6/100,000 persons/year between 1980 and 1984 (Codd et al., 1989). An update on frequency is lacking.

Cardiac enlargement and systolic dysfunction of one or both ventricles are the main clinical findings associated with progressive congestive heart failure (Gavazzi et al., 1995; Baroldi et al., 1998c). The early phase and predictors of this disease are still not recognized, and often, dilatation and pump dysfunction remain asymptomatic for years. In particular, equally severe pump dysfunction with an adverse prognosis identical to that with marked dilatation may occur without or with minimal dilatation. By Holter monitoring, ventricular arrhythmias are observed with ventricular tachycardia occurring in 20 to 60% of patients. Systemic or pulmonary embolism was reported in 3 to 18% of cases, mainly in more compromised patients. Left ventricular thrombi were detected by echocardiography with a frequency of 8 to 60%.

The familial occurrence of this cardiomyopathy once considered rare, shows an incidence of 20 to 40%, mainly as an autosomal dominant pattern of inheritance.

By this method, conflicting results are reported in evaluating morphologic predictors of outcome. Some studies found that fibrosis, reduced myofibrillar volume, or myofibrillar loss has a predictive role, but these results are not confirmed by others (Baroldi et al., 1998c).

At autopsy, often all four heart chambers are dilated (Figure 53), with an increased weight (>500 g) despite a normal or reduced cardiac wall thickness (Table 36), which, in turn, corresponds to a normal or reduced diameter of myocytes (*weight/size paradox*).

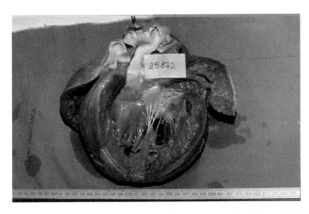

Figure 53   Dilated cardiomyopathy. A heart of 560 g with dilated left ventricular cavity, normal thickness of cardiac wall, normal valve, and no gross evidence of myocardial fibrosis.

Attenuation, absence of distension, cloudy swelling edema, and loss of myofibrils (colliquative myocytolysis) prevail in the luminal half of the left ventricular wall (Table 7; Figure 18). There is minimal lymphocytic infiltration (Table 35) and minor myocardial fibrosis (Table 6) that is mainly interstitial and with an undulating aspect (Figure 38).

Both extramural and intramural coronary arteries are normal. Other findings, such as catecholamine necrosis, a comparison with controls, and further discussion are found in the section entitled "Congestive Heart Failure."

Hypertrophic

Gross anatomy is marked by a thickened left ventricular wall. It is rarely symmetric; rather, the majority of cases (95%) present asymmetric hypertrophy with a maximal increase in the interventricular septum (interventricular septum/free wall ratio = 3:2), which, as a result, protrudes into the left ventricular cavity, drastically reducing it (Figure 54). This obstruction, well seen by Doppler echocardiography imaging, generally involves the median-superior portion of the interventricular septum. The whole septum or its apical region or the right ventricle rarely protrude. Other findings may include endocardial fibrosis marked by a white thickening of both the endocardium where the anterior mitral leaflet beats against the septal wall and the ventricular aspect/chordae tendineae of the anterior mitral valve (75%), atrial dilatation (100%), and a prominent septal band connected with the superventricular crista in the right ventricle. In cases involving congestive heart failure, the heart has a gross aspect typical for dilated cardiomyopathy (Baroldi et al., 1998b).

Histologically, the pathognomonic change is a loss of the usual parallel alignment of myocardial cells. They assume a star-like disposition, defined as *disarray*. Different patterns of disarray were described according to the oblique or perpendicular orientation of bundles or single myocytes (Maron et al., 1992), disposed in any direction (Figure 55) with myofibrillar connections, seen at the *ultrastructural level* (Ferrans et al., 1972; Figure 56), associated with anomalous irregular Z lines. The latter are a possible indication of sarcomerogenesis, similar to that seen in the embryonic development of the myocardium from myoblasts. Between the disarrayed myocytes is an increased interstitial matrix and a myofibrous hyperplasia of arterioles (Figure 57).

Table 36 Frequency and Extent of Myocardial Disarray in Different Pathologic and Normal Conditions

Disease	Sudden/Unexpected Death		Intracranial Hemorrhage	Transplanted Hearts	AIDS	Congestive Heart Failure	Cocaine Abuse	Carbon Monoxide[a]	Electro-cution[a]	Head Trauma Death[a]	
	Coronary	Chagas'								Instantaneous	<1 h
Cases	25	34	27	46	38	144	26	26	21	26	19
Disarray											
Present (%)	12(48)	9 (26)	15(56)	21(46)[b]	4(11)	20 (14)	4 (15)	—	—	—	—
3 sites	4	8	5	6	2	17	4	—	—	—	—
>3 sites	8	1	10	15	2	3	—	—	—	—	—
Left ventricle	12	3	14	19	4	19	4	—	—	—	—
Right ventricle	8	6	8	13	2	13	—	—	—	—	—
Interventricular septum	9	4	14	17	4	17	—	—	—	—	—
Heart weight (g)											
<500	4	9	15	17	4	4	4	—	—	—	—
500	8	—	—	4	—	16	—	—	—	—	—
Myocardial fibrosis (%)											
<20	6	—	8	4	—	1	—	—	—	—	—
20	2	1	—	1	—	2	—	—	—	—	—
Contraction band necrosis × 100 mm² SE											
Foci	27 ± 10	3 ± 1	26 ± 7	36 ± 9	4 ± 2	2 ± 0.3	4 ± 4	1 ± 1	5	0.5	12 ± 18
Myocells	185 ± 48	34 ± 16	67 ± 21	262 ± 47	13 ± 5	11 ± 14	11 ± 14	5 ± 2	8	35	21 ± 23

Note: AIDS, acquired immunodeficiency syndrome. Disarray extent is expressed in number of the 16 sites in each heart (free wall, anterior, lateral, posterior left and right ventricles, interventricular septum, anterior and posterior sampled from one superior and one inferior total heart slices) in which disarray was present. Only areas normally without disarray were considered.

a Normal subjects without resuscitation attempts.

b Absence of disarray in nine hearts with less than 1 week survival.

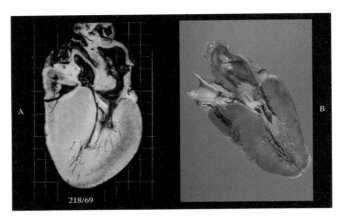

Figure 54   Hypertrophic cardiomyopathy. Asymmetric and symmetric forms (B) both with markedly reduced left ventricular cavity. (Part A: Courtesy of Dr. M.D. Silver.)

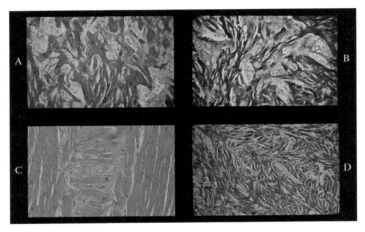

Figure 55   Myocardial disarray. Different aspects (A–D) with increased interstitial fibrosis.

From a practical standpoint, pathologic disarray can be diagnosed when it involves more than 20% of a histologic area (Maron et al., 1979). However, myocardial disarray, if pathognomonic for hypertrophy cardiomyopathy, is also found in specific areas of a normal heart at the junction of the free wall with the septum, at the origin of papillary muscles, around a myocardial scar, in congenital cardiac malformations, and in some diseases (lentiginosis, Friedreich's ataxia, Turner's syndrome, hyperthyroidism).

The clinical pattern is explained by the complex architecture of the myocardium. However, the pathogenesis of many symptoms is still obscure. Subaortic stenosis, due to protrusion of the asymmetric septum into the left ventricle, is not a stable anatomic condition, because pressure gradients change in the same patient. Furthermore, ejection occurs during the first part of systole, leaving a very small end-diastolic volume. Nevertheless, the ejection fraction is 80% higher than normal with an aortic flow unrelated to the pressure gradient that can be obtained in a normal heart by a strong inotropic stimulation or acute restrictive hypovolemia. All of these facts indicate a possible relaxation defect of myocardial cells. Other symptoms such as angina or signs such as ECG evidence

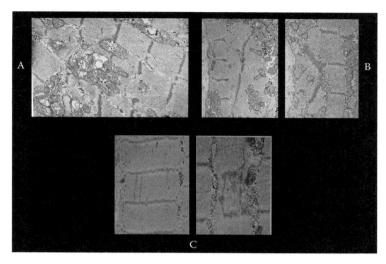

Figure 56 Myocardial disarray. Ultrastructural view of the disorganized myofibrils (A) and irregular Z lines, sign of a disorganized sarcomeric neogenesis (B, C). (Parts B, C: Courtesy of Dr. V. Ferrans.)

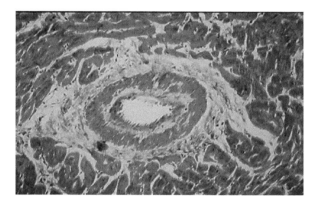

Figure 57 Hyperplastic tunica media in an arteriole within disarrayed myocardium.

of ischemia are interpreted as being consequent to coronary insufficiency relative to the markedly increased myocardial mass. This concept is contradicted by the long-lasting, asymptomatic performance of markedly hypertrophed hearts in cor pulmonale (see above section entitled "Relative Coronary Insufficiency"). Finally, reentry in disarrayed myocardium or focal myocardial fibrosis or reduced diastolic filling of the small left cavity or obliterative thickening of the arterioles are proposed to explain sudden death in this condition (Maron et al., 1979; Saumarez et al., 1992) — all unproved hypotheses (see below).

An essential contribution to understanding the etiopathogenesis of this autosomic dominant familial disease is provided by the recognition of genetic defects. They are reported in Chapter 9. I note that it is equally essential to know how causal and predisposing genetic factors may act (i.e., pathogenesis). Persistence of fetal "disorganized" myocardium, amartia, dismetabolism of contractile myocytes, small vessel disease, antigenic histoincompatibility, and adrenergic disorder, are among the more frequently advocated pathogenic mechanisms (WHO, 1984). Adrenergic disorder will be discussed in Chapter 10.

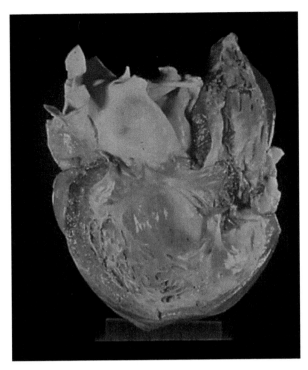

Figure 58   Restrictive cardiomyopathy. Marked endocardial fibrous thickening and left ventricular dilatation.

Restrictive

Long ago defined as endomyocardial fibrosis in tropical areas or Löffler hyperplastic endocarditis, this form is an endocardiopathic condition with reduced diastolic relaxation of the myocardium. The thickness of the endocardium may reach several millimeters, with the involvement of both ventricles occurring in 50 to 70% of these cases and with characteristic retraction of the right ventricle. Fibrous septa may join the endocardium with the epicardium, and the cardiac cavities show a reduced diameter (Figure 58). Histologically, the thickened endocardium is the result of superimposed layers of fibrous tissue, the deepest being formed by granulation tissue, blood vessels, and inflammatory cells, including eosinophils. Such an inflammatory reaction can be seen in the intramural epiendocardial septa mentioned above (WHO, 1984).

Arrhythmogenic

This rare entity, apparently more frequent in a small geographic area of Northern Italy (Corrado et al., 2000) and referred to as arrhythmogenic cardiomyopathy of the right ventricle (WHO 1995), may occasionally involve the left ventricle. Histologically, it is characterized by substitution of the myocardium with adipose or fibroadipose tissue associated with lymphocytic infiltrates that may extend transmurally (Figure 59). Four different clinical patterns have been distinguished: occasional arrhythmogenic episodes; severe malignant arrhythmia; right ventricular insufficiency; or biventricular insufficiency (Corrado et al., 2000). The nature of this cardiomyopathy is still obscure (Thiene and Basso, 2001).

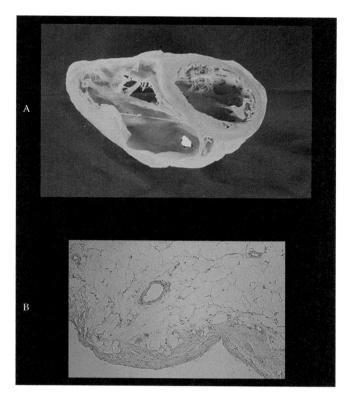

Figure 59 Right arrhythmogenic cardiomyopathy. Gross (A) and histological (B) views showing an almost total substitution of the myocardium with adipose tissue.

Sudden Death in Cardiomyopathies

The highest frequency (50%) of sudden death was observed in hypertrophic cardiomyopathy (WHO, 1984). In dilated cardiomyopathy, the frequency is lower (14%; Fieguth et al., 1994), while sudden death is rarely reported in restrictive cardiomyopathy. In arrhythmogenic cardiomyopathy, a high frequency of sudden death is reported without a percent figure offered (see frequency in infants and children in Chapter 7).

A more extensive quantitative study of myocardial functional lesions in all of these cardiomyopathies could help in understanding the cause of cardiac arrest at autopsy. For instance, in the few cases of arrhythmogenic cardiomyopathy that we examined, foci of contraction band necrosis were present. In this respect, an increase in the number and length of intramyocardial nerves in about half of the hearts excised at transplantation from patients with different cardiac diseases and with a clinical history of arrhythmia (Cao et al., 2000) is worth mentioning.

Orthotopic Transplanted Human Heart

This is another interesting experimental model in humans, because the transplanted heart has a complex natural history in which many factors have a role in its outcome. For example, the status of a donor's heart, surgical complications, immunosuppressive therapy, rejection, allograft coronary arteriopathy, psychological reactions, infectious diseases, tumors, and allograft failure, all interplay. This suggests a review of pathologic findings related to transplantation:

> *Donor hearts*: The majority originate from normal young subjects who die by accidental head trauma. In 45 instances of this type of death, the unique postmortem findings in the heart were occasional small atherosclerotic plaques in coronary arteries with minor (50%) lumen reduction and foci of contraction band necrosis present in 42% of 19 subjects who survived 1 h and absent in 26 who died almost instantaneously. When present, the necrosis was of minor extent (12 ± 18 foci and 21 ± 33 myocytes × 100 mm², Table 6). The conclusion was that a donor heart from subjects dying from head trauma had no ischemic or anoxic changes and that the few adrenergic lesions seen would not compromise myocardial function (Baroldi et al., 1997a). This is a conclusion supported by the excellent function of most transplanted hearts once circulation is restored. A final note is that to define ischemic, to classify both the donor's and the recipient's excised hearts as ischemic (Winters et al., 2001) would be questionable because these are arrested, nonbeating hearts.
>
> *Infectious diseases*: Likely related to immunosuppressive therapy, infectious diseases caused death in 17% of 186 orthotopic transplant recipients; surgical complications in 5%; tumors in 7%; acute rejection in 10%; graft vasculopathy, often defined as chronic rejection, in 15%; graft failure in 39%, including primary pump failure 19%, multiorgan failure 11%, and pulmonary hypertension 9%; and others in 7% (Gallo et al., 1993). In this study, the frequency of sudden death was not reported.

In 46 personally examined transplanted hearts (Baroldi et al., 2003a), an acute rejection so massive as to justify pump failure was seen in only one instance (2%) in contrast to that occurring in three of ten reviewed hearts in the precyclosporine era (30%) (Figure 60). Allograft vasculopathy (Figure 27A; already described in the chapter on nonatherosclerotic extramural coronary arteriopathies) was detected in extramural coronary arteries and their extramural branches of any size in four cases and extramural coronary atherosclerosis in 12 cases, all with a survival of more than 1 yr. No involvement of any type was observed in the intramural vessels. Infarct necrosis was present in seven cases, five of the patients survived less than 30 days with normal coronary arteries, one survived 202 days without coronary disease, and one survived 3 yr with a >70% lumen reduction due to allograft coronaropathy with superimposed atherosclerosis. Medial neuritis was seen only when atherosclerosis was present. In contrast, contraction band necrosis, erroneously interpreted as due to ischemia (Day et al., 1995; Fyfe et al., 1996), was detected in 85% of the 46 hearts with an extent significantly higher than in other controls (Table 37), and a frequency of 100% was detected when survival was less than 30 days, 85% when survival was 30 to 365 days and 64% when more than 365 days, respectively. These findings, which included early to healing contraction band lesions, suggest a possible role of myocardial supersensitivity to catecholamines (see above) in graft failure affecting these globally denervated hearts and a revision of the cause of death in the presence of other pathologic patterns. According to our findings, the coagulative necrosis (Day et al., 1995; Fyfe et al., 1996) corresponds to the early eosinophic hypercontraction of catecholamine necrosis, and posttransplant

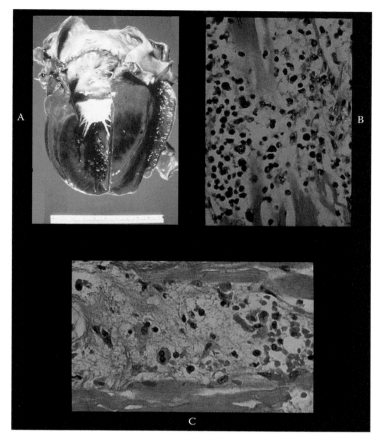

Figure 60 Transplanted heart. Massive acute rejection: (A) gross view of a red, edematous myocardium with (B, C) marked interstitial exudation and diastasis of myocardial cells. This pattern may explain acute cardiac insufficiency. (Courtesy of Dr. D. Cooley.)

lymphoproliferative disorder (Winters and Schoen, 2001) is the early repair stage of this necrosis (Figure 50 and Figure 51) characterized by numerous macrophagic monocytes.

Chronic rejection is generally interpreted as a result of chronic ischemia secondary to allograft vasculopathy or allograft vasculopathy combined with atherosclerosis or to atherosclerosis alone. As indicated above, our findings do not support this concept. Finally, all of the patients in our cases died in the hospital, and death was not sudden. The frequency of the latter in heart-transplanted patients is still uncertain.

Congestive Heart Failure

Remodeling

In the past few years, remodeling has become a popular term used to signify all heart changes that occur in congestive heart failure related to increased heart volume and mass with deterioration of cardiac function due to cardiac cell lengthening, ventricular wall thinning, continuous infarct expansion rather than extension, inflammation or reabsorption

Table 37   Catecholamine Necrosis (or Coagulative Myocytolysis or Contraction Band Necrosis) in Human Transplanted Hearts and Controls

Cases	Human Transplanted Hearts				AIDS	Chagas'	Intracranial Hemorrhage	Head Trauma	
Total	9	10	13	14	38	34	27	26	19
Survival	<9 days	7–30 days	31–364 days	365 days	Months	Seconds	Hours	Seconds	1 h
		Catecholamine necrosis							
Frequency (%)	100	100	85	64	66	100	89	4	42
Extent (Number × 100 mm^2)									
Foci (ES)	95 ± 80	22 ± 31	15 ± 30	15 ± 19	4 ± 11	3 ± 6	26 ± 34	0.5	12 ± 18
Myocells (ES)	316 ± 205	306 ± 357	215 ± 264	217 ± 354	13 ± 27	30 ± 51	67 ± 104	35	21 ± 23
Histologic pattern[a]									
Early bands	9	—	1	4	19	8	13	1	8
Alveolar	—	8	6	3	6	5	9	—	—
Healing/healed	—	2	4	2	—	4	2	—	—

Note: ES = error standard.

[a] Advanced stages are usually associated with early ones.

of necrotic tissue, scar formation, dilation and reshaping of the left ventricle, myocardial cell hypertrophy or ongoing loss (apoptosis), and excessive accumulation of collagen in the cardiac interstitium. A sequence of events, mainly consequent to absolute or relative ischemia that acts by immediate diastolic and systolic dysfunction that can recover spontaneously (myocardial stunning) or postreflow (myocardial hibernation) or be irreversible (myocardial necrosis or apoptosis); by oxidative stress through oxygen radical formation through proinflammatory cytokines and tumor necrosis factor, produced in quantities in failing hearts (Feldman et al., 2000a); and increased neurohormonal activation, an increased level of norepinephrine being correlated with prognosis. This exhaustive consensus (Cohn et al., 2000), like any consensus, requires revision, beginning with the word "remodeling," which is synonymous with "rebuilding, reconstruction, restructuring, reassembling" (*Roget's International Thesaurus*, 1992) indicating a change in structure without deterioration or with amelioration of function, which is not the case for all destructive or repair processes included in the previous definition.

Pathogenetic Factors

In the long list of cardiac diseases with a progressive, intractable loss of cardiac pump function ending in cardiac arrest (Silver and Freedom, 1991), factors such as myocardial fibrosis, of ischemic or myocarditic nature, proliferation of collagen matrix, and apoptosis are considered the main causes of this dysfunction. In a systematic quantitative study of histologic variables measured as a total number times square millimeters or as a percent of the total histologic examined area (16 total wall samples × heart), the significance of different parameters was estimated in hearts excised at transplantation for irreversible congestive heart failure and controls (Baroldi et al., 1998c). All clinical parameters were equivalent in the former groups, and the mean heart weight was increased, while anterior left wall thickness was within normal values (Table 38). The number of focal lymphocytic infiltrates and lymphocytes per focus were insignificant and similar in all excised hearts and controls but the Chagas' group, with the exception of an increased frequency of these infiltrates in normal controls who survived longer (Table 35), perhaps signifying an activation of dendritic monocytes as an expression of increased catecholamine necrosis (Table 6 and Table 37). The frequency of contraction band necrosis was 94% in ischemic, 84% in

Table 38 Heart Weight and Anterior Left Ventricular Wall Thickness in Congestive Heart Failure and Controls

Source	Cases	Men (%)	Age (Years)	Heart Weight (g)	Transverse Diameter (mm)	Anterior Left Wall (mm)
Congestive heart failure	144	89	47 ± 1	473 ± 10		
Coronary heart disease[a]	63	97	51 ± 8	565 ± 108	137 ± 14	12 ± 4
Dilated cardiomyopathy[a]	63	86	42 ± 12	639 ± 155	148 ± 15	15 ± 4
Chronic valvulopathy[a]	18	83	46 ± 8	827 ± 179	147 ± 15	16 ± 4
Hypertrophic cardiomyopathy[a]	31	65	38 ± 12	659 ± 174	—	18 ± 4
Silent Chagas' disease	34	76	49 ± 13	464 ± 110	114 ± 18	15 ± 3
Acquired immunodeficiency syndrome	38	87	31 ± 10	368 ± 67	91 ± 21	13 ± 3
Brain hemorrhage	27	22	58 ± 12	434 ± 91	107 ± 15	18 ± 7
Head trauma in healthy individuals	45	82	42 ± 17	364 ± 47	104 ± 15	12 ± 5

[a] Hearts excised at transplantation: the weight of excised hearts was adjusted by adding theoretical atrial weight (Reiner 1968) — actual heart weight × 100/75.

dilated cardiomyopathic, and 73% in valvulopathic hearts associated with congestive failure. A similar frequency was observed in controls without congestive heart failure, in those with transplanted hearts, and in brain hemorrhage patients with a maximum (100%) in Chagas' patients dying suddenly and unexpectedly. However, the number of foci and myocardial cells with catecholamine necrosis in all stages of this lesion, from early crossbands to alveolar and scarring, was very low (Table 6). In contrast, colliquative myocytolysis showed maximal frequency and extent in failing excised hearts. A higher degree of dilatation and myocardial clinical dysfunction did not show a more severe pattern of this lesion, indicating that the latter was a consequence rather than a cause of congestive heart failure. To test the amount of viable myocardium that persisted, the extent and type of fibrosis were calculated in percent in each histological area by an orthogonally bisected ocular — the only reliable method, because fibrous scar tissue often transforms into adipose tissue (Figure 38) (Baroldi et al., 1997c). On that basis, a fibrous index (total fibrotic area/total histologic area $[mm^2] \times 100$) showed that even in the coronary heart disease group, the amount of histologically viable myocardium greatly exceeded (80%) fibrous tissue (Table 6). This was a finding confirmed *in vivo* by thallium uptake (De Maria et al., 1996). Apart from the central portion of a healed infarct with dense, compact, and straight collagen fibers, the myocardial fibrosis showed a corkscrew aspect (Figure 38). This means that in hypertrophy, the development of fibrous tissue — subsequent to myofibrillogenesis (Weber, 1989) — occurs in a beating myocardium with the corkscrew morphology being a functional adaptation that cannot be the cause of myocellular dysfunction, no matter whether collagen synthesis relates to the fibrillar collagen matrix or reparative fibrosis. The progressive loss by adrenergic myotoxicity of a few myocytes substituted by microcollagenization may result in interstitial or intermyocellular fibrosis erroneously interpreted as an increased fibrillar matrix.

There is no pathological or clinical (lack of energy deficiency as lactate production, reduction of coronary sinus oxygen, pH, and symptoms of ischemia) (Poole-Wilson, 1993) evidence to support the concept that congestive heart failure is caused by ischemic myocardial fibrosis or virus myocarditis; blood flow reduction is not related to the extent of myocardial fibrosis (Parodi et al., 1993). Furthermore, Linzbach's concept of myocellular slippage and myocell neogenesis, more recently proposed by others (Beltrami et al., 1994; Weber, 1989; Cohn et al., 2000), lacks convincing support. The size and weight paradox (i.e., a normal cardiac wall and myocardial cell thickness despite a pathological increased heart weight [500 g]) was explained by myocytes stretching — not confirmed by Linzbach (1960) or by us (Figure 18) (Baroldi et al., 1998c) — or by longitudinal cleavage of hypertrophed myocytes when the pathological heart weight is reached (Linzbach, 1947). This longitudinal cleavage was not clearly demonstrated, and the neogenesis of myocardial cells as shown by mitosis in congestive heart failure (Beltrami et al., 1994) or in 4% of myocardium at the border of a human acute infarction and in 1% in normal regions (Beltrami et al., 2001) needs comment. In the latter condition, the ratio of normal to mitotic myocells was 0.015, considered sufficient if sustained, mitosis could form 150 g of myocardium in less than three months. This is a typical theoretical calculation without solid background. The concept that some parenchymal elements, as for instance myocardial cells in the heart or neurones in the brain, are unable to reproduce in adult age, was challenged by their culture *in vitro*. However, it is likely that the demonstration of their duplication *in vivo* and their eventual reproduction from injected stem cells may have little reparative significance. What needs to be proven is their integration within a functioning tissue. Both

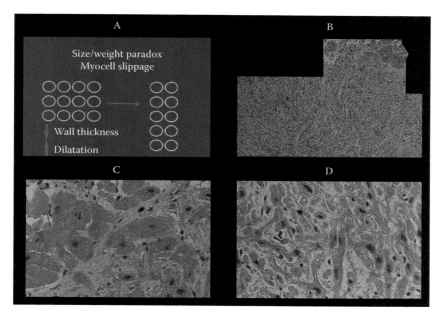

Figure 61   Weight/size paradox in congestive heart failure. The slippage of myocardial cell to explain a normal wall and myocellular size despite a heavy cardiac weight (A) does not consider the myofibrillar bridging between myocells and fibrillar collagen connections. In the slippage of myocells, capable of reducing, for instance, a 3 cm cardiac wall to 1.5, one should imply an extensive destruction of all interstitial structures (vessels, including lymphatics, and nerves) and consequent widespread tissue damage never seen in this condition. Even a neogenesis of the myocardial cell was never observed. Only once in an endomyocardial biopsy at a previous site of sampling in a transplanted heart, did we have the opportunity to see new myocells forming a focus of small elements assuming the aspect of atrioventricular node (B–D). Just to emphasize that when a neogenesis exists, it can be seen apparently without integration in the functioning myocardium.

myocardial cells and neurons have complex interdigitations and connections, and beside any theoretical calculation, the question is whether connections occur in the adult, otherwise, duplication has no compensatory meaning. Facts speak against the latter supposition. At least in the myocardium, small foci of one or a few necrotic myocytes repair by fibrous replacement without any evidence of myocardial cell neogenesis. Observed only in one instance, in an endomyocardial biopsy at a previous site of sampling, were small, newly formed, myocardial cells with an atrioventricular-like node feature not connecting with normal surrounding myocytes (Figure 61), confirming that when they exist, they are recognizable without any connection with normally functioning myocytes.

Nuclear amitosis is visible in normal and atrophic hearts (Baroldi et al., 1967) and cannot be considered an index of myocardial cell neogenesis; this may distort any conclusion based on nuclear frequency and volume. Finally, the hypothetical concept of interpenetration (slippage) of myocytes with thinning and elongation of the cardiac wall is also questionable. First, how can a hypertrophied cell reassume a normal size despite the persistence of the hypertrophying stimulus, and second, how can slippage occur when each myocyte is connected to adjacent ones by numerous and conspicuous myofibrillar bridges, collagen matrix network, and an interstitium full of arterial, venous, and lymphatic vessels and nerves. An interdigitation of myocytes capable of reducing, for example, a wall thickness

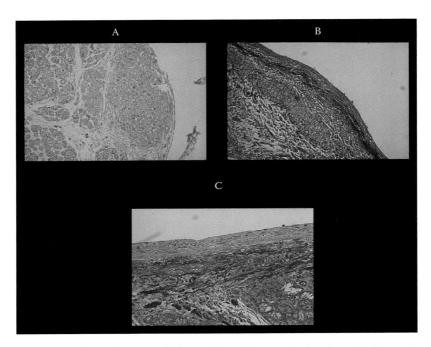

Figure 62   Endomyoelastofibrosis of the endocardium. Thickening of the endocardium is considered consequent to intracavitary thrombi organization. The latter is a relatively rare event. The majority of endocardial thickening is due to the formation of smooth muscle cell bundles (A) with hyperelastosis (B) ending in fibrous replacement (C). This is a process similar to that shown in the early pathogenesis of an atherosclerotic plaque.

from 3 cm to 1.5 cm (Figure 61) should destroy all interstitial structures, with severe global damage of the myocardium, a pattern never described.

The endocardial thickening, generally interpreted as healed endocardial thrombosis, in all hearts with congestive heart failure is mainly an endocardial myofibroelastosis, which starts with nodular proliferation of smooth muscle cells within the basal lamina of the endocardium followed by elastic tissue hyperplasia and final collagenization (Figure 62), a change that recalls the first stage of the atherosclerotic myohyperplastic plaque. The pattern is also present in control hearts, suggesting a pathogenesis not related to myocardial thrombi and congestive failure only (Baroldi et al., 1998c).

A last note concerns apoptosis as a cause of congestive failure. This is a hypothesis contradicted by the heavy hearts we observe in this condition. If apoptosis, which by definition is a loss of myocytes without fibrous repair, could explain failure by a loss of myocardial cells, instead of a heavier-weight heart, we should see normal if not atrophic hearts. Finally, the claim that the increased weight is due to myocardial cell neogenesis should consider that the latter is incompetent to contrast failure.

Sudden Death in Other Cardiac Conditions

In the first part of this section on coronary heart disease, a definition of sudden death was given, and findings on sudden and unexpected death associated with coronary atherosclerosis were reported and compared with other patterns of coronary heart disease. In this section, other cardiac pathologic conditions associated with sudden unexpected or expected death are examined, recalling that any cardiac disease may have that outcome.

Extramural Coronary Arteries

Congenital Anomalies

Whether a single coronary artery is a potential death threat is controversial. In this rare condition, the concern is for a possible obstruction of the ostium or main stem of the single vessel, even if a variety of combinations were reported at autopsy, as in the following cases with a single coronary artery: an 80-year-old man dying from noncardiac disease (Blake et al., 1964); a 79-year-old man with a myocardial infarct but without coronary occlusion (Tremouroux et al., 1959); a sudden death in a 52-year-old man with severe angina, coronary atherosclerosis of the two main coronary artery branches, and a myocardial scar (Spring and Thomsen, 1973); cardiac failure in a 27-year-old gravida with normal coronary artery and extensive myocardial fibrosis without a history of coronary heart disease (Halperin et al., 1967); and sudden death in athletes following stressful exercise (Blake et al., 1964; McClellan and Jokle, 1968).

When both coronary arteries originate from the anterior aortic sinus, the anatomic anomaly presents with angulation and a slit-like ostium plus compression between aorta and pulmonary arteries of the left coronary artery. These findings were related to a possible ischemic mechanism able to explain sudden death, which is frequently observed during exercise in this situation. In contrast, the origin of both coronary arteries from the left aortic sinus is not related to sudden death (Roberts et al., 1982b; Taylor et al., 1992a; Cheitlin et al., 1974).

Inflammatory Processes

The precise frequency of sudden death in the course of extramural coronary arteritis is not known for two main reasons. Their incidence is generally low, and their presence, especially when death is sudden or unexpected, is easily missed if autopsy or histologic studies are not done. In some specific conditions, such as polyarteritis nodosa and the mucocutaneous lymphonode syndrome or Kawasaki disease, coronary artery aneurysms occur. Their rupture may determine a sudden demise due to hemorrhage with cardiac tamponade (Sinclair and Nitsch, 1949; Kegel et al., 1977). Recently a 21-year-old subject with Kawasaki disease died suddenly and unexpectedly without rupture of two large coronary aneurysms (Fineschi et al., 1999b). Another case affected a 2-year-old child with aneurysms of both coronary arteries associated with obliterative intimal thickening and thrombosis (Burns and Manion, 1969). And there was a case of isolated periarteritis in a young pregnant woman (Ahronheim, 1977). Both died suddenly and unexpectedly.

Coronary Ostial Stenosis

A percentage (16%) of sudden deaths occurred among 233 cases of syphilitic aortitis, with 37 cases presenting obstruction of one or both coronary ostia. Of the latter 37 cases, 17 (47%) died suddenly (Scharfman et al., 1950). From the files of the Armed Forces Institute of Pathology, of 11 cases with bilateral ostial (sub)occlusion, none had a myocardial infarct, and only one died suddenly (Baroldi and Silver, 1995).

Coronary Embolism

At present, coronary embolism is a rare cause of sudden death in Western countries. In a previous report (Wenger et al., 1958), the incidence of sudden death was 60%.

Coronary Dissecting Aneurysm

Sudden death is frequent in cases of coronary dissecting aneurysm. Of 24 cases, 18 cases presented patients (15 women and 3 males) who died suddenly and unexpectedly (Claudon et al., 1972).

Mural Coronary Arteries

A myocardial bridge encircling a coronary artery, especially the left anterior descending branch, was the main coronary finding in three subjects (Morales et al., 1980) who died suddenly: two, a 34-year-old man and a 17-year-old girl, had no CHD history; and a 54-year-old man had episodes of precordial pain for many years associated with a "milking effect" at coronary cineangiography of the left descending branch without other changes. This raises the question of whether myocardial bridges are a normal anatomic variant or an obstructive impairment (Noble et al., 1976). Surgical debridging has been successful in CHD patients (Faruqui et al., 1978).

Intramural Coronary Arterial System

The only pathological findings of the intramural arterial system reported in the literature associated with sudden death are platelet aggregates and fibrin-platelet thrombi or emboli (Jorgensen et al., 1968; Haerem, 1972; Frinck, 1978; El-Maraghi and Genton, 1980; Falk, 1985; Davies et al., 1986). These lesions were considered responsible for acute ischemia causing sudden death. Emboli with cholesterol crystals were demonstrated in 4 of 25 (Falk, 1985) and in 2 of 90 (Davies et al., 1986) cases.

Only one study in the literature investigated the relationship between coronary collaterals and sudden death by postmortem coronary injection of calibrated microspheres 40 to 75 μm in diameter, followed by radiopaque barium in gelatin. Anastomotic channels were never seen in 76 normal subjects and were documented in 10 of 13 sudden death cases associated with healed infarct and in 1 of 16 sudden death cases without an infarct. The conclusion was that a previous infarct was essential for functioning collaterals to develop, while their absence could explain an ischemic sudden death in the absence of a previous infarct (Spain et al., 1963) (see section on Coronary Heart Disease).

Heart Muscle and Cardiac Valves

Congenital Anomalies of the Heart

Among the many complications associated with congenitally malformed hearts, sudden death is a rare event (Edwards, 1968). A claim is that the latter is more frequent in any condition producing obstruction of the left ventricular outflow tract. However, in 199 cases of congenital aortic stenosis in children, only two died suddenly (Glew et al., 1969). (See Chapter 7.)

Heart Tumors

In an extensive review of heart tumors, the frequency of sudden and unexpected death was 5 of 130 cases with myxoma, 3 of 42 with papillary fibroelastoma (13 from aortic cusps occluding a coronary ostium), 6 of 10 fibroma of interventricular septum (often in young persons), 2 of 32 lipoma of the interatrial septum, 7 of 15 mesothelioma of the atrioventricular node, and 2 of 14 teratoma. Other benign tumors, such as 36 rhabdomyoma, 7 fibroma and 5 lipoma of the free cardiac wall, 15 hemangioma, and 89 bronchogenic pericardial cysts had no history of sudden death, with the same being true of malignant tumors (McAllister and Fenoglio, 1978). The case of an enormous fibroma of the left ventricle in a 12-year-old girl athlete (Figure 36) who died suddenly and unexpectedly following a swimming race is worth mentioning (Heath, 1969).

Conduction System

In Chapter 3, we suggested a method that permitted an acceptable evaluation of the conduction system. Any time a sudden death is unexplained and there are indications in the clinical history of preexistent arrhythmia or interventional therapy involving partial ablation of the conduction system, a serial section study of this system should be considered. In an apparently healthy 29-year-old man who died suddenly, the unique postmortem finding was a hemorrhagic infarct (mainly, contraction band necrosis as shown by microphotos) limited to the His bundle without significant vessel changes (James and Riddick, 1990). The underlying diseases and the pathological findings from a series of 77 case reports in which a serial section study was performed (James and Jackson, 1977; James and Puech, 1974; James et al., 1975d, 1976b, 1978a; Brechenmacher et al., 1976, 1977) are reported in Table 39 and Table 40, respectively.

In another study of 49 patients dying suddenly, 39 had an acute myocardial infarction with severe obstructive atherosclerotic coronary disease and previous infarct in about half of the cases. The histologic findings in the conduction system were intimal obliterative thickening of the sinoatrial branch in 26%, atrioventricular node branch in 52%, fibrosis or fatty replacement of both nodes in 22%, or His bundle in 45%, and bundle branches in 51% (Lie, 1975). In another study, arterial dysplasia of the atrioventricular node was present in 44% of 27 sudden death cases and in 6% of controls (Burke et al., 1993). Degeneration and inflammation of nervous structures of the conduction system were

Table 39   Serial Section Study of the Conduction System — Underlying Disease in 77 Cases Reported by T.N. James

Disease	Number of Cases
Hypertrophic cardiomyopathy	22
Scleroderma	8
Long Q–T syndrome	8
Rupture of infarcted IV septum	5
Pheochromocytoma	3
Type A Wolf-Parkinson-White syndrome	2
Persisting superior vena cava	2
Rheumathoid arthritis	2
Pickwickian syndrome — Homocystinuria	
Familial congenital heart block – Whipple's disease	
Ankylosing spondilitis — Sarcoidosis	
Coarctation of aorta — Metastatic hypernephroma	1 each
Idiopathic conduction disturbances	7
No history	10

Table 40 Serial Section Study of the Conduction System — Pathological Findings in 77 Cases Reported by T.N. James

Pathological Findings	Number of Cases
1. Benign tumor (fibroma, Purkinje's cell, polycystic AV)	4
2. Fibromuscular medial displasia sinus/AV node arteries	
No underlying disease	6
Hypertrophic cardiomyopathy	13
Scleroderma	7
Pheochromocytoma	3
IV septum rupture (infarct)	3
Homocystinuria-rheumatoid arthritis	
Myocarditis-sarcoidosis-ankylosing spondylitis	
Coarctation of aorta	6 each
3. Anomalies of conduction system (persistent fetal dispersion or malformations AV node/His bundle, unusual, connections, venous lacunae, etc.)	25
4. Focal neuritis and neural degeneration (long Q–T syndrome)	8
5. Focal or diffuse replacement by adipose tissue	12
6. Focal degeneration or fibrosis	35
7. Panarteritis (Whipple's disease)	1
8. Disseminated intravascular coagulation	1
9. Platelet aggregates (pheochromocytoma, homocystinuria)	4
10. Fibromuscular polyploid mass occluding SN artery	1
11. Severe coronary atherosclerosis	10
12. Atresia left main trunk with diffuse myocarditis	1

observed in cases of sudden death (James, 1986; Rossi, 1985; Shvalev et al., 1986). Unspecific findings (mononuclear cells in and around the sinoatrial node, fibrosis of the atrioventricular node, etc.) were reported in young obese subjects who died suddenly (Bharati and Lev, 1995).

Physical Strain

The risk of strenuous exercise causing sudden death in coronary heart disease is controversial (Gibbons et al., 1980). The sudden and unexpected terminal episode occurred at rest or while sleeping in 69% of our 2324 reviewed cases. The remaining 707 were at work without evidence that strenuous physical exercise preceded the outcome (Baroldi and Silver, 1995). Only in the three previously quoted cases with mural coronary arteries did sudden death occur concomitantly with physical effort (Morales et al., 1980).

On the other hand, sudden death in young athletes and joggers presents a unique case. Among 78 young athletes (Heath, 1969; Opie, 1975; Noakes et al., 1979; Maron et al., 1980; Morales et al., 1980; Waller and Roberts, 1980; Tsung et al., 1982; Virmani et al., 1982; Voigt and Agdal, 1982; Thiene et al., 1985), coronary atherosclerosis was considered the cause of sudden death in 3% of subjects younger than 20 years, 40% between 20 and 29 years, 67% between 30 and 39 years, and 100% after 40 years. In 36 joggers (Morales et al., 1980; Virmani et al., 1982; Thiene et al., 1985), 92% had severe coronary atherosclerosis (Table 41).

Among 1276 sudden death cases reviewed from the literature (Baroldi and Silver, 1995) a pathological heart weight (500 g) was present in 46%. In our experience, 43% of 133 sudden and unexpected death cases without extensive myocardial fibrosis had hypertrophied hearts in contrast to 76% of 75 cases with extensive fibrosis. It is interesting to note

Table 41 Cardiovascular Pathology in 78 Athletes Who Died Suddenly Related to Their Age (Reports in the Literature)

Pathology	Age (Years)					
	<20		20–29		30–39	
Anomalous origin coronary arteries	5 (17%)	2 (8%)	—	—	7 (9%)
Hypoplasia right coronary artery	1 (3)	—	—	—	1 (1)
"Mural" left anterior descending branch	1 (3)	—	—	—	1 (1)
Hypertrophic cardiomyopathy	10 (34)	4 (16)	1 (17)	—	15 (19)
Floppy mitral valve	1 (3)	2 (8)	—	—	3 (4)
Idiopathic left ventricular hypertrophy	3 (10)	1 (4)	—	—	4 (5)
Heart tumor	2 (7)	—	—	—	2 (3)
Ruptured aorta	1 (3)	1 (4)	—	—	2 (3)
Lung thrombo-embolism	—	1 (4)	—	—	1 (1)
Anomaly conduction system	1 (3)	—	1 (17)	—	2 (3)
Right ventricular dysplasia	3 (10)	4 (16)	—	—	7 (9)
Atherosclerosis coronary	1 (3)	10 (40)	4 (66)	18 (100)	33 (42)
Total	29 (100)	25 (100)	6 (100)	18 (100)	78 (100)

that among 115 young soldiers who died suddenly (Moritz and Zamcheck, 1946), none had a pathological heart weight. In 40 athletes at the peak of training, echocardiography showed a left ventricular enlargement (>60 mm) and wall thickness increase (>13 mm). After a deconditioning period ranging from 1 to 13 yr, in 31 athletes, the sizes returned to almost normal parameters — in 50%, increased body weight and recreational physical activity — but 9 athletes had a persistent 60 mm cavity despite a normal wall thickness (Pelliccia et al., 2002).

Endocardial Lesions and Sudden Death

6

MALCOLM D. SILVER

Contents

Introduction

In the past, before cardiac surgery and interventional procedures were introduced to treat valvular heart disease, it usually led inexorably to fatal heart failure, a situation that still prevails in countries where such treatments are not readily available. Nevertheless, some afflicted individuals died suddenly. Where health care is universal, most of the conditions mentioned in this chapter will be diagnosed and treated. Nonetheless, some subjects with them will die suddenly. Furthermore, the indigent, the elderly living alone, psychotics, alcoholics, recent or illegal immigrants, and others at the fringe of society may slip through the health care net, develop unrecognized valvular disease, and present as cases of sudden death.

Endocardial and valvular diseases are not, per se, common causes of sudden death. Also, their frequency and determinants vary with a pathologist's practice and the population examined (Myerberg and Castellanos, 1992; Roberts, 1993; Corrado et al., 1998; Thiene, 1998; Thiene et al., 2001; Virmani et al., 2001). In my experience, for example, the cases most frequently encountered were associated with aortic valve stenosis or mitral valve prolapse. Problems associated with prosthetic heart valves were next most frequent.

Those endocardial lesions that cause sudden death usually affect heart valves rather than the endocardium lining heart chambers. Where pertinent, those affecting both natural and artificial valves will be discussed. Even so, some lesions mentioned in this chapter arise on mural endocardium. With the exception of certain valvular lesions that permit survival to adulthood and are associated with sudden death, I will not discuss congenital cardiac malformations or sudden death that may follow their repair. Rather, the reader is referred to Chapter 7, where these topics are presented. When discussing heart valve lesions, conditions that may affect any valve will be presented first, followed by those that affect individual valves. These sudden deaths have several potential mechanisms (Table 42).

When valvular regurgitation or valve stenosis or occlusion causes sudden death, it is by acute heart failure, and the pathology is more often related to left-sided than to right-sided heart valves. The left heart works at higher pressures than the right. Thus, sudden left-sided valvular incompetence or obstruction can induce potentially fatal acute pulmonary edema. Because of the lower pressures, sudden right-sided valve incompetence or obstruction is better tolerated. Thus, sudden death is uncommon when a tricuspid or pulmonary valve is affected. Indeed, valve excision without replacement by a prosthesis was used to treat tricuspid valve infective endocarditis (Yee and Khonsari, 1989). Depending on the duration of heart failure preceding the sudden valvular event, secondary pathologic changes may be observed in lungs, liver, or heart chambers with dilatation and hypertrophy affecting the latter. Remember that stenosis in the vicinity of the aortic valve also interferes with coronary artery perfusion and raises the possibility of sudden death.

Infective Endocarditis

This disease is characterized by an infection of the valvular endocardium associated with infected thrombotic vegetations (Figure 63). Generally, it involves a single, native, left heart valve (usually the mitral) more often than a right-sided one, but multivalvular infections of left heart valves or of both right-sided and left-sided ones occur under certain conditions. Examples are an infection developing in the presence of multivalvular rheumatic disease (aortic and mitral valve commonly infected) or if an infection is associated with the intravenous administration of drugs by addicts (Figure 63A). Sometimes, a congenital cardiac anomaly (Figure 63B), preexisting valvular disease (Figure 63C), preceding medical condition, or a therapeutic, surgical, or dental procedure is associated with a heart valve infection, but often that is not the case. Infective endocarditis can develop on a "normal" heart valve that shows no underlying pathology. Often involved in the latter instance is

**Table 42   Mechanisms of Sudden Death in Infective
Endocarditis Affecting Either Native or Prosthetic Valves**

Fatal Cardiac Arrhythmia
Valvular stenosis or occlusion
Valvular regurgitation (sudden or progressive) due to
   Cusp/leaflet ulceration
   Cusp/leaflet perforation
   Cusp/leaflet aneurysm, +/– its rupture
   Mural spread with chordal or papillary muscle rupture
   Prosthetic valve occluder entrapment
   Simulated valvular incompetence resulting from paravalvular leak
   Prosthetic valve dehiscence
   Sinus formation
Other
   Septicemia
   Direct spread of aortic valve vegetation occluding a coronary artery
   Bland or infected emboli causing
      Vascular occlusion
      Organ infarction
      Organ inflammation, for example, focal myocarditis
      Organ abscess(es), +/– rupture
      Arteritis with complicating occlusive thrombus
      Infected (mycotic) aneurysm with rupture
   Annular abscess producing
      Pericarditis
      Coronary artery arteritis with local mural or occlusive thrombus
      Rupture with hemopericardium and cardiac tamponade
      Aortic rupture
      Perforation of the membranous interventricular septum
      Destruction of the atrioventricular node
      Paravalvular leak

the aortic valve or, where drugs are self-administered intravenously, the tricuspid valve (Figure 63A).

In the gross, vegetations are usually attached to the atrial aspect of an atrioventricular leaflet or to the ventricular surface of a semilunar cusp and are related, in either instance, to the lines of leaflet/cusp closure on their flow surfaces (Figure 63A and Figure 63C). The valvular infection can remain localized to one cusp/leaflet or spread to a contiguous one ("kissing" lesions, Figure 63C) or to chordae tendineae or papillary muscles and cause their rupture. Sometimes, vegetations are located away from these usual sites. If that occurs, a pathological cause is usually apparent. For example, those found on the outflow surface of the anterior mitral valve leaflet near its base or midpoint may represent a mural infection if an infected aortic valve is incompetent, or result from the extension/rupture of an annular abscess burrowing from the aortic valve (see below). Vegetations observed near the base of the posterior mitral leaflet on its inflow surface develop when the infection is associated with a calcified mitral valve annulus. Vegetations can be gray-pink (Figure 63C), soft, and friable, or gray, yellow-brown, and firm. They vary in size with those associated with fungal infections said to be more voluminous. Small ones, or those remaining after embolization of a part, may be overlooked, because they only thicken or cause an irregularity along the line of closure. Microscopically, vegetations are composed of intermingled thrombus, inflammatory cells, and microorganisms, the latter sometimes in colonies (Figure 63D).

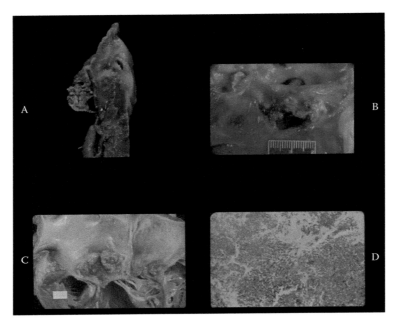

Figure 63 (A) Candidal infective endocarditis affecting the posterior leaflet of a "normal" tricuspid valve in a drug addict who used the intravenous route of administration. This section through the valve annulus shows the vegetation attached to the leaflet. Note the close association of the right coronary artery to the valve annulus. It (and the circumflex coronary artery with mitral valve infections) may thrombose if an annular abscess develops and spreads laterally. (B) Staphylococcal infective endocarditis on a congenitally bicuspid, calcified, aortic valve. In this instance, the disease has perforated a cusp. An annular abscess, the ostium of which is directly posterior to the cusp's upper margin, also developed. (C) Streptococcal infective endocarditis on a mitral valve showing rheumatic valvular disease. Vegetations are aligned along the lines of closure with that on the anterior leaflet, in becoming exuberant, spreading onto its body. The juxtaposition of vegetations raises a question of whether they represent contiguous spread and "kissing lesions." (D) Histology of thrombotic vegetation illustrated in part A showing Candida organisms (PAS stain).

The organisms may be visualized in Gram or Gomori methanamine silver stain, but that is not always the case. Some that might be expected to take up these stains do not, and others need different histological stains to demonstrate them. Both the vegetations and adjacent heart valve show an inflammatory reaction that may be acute or chronic. Where an infection has been present a protracted period healing changes may be obvious, and parts of a vegetation may calcify.

Any microorganism or blood-borne parasite has the potential to induce infective endocarditis. In practice, Gram-positive cocci cause it much more often than Gram-negative organisms, and bacteria much more often than fungi. Usually, one organism induces the disease, but sometimes more than one is responsible (superinfection). Nosocomial infections may cause infective endocarditis.

This valvular infection affects males more frequently than females. Depending upon the location of a pathologist's practice in the world, it primarily affects young adults with preexisting valvular disease (mainly rheumatic) or older patients, many of whom have either degenerative valvular disease or normal valves. I stress these geographic differences particularly because of the speed of travel nowadays. For example, Datta and colleagues (1982), reporting an autopsy series of 120 cases from northern India, noted that where a route of infection was known, puerperal sepsis was a common source; affected individuals were of younger age, these cases exhibited significant rheumatic valvular disease, and there was an absence of both narcotic addiction and of degenerative heart disease among those studied. Also, Naidoo and coworkers (1990) described six cases of treated tricuspid valve endocarditis in young women from South Africa and recorded that four were a result of genital sepsis.

Clinical manifestations are the result of microorganisms proliferating in the bloodstream or are caused by local or other systemic complications of the disease. In the particular circumstances where valve leaflets or cusps are suddenly perforated producing acute valvular insufficiency, clinical manifestations are usually acute and causative organisms are regarded as "virulent." In these circumstances, the valve may not show underlying pathology. On the other hand, the infection can be indolent if a microorganism is less virulent, with muted clinical signs and symptoms. Such infections often involve previously damaged heart valves. Nevertheless, clinical manifestations produced by a particular microorganism can vary with circumstances, and a less virulent infection may cause pathological changes that are described below and cause sudden death.

We will now discuss some of the causes of sudden death in infective endocarditis, as listed in Table 42.

Often the pathology found at autopsy provides a plausible explanation for a fatal arrhythmia, for example, a myocarditis or myocardial infarction or the destruction of or encroachment upon the atrioventricular node by an annular abscess. In other instances, no explanative pathology is obvious.

When native valve orifice stenosis occurs as a result of enlarging vegetations, a changing heart murmur is the usual clinical manifestation. However, acute mitral valve obstruction also occurs and may cause sudden death (Prasquier et al., 1978). Infected vegetations also spread across the orifice of a prosthetic heart valve and can seriously stenose or occlude it (Figure 64A and Figure 64B). In doing so, they may interfere with occluder seating and valve closure or, by their increasing volume, entrap an occluder in a partially open position rendering the valve incompetent. If vegetations stop occluder movement in its closed position, the orifice is effectively blocked. Those located at the pivot point of disc or bileaflet occluders also hinder or stop their movement. Infected vegetations may also encase tissue valve cusps causing severe incompetence. Valvular regurgitation that is either acute or insidious in onset is caused by the infection ulcerating a valve's free margin or by the infection destroying its substance and inducing perforation (Figure 63B). Tissue destruction may also lead to a cusp/leaflet aneurysm that interferes with cusp/leaflet closure and may rupture immediately or subsequently. These changes affect both native valves, usually those of the left heart, and the cusps of tissue valve prostheses.

As mentioned above, the clinical manifestations of infective endocarditis are the result of infective organisms proliferating in the bloodstream and of generalized or local complications of the disease. It is rare for a patient to be overwhelmed by septicemia and die suddenly.

In discussing other complications of the disease, the vegetations attached to a heart valve provide a source of bland or infected emboli. Depending upon their size and potential for infection, such emboli may induce the following:

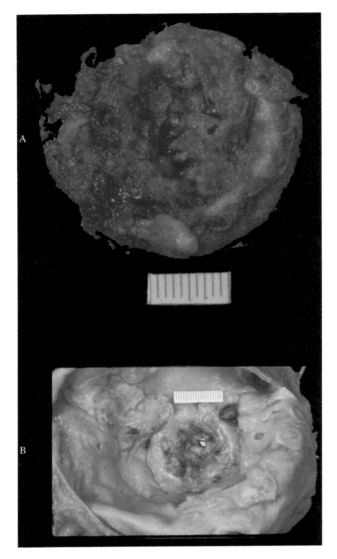

Figure 64 (A) Aortic tissue valve prosthesis affected by endocarditis showing complete occlusion of its lumen by overgrowth of infected vegetations. (B) Mitral Bjork–Shiley prosthesis viewed from the left atrium. The orifice is almost completely occluded by infected vegetations. The artificial valve's disc was also "entrapped" and immobilized.

1. Acute lethal effects, for example, arterial ostial occlusion or vascular embolism producing a stroke or acute myocardial infarction
2. A distal arteritis with an associated complicating occlusive thrombus
3. Infected (mycotic) aneurysms that form especially on peripheral arteries; the latter may rupture months after an infection, causing sudden death, even where a patient had apparently curative medical treatment
4. Abscesses in tissue

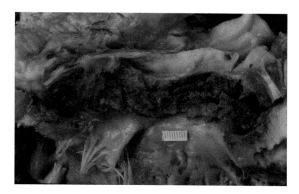

Figure 65   Circumferential annular abscess associated with infected tissue valve prosthesis excised at autopsy and illustrated in Figure 64A. At cut edges, lateral extension of the abscess is obvious. This valve had partially dehisced.

A serious focal myocarditis with or without abscess formation is often found associated with an aortic infective endocarditis, while a brain abscess may develop. Anguera et al. (2001) reported the sudden death of a patient with *Staphylococcal aureus* endocarditis that affected native aortic and tricuspid valves resulting from perforation of a myocardial free wall abscess. A patient with tricuspid valve endocarditis, especially if caused by a staphylococcal infection, may die suddenly of an acute pneumonia or as a result of complications arising from a lung abscess. Immune-mediated complications associated with microorganisms proliferating in the blood rarely kill patients suddenly.

Then, infection may spread locally. This can induce chordal or papillary muscle rupture or lead to an annular, perivalvular, or ring abscess (Figure 63B and Figure 65). Such abscesses are more common with aortic valve infections and are associated with both native and prosthetic valves, the latter more likely a mechanical than a tissue valve. An abscess may subsequently burrow into adjacent tissue proceeding laterally, proximally, or distally. Here it is vital to understand the associated anatomy of each heart valve. For example, I have seen annular abscesses of the tricuspid, mitral, and aortic valves (associated with the left cusp) spread laterally and either rupture acutely into the pericardial cavity or produce a pseudoaneurysm. With either an acute or late rupture, the latter may cause fatal cardiac tamponade. Lateral spread, particularly in aortic and mitral areas, also produces a pericarditis or affects the right or circumflex coronary artery with compression or occlusive thrombosis as sequelae. Lateral spread from other parts of the aortic annulus may impinge on the right atrial wall; the right ventricle near the membranous interventricular septum with risk of perforation and damage to the conducting system (producing complete heart block); or the outflow tract of the right ventricle or the main pulmonary trunk. Such extensions, if they break through to a related heart chamber or blood vessel, for example, the coronary vein, produce fistulae, and add to the blood volume that a ventricle must handle, resulting in severe heart failure. An annular abscess in the aortic area spreading proximally can induce aortic rupture, while proximal spread at that site or in the mitral area may produce a fistula that allows blood to bypass the valve annulus. Such paravalvular leaks have the same clinical effects as valvular regurgitation. Alternately, if a patient has a prosthetic valve *in situ*, the annular abscess or its spread may allow valve dehiscence. Distal spread of a right aortic cusp abscess may be into the right or left ventricular myocardium allowing subsequent rupture into a cardiac cavity or pericardium, while those related to the left and posterior aortic cusps may spread into the base of the anterior mitral valve leaflet inducing secondary infection or its rupture, destroy the atrioventricular node, or perforate the interventricular septum. Chan (2002) discusses the early clinical course and outcome among 43 patients with infective endocarditis complicated by perivalvular abscess. Dowling and Buja (1988) reported an unusual case of local spread causing sudden death in a 44-year-old man. In this instance, infected vegetations from a *Streptococcus viridans* endocarditis affecting the left cusp of a normal aortic valve extended into and obstructed the ostium of the left main coronary artery; both coronary vessels arose from the left coronary sinus.

Pathologists should keep in mind that some cases of sudden death present healed lesions of infective endocarditis, for example, cusp/leaflet ulceration, perforation, or aneurysm; chordal or papillary muscle rupture; or sinuses produced by spreading annular abscesses. All are liable to produce severe continuing heart failure.

Valvular or Mural Thrombotic Lesions

Infectious Thrombotic Lesions

Much less common than a valvular infection is one located on the endocardium of a heart chamber. This can occur where microorganisms in the bloodstream attach to an endocardial thrombus resulting from an intracardiac blood jet impinging on the surface or, if the jet is long standing, to an endocardial patch of fibrosis, both being called a "jet lesion." They are found, for example, in the right ventricle associated with a ventricular septal defect, on the septal endocardium of the left ventricle and the anterior mitral leaflet associated with aortic incompetence, or on the left atrial wall, especially the posterior one in mitral regurgitation. Alternatively, such endocardial infections, all termed "mural endocarditis," develop when a bacterial or fungal myocardial infection extends to the endocardium.

Thrombus or pseudointima that attach to valvular prostheses — patches introduced to occlude defects or conduits used to repair congenital anomalies — are sites of potential infection. So, too, are thrombi that form along catheters introduced to monitor patients (e.g., central venous lines, or for therapeutic purposes, for example, intravenous alimentation) or thrombi that form on pacemaker leads. Such infections in vessels produce "infective intimitis." All are uncommon. Rare, too, is an infective intimitis observed in the aorta distal to a coarctation. Like those in infective endocarditis, all of these lesions produce a similar range of clinical features and complications. Keep in mind that infections also develop in association with mural thrombi present in cardiac or vascular aneurysms. They, too, are uncommon, and *Escherichia coli* is often the infecting organism. Bland or infected emboli may arise from the infected aneurysm or the heart or vessel rupture, through erosion of its wall.

Bland Thrombotic Lesions

Bland thrombi attached to a heart valve, especially left-sided ones, as in nonbacterial thrombotic endocarditis (NBTE), can be a source of fatal distal emboli (Figure 66).

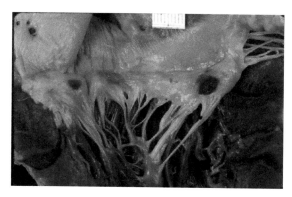

Figure 66   Nonbacterial thrombotic endocarditis of the mitral valve in a patient with leukemia. Similar lesions on the aortic valve had embolized to the left anterior descending coronary artery, causing sudden death.

Histologically, the vegetations consist largely of platelets and fibrin with few polymorpho-nuclear leukocytes, and the underlying valve is not inflamed. NBTE lesions occur in patients with many different clinical conditions. Often they have a hypercoagulable state or disseminated intravascular coagulation. The condition is also linked to the antiphos-pholipid syndrome, while other patients have malignant disease, especially a mucin-producing adenocarcinoma of lung, pancreas, or colon or a lymphoma. Steiner (1995) compared the findings in cases of infective endocarditis and NBTE occurring in a necropsy study of 320 cases. (I point out that distal embolism causing sudden death may also be the first manifestation of a mural thrombus associated with a "silent" myocardial infarction.)

NBTE lesions are a common autopsy finding on right heart valves where a patient had a central venous catheter inserted. They have the potential for both embolism and infection. Tiny emboli are often present in the lungs in such cases but do not usually cause sudden death. However, a case of pericardial tamponade that caused sudden death following coronary occlusion and myocardial infarction was reported (Laurain and Inoshita, 1985). The lesions of NBTE rarely become infected.

I described above a range of complications that infected vegetations can produce in artificial heart valves. Comparable complications occur if bland thrombus accumulates on such valves.

A pathologist is never certain whether a vegetation attached to a heart valve is bland or infected. Both histologic and microbiologic examination are necessary. In the latter instance, culturing a fragment of the vegetation will likely yield better results than a swab taken from it. When making a diagnosis of endocarditis, the pathologist must be circumspect. In this, it is best to adhere to the redefined major and minor criteria employed by clinicians. They confirm the diagnosis of infective endocarditis if microorganisms are recovered from blood cultures or are demonstrated by polymerase chain reaction techniques; serologic changes occur; or microorganisms are demonstrated in vegetations, abscess tissue, or in emboli that are examined histologically. If microorganisms are not recovered after repeated blood sampling and prolonged culture, presence of the disease may be inferred by echocardiographic findings revealing vegetations; or sinuses or fistulae induced by annular abscesses, as demonstrated by Doppler flow techniques (Bayer et al., 1998). If such criteria had been used by Edwards and colleagues (1986) in reporting a supposed case of infective endocarditis affecting the Eustachian valve in the right atrium, the diagnosis would be regarded as not proven.

Myxomatous Degeneration of Heart Valves

In this condition, an excess proteoglycan material is observed histologically in a valve cusp or leaflet, most noticeably in the fibrosa. When the change is severe, the deposit is generalized, extending toward the base of the fibrosa and into the remainder of a valve tissue (Figure 67A and Figure 67B). Annular tissue and, in atrioventricular valves, chordae tendineae are also involved. The deposit in the valve components associated with degeneration and loss of elastic tissue and collagen is obvious in a Movat stain but best demonstrated by electron microscopy (Akhtar et al., 1999). In making a histologic diagnosis several points must be kept in mind. They include the histologic structure of normal heart valves. Here McDonald and colleagues (2002) using digital imaging technology made an important contribution to understanding normal mitral and aortic valves. Histologically,

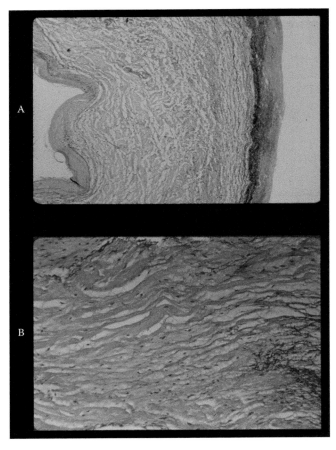

Figure 67 (A) Histologic section of a myxomatous mitral valve showing increased proteoglycan material in its substance. (B) Higher power to demonstrate the green-stained proteoglycans. Note the fibroelastic thickening on the atrial side of the leaflet, a not unusual finding in this condition (Movat stain).

valve cusps and leaflets present four tissue layers covered on either side by endothelial cells (Veinot et al., 2001). Those of the atrioventricular leaflets, proceeding from atrial to ventricular surfaces, are the auricularis, spongiosa, fibrosa, and the ventricularis. Semilunar cusps have similar layers if named, proceeding from ventricle to great vessel, the ventricularis, spongiosa, fibrosa, and arterialis. The spongiosa normally consists mainly of proteoglycans in which a few elastic fibers, collagen, and connective tissue cells such as fibroblasts and primitive mesenchymal cells are found. This thin layer extends the full length of atrioventricular leaflets with the valves presenting in their distal third fibrosa and spongiosa layers covered by endocardium. In addition, the spongiosa of the proximal third of mitral leaflets contains cardiac myocytes and capillaries extending into it from the left atrial myocardium. In semilunar cusps, the spongiosa is prominent only in their basal third. Thus, when the Movat stain is used, and it is the best available to define proteoglycans in the cardiovascular system, normal atrioventricular leaflets present a narrow band of green material that widens at either end of the valve, more so distally than proximally, while semilunar cusps have a green layer at their base. It must be appreciated that focal myxomatous accumulations can be found in heart valves. Here, myxomatous tissue formation might be regarded as a response to local "injury" or adaptive changes. These findings must not be mistaken for pathological myxomatous degeneration. Rather, by Movat stain, the latter presents far more abundant deposits of green proteoglycans that in severe cases may thicken all layers of an affected valve throughout its length (Figure 67B) as well as be obvious in other valvular components.

Myxomatous degeneration may be regarded as having many causes and to occur as a focal, secondary, or primary process. In the first instance, the affected valve is likely to be from an elderly person or show evidence of preexisting pathology, for example, rheumatic valvular disease with the focal deposits of myxomatous material often related to scar tissue. In such circumstances, valve prolapse and sudden death are unlikely. Injury or local adaptive changes affecting a valve may also explain secondary myxomatous degeneration. It is often focal but can be generalized. The latter may be the reason for myxomatous deposits in, for example, tricuspid or pulmonary valves associated with pulmonary hypertension or the mitral valve associated with the Eisenmenger's complex. In these instances, only one valve usually shows the change with the mitral valve most often affected. Generalized secondary myxomatous change may lead to valve prolapse and sudden death. Primary myxomatous degeneration also produces localized or generalized changes and is also associated with valve prolapse and sudden death. Here the condition is often related to recognized genetic and chromosomal syndromes, some having far-reaching effects and others affecting heart valves alone. In the former case, myxomatous degeneration may be part of, for example, Marfan, Ehlers-Danlos, or Turner's syndromes with the patient showing dysmorphism in other systems, or formes fruste. As some authors emphasize, either a pathologic or gene-based diagnosis of a heritable disease should trigger family studies (Seidman and Sampson, 2001; Thiene et al., 2001). Thus, for public health reasons, forensic pathology units must be able to do diagnostic gene analyses or be associated with centers where such tests are done.

The pathologic complications of myxomatous degeneration are listed in Table 43.

In discussing them, the following points are made. Myxomatous deposits also occur in other structures important in valve closure; thus, annular dilation might be expected, whereas aneurysm, shredding, and perforation are most likely to affect aortic valve cusps due to deposits thinning them. Associated weakening of valvular structure and chordal

Table 43　Complications of Myxomatous Degeneration of Heart Valves

Annular dilation or calcification
Cusp/leaflet
Aneurysm
Shredding or perforation
Prolapse
Chordal rupture

changes can induce mitral valve prolapse. Chordal elongation also occurs, with their rupture possible. Further comments are presented under individual valves.

Rheumatic Valvular Disease

Rheumatic valvular disease is the end product of repeated attacks of acute rheumatic fever, a cross-reactive, immunologically mediated, inflammatory disease that affects the tissues of many organs and occurs within 10 days to 6 weeks of a pharyngitis caused by certain strains of group A (beta hemolytic) streptococci (Guilherme et al., 2000; Galvin et al., 2000; Roberts et al., 2001; Quinn et al., 2001; Galvin et al., 2002). Children after the age of 5 are most often affected, but the disease can first appear in middle or later life. Once afflicted, an individual is prone to further attacks, with each bout of pharyngitis likely to produce the same clinical manifestations as the original attack, although a recurrence in an older individual may produce unusual ones. The frequency of episodes usually diminishes with increasing age. Patients rarely die during an acute attack of rheumatic fever, although I have seen two sudden deaths in such circumstances affecting migrants to Canada, one a teenager from the West Indies and the other a middle-aged East Indian woman. Both deaths were associated with a florid rheumatic myocarditis. Josselson et al. (1984) describe a similar case, but in that instance, the 18-year-old male had no history of acute rheumatic fever. Sudden death may also occur if mitral valve chordae rupture during an acute attack.

Gross changes found in rheumatic valvular disease develop as a result of the organization of minute, thrombotic, endocardial vegetations found along the lines of closure formed during acute attacks of rheumatic fever, and the inflammation induced by the attacks in various parts of the valve. This scarring distorts valve components. The process may also be aided by the deposition and organization of minute thrombi resulting from altered hemodynamics as valvular distortion evolves. Thus, valve damage is cumulative. The net effects can shorten and thicken cusps/leaflets, inducing valvular regurgitation (Figure 68A and Figure 68B); fuse adjacent sides of cusps/leaflets at commissural areas producing luminal stenosis (Figure 68B and Figure 68C) — this tends to be the principal change in tricuspid stenosis, producing a diaphragm-like orifice with only minor associated chordal changes; fuse, thicken, or shorten chordae tendineae of the mitral valve with leaflet tethering and valve regurgitation (Figure 68A); or fuse chordae to cause a mitral subvalvular stenosis (Figure 68D). Thus, clinically, a rheumatic valve may be purely stenotic, purely regurgitant, or exhibit both functional abnormalities.

In Western countries, it usually takes 20 or more years for such valvular distortion to evolve. However, in other countries where children are subject to repeated attacks of

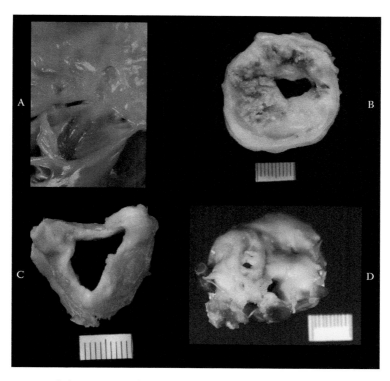

Figure 68   Lesions of rheumatic valvular disease seen in the gross. (A) Mitral valve showing shortened and thickened leaflets near a commissural area. In this instance, these changes are also associated with fusion and marked shortening of chordae tendineae so that the leaflets are tethered to the apicies of the papillary muscles, and severe mitral regurgitation resulted clinically. (B) Surgically excised mitral valve viewed from its atrial aspect showing stenosed orifice resulting from fusion of adjacent sides of leaflets near commissural areas. A similar process obliterated the clefts of the posterior leaflet (uppermost). Note the secondary calcification affecting this valve with some nodules ulcerated. (C) Surgically excised aortic valve viewed from the aortic aspect. Cusps are shortened and thickened by both fibrous tissue and calcium nodules. The fusion of their adjacent sides near commissural areas produced the stenosed, triangular orifice. (D) Marked chordal thickening and fusion illustrated in a surgically excised mitral valve viewed from its ventricular aspect. This produced a subvalvular stenosis.

rheumatic fever and, perhaps, suffer from other conditions (e.g., malnutrition), the process is speeded up, with children and adolescents presenting with severe mitral stenosis (juvenile mitral stenosis) caused mainly by fibrosis, although Chopra and Bhatia (1992) in their series reported from Delhi found various degrees of calcification in 36% of their patients. The authors also commented that autopsy in juvenile patients revealed moderate to marked hypertrophy of medium-sized pulmonary arteries with dilation lesions seen in a few cases.

The histologic features of a distorted valve may vary, but generally, cusps/leaflets are diffusely fibrosed, often with obliteration of their original architecture, and show increased vascularization (small, thick-walled vessels); focal lymphocytic infiltrates exist, often near the cusp/leaflet base — Aschoff's nodules, per se, are not found in cases that have taken many years to reach this stage but are, on rare occasions, observed in the valves of people with active rheumatic carditis (Freant and Hopkins, 1997) or where stenosis developed rapidly; foci of myxomatous degeneration are obvious within cusps/leaflets; and chordae are thickened by fibrous tissue and may be fused together. Secondary morphologic changes often encountered include valvular, annular commissural, and chordal calcification. This complication is less likely where valve stenosis developed rapidly. Such calcium nodules further distort valve components and may ulcerate and embolize contents, cause hemolysis, or be a site of thrombus deposition with subsequent embolization or infection. Lambl's excrescences on a distorted valve are also a secondary morphologic change that can be encountered.

Other pathological changes include myocardial hypertrophy and dilation of heart chambers caused by associated heart failure. With mitral stenosis, either free-floating (ball) or mural thrombi may develop in the left atrium. Both can calcify with time. Rarely, a ball thrombus obstructs the valve orifice. Left atrial thrombi may give rise to emboli. Pulmonary vascular changes can develop and, with right heart failure, liver congestion with or without associated fibrosis.

Rheumatic valvular disease affects different valves under different circumstances. No gender difference exists in cases of acute rheumatic fever, yet mitral stenosis affects women twice as often as men. Indeed, the majority of cases present with only the mitral valve affected; mitral and aortic valve involvement is next most frequent, with a lesser number of cases having both these valves and the tricuspid valve afflicted. Generally, rheumatic valvular disease of the pulmonary valve is uncommon except in places with a high altitude, for example, Mexico City. In those circumstances, only the pulmonary and tricuspid valves may be affected.

Patients with rheumatic valvular disease may die suddenly of an arrhythmia or heart failure induced by mitral or aortic stenosis (see below) or, in considering other mechanisms, from embolism or infective endocarditis.

Sudden Death in Patients Bearing Heart Valve Prostheses

A minority of those who die suddenly after having had one or more heart valve prostheses inserted years previously do so of conditions not directly related to their prostheses; but, the majority of such deaths are directly related to them. Schoen (2001) discusses the topic of postoperative mortality. Among the first group, one finds individuals in whom coronary artery disease had progressed. Their sudden deaths are likely due to an arrhythmia induced by ischemia or are a result of myocardial infarction or its complications. However, others who die of an arrhythmia do not have severe coronary artery disease. Still others die suddenly of causes that may be related to myocardial changes wrought by the original valvular disease or its progression, despite prosthesis insertion; complications of bygone cardiac surgery; or intercurrent, noncardiac diseases.

Complications related to the operative procedure rarely kill many years after operation, but the examining pathologist must look for evidence of disproportion between the prosthesis and its surrounds, pseudoaneurysm formation in annular areas, particularly related to the mitral valve or of the left ventricle from tissue damage induced during removal of a diseased valve, and for the possibility of vascular injury affecting the aorta or coronary arteries. Sutures can entrap parts of a prosthesis leading to dysfunction (see Chapter 13).

The accumulation of bland thrombus and the development of infectious endocarditis are continuing risks for a patient who bears a prosthetic heart valve, especially a mechanical one. The sequelae of such developments are discussed above. Thrombus or the healed vegetations of infected endocarditis may become organized, forming pannus that stenoses a valve orifice or interferes with tissue valve cusp movement or either the movement or seating of a mechanical valve occluder. Most patients bearing a mechanical prosthesis and some with tissue valves, for example, those with atrial fibrillation or a history of an episode of thromboembolism, receive anticoagulant therapy. They face the danger of anticoagulant-related hemorrhage if dosage is not properly monitored and maintained.

Many patients bearing a heart valve prosthesis now survive for decades. Nevertheless, with time, the risk of structural failure increases in tissue valves due to deterioration or calcification of their collagen components. Thus, attachments can separate from the valve stent; ulcers and perforations may appear in cusps, or they can stretch, while root dilation occurs with unstented valves. All of these changes lead to prosthesis incompetence that, if rapid in onset, may kill a patient suddenly. Calcification is another late complication of tissue valves. This may stenose their orifices or cause cusp perforation. Stent "creep" also causes valve stenosis.

The components of mechanical valves may be subject to wear or metal fatigue, as they are called upon to meet the requirements of a functioning heart. In the past, these problems caused occluders to escape their housings either following a direct change in the poppet/disc substance or fracture of the metal or pyrolytic carbon components that restricted poppet/disc movement. As each defect was defined, the affected prosthesis was either withdrawn from the market or modified. Today's prostheses are designed to reduce the risk of these complications by modifying components or by reducing the risk of tissue valve calcification. Even so, man designs, but nature decides. Therefore, the possibility of new problems arising, as models are *in situ* for even longer periods, must be kept in mind and searched for.

Other Endocardial Lesions Associated with Sudden Death

The following endocardial conditions are merely mentioned here as rare associations with sudden death. Thus, patients with endomyocardial fibrosis may die suddenly (Balakrishnan et al., 1993; Guimaraes, 1993). Among cardiac tumors, I have seen a large left atrial myxoma suddenly obstruct the mitral valve orifice and another cardiac myxoma give rise to a tumor embolus that obstructed a coronary artery, both events killing the affected individuals suddenly. Found in the literature are reports of papillary fibroelastomas that arose on aortic valve cusps and either blocked a coronary ostium with their fronds (Butterworth and Poindexter, 1973) or embolized a part to a coronary artery, thereby inducing sudden death (Harris and Adelson, 1965). These tumors are also a source of thromboemboli and are now considered less clinically benign than was once thought, especially if they are located in the left heart (Valente et al., 1992; Hynes et al., 2002).

Lesions of the Tricuspid Valve

Congenital Lesions

I examined a 37-year-old patient who died suddenly and presented with a congenitally corrected transposition of the great vessels unknown before autopsy as the only cardiac anomaly. Ebstein's anomaly, discussed in part in Chapter 7, is also associated with sudden death in adolescents and young adults due to a risk of supraventricular tachycardia, which can have a variable pathologic anatomy. Fused or malformed tricuspid valvular tissue usually affecting the posterior and septal leaflets is displaced into the right ventricular cavity with the basal attachments of the leaflets adhering to the right ventricular wall distal to the true valve annulus. The valve tissue is often redundant, but the anomaly may be mild, not associated with any hemodynamic dysfunction, and manifest microscopically as a mild lowering of the attachment of the tricuspid septal leaflet (Thiene et al., 2001).

Acquired Lesions

Almost any of the acquired valvular lesions mentioned above in affecting the tricuspid valve could be associated with sudden death, but this would be an extremely rare occurrence. More likely, in such instances, tricuspid valve involvement is part of a condition that also affects left heart valves with problems there causing the sudden death. Burke et al. (1997c) discuss sudden death in intravenous drug addicts which is associated with acute infectious endocarditis located solely on the tricuspid valve. The authors also emphasize the frequency of healed infective endocarditis lesions of this valve observed among their cases. Infection of the tricuspid valve also poses a risk of pulmonary abscess and of paradoxical embolism. Dickens et al. (1992) reported a case where a cardiac metastasis of a primary adrenal cortical carcinoma to the endocardium of the right atrium produced an intracavitary mass that the authors believed caused sudden death by blocking the tricuspid valve orifice. Carcinoid heart disease, which most often affects right heart valves, is not usually associated with sudden death.

Lesions of the Pulmonary Valve

Right ventricular outflow tract obstruction as part of tetralogy of Fallot and sudden death are discussed in Chapter 7. Isolated pulmonary valvular stenosis may be congenital or acquired but is not usually associated with sudden death.

Lesions of the Mitral Valve

Many mitral valve lesions that induce either acute valvular regurgitation or stenosis/obstruction and that are associated with sudden death have been described above.

Mitral Valve Prolapse

Mitral valve prolapse syndrome has many synonyms and is one of the more prevalent cardiac valvular abnormalities. Clinically, it may be associated with acute, chronic, or acute on chronic mitral regurgitation, stroke, or sudden death. It has many causes, with myxomatous degeneration a frequent one. Other potential mechanisms include an anomalous arrangement/insertion of chordae tendineae (van der Bel-Kahn et al., 1985; Virmani et al., 1987), while Hutchins and colleagues (1986) suggested that the floppy valve develops from hypermobility of the valve apparatus secondary to annulus fibrosis dysjunction, an anatomic variant in its morphology.

Chordal rupture leading to prolapse may follow trauma (see Chapter 12), infective endocarditis, or fraying caused by calcium nodules in the mitral annulus. However, the most frequent culprit is myxomatous degeneration occurring within the chordae. Chordal rupture probably does not occur as a result of an acute myocardial infarction, per se. Rather, chordae avulse from the tip of an infarcted papillary muscle. In these cases, careful gross examination will reveal a tiny nub of infarcted myocardium or of fibrous tissue from the apex of the papillary muscle attached to the one or two detached chordae.

Mitral valve prolapse is also a consequence of papillary muscle rupture following myocardial infarction.

Gross morphology will vary with the cause of valve prolapse, as will the mechanism of sudden death. Where prolapse developed suddenly, for example, following chordal rupture or rupture of the body of an infarcted papillary muscle, leaflets may be normal if pathology is obvious in chordae or the papillary muscle. In these instances, acute pulmonary edema may have killed the patient.

I warned above about overdiagnosing myxomatous degeneration histologically. A similar warning must be made to distinguish the "hooding" of mitral valve leaflets from mitral valve prolapse. The free margin of a leaflet between chordal insertions may bulge slightly toward the left atrium. Edwards (1988) called this "interchordal hooding"; the change is not associated with clinical symptoms. Edwards considered that hooding fell within the range of normal morphology, it affected both the rough and clear zones of a leaflet (see Ranganathan et al., 1970, for definitions); the bulge was less than 4 mm high; and the change involved only one third or less of a leaflet's free margin. He diagnosed prolapse when such hooding was greater than 4 mm in height and involved a leaflet's rough and clear zones and at least half an anterior leaflet or two thirds of a posterior one. Figure 69 demonstrates the gross morphology of some prolapsed mitral valves. The extent of leaflet prolapse varies from case to case.

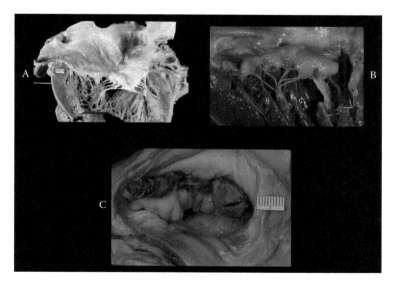

Figure 69   Gross morphology of myxomatous mitral valves. (A) Part of an anterior leaflet bulging toward the left atrium. (B) Close-up of prolapsed scallops of the posterior leaflet. Note that some chordae appear elongated and some thickened. (C) Left atrial view of most of a prolapsed posterior leaflet. Note the thrombus deposited in the angle between the base of the prolapsed leaflet and the atrial wall. Such a thrombus can be a source of emboli.

Where prolapse is caused by myxomatous degeneration complicated by acute chordal rupture, acute pulmonary edema follows. However, these patients also die suddenly of an arrhythmia. Although the exact cause of the arrhythmia is uncertain, Burke et al. (1997b) concluded that arterial dysplasia may contribute through ischemic ventricular scarring. Their histologic study of the hearts from 24 young adults with valve prolapse who died suddenly revealed fibromuscular dysplasia stenosing the atrioventricular nodal artery in 18 (and in four of 16 control hearts), with the morphologically measured degree of luminal narrowing being significantly greater in hearts with valve prolapse. The authors also found that the degree of fibrosis at the base of the ventricular septum, as calculated by computerized morphometry, was greater in hearts with mitral valve prolapse.

In myxomatous degeneration, the prolapsed valve or part thereof is usually gray-white, thickened, and bulges with a hooded appearance toward the left atrium. The latter is best appreciated if an affected valve is viewed from the left atrial aspect before opening the left ventricle (Figure 69C). The endocardium on the leaflet flow surface is usually thickened by fibromuscular tissue, and tiny thrombi may be attached to both leaflet and chordae. Linear thrombus can also form at the base of a prolapsed leaflet in the angle between it and the adjacent atrial wall. The angle becomes obliterated by fibrous tissue if, over time, such thrombus is organized. Cerebral ischemic events prevalent in younger individuals were related to prolapsed mitral valves and such thrombi (Barnett et al., 1980), but subsequent clinical studies produced controversy (Zenker et al., 1988; Gilon et al., 1999). Figure 69C illustrates a flamboyant example of thrombus at the base of a prolapsed myxomatous valve. It would be hard to believe that it could not give rise to emboli that induce cerebral, as occurred in this case, or other embolic pathology.

The cut surface of a prolapsed leaflet is thickened and has a typical, grayish, gelatinous appearance. Chordae are often elongated (Figure 69B) and have a reduced diameter. When thinned in this way, they can rupture with their free ends apparent in the gross. In other instances, a ruptured chorda can recurve upon itself and reattach to a leaflet's ventricular surface, presenting as a loop. Alternatively, chordae may be thickened (Figure 69B). When this occurs, they do not fuse together and commissural fusion is not obvious, which are morphological differences that help differentiate this condition from rheumatic valvular disease. One must search for other pathological complications known to occur with myxomatous degeneration, such as annular calcification, which develops in long-standing cases, or infective endocarditis (Figure 70). The latter may also cause sudden death.

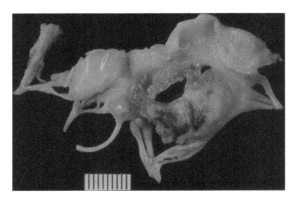

Figure 70   Infectious endocarditis affected a myxomatous anterior mitral leaflet with perforation illustrated here in a surgically excised specimen.

Periaortic Lesions

Many conditions that induce acute valvular regurgitation and sudden death by pulmonary edema are discussed above, and others will be mentioned here. When considering stenotic lesions in the aortic area, subvalvular and valvular lesions will be discussed, with supravalvular ones presented elsewhere (see Chapter 7).

Any obstruction to the left ventricular outflow tract induces a self-sustaining and slowly progressive effect. The heart copes with the worsening pressure gradient across the obstruction through progressive hypertrophy of the left ventricle, with, in most instances, that adaptation allowing cardiac function to be sustained for many years. Once the left ventricle fails to compensate, symptoms of angina pectoris, syncope, and dyspnea manifest, and a patient's prognosis worsens unless the stenosis is relieved. Sudden death may occur in those who have such symptoms, but it may also happen in an asymptomatic individual. The combined effects of a reduced coronary flow resulting from the stenosis and increased oxygen demand by the hypertrophied myocardium are the likely explanations for angina pectoris, syncope, and sudden death. The exact mechanism of the latter is obscure, but patients with aortic stenosis are prone to complex ventricular arrhythmias.

Subvalvular Aortic Stenosis

Membranous subvalvular aortic stenosis is discussed in part in Chapter 7. It is more common in males than in females. Occasionally, the condition produces minimal or no clinical symptoms, and teenagers or young adults with it die suddenly, often during exercise. It can have a familial occurrence, so first-degree relatives of sentinel cases who die suddenly should be evaluated medically (Urbach et al., 1985).

The lesion presents as gray-white fibroelastic tissue in the outflow tract of the left ventricle affecting an area 1 to 2 cm distal to the aortic valve cusps and in the plane of the mitral valve's annulus fibrosus. The tissue is from 2 to 4 mm wide, most obvious on the membranous interventricular septum, and may appear as a discrete band, an accumulation of several bundles, or, most commonly in adults, as a diffuse ridge (Figure 71). Fibroelastic tissue on the septal wall sometimes extends toward the right aortic valve cusp. The membranous obstruction may be crescentic and extend across the anterior mitral valve leaflet

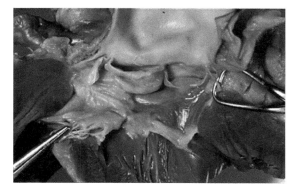

Figure 71 Membranous subaortic stenosis causing sudden death in a 21-year-old male. Note the distinct ridge of fibroelastic tissue on the septal wall of the left ventricle proximal to the aortic valve cusps. In this instance, the membrane extended onto the ventricular surface of the anterior mitral valve leaflet.

with the ends inserting into the ventricular aspect of that leaflet's base. The distal part of the leaflet here is usually normal and not thickened as it is in hypertrophic cardiomyopathy accompanied by systolic anterior movement of the leaflet. More commonly in adults, the membrane forms a distinct circumferential collar. This lesion can be overlooked, with observed changes dismissed as endocardial thickening. A better impression of the degree of stenosis is gained by reconstructing the outflow tract, if opened at autopsy, and sighting along it toward the aortic valve. Except at the membrane, the outflow tract of the left ventricle is not narrowed as it is in other forms of subvalvular stenosis. One may find the area between membrane and aortic valve dilated. The latter usually has three cusps that may be normal or become thickened by a blood jet if the subvalvular stenosis is severe; rarely the cusps calcify. These damaged cusps are prone to infective endocarditis. Other forms of subvalvular stenosis causing sudden death are discussed elsewhere.

Valvular Aortic Stenosis

Stenosis at the level of the aortic valve may be congenital or acquired. Schoen and Edwards (2001) discuss these conditions. Acquired lesions result from cusp stiffening or immobility due to fibrosis, the deposition of calcium or other materials within the cusp substance, or by both processes; fusion of adjacent sides of cusps near their commissural area (commissural fusion); or a combination of these changes. Irrespective of the initial cause, further cusp sclerosis and dystrophic calcification occur with time and worsen the stenosis so that the end result may be a severely stenosed calcified valve, that is, calcific aortic stenosis. Of itself, the latter is not a satisfactory diagnosis, because careful gross examination can usually determine whether initial changes were postinflammatory (rheumatic), degenerative, or congenital. However, in a small percentage of cases where changes are very marked, this may not be possible, and a diagnosis of calcific aortic stenosis must suffice. In practice, stenotic lesions that affect the valve are the ones most commonly encountered in cases of sudden death. It may occur at any age but is most prevalent among middle-aged and older patients.

A high-pressure blood jet exiting from a severely stenosed aortic valve can induce degenerative changes in the ascending aorta, with disruption of medial elastic tissue, increased accumulation of glycosaminoglycans, and focal cystic medial degeneration. These may result in a local aneurysm where the jet impacts or a more diffuse poststenotic dilatation. The former may rupture, whereas aortic medial changes make the ascending aorta prone to intimal tears and aortic dissection, the latter seemingly more commonly associated with congenitally anomalous valves than stenosed ones having three cusps (Larson and Edwards, 1984).

Elderly patients who have valvular aortic stenosis may hemorrhage from an arteriovenous malformation (angiodysplasia) of the gastrointestinal tract, most often affecting the right colon. The mechanism of this association is unknown.

Congenital Lesions Causing Aortic Stenosis

Within the range of congenital anomalies, an aortic valve may present as a dome-shaped structure with no commissures, with only one commissure, or be bicuspid. In this group, males are affected more often than females, with the male-to-female ratio in congenitally bicuspid valve being 4:1. Rarely, an aortic valve has four cusps.

Acommissural and Unicuspid Aortic Valve

Both acommissural and unicommissural valves are frustum shaped. The former has a domed appearance, like a volcanic cone, and may show no obvious commissures or have ridges marking the site of aborted ones (raphes) passing from the sinus wall to the cusp in the depth of the sinus (Figure 72A). This type of congenital abnormality causes a marked degree of valvular stenosis from birth; thus, it may be lethal in infancy. The two cases of sudden death associated with this anomaly that I have seen occurred in preteen boys engaged in sport. In both instances, the cusp was thickened at its free margin but not calcified.

Roberts (1979) likened the ostium of a unicuspid aortic valve to an exclamation mark. This is because its free margin, which is attached to the sinus wall at the site of an aortic commissure, extends toward the opposite sinus wall and then returns to reattach at that commissure without forming any other commissure (Figure 72B). Again, raphes may be found within the depths of its sinus. Unicuspid valves are much less frequent than congenitally bicuspid ones. In my experience, subjects with unicuspid valves died suddenly as teenagers or young adults, whereas those with congenitally bicuspid valves died suddenly in their thirties or older. However, this is not an immutable rule, and the age at death may overlap. The unicuspid valve's free margin and substance are usually thickened by fibrous tissue, with calcification found in the valves of older individuals.

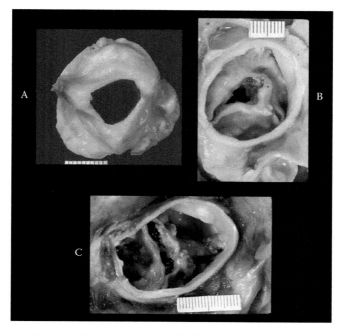

Figure 72 Congenital anomalies producing valvular aortic stenosis. (A) Dome-shaped acommissural valve here illustrated in a specimen surgically excised from a 15-year-old male. Flecks of calcium appear as yellow streaks. (B) Unicuspid aortic valve from a 21-year-old male who died suddenly with no other obvious cardiac lesions. Note cusp thickening and raphes in sinus. (C) Sudden death in a 55-year-old female with a congenitally bicuspid valve. Note raphe in sinus and sclerosis and calcification of cusps.

Bicuspid Aortic Valve

A biscuspid aortic valve is its most common congenital anomaly. It occurs in males three to four times more often than women (Waller et al., 1973) and may be an isolated phenomenon or be associated with other congenital cardiac anomalies, for example, coarctation of the aorta. If both cusps are of equal size, the commissures are opposite each other, and the patient may experience no valvular dysfunction during life. However, most bicuspid valves have a larger conjoint cusp because the circumferential amplitude of the commissures is more than 180°. Such bicuspid valves usually become dysfunctional with time, producing either valvular stenosis or incompetence, both of which may cause sudden death.

Among the bicuspid valves they examined, Angelini and colleagues (1989) found approximately two thirds had cusps located medially and laterally with a coronary artery ostium arising from each sinus, the other third had anterior and posterior cusps with coronary artery ostia arising from the anterior sinus. Left coronary artery dominance is more common in patients with a bicuspid valve, and the left coronary artery ostium is more likely to arise from the ascending aorta distal to the sinotubular junction. Again, about 60% of bicuspid valves have a raphe observed as a ridge of tissue in the sinus of the larger conjoint cusp. This represents the aborted development of one commissure or fusion of two cusps during fetal life. If present, the raphe may vary in height (from sinus floor toward the cusp's free margin), length (extension from sinus wall toward the free margin), and degree of differentiation (separation) of its free margin (Waller et al., 1973). Rarely, instead of being a continuous solid ridge, a raphe presents as a cord-like structure 1 mm in diameter or less extending from a commissural mound to the conjoint cusp's free margin. This results in a fenestration between the cord, sinus side of the conjoint cusp, and sinus wall (Walley et al., 1994). The height of a conjoint cusp is often minimal at its midpoint or, as another variant, presents a distinct notch at this location; both favor the development of valvular incompetence.

With time, a bicuspid valve is prone to calcification (Figure 72C). This change, like that occurring in the cusps of valves with three cusps with increasing age, develops in areas of increased mechanical stress, first along the raphe of the bicuspid valve and in both bicuspid and tricuspid valves at their attachments and basal halves. Usually most obvious on their sinus sides, the individual nodules of calcification may subsequently fuse or ulcerate and may show bony metaplasia. Occasionally, an associated bar of calcium extends from the aortic annulus into the anterior mitral valve toward its midpoint. These cuspal changes, with associated fibrosis, both thicken and stiffen the cusps, leading to valvular stenosis. Bland thrombus may develop on the calcium nodules, or they may ulcerate. Stenosed aortic valves are prone to infective endocarditis, and if the stenosis is severe, it may induce hemolysis marked by hemosiderosis of renal tubules.

Patients with a congenitally bicuspid valve who die suddenly usually have a severely calcified stenosed valve. The degree of stenosis observed in elderly patients who have little physical activity may be remarkable.

Quadricuspid Aortic Valve

The quadricuspid aortic valve is a rare anomaly. Cases of calcific aortic stenosis affecting them have been reported, if not instances of sudden death associated with the change.

Other Congenital Conditions Associated with Calcific Aortic Stenosis

Materials may be deposited in cusp and annular tissue as a result of a genetic abnormality. They, with subsequent calcification, induce aortic stenosis. For example, I have seen cases of sudden death where one patient had calcific aortic stenosis associated with Fabry's disease, and another had gouty deposits in cuspal and annular tissue. Rarely, type II hyperlipidemia, glycogen storage disease, and ochrononis may all be associated with calcific aortic stenosis, with the latter a potential cause of sudden death. However, I have not seen such cases.

Acquired Lesions Associated with Aortic Stenosis

Some of these have been discussed above. A stenosed bicuspid valve is usually distinguishable from one calcified and stenosed following rheumatic valvular disease by a lack of commissural fusion. The latter often produces a triangular stenosed valve orifice (Figure 73A), although acquired bicuspidization is possible. Women are often affected, and the mitral valve may show rheumatic damage. Calcification, mainly at the bases of a valve with three cusps and not associated with marked cusp sclerosis or commissural fusion, helps differentiate aortic stenosis resulting from degenerative changes (Figure 73B). In addition, the latter cases usually occur in elderly individuals, predominantly men, who are in their sixties or seventies.

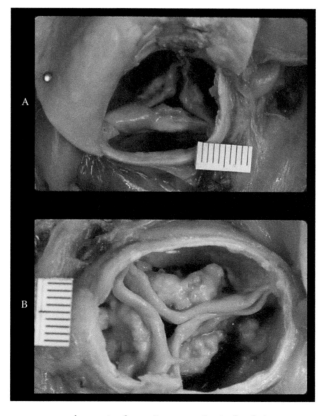

Figure 73 Common causes of acquired aortic stenosis, in both instances associated with the patient's sudden death. (A) Rheumatic valvular disease. Note cusp sclerosis and commissural fusion that produced a triangular stenosed orifice. (B) Degenerative calcification in the valve from a 75-year-old man. Note lack of commissural fusion and calcium deposits toward the base of the cusps, leaving the free margin relatively unaffected.

Acute Aortic Incompetence Causing Sudden Death

The chord-like variant of a raphe in a congenitally bicuspid valve (described above) may rupture, allowing the conjoint cusp to prolapse.

Many of the lesions inducing acute cusp perforation, ulceration, or prolapse of native, tissue, or mechanical heart valve prostheses were described above. Calcium nodules may fray and perforate a cusp. Cusps are also perforated both by direct penetrating trauma and by back pressure in crushing injuries, where an intraaortic pressure wave or water-hammer stress is created, rupturing them at their bases (Lobo and Heggtveit, 2001). An intimal tear in the ascending aorta can involve an aortic valve commissure and release it. Similar effects occur when the commissural attachments of a tissue valve separate from their stent. Nowadays, most of these conditions are recognized clinically and dealt with surgically, but all have the potential to cause sudden death.

7

Sudden Cardiac Death in Infants and Children

MEREDITH M. SILVER

Contents

Introduction

Childhood is commonly accepted as the first two decades of life. Its stages are defined as follows:

Infant	**Birth to 2 years**
Newborn	Birth to 1 month
Young infant	1 month to 1 year
Older infant	1 to 2 years
Child	**2 to 20 years**
Young child	2 years to puberty
Adolescent	Puberty to 20 years

Before discussing the incidence and causes of unexpected cardiovascular sudden death (SD) in childhood, I will review those pertaining to all causes of natural SD that affect two distinct age groups: newborn and infants to the age of 1 year, and older infants and children. In subsequent discussion, the second category will often be further subdivided into pre- and postpubescent children. Indeed, adolescent children share the same cardiovascular problems as young adults until the age of 35 or 40, beyond which coronary atherosclerosis becomes the most common finding associated with SD. Because of improved survival in subjects with treated congenital heart disease and because many congenital syndromes associated with SD are seen in adults, I will review those childhood causes that also act in young adult life.

A cause of natural SD in children, as in adults, is virtually always based in the heart, lungs, or brain. Those associated with the heart are the most prevalent, excepting only in young infants in whom sudden infant death syndrome is, in industrialized countries, the most common pattern. An overlap exists between the three organ systems important in SD, heart, lungs, brain, and the mechanism of death. Common chronic conditions such as epilepsy and asthma may well involve a terminal cardiac dysrhythmia. The same mechanism would explain SD in most cases of congenital heart disease and in the great variety of structural and functional heart diseases that are the focus of this chapter.

The time interval chosen as "sudden" between 1 h and 1 day has been questioned. For example, children found dead in their beds having seemed normal several hours previously are classed as cases of SD. The younger the subject, the more likely is this scenario. Some authors exclude any case of dying after admission to hospital, even if the death occurs within 24 h of initial collapse, because hospital treatment alters the natural course of the disease (Wren et al., 2000).

Choosing a time interval as long as 24 h between initial collapse and death means, in infants and young children who are totally dependent on caregivers, that a high proportion of natural SD cases will be due to acute infections, asthma, and central nervous system diseases. The mode of death in such a case is a lot less sudden than in a child observed to drop dead while playing at school or in an adolescent who dies suddenly while actively engaging in a sport. Hence, the papers cited below (that use the 24 h interval unless stated otherwise) give prominence to relatively protracted deaths associated with infection or, to put it another way, make us realize that instantaneous, presumptively dysrhythmic, SD is truly rare in childhood.

Sudden Death between Birth and One Year

Sudden death occurring in childhood is most common in infants less than 1 year old. Until recently and in developed countries, sudden infant death syndrome (SIDS) or crib death accounted for approximately 3 deaths per 1000 live births. Lately that incidence has fallen with the general adoption of a supine infant sleeping position. During the first year of life, and specifically from the age of 1 to 6 months, SIDS is by far the most common pattern of SD, accounting, in Canadian studies, for approximately 80% of cases (Czegledy-Nagy et al., 1993; Côtè et al., 1999). The next most common pattern after SIDS is infection, which is more common than cardiovascular causes (Norman et al., 1990; Côtè et al., 1999). The most common cardiovascular cause of SD in this age group is congenital heart disease, but other anatomic and functional heart diseases can, in rare instances, cause SD in this age range (Table 44).

Sudden Infant Death Syndrome (see also Chapter 8)

Although numerous heart, lung, and brain conditions as well as local and systemic infections and toxemia have been named to cause SIDS, as well as unnatural causes (smothering, poisoning), its etiology remains unknown. Even after excluding all causes of SIDS-like death, it is highly doubtful if SIDS is a single entity. A pathologist classifies a case of natural death as SIDS when a previously well infant dies while sleeping and no cause of death is found through a comprehensive investigation of the death scene and a detailed autopsy that includes microbiologic studies and screening for drugs and inborn metabolic errors. How rigorously these conditions are fulfilled varies a good deal from one jurisdiction to another. Many are beyond the control of the pathologist who does the autopsy. Also, the age range accepted for a SIDS death varies with local forensic guidelines and individual opinions held by pathologists. Finding another cause of death at autopsy precludes a diagnosis of SIDS, but the pathologist may and commonly does find one or more minor abnormalities, such as a respiratory infection. Also, a small proportion of these patients are born prematurely (Czegledy-Nagy et al., 1993). As investigative methods improve, several distinct

Table 44 Findings in Cardiac Sudden Death between Birth and One Year of Age

Anatomic
Congenital heart disease, especially ductus-dependent kind, with or without surgery
Diseases of coronary arteries and ostia, congenital and acquired
Myocardial diseases, congenital and acquired
Acute myocarditis
Dilated cardiomyopathy with endocardial fibroelastosis
Metabolic cardiomyopathies
Cardiac tumors and hamartomas

Functional
Congenital long Q–T syndrome

Conduction System
Congenital complete heart block due to maternal antiphospholipid antibodies

diseases have been found to mimic a SIDS death. For example, inborn metabolic errors such as medium-chain coenzyme A dehydrogenase deficiency, discussed below, may cause a SIDS-like death in infancy. This becomes more likely when SIDS repeats within a sibship (Vockley, 1994). Another condition that frequently mimics SIDS is an enlarged heart with endocardial fibroelastosis. We have seen several such cases in our autopsy service and in infant hearts referred from other centers. The presumptive anatomic diagnosis with such findings would be dilated cardiomyopathy, and because this condition may be familial, the pathologist should suggest investigations in the relatives. Such cardiac findings are sometimes either overlooked at autopsy or not considered significant, and the case is labeled as SIDS. We know this because, having made the diagnosis of dilated cardiomyopathy in a new hospital autopsy case, we then reviewed the referred autopsy findings in a previously born sibling said to have died of SIDS and found an enlarged heart with biventricular endocardial fibroelastosis (Figure 74).

Because the cause of a SIDS death is not known, one cannot even be sure that it is natural. An adequate drug screen should uncover deliberate or accidental poisoning. Homicide by smothering may be suspected by officials at the scene, but this will be neither confirmed nor denied by autopsy. Adequate microbiologic studies should, after correlation with gross and microscopic findings, reveal an overwhelming infection. The term "SIDS" was introduced to spare parental anguish and guilt but has the disadvantage of implying that it is a single disease entity rather than a classification for deaths of unknown cause occurring in apparently healthy infants. Parenthetically, the old terms "crib death" or "cot death" had the advantage of admitting complete ignorance of the cause of death. Whether or not the majority of SIDS cases are due to a specific respiratory (or central nervous system, or cardiac) developmental problem remains unproven. A heritable dysrhythmic heart condition, the long Q–T syndrome, is commonly cited to cause SIDS or a SIDS-like death and will be discussed below.

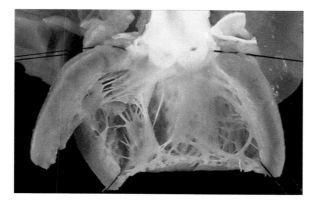

Figure 74 A 3-month-old infant was found dead in his crib. The heart was three times heavier than expected, and the left ventricle, seen here opened through the outflow tract, was dilated and lined by a milky layer of endocardial fibroelastosis. These cardiac findings exclude a diagnosis of SIDS but did not explain the sudden death. A female sibling previously autopsied in this department was stillborn and hydropic and had similar although milder gross cardiac findings, suggesting a familial dilated cardiomyopathy.

Sudden Death between One Year and Adult Life

This age range, starting at 1 year and thereby excluding SIDS cases, has been chosen by authors in many studies, the upper age limit being either 19, 20, or 21 years. After the age of 1 year and up to adult life, SD is very rare, a hundred times rarer than in early infancy. A large epidemiologic study from the U.S. gave an incidence of 4.6 per 100,000 population per year (Neuspiel and Kuller, 1985). In a U.K. study, the SD rate was 3.3 per 100,000 of population per year (Wren et al., 2000). From a study in northern Spain in which natural SD patients died less than 6 h from onset of symptoms, a mortality rate of 1.7 per 100,000 population per year was found (Morentin et al., 2000). Once unnatural deaths were excluded, these three studies divided cases according to whether the subject had a previous recognized disorder such as epilepsy, a chronic heart disease, or asthma, which collectively make up a large proportion of total SD cases in childhood. For example, in the U.K. study, half had a known chronic disease including epilepsy, cardiac disease, or asthma (Wren et al., 2000). Apart from a high proportion of cardiovascular causes, including both congenital and acquired forms, the most common causes of death were infections, epilepsy, cerebral diseases, and asthma. Males consistently outnumber females in these deaths, the rate for boys being at least twice as high as that for girls (Molander, 1982; Driscoll and Edwards, 1985; Kitada et al., 1990; Morentin et al., 2000). In a study of high school students participating in sports at the time of their SD, the male-to-female ratio was 5:1 (Van Camp et al., 1995).

Neuspiel and Kuller (1985) reported that the rate was highest in the second year of life, partly due to greater vulnerability to infections at that age; it declined to a nadir from the ages of 4 to 14 years and then rose moderately until adult life. Those authors suggested that the group of unexplained deaths in older infants and young children might represent an etiologic continuum with SIDS. However, I reemphasize that SIDS may not be a single disease. Inherited defects of mitochondrial fatty acid beta-oxidation, such as medium-chain acyl coenzyme A dehydrogenase deficiency, discussed below, that cause SIDS-like deaths in infancy, also cause Reye's-syndrome-like death in older infants and young children and tend to recur within a sibship (Vockley, 1994). These inborn errors of metabolism affect tissue metabolism, as evidenced by fatty changes seen histologically in liver, myocardium, and proximal renal tubules (Iafolla et al., 1994).

Other reports do not take the entire population into account but review consecutive autopsies on SD victims and, thereby, calculate the proportion attributable to various causes. A 6-year study of children between 1 and 20 years of age included 31 cases of SD, half due to infection and 7 (23%) to a cardiac disease (Molander, 1982). Keeling et al. (1989) found 169 SD cases in persons aged 2 to 20 collected over a 20-year period; 92 (54%) had recognized disorders, usually congenital heart disease, asthma, or epilepsy, and deaths in the remaining 77 children were most often due to infection.

All these studies contain a group of SD cases in which no cause was found at autopsy and the heart was anatomically normal (unexplained death or presumed natural SD cases) — the incidence varied from 6.5% (Keeling et al., 1989) to 32% (Morentin et al., 2000). A sudden disturbance of cardiac rhythm would be considered likely in many cases, especially if the death was observed and associated with emotional stress or physical exertion. Among a group of schoolchildren aged 5 to 19 years, unexplained cases were thought likely to be cardiac, based on the circumstances surrounding the death, and were equal in number (19 of 64: 29.7%) to those attributed to underlying anatomic heart disease

Table 45 Findings in Cardiac Sudden Death between One Year of Age and Adult Life

Anatomic

Congenital heart disease with or without surgical palliation or repair
Diseases of coronary arteries and ostia
    Congenital anomalies in origin and initial course
    Acquired coronary ostial/arterial diseases with stenosis or thrombosis
Diseases of proximal great vessels and heart valves, congenital and acquired
    Aortic dissecting aneurysm
Myocardial diseases, congenital and acquired
    Myocarditis, acute and healing
    Cardiomyopathies, familial and unexplained
    Metabolic cardiomyopathies
Cardiac tumors and hamartomas

Functional

Primary disorders of heart rhythm

Conduction System

AV conduction block and sinus node dysfunction, congenital and acquired

(Kitada et al., 1990). A retrospective diagnosis of long Q–T syndrome is possible by reviewing an electrocardiogram of a dead subject or of relatives, as happened in 2 of 41 previously unexplained cases in one study (Wren et al., 2000).

Cardiac malformations vary greatly in severity, and only some are significantly associated with SD. One may presume that its mechanism is the same in a child with known congenital heart disease as that in another with an unsuspected or undiagnosed cardiac malformation. One may further presume that the mechanism is a cardiac arrest, starting either as asystole following bradycardia or as ventricular fibrillation. One of these dysrhythmias would be the final common pathway for many rapidly fatal conditions that start in the head (epilepsy, intracranial hemorrhage), the respiratory system (bronchopneumonia, acute epiglottitis), or the circulatory system (septicemia, toxemia). A child with epilepsy may die during an observed seizure or be found dead in bed and an autopsy may not demonstrate a cause of death; SD would be linked to epilepsy, even though the mechanism of the SD may not be understood. Some electrophysiologic and morphologic studies suggest that the mechanism in epilepsy may be due to a sudden cardiac dysrhythmia (Natelson et al., 1998; Nei et al., 2000; Opeskin et al., 2000).

Summarizing reported studies of SD in this age range of childhood, one can say it is very rare and that cardiovascular findings, which are of great variety (Table 45), make up approximately one third of the total. Compared to their incidence in early infancy, cardiovascular SD cases in older infants and children make up a much larger proportion of total natural SD cases.

Classification of Cardiovascular Findings Associated with Sudden Death in Children

It is conventional to divide congenital and acquired conditions considered to cause cardiovascular SD into those associated with cardiac or vascular pathology (anatomic or structural changes) and those with an essentially normal heart and vessels when examined at autopsy (functional or dysrhythmic disorder). As we will see, this division is arbitrary. The various anatomic and functional factors in infants and childhood that will be discussed below are listed in Table 44 and Table 45. Abnormal pathologic findings in the cardiac conduction system occupy a gray area between the two groups. That is because it is the exception rather than the rule to include conduction system morphology in a pathologic examination of the heart, many pathologists do not yet fully understand the normal structure and development of that system, and abnormal changes in the conduction system may be secondary rather than primary. Of course, many pathologists have reported interesting morphologic findings in the conducting system in cases of SD (see Chapter 5, section entitled "Conduction System" and Chapters 8 and 9).

Anatomic Causes

Anatomic heart diseases that cause SD in these age groups as well as in young adults include both congenital and acquired lesions. These paragraphs include some general comments with specific lesions subsequently discussed in greater detail.

Congenital cardiac diseases, whether treated medically or following surgical palliation or correction, are associated with SD in all age groups. As early surgical repair becomes more common and successful, SD occurring two or more decades after surgical correction has become relatively common.

In newborns, severe kinds of cardiac malformation may be lethal, but the classification of such deaths will depend on the circumstances. A newborn infant with ductus-dependent congenital heart disease will sometimes die at home or in an emergency room, because new parents do not appreciate that the infant is very ill: let us call the patient in such a case "infant A." Other newborns diagnosed with congenital heart disease are likely to be in a hospital specializing in pediatric cardiology and be treated early with palliative and corrective cardiac surgery. A newborn with the same pattern of disease as infant A may well die suddenly in the hospital: our hypothetical "infant B." This death would not be classed as unexpected, and the pathologist's anatomic findings, perhaps identical to those for infant A, would be directed to confirming or denying cardiologic findings rather than explaining the cause of the SD.

Isolated anomalies of the coronary ostia or proximal coronary arteries are associated with SD in all age groups, with the younger deaths associated with more serious anomalies. For example, origin of the left coronary artery from the pulmonary artery leads to death by myocardial infarction in early infancy, whereas acute angulation of the ostium or ectopic origin from the wrong aortic sinus are more likely in cases of SD in children beyond infancy and in young adults. Such SD is often associated with physical exertion, possibly because this leads to increased myocardial oxygen demand or adrenergic stress (see Chapter 10).

After coronary artery anomalies, especially in adolescents and young adults, exercise-related SD is caused most often by either ruptured aortic aneurysm or hypertrophic cardiomyopathy. The latter is rare in childhood: in an epidemiologic study covering over 8 million people in the 1- to 20-year-old age group, the incidence of death from this condition was 1 in 1.35 million (Wren et al., 2000). Even less common is SD from arrhythmogenic right ventricular cardiomyopathy. This event, as well as affecting adolescent males in particular, is also exercise or stress related. Sudden death from either hypertrophic or right ventricular cardiomyopathy is not clearly related to exercise until after pubescent growth of body and heart muscle plateaus. These two types of familial cardiomyopathy are also associated with SD in infants and young children, although at a much lower rate and affecting both sexes. Nevertheless, exercise-related SD also occurs in childhood, which is, for most of us, the most physically active stage of life.

Myocarditis is the most common acquired heart disease causing SD in this age group (infants, children, young adults). Although attributed to a viral infection, identification of a causative virus is often lacking. Deaths in this condition usually follow a short or more prolonged clinical illness with heart failure. Deaths that occur before the illness is overt are attributed to a cardiac dysrhythmia. Sudden death can also occur in the healing phase, when the heart condition can evolve into a dilated cardiomyopathy. Acute Kawasaki coronary arteritis may cause death in infants. Older children can die suddenly from ruptured coronary artery aneurysms that are likely the result of healed Kawasaki disease. Other types of myocardial lesions, such as tumors and hamartomas, are very rare causes of SD in childhood.

All of the above cardiovascular conditions may be associated with a SD, or with death that appears sudden due to failure to recognize clinical illness in a young infant. The more likely mode of death in all of the above anatomic conditions would be heart failure leading to an anticipated death.

Functional Causes

An example of a functional cause in the newborn is congenital complete heart block, resulting most often from maternal antiphospholipid syndrome but which may be part of complex congenital cardiac malformation. Other types of dysrhythmia may be lethal in the very young, for example, sustained supraventricular tachycardia, chaotic atrial rhythm, and junctional ectopic rhythm, but deaths in these conditions would not likely be sudden or unexpected, because the infants would have severe heart failure. Congenital long Q–T syndrome is recognized in the newborn and is said to be highly lethal at this age.

In general, functional disorders in children are rare and include the two genetic forms of the long Q–T syndrome, familial polymorphic ventricular fibrillation and Wolff-Parkinson-White syndrome complicated by atrial fibrillation. Complete heart block is a possible basis for SD but is less likely than in the very young and is more likely to be iatrogenic than congenital.

Sudden Death from Congenital Heart Disease

The determination of sudden death from congenital heart disease constituted an important group in Toronto studies (Thornback and Fowler, 1975; Harrison et al., 1996; Oechslin et al., 2000), being the most common mode of death (26%) among 2609 consecutive adult cases followed at this center (Oechslin et al., 2000). However, the figures are biased by case selection, because children treated for congenital heart disease at the Hospital for Sick Children are followed as adults in the Toronto Congenital Cardiac Centre. A population-based view is given in two U.K. studies where all autopsied childhood (2 to 21 years of age) cases of SD during a time period were reviewed retrospectively. During a 20-year period in Oxford and Southern Oxfordshire, 26 of 169 patients aged 2 to 20 years had SD attributed to congenital heart disease; in only one instance was the latter not diagnosed before autopsy (Keeling and Knowles, 1985). Another similar 10-year study from the U.K. Northern Health Region revealed that 31 of 229 SD cases with patients aged 1 to 21 years had congenital heart disease, and in 9 of these patients, this was not known until autopsy (Wren et al., 2000). A review that tabulated 469 SD cases from nine population studies (Liberthson, 1996) reported that congenital heart disease (usually aortic stenosis) caused SD in 63 cases (10%); the numbers in each tabulated study varied widely, and 4 had no cases of congenital heart disease, possibly due to case selection.

The types of congenital heart disease associated with SD have changed dramatically during recent decades in which surgery evolved. Improved methods of medical control, more aggressive surgery, early surgical correction rather than palliation, implantation of defibrillators, and radiofrequency ablation of dysrhythmic foci are all helping to improve survival rates.

Congenital Heart Disease in Newborn and Young Infants

Severe cardiac malformations are likely to cause early death and in some such cases it will be sudden. Mortality is highest in newborns, especially in the first few days of life, and is due to the ductus arteriosus closing in cases of ductus-dependent malformations, such as hypoplastic left heart syndrome, critical aortic stenosis, complete transposition of great vessels, and aortic coarctation. Mortality and the incidence of SD goes down with survival into early infancy. In our institute, cardiac malformations caused less than 10% of SD in young infants (Czegledy-Nagy et al., 1993; Silver, 1998); similar numbers are reported from other Canadian centers (Norman et al., 1990; Côtè et al., 1999).

Those with congenital heart disease that suffer SD in early infancy usually present following an observed cardiovascular collapse preceded by minimal signs of illness, for example, problems with feeding. In ductus-dependent lesions, such as hypoplastic left heart syndrome, it is easy to understand circulatory collapse and death in newborns. At this age, it is highly unlikely for SIDS to cause SD. However, in supposedly well infants, we have seen SD occur at several weeks of age resulting from critical aortic stenosis or severe aortic coarctation. Hence, the age range overlaps with that of SIDS (Figure 75). For example, SD reported in four infants was attributed to total anomalous pulmonary venous connection; two of the four were found dead in their cribs, thus resembling SIDS cases (Byard and Moore, 1991). Newborns and infants with total anomalous pulmonary venous connection appear to be prone to SD: 8 of 52 (16%) autopsy cases from a two-center study presented in this manner despite an absence of significant antemortem symptoms and signs (James et al., 1994). Ventricular septal defect caused SD in four infants between the ages of 1 week and 3.5 months — all had cardiomegaly (Cohle et al., 1999). A 3-month-old infant died suddenly, and a subarterial ventricular septal defect was found at autopsy (Byard et al., 1990a). Sudden death is reported in children with unsuspected or known Ebstein anomaly, but infants with a severe form of this condition tend to die in the first week of life of severe cyanotic congestive heart failure (Gentles et al., 1992).

Cardiac malformations in the newborn are becoming rare in some countries, because diagnoses are now being made in early fetal life. Sometimes termination of the pregnancy follows these diagnoses. Most newborns with prenatally diagnosed malformations are treated at birth and often receive early surgical palliation or corrective surgery. Parenthetically,

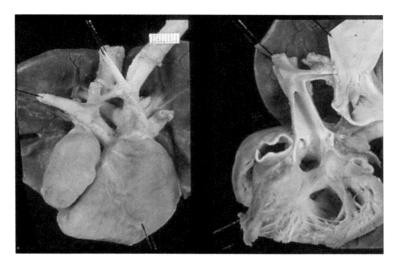

Figure 75 A 10-week-old infant was found dead in his crib; SIDS was suspected. The heart weighed twice the expected weight and was both globally dilated and hypertrophied. An obstruction lay at the aortic isthmus, seen here in both closed (left) and open views. In the latter, the left common carotid artery is retracted upwards and to the left to show the narrowest part of the aorta just proximal to its descending segment (also retracted). The main pulmonary artery is transected at its root and was opened distal to its bifurcation through the ductus arteriosus into the descending aorta. The latter corresponds to the grooved segment adjacent to the aortic isthmus. The grooved appearance of the ductus is caused by medial contraction.

congestive heart failure is a far more likely mode of presentation in such an infant than is SD. Again parenthetically, sudden dysrhythmic deaths are common in subjects with congestive heart failure at any age, but such deaths would be expected.

Congenital Heart Disease in Children and Young Adults

Among children with cardiac malformations, the reported incidence of SD depends on the population studied. For example, a registry for hearts referred for pathologic study by coroners and pathologists reported isolated abnormalities of coronary artery origin or takeoff as the predominant cause of SD in children; the authors saw virtually no cases of cardiac malformation and, in fact, used ventricular septal defect cases as their control group (Steinberger et al., 1996). In contrast, another study done by reviewing death certificates of children who died suddenly between 1985 and 1994 in one English health region found that 28 of 33 SD cases had known congenital heart disease; the single case of anomalous origin of a coronary artery in this study was discovered at autopsy (Wren et al., 2000).

In my experience at a tertiary pediatric referral center, congenital heart disease cases (usually late postoperative cases) comprised at least one quarter of childhood SD cases. It should be remembered, however, that deaths in congenital heart disease subjects at any age are far more likely to be expected and protracted rather than sudden. My autopsy experience is that most childhood subjects with congenital heart disease and SD had received corrective cardiac surgery several years previously and, subsequently, were followed as patients at cardiology clinics. Such children appeared healthy and well grown and were physically active.

Genetic and Chromosomal Associations of Congenital Heart Disease

In the past, it was thought that most of these cases were sporadic rather than genetic. A well-known exception was the autosomal dominant Holt-Oram syndrome, in which subjects have abnormal radial ray bones ("fingerized thumb") as well as cardiac malformations, usually an atrial septal defect (Basson et al., 1994). Complete heart block is reported in this syndrome (Newbury-Ecob et al., 1996). Other autosomal dominant syndromes, with or without dysmorphism, link a secundum atrial septal defect with conducting tissue disease and the possibility of SD (Mandorla and Martino, 1998).

With increasing frequency, single gene defects are recognized to cause certain forms of cardiac malformations. In most affected families, transmission is autosomal dominant (Payne et al., 1995). This is important because SD may be expected to affect family members. Examples include Ebstein anomaly, atrial septal defect, atrioventricular septal defect, aortic valve stenosis/bicuspid aortic valve, and total anomalous pulmonary venous connection. Tetralogy of Fallot, the most common cardiac malformation in adults, is often associated with a deletion in the long arm of chromosome 22; these subjects may have familial or sporadic cardiac malformation and minor facial dysmorphism (Amati et al., 1995). Monosomy of chromosome 22 is thought to cause 5% of surgically treated cardiac malformations, including aortic arch and conotruncal defects (Payne et al., 1995).

Cardiac malformation has a known association with abnormal chromosomes. For example, an atrioventricular septal defect or ventricular septal defect is often seen in Down syndrome (trisomy 21) and also occurs as an autosomal dominant familial defect (Payne et al., 1995). Cardiac malformation is observed in many dysmorphic syndromes caused by chromosomal or single gene defects, for example, aortic coarctation in Turner's syndrome (monosomy X) and supravalvular aortic stenosis in Williams syndrome (elastin gene defect). Dysmorphic syndromes that have a known association with SD will be mentioned below in relation to the specific cardiac malformations with which they are associated.

Late Sudden Death after Surgical Correction

SD occurring two or more decades after surgical treatment of, for example, left ventricular outflow tract obstruction, transposition of the great vessels, and tetralogy of Fallot, accounts for approximately one third of SDs in persons less than 35 years of age (Basso et al., 1995). A study of 41 cases from a follow-up group of 3589 postsurgical patients with a variety of cardiac defects found that the risk of late SD appeared to be time dependent and increased incrementally 20 years after operation for tetralogy of Fallot, aortic stenosis, or coarctation (Silka et al., 1998). The risk among patients in this follow-up group of 41 cases was 25 to 100 times greater than that in an age-matched control population, with this increased risk primarily related to those with cyanotic or left heart obstructive lesions. Sudden death was attributed to arrhythmia in 30 of the 41 patients, circulatory (embolic or aneurysm rupture) in 7, and acute heart failure in 4.

Paroxysmal atrial flutter is a common and dangerous finding among children surviving surgical repair or palliation of an anomaly. Of 380 consecutive patients with atrial flutter, 10 died suddenly in one multicenter study (Garson et al., 1985). Supraventricular tachycardia, usually atrial flutter, seemed the most dangerous arrhythmia in repaired tetralogy of Fallot and in both the Mustard and Senning atrial baffle operations for transposition of the great vessels. Hence, the type of arrhythmia that underlies SD in these cases may differ from that seen in adults dying suddenly from coronary artery disease. Nevertheless, electrocardiographic studies of children with aortic stenosis point to ventricular arrhythmias that presumably start on the basis of myocardial ischemia (Wolfe et al., 1993; Yilmaz et al., 2000).

In the following discussion, I will deal with the natural history of the most common forms of congenital heart disease associated with SD and, where pertinent, with SD occurring late after surgical repair of that defect.

Tetralogy of Fallot

The natural history of tetralogy depends on the severity of right ventricular outflow tract obstruction (Figure 76). Those with mild obstruction show little or no cyanosis in childhood, the so-called pink tetralogy, but most present with cyanosis and have significant obstruction and consequently a right to left shunt through the ventricular septal defect; systemic to pulmonary arterial anastomoses bring more blood to the undersupplied lungs. Palliative or reconstructive surgery at this stage allows prolonged life.

We have seen SD in children and teenagers many years after a successful repair of tetralogy of Fallot; these patients had both right and left bundle branch block. Such subjects are more likely to die during sporting activities than during sleep. One study found a significant association of SD with older age at the time of surgical correction (Johnsson et al., 1995). The mechanism is controversial. Based on morphologic examination of the heart and conduction system, ventricular arrhythmia was considered more likely than atrioventricular block (Deanfield et al., 1983). Gillette and Garson (1992) favored sudden complete heart block. Retrospective analysis in another study showed a significant association with complete heart block that persisted beyond the third postoperative day (Hokanson et al., 2001). Agonal bradycardia due to advanced atrioventricular block was documented in 2 of 8 patients who died suddenly late after surgical correction; both had transient postoperative block and residual bifascicular block (Silka et al., 1998). One study reported that late ventricular tachycardia and

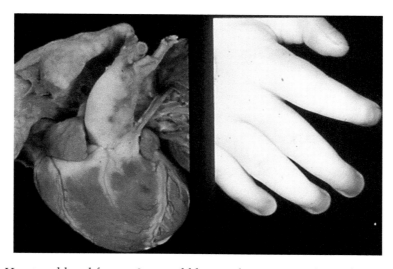

Figure 76 Heart and hand from a 2-year-old boy with severe tetralogy of Fallot, who died in the hospital soon after his first presentation with severe cyanotic congestive heart failure. In the heart, note the enlarged right ventricle giving off a diminutive main pulmonary artery, seen to the right of the aorta. Pulmonary blood supply was through large collaterals anastomosing spinal with pulmonary arteries. Note also the clubbed and cyanosed fingertips.

SD were associated with pulmonary valve regurgitation, whereas atrial flutter or fibrillation was associated with tricuspid valve regurgitation (Gatzoulis et al., 2000). Therrien et al. (2001) reported that replacement of regurgitant pulmonary valves in 70 patients together with intraoperative cryoablation of arrhythmic foci in some led to stabilization of the electrocardiographic QRS interval and a reduction in atrial and ventricular arrhythmias.

Surgical techniques have both changed and been refined during the several decades during which surgical correction of tetralogy has been performed. For example, Dietl et al. (1994) found that cases in which the right ventricular approach was employed had more dysfunction and pulmonary valve regurgitation than in those in whom a right atrial approach was used, the latter having a significantly reduced risk of ventricular arrhythmia. Also, right ventricular scarring after repair via a ventriculotomy approach is a source of postoperative arrhythmias (Garson et al., 1983).

Transposition of the Great Vessels

Transposition of the great vessels is incompatible with life unless a shunt exists or is created surgically to allow blood to cross from the pulmonary to the systemic circulation. Hence, the natural history of the lesion is for death in a newborn when the ductus arteriosus closes. Atrial baffle procedures (Mustard, Senning, and others) brought the pulmonary blood into the right ventricle, supporting the systemic circulation, and made the left side of the heart support the pulmonary circulation, thus allowing prolonged survival.

At this center, SD was the most frequent cause of late deaths among 534 patients followed up to 30 years after undergoing a Mustard atrial baffle procedure (Gelatt et al., 1997; see also Figure 77). Also in Toronto, of 86 consecutive adult patients followed an average of 23 years, nearly half had supraventricular tachycardia, usually atrial flutter, and 2 died suddenly (Puley et al., 1999). Supraventricular tachycardia has been documented in other studies of patients with either Mustard or Senning atrial inflow correction procedures and appears to be a significant risk factor for late SD (Birnie et al., 1998). Of six SD subjects dying late after the Mustard operation, five succumbed during active physical exertion, and polymorphic ventricular tachycardia or ventricular fibrillation was documented in four (Silka et al., 1998). Radiofrequency ablation has been used to treat intraatrial reentry tachycardias following Mustard and Senning procedures (Van Hare et al., 1996). Pacemakers may prevent dysrhythmic deaths caused by the extensive intra-atrial damage induced surgically (Hayes and Gersony, 1986; Puley et al., 1999). Such atrial inflow corrective operations have now been replaced by arterial switch procedures (Hayes et al., 1986; Planche et al., 1988). Thus, late postoperation dysrhythmic SD should become less common. When coronary arteries are properly reimplanted, the danger of late SD is reduced. One early arterial switch operation that involved single-orifice reimplantation of both coronary artery ostia was rapidly abandoned because coronary artery obstruction occurred in 11 of 35 patients (Bonhoeffer et al., 1997).

Sudden Death in Left Ventricular Outflow Tract Obstruction

Obstruction of the left ventricular outflow can occur proximal to the aortic valve (subaortic), at the valvular level, distal to the valve (supravalvular), or in the aorta (coarctation). Late SD following surgery is reported in some of these lesions.

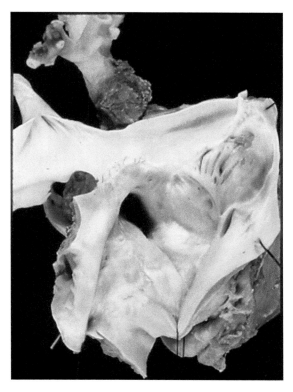

Figure 77   This 8-year-old-boy died suddenly 2 years after undergoing a Mustard atrial baffle operation for transposition of the great vessels. In this view of the right atrium, note four pulmonary veins with suture lines still visible at the anastomoses. In cases that survive longer, the atrial lining becomes more sclerotic. I have seen suture granulomas in the region of the atrioventricular node in similar cases. The conduction system was not examined in this case because no dysrhythmias were documented.

Subaortic Stenosis

A distinct fibroelastic membranous shelf in the outflow tract that is presumed to be congenital may cause subvalvular stenosis (see Chapter 6). We saw an example in an 18-year-old woman who suddenly died immediately postpartum. In the newborn, the stenosis often affects the outflow tract diffusely and is associated with aortic valvular stenosis. As in the latter condition, symptoms tend to progress rapidly (Kugelmeier et al., 1982). In the adult, subvalvular stenosis is often due to diffuse endocardial fibrosis affecting the outflow tract and is only slowly progressive. Also among adults, the condition may be an acquired deformity rather than a malformation, because it occurs in cases of repaired congenital heart disease (Oliver et al., 2001) or where there is a preexisting abnormality of the mitral valve apparatus or the interventricular septum (Lampros and Cobanoglu, 1998). Sudden death is reported occasionally in the adult disease (Trinchero et al., 1988). A familial occurrence has been noted (Onat et al., 1984; Petsas et al., 1998).

   Muscular stenosis at this level is, of course, typical of hypertrophic cardiomyopathy (see below).

Aortic Valvular Stenosis Including Late Sudden Death after Surgery

Before surgical treatment of left ventricular outflow tract obstruction became common, congenital heart disease was the leading association with SD in children, and the most common lesions obstructed the left ventricular outflow tract (Thornback and Fowler, 1975). Typically, it was caused by stenosis at the valvular level and associated with SD during physical exertion, the mechanism being a lethal dysrhythmia probably triggered by acute myocardial ischemia in the hypertrophied left ventricle (Lambert et al., 1974). Since surgical correction has become the norm, studies that mention mortality are mainly concerned with the timing of surgery relative to the patient's age and the degree of stenosis. Severe stenosis (critical aortic stenosis clinically) is highly lethal in the newborn (Turley et al., 1990). Even when asymptomatic, the stenosis is rapidly progressive in infants (Anard and Mehta, 1997), but deaths in newborns and infants with aortic stenosis, with or without surgery, would rarely be unexpected. Nevertheless, we have seen cases of both unsuspected aortic stenosis and coarctation at autopsy, the preautopsy diagnosis being SIDS (Figure 75).

Exercise testing is done in aortic stenosis subjects of any age to evaluate ST segment depression and the development of symptoms, ventricular arrhythmia, or an inadequate rise of systolic blood pressure during exercise. Among 66 consecutive asymptomatic patients with severe isolated aortic stenosis, 4 that had a positive exercise test subsequently died suddenly (Amato et al., 2001). A study of 40 asymptomatic children with mild or moderate aortic stenosis found that they had a significantly longer Q–T interval than healthy children, as well as increased heart rate during treadmill exercise, the latter being interpreted as a sign of early myocardial ischemia and indicating vulnerability to SD (Yilmaz et al., 2000). An electrocardiographic study of subjects with cardiac malformations revealed that serious ventricular arrhythmias were most frequently noted in patients with aortic stenosis; they also had the highest incidence of SD (Wolfe et al., 1993).

An autopsy examination of 182 young (less than 35 years) cardiovascular SD cases revealed that 2 had isolated bicuspid aortic valve, and 3 had bicuspid aortic valve as well as aortic isthmic coarctation (Basso et al., 1995). Aortic valvular stenosis is significantly associated with Turner's syndrome, as are a bicuspid aortic valve, the infantile type of aortic coarctation (tubular hypoplasia of the aortic isthmus), and total anomalous pulmonary venous connection. Cardiac malformations vary with the genotype (Moore et al., 1990; Mazzanti et al., 1998; Prandstraller et al., 1999). Sudden death in Turner cases is, however, usually due to a ruptured dissecting aortic aneurysm (see below) rather than left ventricular outflow tract obstruction.

Among 27 patients who had an aortic valvotomy when that operation was new, 3 late cases of SD occurred (Stewart et al., 1978). They were significantly associated with postoperative left bundle conduction defects (Thomas et al., 1982). Mortality in surgically treated aortic stenosis, including arrhythmic deaths, declined rapidly with improved surgical techniques, cardiologic methods used to define the need for surgery, and better anesthetic techniques. A 1980 study indicated that SD risk was minimal and irreversible myocardial damage was unlikely if the indications for operation were moderate or severe aortic stenosis, together with restrictive symptoms or ST and T wave changes on the electrocardiogram (Hossack et al., 1980). One case of late SD was reported among infants and children treated by palliative valvotomy, whereas four other late deaths were due to nondysrhythmic causes (Kugelmeier et al., 1982). In a surgical follow-up study of aortic stenosis in children who were managed medically or surgically, researchers found that more than half of the cardiac deaths were sudden and unexpected and correlated retro-spectively with an outflow gradient greater than 50 mm Hg (Keane et al., 1993). Residual valvular stenosis or insufficiency was present in 3 of 4 patients dying suddenly due to cardiac arrhythmia late after repair (Silka et al., 1998).

Supravalvular Aortic Stenosis Including Williams Syndrome

Supravalvular aortic stenosis is seen in Williams syndrome. The latter may be familial (autosomal dominant) or sporadic and results from mutations in the elastin gene on the long arm of chromosome 7 (Keating, 1995). It may be associated with SD and coronary artery stenosis and severe biventricular outflow tract obstruction (Bird et al., 1996). Mor-phologically, the medial dysplasia at the sinotubular junction region of the aorta may be discrete or diffuse and, especially in the latter variant, extend into the proximal coronary arteries, causing stenosis and myocardial ischemia (van Son et al., 1994).

Supravalvular aortic stenosis may also be familial (autosomal dominant) or sporadic but nonsyndromic. In these instances, the dysplasia found in the aortic root is similar to that seen in Williams syndrome (O'Connor et al., 1985). The pathology of the diffuse variant is described in children who did not have Williams syndrome, one of the four children had a brother die suddenly (Vaideeswar et al., 2001). The stenosis may be asso-ciated with dysplastic aortic valve cusps that isolate one or other coronary ostium with such subjects prone to SD during exercise (Thiene and Ho, 1986; Debich et al., 1989; Sun et al., 1992). A similar isolation of the left coronary artery ostium in a dysplastic aortic root caused SD in adolescent males (Kurosawa et al., 1981; Debich et al., 1989). Other complications of supravalvular aortic stenosis that might lead to SD include severe coro-nary artery stenosis (van Son et al., 1994) and aortic aneurysm (Beitzke et al., 1986).

We saw this form of stenosing aortopathy in an infant who died with heart failure and in a 2-month-old male with Williams syndrome and a SIDS-like death. I observed a similar lesion causing heart failure and hydrops in a midgestation fetus, and a similar lesion was reported in a stillborn (Cagle et al., 1985).

Aortic Coarctation

The two types of aortic coarctation, infantile and adult, refer to different morphology of the obstruction. The adult type, which may present at any age, is caused by a shelf-like invagination of media from the posterior juxtaductal aortic wall into the lumen; intimal thickening from a jet lesion just distal to the lesion adds to the stenosis (Glancy et al., 1983). The infantile form of coarctation is a tubular hypoplasia of the aortic isthmus as seen in Figure 75. In newborns, coarctation of either type is rarely an isolated malforma-tion, and prognosis often depends on the associated malformations (Barbero-Marcial et al., 1982; Kopf et al., 1986). Apart from unsuspected instances causing SD in newborns and young infants (Figure 75), undiagnosed coarctation may be linked with SD in child-hood or adult life from the complications of chronic hypertension, including dissecting aortic aneurysm and ruptured berry aneurysm.

Late after repair of isolated aortic coarctation, SD is rather rare, affecting only 2 of 138 patients in one report (Sorland et al., 1980). Dissecting aneurysm of the aorta was proven as the main cause of late SD in an earlier study (Forfang et al., 1979), but ischemic heart disease is also likely in adult patients because of chronic hypertension (Toro-Salazar et al.,

2002). Coronary artery disease caused death in 32 of 645 patients, with SD being the second most common type of late death in that study (Cohen et al., 1989). Arrhythmic SD occurring late after aortic coarctation repair was associated with severe ventricular hypertrophy in 12 subjects (Silka et al., 1998).

Repair of an aortic coarctation using an orthoptic conduit may be complicated by late dehiscence at the suture line or false aneurysm formation, both with a risk of late SD from massive hemorrhage (Emmrich et al., 1982). Aortoplasty with a Dacron patch also carries a risk of aneurysm formation that may rupture to cause SD, particularly in pregnant women (Parks et al., 1995).

Ebstein Anomaly of the Tricuspid Valve

An arrhythmia is the most common initial presentation of Ebstein anomaly in adolescents and adults. The arrhythmias documented in all age groups include right bundle branch block, supraventricular tachycardia, and Wolff-Parkinson-White (WPW) syndrome (Tuzcu et al., 1989). Among 45 adolescents and young adults with the anomaly and an arrythmia, the latter was supraventricular in 24, atrial flutter or fibrillation in 12, WPW syndrome in 8, and ventricular tachycardia in 1 (Chauvaud et al., 2001). The association of Ebstein's anomaly with dysrhythmias is discussed further below.

Newborns and infants with Ebstein's anomaly are likely to die of cyanotic congestive heart failure, whereas children and adult subjects, whether treated medically or surgically, are prone to SD (Tuzcu et al., 1989; Celermajer et al., 1994; Attie et al., 2000; Chauvaud et al., 2001).

Sudden death is reported after surgical repair of Ebstein tricuspid valve especially in those with preoperative or perioperative dysrhythmias (Oh et al., 1985; Tuzcu et al., 1989; Chauvaud et al., 2001). Because SD correlated with increased heart size and atrial fibrillation in their study, Gentles et al. (1992) proposed that right atrial dilation and hypertrophy might explain the atrial dysrhythmias in such subjects.

Other Surgically Treated Congenital Heart Diseases

A report of 380 children and young adults between 1 and 25 years of age indicated that the most common association of atrial flutter was repaired congenital heart disease; 10% of the 380 subsequently died suddenly (Garson et al., 1985). An electrographic study found that those with ventricular septal defect had serious arrhythmias (multiform premature ventricular contractions, ventricular couplets, and ventricular tachycardia) and a tendency for SD, and only those with aortic stenosis had a higher frequency of these complications (Wolfe et al, 1993).

Among the various septal defects, if repair is delayed until the right ventricle shows dilation and hypertrophy from chronic volume overload, the latter might potentiate dysrhythmias. The ventriculotomy approach to close a ventricular septal defect has been abandoned because right ventricular scarring was thought to promote arrhythmias (Garson et al., 1983).

Surgical series dealing with various forms of congenital heart disease contain occasional cases of late SD. The most common defects, apart from those described above, are ventricular, atrial, and atrioventricular septal defects, in that order. This may reflect the relative

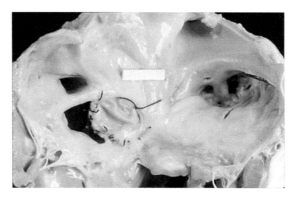

Figure 78 This 18-year-old male had a ventricular septal defect repaired 5 years earlier through a ventriculotomy approach. He died suddenly while at school. No electrocardiograms were available, but he had a recent history of syncope. In this view of the right ventricular inflow tract, a 2 cm nodular area is seen below the centimeter marker, deep to the septal leaflet of the tricuspid valve and just to the left of the coronary sinus. Histologically, this area presented suture granulomas. Grossly, it occupies the apex of Koch's triangle (Todaro tendon above, coronary sinus to the left, tricuspid valve base below) used to localize the atrioventricular node prior to histologic study.

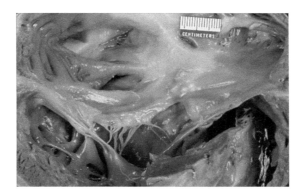

Figure 79 A ring of sutures marks the patch closure of a septum primum atrial septal defect. As in an atrioventricular (AV) septal defect (AV canal, endocardial cushion defect), the AV conduction system would be congenitally distorted (see also Figure 98). The heart is from a 12-year-old girl who died suddenly 3 months after operation, done before electrophysiological mapping of the conduction system became the norm.

frequency of these conditions and of their surgical correction rather than any inherent tendency for such surgical repairs to cause late arrhythmic SD. However, the proximity of septal defects to the conduction system and its vulnerability to surgical damage may underlie dysrhythmic deaths occurring after surgical repair. Surgical damage to the conduction system is discussed below. Figure 78 shows a repaired ventricular septal defect, and Figure 79 shows a repaired primum-type atrial septal defect.

   Complex cardiac malformations that require multiple staged operations have an associated high mortality. This would include occasional cases of SD from either arrhythmia or heart block (Razzouk et al., 1992; Najm et al., 1997; Gates et al., 1997; Jahangiri et al., 2001), but such deaths would rarely be deemed unexpected.

An actuarial study of patients who had heart valves replaced revealed that SD occurring either early or late following the procedure was the second most common mode of death, after heart failure (Blackstone and Kirklin, 1985). Late SD is rare after implantation of mechanical heart valves in children and may be the result of their complex malformations (Yamak et al., 1995; Lubiszewska et al., 1999).

Insertion of Blalock-Taussig shunts to palliate obstructed blood flow to the lungs, such as in tetralogy of Fallot, may be followed by SD not associated with blockage of the shunt (Fermanis et al., 1992). The use of transcatheter intracardiac devices also carries a risk of SD, 1.2% among 777 patients. This risk was higher in those with more severe underlying heart disease or multiple devices (Perry et al., 2000). Because intravascular stents are used in patients with severe malformations, it is not surprising that SD occurs occasionally when they are inserted, 1 in 85 patients in one study (O'Laughlin et al., 1993).

Eisenmenger's Syndrome and Pulmonary Hypertension

Whereas cardiac dysrhythmia would account for the majority of SD cases in congenital heart disease, suprasystemic pulmonary pressure can also be an important cause. When the Eisenmenger's syndrome supervenes with a left to right shunt via a septal defect or patent ductus arteriosus, it is a harbinger of death. However, most patients survive until early adult life (Vongpatanasin et al., 1998). In a study of SD in young adults following repair of ventricular septal defect, 9 of 23 had this shunt reversal with reactive pulmonary hypertension (Harrison et al., 1996). Sudden death is common in subjects with the Eisenmenger's reaction (Daliento et al., 1998). It is most often due to pulmonary hemorrhage or a ruptured great vessel and is typically nontachyarrhythmic (Niwa et al., 1999). It is difficult to conceive how such a SD would be classed as unexpected, except perhaps in young infants in whom malformation was unsuspected prior to autopsy, or when SIDS was suspected (Cohle et al., 1999; Byard et al., 1990a). Eisenmenger's reaction is also seen in truncus arteriosus and the univentricular heart. Pulmonary hypertension, whether secondary to malformation or unexplained, is a risk factor for SD, especially during cardiac catheterization (Fuster et al., 1984). We observed a child who died suddenly during cardiac catheterization. She had a known atrial septal defect and partial anomalous pulmonary venous connection, the latter first identified during the catheterization; severe pulmonary hypertensive arteriopathy was demonstrated at autopsy (Figure 80).

Pathology of the Heart in Cases of Sudden Death Associated with Congenital Heart Disease

A pathologist performing an autopsy in a case of SD must look for previously unsuspected cardiac malformation. Should the latter be found, it must be decided whether it caused the SD. A history of syncope will be important in this decision, as well as electrocardiographic records and family history of cardiac malformation or SD. A literature search may show case reports linking the malformation to SD, but such a link must be considered anecdotal. In cases where death was observed, such as a child dropping dead while playing at school, or if ventricular fibrillation was documented during resuscitation, then a dysrhythmic death would not be doubted. Finding a previously unsuspected ventricular septal

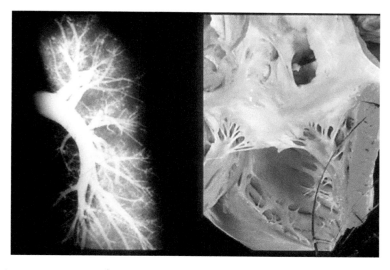

Figure 80 A postmortem pulmonary angiogram (left) shows marked dilation of proximal pulmonary arteries with peripheral pruning, consistent with severe pulmonary hypertension during life. The opened heart, left inflow view, from this 6-year-old girl shows a secundum-type atrial septal defect. She died suddenly during cardiac catheterization. Partial anomalous pulmonary venous connection was discovered during that examination. I attributed her sudden death to unsuspected pulmonary hypertension. Similar postmortem pulmonary angiograms are found in patients dying with Eisenmenger's reaction or unexplained pulmonary hypertension

defect in a normal-sized heart at autopsy would not explain a dysrhythmic death. If the conduction system were competently examined histologically in such a case, perhaps abnormalities would be found that might explain the SD. Looking into the family history and instigating a search for a familial dysrhythmic heart condition, such as the long Q–T syndrome, should be done even if, for example, a ventricular septal defect is found at autopsy. In addition to documenting anatomic and surgical lesions and correlating them with clinical findings during autopsy, a pathologist will seek morphologic evidence of an anatomic substrate for any suspected arrhythmia. For example, a finding of ventricular hypertrophy should correlate with outflow obstruction for that ventricle and evidence of recent or old myocardial ischemia in the hypertrophied myocardium. Sometimes one finds scarring and stitch granulomas in the region of the sinus atrial node (for example, in a Mustard repair for transposition of the great vessels) that would seem to explain a clinical sick sinus syndrome.

Sudden Death Associated with Coronary Artery Disease in Children

The most common malformation in this category is an anomalous origin or course of the subepicardial coronary arteries. All other lesions discussed in this section are rare.

Anomalous Origin or Course of Proximal Coronary Arteries

Coronary artery anomalies may be minor or major. Minor anomalies such as variations in the number and location of ostia usually have no significance. Major anomalies of ostia or proximal coronary arteries, on the other hand, may cause heart failure or SD via a

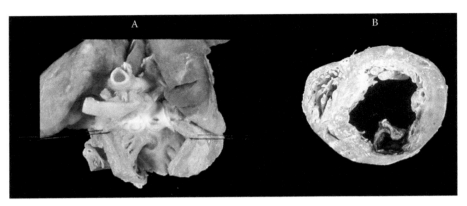

Figure 81 (A) This female infant, 5 months old, was brought to the hospital with asystolic cardiac arrest. Her right coronary artery arose from the sinus related to the right pulmonary cusp, seen above the opened muscular pulmonary infundibulum. (B) A transverse section of the ventricles from the case described in Figure 80 showing a recent anterolateral infarct in the left ventricle. The white dots and streaks in the myocardium were caused by barium injected into the right coronary artery (which arose normally from the aorta) at postmortem coronary angiography. Some white subendocardial areas represent myocardial scars.

dysrhythmic mechanism initiated by acute myocardial ischemia. The more serious the anomaly, for example, the left coronary artery arising from the pulmonary artery, the more likely death will occur at an early age; autopsy will usually show previous and recent myocardial necrosis in such cases. The most lethal anomaly of all, killing 60% of those with the condition before 2 weeks of age, is for both coronary arteries to arise from the pulmonary artery (Heifetz et al., 1986). Anomalous origin of the left coronary artery usually causes death in the first year of life, whereas subjects with a right coronary artery arising from the aorta usually survive longer (Figure 81). If diagnosed early, surgical relocation of an anomalous artery is life saving (Kakou Guikahue et al., 1988).

Among young infants, SD initially thought to be SIDS has been attributed to slit-like or stenotic coronary artery ostia and acute angulation of takeoff or ectopic origin from the aorta (Lipsett et al., 1991; Steinberger et al., 1996). Such autopsy findings can easily be overlooked unless careful examination of the proximal coronary tree is done. If an acute myocardial infarction was present, attributing a SD to such an anomaly would be acceptable. Anomalies such as ectopic origin from the aorta may be lethal in later infancy, childhood, or adult life. In many cases of SD in children and young adults with anomalous coronary arteries, it was recorded that they died during or after exertion (Mahowald et al., 1986; Land et al., 1994; Ohshima et al., 1996; Garfia et al., 1997). In larger series that included children, SD and exercise-related deaths were most common when the left main coronary artery arose from the right coronary sinus or the left main coronary artery from the pulmonary artery (Taylor et al., 1992a; Frescura et al., 1998).

Absence of the proximal right coronary artery and severe hypoplasia of that vessel have also been described in childhood SD (Byard et al., 1991b; Moore and Byard, 1992; Ohshima et al., 1996). One boy with a single (absent proximal right) coronary artery who died during a period of physical and mental excitement had a mother and a grandmother who died suddenly (Ohshima et al., 1996).

Analysis of 7857 pediatric autopsy cases from four tertiary centers gave an 0.5% incidence of anomalous coronary arteries, with ectopic origin from the aorta being most frequent (43%), followed by origin from the pulmonary trunk (40%); the mean age at death was 2.2 years (range 4 h to 14 years) (Lipsett et al., 1994). In an examination of hearts referred to the authors because of sudden out-of-hospital death of previously healthy subjects, 22 cases aged between birth and 21 years had anomalous coronary arteries (Steinberger et al., 1996). The most common coronary artery finding was a combination of ectopic origin from the aorta with acute angulation at origin (10 cases), followed by ectopic origin alone (6 cases) and acute angulation alone (5 cases), and ostial stenosis in a single case. Another review that included children (150 consecutive cases of SD within 6 h of the onset of symptoms) revealed 8 males and 4 females, age range 2 to 35 years, mean 24.2, who had congenital coronary artery anomalies: a deep intramyocardial course in six, origin from the wrong sinus in three, and ostial obstructions in three (Corrado et al., 1992). Frescura et al. (1998) reported an isolated anomalous origin of coronary arteries in 27 of 1200 specimens (2.2%) including the left coronary artery from the pulmonary trunk in 5, origin from the wrong aortic sinus in 12 (both coronary arteries from the right sinus in 4, from the left sinus in 7, left coronary artery from the posterior sinus in 1), left circumflex branch from right aortic sinus or from the proximal right coronary artery in 3, high takeoff of the right coronary artery in 3, and stenosis of the coronary ostia attributable to a valve-like ridge in 4. In that study, 16 cases of SD, with this occurring in all cases of left coronary artery origin from the right aortic sinus, were precipitated by exercise. Often, SD was the initial presentation (Frescura et al., 1998).

A deep intramural course of the proximal left main coronary artery also appears to predispose one to SD through the mechanism of myocardial ischemia, especially during exercise (Burke et al., 1991a). Among 39 hearts, each with an intramural course of the left anterior descending coronary artery, 22 had demonstrated myocardial ischemic lesions in the distribution of that artery. Thirteen of the 22 hearts were from victims of SD, with 6 of the 13 having died during vigorous exercise; hence, a deep intramural arterial course was considered abnormal rather than a simple anatomic variant (Morales et al., 1993).

Coronary ostial stenosis is sometimes associated with an anomalous origin or course of coronary arteries. Angiographic study indicated that two of five children with coronary ostial stenosis had an aberrant left coronary artery arising from the contralateral aortic sinus (Jureidini et al., 2000). Ostial stenosis may also be an isolated congenital anomaly (Amaral et al., 2000).

Although coronary artery anomalies are often isolated, they also occur in cases of congenital heart disease. An echocardiographic study of 62,320 children revealed that isolated anomalies of the origin/initial course of coronary arteries were present in 14, another 8 had anomalies of the coronary arteries plus 4 had tetralogy of Fallot, 1 had truncus arteriosus, and 3 had transposition of the great vessels (Werner et al., 2001). That study also identified four coronary fistulas. The latter are usually associated with cardiac malformations, for example, in pulmonary atresia with intact septum. An angiographic study of 119 cases of tetralogy of Fallot found that 11 children had anomalous coronary arteries (Dabizzi et al., 1980). Another reported that the incidence of abnormal coronary arteries in congenital malformed hearts, especially tetralogy of Fallot, was 2% or more, whereas such anomalies were an incidental finding in 0.3 to 0.8% of conventional coronary angiography (Felmeden et al., 2000). In a pediatric autopsy study, anomalous coronary arteries were associated with other anomalies in 57% of cases, 43% of these being cardiac malformations (Lipsett et al., 1994).

Fibromuscular Dysplasia

Although this stenosing vascular lesion typically affects renal vessels and sometimes carotid and iliac arteries, a morphologically similar lesion occurs in the aorta and coronary arteries and, in some cases, in both coronary and renal arteries. The term is used for a nonathero-sclerotic, noninflammatory, stenosing arterial disease predominantly affecting the media, with disorganized hyperplasia of medial smooth muscle cells. The condition was classified morphologically in renal arteries and distinguished from intimal fibroplasia and perarterial fibroplasias (Harrison and McCormack, 1971). Whether the lesion reported in coronary arteries is the same disease as in renal vessels is not known. Furthermore, whether the morphologic subtypes described in the kidney also occur in the heart is uncertain. The literature concerning this coronary vascular lesion consists largely of case reports but has the advantage of being based on autopsy findings. For example, a 3-month-old male who died suddenly with a myocardial infarct due to fibromuscular dysplasia in the left anterior descending coronary artery also had widespread, fibromuscular dysplasia in the aorta and its large branches (Imanura et al., 1997). Three infant siblings with fibromuscular dysplasia in the aorta, its main branches, and the main coronary arteries had SIDS-like deaths due to myocardial infarction (Dominguez et al., 1988). Both of these case reports are unusual in the patients' age, because most subjects die in adolescence or young adult life. Also, familial occurrence is uncommon, although it has been reported (Burke et al., 1993).

One wonders whether an overlap exists between fibromuscular dysplasia affecting the aorta and its proximal branches including main coronary arteries and the stenosing aorto-pathy in the diffuse form of supravalvular aortic stenosis found in Williams syndrome. Young age of onset and familial occurrence are usual in that syndrome. Severe stenosis of all proximal coronary arteries was described in four children with Williams syndrome: the lesion involved all three layers of the arterial wall with intimal hyperplasia, fibrosis, and disorganization; disruption and loss of the internal elastic membrane; indistinct intimal-medial junction; medial hypertrophy and dysplasia; and adventitial fibroelastosis (van Son et al., 1994).

In our autopsy service, we had just such a problem with this differential diagnosis. Sudden death in a previously healthy 13-month-old boy was associated with severe medial thickening at the roots of both the aorta and main pulmonary artery. Subsequently, a 3-month-old male sibling died just after completion of cardiac catheterization, done because of the history in his brother; heart block developed during the procedure. Autopsy revealed similar stenoses of both proximal great vessels. The media of the ascending aorta and the main pulmonary artery were approximately twice as thick as normal for the child's age, with many more layers of smooth muscle, collagen and elastic, but medial histologic organization appeared normal, and the coronary arteries were not involved.

The papers cited hereafter concern older children and young adults. In three reported cases, fibromuscular dysplasia was confined to intramural coronary arteries (Veinot et al., 2002). Other cases had the lesion restricted to the arteries supplying either the sinus or atrioventricular nodes (Anderson et al., 1981a; Burke et al., 1993; Zack et al., 1996; Jing and Hu, 1997; Michaud et al., 2001) or even confined to the branch supplying the bundle of His (James and Riddick, 1990). A histologic study of face sections from serial blocks of the conduction system in 381 cases of SD in patients up to 40 years of age found that seven had dysplasia of the atrioventricular node artery (Cohle et al., 2002). Another morphologic study of the conduction system in 27 cases of SD and 17 control subjects who died from trauma reported fibromuscular dysplasia with acid mucopolysaccharide deposition in the atrioventricular node artery of 12 of the SD group and in only one control heart (Burke et al., 1993). Morphometrically in that study, the atrioventricular node arteries in 10 SD subjects were significantly narrower than the control value; 5 of the 10 cases died during exercise (Burke et al., 1993).

Fibromuscular dysplasia of the atrioventricular nodal artery was also observed in cases of prolapsed mitral valve (Burke et al., 1997b; Veinot et al., 2002). Burke and colleagues found by morphometry that arterial luminal narrowing and myocardial fibrosis at the base of the interventricular septum were significantly greater in mitral valve prolapse cases than in control hearts and proposed that these lesions contribute to SD in mitral valve prolapse subjects (Burke and Virmani, 1998). Because 4 of the 16 control hearts in this study showed the same lesion, the latter must sometimes be an incidental finding, also evident from a large morphologic review (James, 1990). Its cause is unknown but it appears, from the above, that the morphologic lesion may have numerous causes.

Intimal fibroplasia affecting coronary arteries may be akin to that described in renal arteries by Harrison and McCormack (1971). In the former vessels, intimal stenosing fibroelastosis due to localized high pressure is seen, for example, in congenital heart disease, such as a newborn with pulmonary stenosis and intact interventricular septum. We would classify this as a form of obliterative endarteritis, but it is termed "fibromuscular dysplasia" in some reports.

Intimal fibroplasia with myofibroblasts that stenosed both coronary arteries and the thoracic aorta and its branches was reported to cause SD in a 3-month-old male (Maresi et al., 2001a). A 10-year-old boy who died suddenly when playing had stenosing intimal smooth muscle cell hyperplasia of all main coronary arteries, renal arteries, and a branch of the pulmonary artery (Siegel and Dunton, 1991).

Coronary artery intimal hyperplasia was reported in the extensively disarrayed and fibrotic left ventricular myocardium of two adolescents with right ventricular cardiomyopathy (Smith et al., 1999). Focal intramural fibromuscular dysplasia occurs within the disarrayed septal myocardium in cases of hypertrophic cardiomyopathy (Veinot et al., 2002). Sudden death in four adolescent and young adult cases (associated with exercise in three) was attributed to this severe change in small intramural coronary arteries (Burke and Virmani, 1998); it is reassuring to read that hypertrophic cardiomyopathy was not present in these four subjects.

Segmental mediolytic arterial disease is described in epicardial coronary arteries in neonates as well as in older children and young adults. It is attributed to vasospasm; such lesions might heal and produce either stenosis or aneurysm and on healing be a possible precursor to fibromuscular dysplasia (Slavin et al., 1995).

Early Onset Atherosclerosis

Accelerated atherosclerosis due to certain chronic diseases may severely stenose coronary arteries and cause myocardial ischemia during childhood. In our service, we saw single cases of SD, at ages between 12 and 17 years, from myocardial ischemia and infarction in severe diabetes mellitus, in beta thalassemia, and in progeria. Also, a male aged 5 years died suddenly with a myocardial infarction. The severe coronary artery stenosis caused by foam cells filling the intima proved to be due to familial hypercholesterolemia; he also had xanthomata at the ankles (Figure 82). Hyperlipidemic inborn metabolic errors are associated

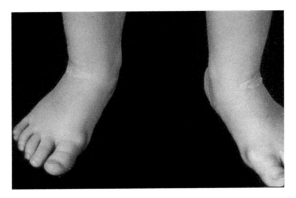

Figure 82   A 5-year-old boy died suddenly while playing. His heart showed a recent myocardial infarct. The coronary tree was severely stenosed throughout by intimal foam cell accumulation that proved to be due to familial hypercholesterolemia. Note the dermal xanthomata around the ankles observed at autopsy.

with acquired stenosis, due to foam cell accumulation in the intima and at the root of the aorta and the major coronary arteries, particularly at their ostia; but death in homozygotes usually occurs during early adult life rather than during childhood (Allen et al., 1980; Haitas et al., 1990; Yamamoto et al., 1989). Rapid onset of atherosclerosis is also seen in coronary arteries damaged by radiation, for example, mediastinal radiation for Hodgkin's disease in childhood (Reinders et al., 1999). An 18-year-old woman, affected by Hodgkin's disease and treated successfully with radiotherapy, died suddenly; autopsy showed an acute septal myocardial infarction and severe focal atherosclerotic lesion of the anterior descending coronary artery (Angelini et al., 1985). Accelerated atherosclerosis also affects transplanted hearts in children (Pucci et al., 1990; Pahl et al., 1994). The consequences of myocardial ischemia and increased risk of SD are like that seen in coronary atherosclerosis of adult life. The latter is reported in young adults and even in older children (Corrado et al., 1988).

Coronary Artery Calcinosis

Idiopathic infantile arterial calcification is a rare disorder of undetermined etiology, characterized by calcific deposits along the internal elastic membrane of large, medium, and small arteries. Intimal fibrous proliferation occurs, and the arterial lumen is narrowed. The constellation of affected organs varies, but coronary arteries are almost always involved, whereas those of the central nervous system are spared. Death from myocardial infarction usually occurs within the first 6 months of life, but occasional cases survive to adulthood (Marrott et al., 1984). Two infants who died suddenly and unexpectedly were found to have idiopathic arterial calcification with widespread fibrointimal proliferation of elastic and muscular arteries and characteristic calcification of the internal elastic laminae (Byard, 1996).

Kawasaki Disease

In the acute phase, usually seen in young infants, Kawasaki disease presents as an acute coronary panarteritis (infantile polyarteritis nodosa) with severe stenosis of all major coronary arteries and focal myocardial infarction. The morphology is similar to acute polyarteritis nodosa, a disease that typically affects young adult males and sometimes involves the subepicardial coronary arteries, causing SD by myocardial ischemia (Swalwell et al., 1991). Since the incidence of rheumatic heart disease has declined in North America, Kawasaki disease is now the leading cause of acquired heart disease in young children.

We reported SD due to ventricular fibrillation affecting a 2-month-old male a few hours after admission to the hospital with a febrile illness (Byard et al., 1991b). We have also seen giant aneurysm formation and death from myocardial infarction in a very young infant, as others have noted (Rowley et al., 1987; Avner et al., 1989). Sudden death from cardiac tamponade due to rupture of an aneurysm was reported in a 2-month-old male (Maresi et al., 2001b).

It is much more likely that SD in Kawasaki disease will occur in the healed stage of the disease. Then, aneurysms formed in the acute stage may enlarge, and progressive luminal occlusion by mural thrombus may cause myocardial ischemia or infarction, likely precipitated by exercise. We saw three such cases in adolescents: two males who were exercising when they died and a female found dead in bed. One of the males died when riding a bicycle. He had no history of an acute febrile illness in infancy. Many giant aneurysms present along epicardial coronary arteries were extensively occluded by old thrombus. His myocardium showed previous and recent ischemic damage. A similar case in the literature includes pathologic description of the aneurysms seen at the healed stage of Kawasaki disease (Fineschi et al., 1999b). This presentation will likely become rare in the future, because the acute illness is now diagnosed early, and the patient is treated with gamma globulin. Also, coronary aneurysms are treated surgically. Other patients will recover from their myocardial infarction and become candidates either for coronary artery bypass grafting (Mavroudis et al., 1996) or heart transplantation. Among nine cases of healed myocardial infarction in children 2 to 17 years old, a history of Kawasaki disease (27%) was second only to anomalous coronary artery (35%) as the cause of SD (Celermajer et al., 1991).

A survey of 74 cases of presumed late Kawasaki disease in adolescent children and young adults included 16% who presented with SD, the remainder had chest pain/myocardial infarction (61%) or arrhythmia (11%), and in the majority, symptoms were precipitated by exercise; 100% had coronary aneurysms (Burns et al., 1996).

Takayasu's Disease

Takayasu's disease is a rare cause of death in childhood, usually affecting females in the adolescent and young adult age range. We reported a case in an 8-year-old girl. She died in ventricular fibrillation a day after cardiac catheterization and angiography, and her death was not unexpected, because she was in the hospital for investigation of a febrile illness and had severe acute heart failure (Chiasson et al, 1990). Autopsy showed a massive recent myocardial infarction due to diffuse fibrostenotic inflammation of all subepicardial coronary arteries; the process also involved the aorta and its main branches (Figure 83).

In other studies of childhood coronary artery Takayasu's disease, diffuse disease of major coronary vessels is rare. Usually the inflammation and scarring involve coronary ostia in direct continuity with aortic disease (Seguchi et al., 1990; Matsubara et al., 1992). Rarely, it is localized in a major coronary artery (Abad, 1995) or causes a coronary aneurysm (Matsubara et al., 1992). Sudden death from aortic regurgitation in Takayasu's disease is reported from Japan (Yajima et al., 1994).

Figure 83   Takayasu's disease of the aortic root here causes marked stenosis of the right coronary ostium. Note the marked aortic intimal thickening and adventitial lymphoid patches. The epicardial coronary arteries were similarly stenosed in this 8-year-old girl who died with ventricular fibrillation while in the hospital being investigated for a recent onset of heart failure. Recent circumferential myocardial infarction of the subendocardial left ventricle was demonstrated as well as scarring due to healed infarcts.

Miscellaneous Vasculitides

Tuberculous aortitis caused coronary ostial stenosis leading to SD in a 12-year-old girl (Chow et al., 1996). Giant cell arteritis associated with thrombosis in the left main coronary artery suddenly killed a 19-year-old female (Cohle et al., 1982).

Ostial Stenosis Due to Surgical Injury, Stents, and Catheters

Coronary arteries and particularly their ostia are vulnerable to iatrogenic damage by catheters and stents or to damage occurring during aortic valve replacement. This also occurs in children with congenital heart disease who have valvular dilation by stent or aortic prosthetic valve insertion.

   We saw a case of SD in an adolescent male. It was caused by coronary ostial stenosis late after removal of obstructive muscle bands from the pulmonary infundibulum during repair of a ventricular septal defect (Figure 84).

Coronary Artery Embolism

Occasional cases of SD in children are due to coronary artery embolism, with the thromboemboli arising on heart valves in both rheumatic fever and congenital heart disease (Stahl et al., 1995). We have seen a thromboembolus coming from a chorionic plate vein in the placenta kill a newborn. This would be termed "paradoxical embolism" in anyone older than a neonate. Others describe newborn cases in which the coronary thromboembolus apparently came from the ductus venosus or umbilical vein, sources that would be accepted only if origin in the placental chorionic plate was excluded. Thromboemboli in stillborn and neonates may also arise from a renal vein thrombus; one such case embolized to the coronary circulation and caused a myocardial infarction (Bernstein et al., 1986).

   As with cardiac myxoma in adults, fragments of a friable tumor growing on a heart valve or mural endocardium may occasionally embolize to coronary arteries and so cause myocardial ischemia/infarction with the risk of SD (Amr et al., 1991; McElhinney et al., 2001).

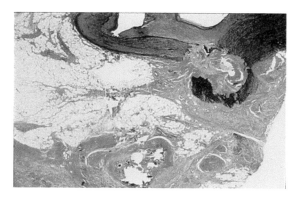

Figure 84   A 14-year-old male died while playing ice hockey. Grossly, the left ventricular outflow tract was extensively scarred due to a remote ventricular septal defect repair and removal of obstructive muscle bands from the right ventricular outflow tract. The left coronary ostia was stenosed, and the right was atretic. In this histologic section, remnants of the right coronary ostium containing old suture material are seen adjacent to the aorta (above). Other suture granulomas are seen in the pericardial fat.

Sudden Death in Diseases of Proximal Great Vessels and Heart Valves

Dissecting aneurysms of the main pulmonary artery occur in subjects with severe pulmonary hypertension and can rupture to induce SD. This was reported in a child with Eisenmenger's syndrome caused by left to right shunting (Walley et al., 1990). Apart from this mention, only dissecting aortic aneurysm will be considered here. Also, among causes of valvular heart disease, only the myxomatous mitral valve will be discussed. A dissection can cause SD by exsanguination. How the valvular lesion causes SD is unknown.

Nonsyndromic Causes of Dissecting Aortic Aneurysm

A predisposing factor for dissecting aneurysm in adults is systemic hypertension; SD from ruptured dissecting aneurysm is also reported in children with chronic renal hypertension (Vogt et al., 1999). Also, there is a known association of dissection with pregnancy (Konishi et al., 1980), with weight lifting (de Virgilio et al., 1990), and with both methamphetamine abuse (Davis and Swalwell, 1994) and crack cocaine abuse (Hsue et al., 2002). All of these associations are possible in older children. The common factor in all instances is the presence of cystic medial degeneration in the aorta, that is, fragmentation of medial elastic fibers. This process, of varying degree, occurs with normal aging (Trotter and Olsen, 1991) but is exaggerated in the conditions mentioned above. Possibly they act in the presence of a genetic predisposition.

An intimal tear has to occur to initiate the dissection. However, both the medial degeneration and intimal tear may be difficult to demonstrate at autopsy. Murray and Edwards (1973) and Silver (1997) discussed the range of pathology that may follow an intimal tear. A review of pathological findings in 204 cases of ruptured dissecting aneurysm identified the usual mechanism of dissection as an intimal tear, although rupture of an aortic branch artery or ruptured vasa vasorum was also demonstrated (Wilson and Hutchins, 1982). Hypertension, Marfan syndrome, and less often traumatic, atherosclerotic, or inflammatory injuries to the aortic media were the factors associated with dissection in that study (Wilson and Hutchins, 1982). However, Silver (personal communication) has the opinion that dissection associated with atherosclerosis occurs only when an intravenous catheter or device accidentally burrows into an ulcerated plaque. Dissecting aortic aneurysm is also noted in adolescents with aortic fibromuscular dysplasia (Gatalica et al., 1992) and in Williams syndrome with supravalvular aortic stenosis (van Son et al., 1994), whereas saccular aortic aneurysm was reported in a male child with familial supravalvular aortic stenosis (Beitzke et al., 1986).

Aortic Dissecting Aneurysm in Marfan Syndrome and Familial Non-Marfan Syndrome

Sudden death due to ruptured dissecting aneurysm in Marfan syndrome is most likely to happen after the third decade, but occasional cases occur in childhood (Tsang et al., 1994). When a death occurs spontaneously in a previously healthy, normally developed young person, usually male and associated with exertion, one thinks of a *forme fruste*, where the disease is expressed only in the aorta (Emanuel et al., 1977). The syndrome varies greatly in severity in different organ systems (Payne et al., 1995), but a family history will help decide if a case is truly one of Marfan syndrome, an autosomal dominant familial disease that affects the cardiovascular system, skeleton, and eye.

The syndrome is caused by a defective fibrillin gene on the long arm of chromosome 15. Fibrillin forms a major component of extracellular microfibrils that cross-link the elastic of arterial media. At least 50 mutations of the fibrillin gene are known (Strauss and Johnson, 1996). Each family usually has a different mutation and varying phenotypes (Payne et al., 1995). Subjects are liable to annuloaortic ectasia due to cystic medial degeneration leading to aortic dissection and rupture with death in adult life; in a high proportion of these cases, death will be sudden. Rupture of an aortic dissecting aneurysm is the primary cause of mortality in Marfan syndrome and is effectively prevented by aortic root replacement (Gott et al., 1996; Westaby, 1999). Aortic elastic fragmentation tends to be severe in these subjects, including *forme fruste* cases, and also in familial dissecting aneurysm subjects (Klima et al., 1983; Trotter and Olsen, 1991). Nevertheless, SD from ruptured aortic dissecting aneurysm will likely occur in young adults rather than in children. Marfan subjects followed by echocardiography had severe dilatation of the aortic root, which usually precedes rupture, in patients as young as 20 years (Hwa et al., 1993).

At least two loci for familial dissecting aortic aneurysm without Marfan syndrome were mapped to chromosomes 5 and 3 (Guo et al., 2001; Milewicz et al., 1998). These patients tend to be older than Marfan subjects at presentation and usually exhibit autosomal dominant inheritance, although both sex-linked and recessive patterns are reported (Coady et al., 1999). Among 93 adults with cystic medionecrosis but not Marfan syndrome, more than half of the 34 deaths were from a ruptured dissecting aneurysm (Marsalese et al., 1990). Familial nonsyndromic aortic dissecting aneurysm is also reported to cause SD in adolescent children (Nicod et al., 1989; Toyama et al., 1989).

We autopsied an 18-year-old policeman-in-training, a large, muscular, and nondysmorphic individual who died of a ruptured dissecting aneurysm. He had no family history of SD. As well as an intimal tear in the ascending aorta, the aorta showed severe cystic medial degeneration. The only dissecting aneurysm we have seen in a 4-month-old infant with Marfan syndrome occurred in the ductus arteriosus (Gillan et al., 1984).

Aortic Dissecting Aneurysm in Turner's, Noonan's, and Ehlers-Danlos Syndromes

Dissecting aortic aneurysm may also kill female subjects with Turner's syndrome who develop aortic root dilatation and cystic medial degeneration in the ascending aorta proximal to an aortic coarctation. Some patients with Turner's syndrome will also have a bicuspid aortic valve that may be stenosed or show a variety of other cardiac malformations, such as anomalous pulmonary venous drainage; in individual cases, malformations vary with the genotype (Nora et al., 1970; Mazzanti et al., 1988, 1998; Moore et al., 1990; Gotzsche et al., 1994; Lin et al., 1998; Sybert, 1998; Prandstraller et al., 1999). Ruptured dissecting aneurysm is reported in subjects with Turner's syndrome without aortic coarctation but with cystic medial degeneration at autopsy (Price and Wilson, 1983; Goldberg et al., 1984; Lin et al., 1986). Among 244 karyotype-proven subjects with Turner's syndrome studied by echocardiography, 3 had aortic dissecting aneurysms; the author noted 42 previously reported cases of aortic dissection in those with Turner's syndrome, all but 5 had coarctation, bicuspid aortic valve, or hypertension (Sybert, 1998).

We autopsied an 18-year-old trainee nurse of short stature who died suddenly of a ruptured aortic aneurysm; the typical Turner's genotype (X monosomy — which precludes fertility) was demonstrated postmortem (Figure 85). Many other genotypes as well as mosaicism are associated with the phenotypic syndrome (Mazzanti et al., 1988; Gicquel et al., 1992; Gotzsche et al., 1994), and many patients with Turner's syndrome can become pregnant. Pregnancy in such women is associated with an increased risk of ruptured dissecting aneurysm, even during adolescence (Lin et al., 1998).

Noonan's syndrome, an autosomal dominant familial disease that also occurs sporadically, has phenotypic similarities to Turner's syndrome, but subjects have a normal karyotype. Heart abnormalities usually affect the right ventricular outflow tract rather than the left (Van der Hauwaert et al., 1978; Burch et al., 1993; Ishizawa et al., 1996). In affected subjects, a small proportion has aortic coarctation as well as other types of cardiac malformation (Marino et al., 1995). Ruptured dissecting aortic aneurysm may occur with this syndrome and also with the LEOPARD syndrome (multiple lentiginosis), but aneurysm rupture rarely happens before adult life. Sudden death in Noonan's patients is more likely associated with hypertrophic cardiomyopathy, discussed below, than with dissecting aneurysm (Ishizawa et al., 1996).

In Ehlers-Danlos syndrome type IV, an abnormal type III procollagen (chromosome 2) forms a major component of the arterial wall, and childhood death may occur from rupture of an aneurysm on a major artery, rupture of an aortic aneurysm or its dissection, or a variety of other vascular catastrophes (Wimmer et al., 1996; Pepin et al., 2000). In this autopsy service, an infant thought to have died of SIDS had a spontaneous subarachnoid hemorrhage due to type IV Ehlers-Danlos syndrome (Byard et al., 1990b).

Myxomatous Mitral Valve

Mitral valve prolapse, or floppy mitral valve caused by this condition, is due to redundant leaflet tissue or to an excessive length or rupture of chordae tendineae, or both. The morphology is described in Chapter 6. While it occurs at all ages, the most severe degree

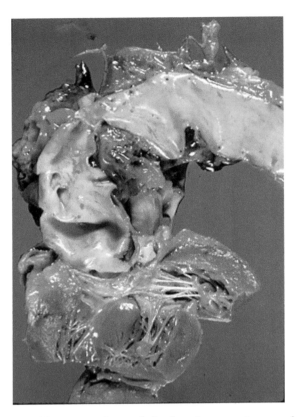

Figure 85   This 18-year-old woman dropped dead at her evening meal. She had ruptured a dissecting aneurysm originating from an intimal tear just distal to the bicuspid aortic valve. Note the isthmic coarctation (tubular hypoplasia) in the aortic arch and compare with Figure 75. She also had streak ovaries. Postmortem karyotype showed X monosomy.

occurs in very young children with Marfan syndrome (Tsang et al., 1994). In them, it is an even more likely cardiovascular complication than an aortic dissecting aneurysm, but only rarely is it associated with SD. However, prolapse can cause severe mitral regurgitation and heart failure. Apart from cases with a known defect in elastic or collagen, such as Marfan and Ehlers-Danlos syndromes, one must presume a weakness in the leaflet structure in this clinically common condition that causes a mitral systolic click on auscultation. Secondary forms may be associated with rheumatic fever (especially after commissurotomy), coronary artery disease with or without ruptured chordae tendineae, and conditions such as atrial septal defect and hypertrophic cardiomyopathy with outflow tract obstruction.

Isolated prolapse of the mitral valve is frequently associated with SD in some reports (Topaz and Edwards, 1985). In other reports, a conspicuous absence of this condition among similar pediatric cases was noted (Garson and McNamara, 1985). Sudden death tends to occur in young women rather than in children and may be familial (Dollar and Roberts, 1991; Chapman, 1994). Ronneberger and colleagues (1998) reported an 8-year-old boy who collapsed while playing soccer. Autopsy revealed a myxomatous mitral valve with lacerations of the posterior leaflet. The authors claimed that this was the youngest subject among some 100 previously reported. Another autopsy study of SD cases from subjects 14 to 35 years of age reported isolated myxomatous mitral valve prolapse in 10%;

most were females with focal fat and fibrosis in the right ventricular outflow tract myo-cardium ("segmental right ventricular cardiomyopathic changes"). The latter might explain a sudden onset of a ventricular arrhythmia (Corrado et al., 1997a). A case reported from the same group observed right ventricular cardiomyopathy in a young woman with mitral valve prolapse and ventricular arrhythmias who died suddenly with documented ventric-ular fibrillation (Martini et al., 1995). A prospective study of 300 patients including children with mitral valve prolapse reported that three cases of SD attributed to ventricular fibril-lation occurred during follow-up (Duren et al., 1988). Other clinical studies documented various cardiac arrhythmias including ventricular premature beats, tachycardia and fibril-lation as well as widening of the Q–T interval (Kulan et al., 1996; Ulgen et al., 1999).

A pathologic analysis of 24 hearts with myxomatous mitral valve prolape from young adults showed a morphometrically significant association with fibromuscular dysplasia in the artery supplying the atrioventricular node. The authors proposed that might explain SD in mitral valve prolapse (Burke et al., 1997b). This study makes one wonder whether subepicardial and intramural coronary arteries are examined adequately at autopsy in cases where SD was attributed to mitral valve prolapse, let alone whether the arteries supplying the conduction system are examined.

Of three patients with mitral valve prolapse who died suddenly, one with heart failure had a normal conduction system, and the other two had accessory atrioventricular connec-tions (Vesterby et al., 1982). Three teenagers who died suddenly had various abnormalities in the conduction system as well as premature changes of aging (Bharati et al., 1983); two also had mitral valve prolapse. Thus, SD in this condition may have many associated factors.

Sudden Death in Myocarditis and Cardiomyopathies

Inflammation of the myocardium includes acute polymorphonuclear myocarditis and abscess formation. These are both secondary to septicemia and are not considered here. Rather, lymphocytic myocarditis, which is diffuse and associated with cardiac myocyte necrosis, is termed acute myocarditis (Aretz et al., 1987; see also Figure 86). Subacute or chronic, myocarditis in this text describes active chronic myocardial inflammation with healing changes. If the latter are absent, one must presume that active chronic myocarditis is of recent onset.

Acute Lymphocytic Myocarditis

In autopsy series, this type of myocarditis was associated with approximately 25% of SDs in childhood cases (Topaz and Edwards, 1985; Neuspiel and Kuller, 1985; Keeling and Knowles, 1989), equal to the incidence of hypertrophic cardiomyopathy (Drory et al., 1991). However, if autopsy cases of SD in children are grouped with those in young adults up to age 35, myocarditis caused death in 7.5%, and hypertrophic cardiomyopathy caused death in 5.5% (Basso et al., 1999). Congestive heart failure is the usual mode of death. It can present with cardiogenic shock, so-called fulminant myocarditis, but this clinical variant actually has a better prognosis than recent-onset myocarditis (McCarthy et al., 2000). Cardiac dysrhythmias, including ventricular tachycardia and atrioventricular con-duction block, may also occur in early stages and lead to SD (Vignola et al., 1984; Balaji et al., 1994; Davis et al., 1996a; Heusch et al., 1996). Among 16 childhood cases in which

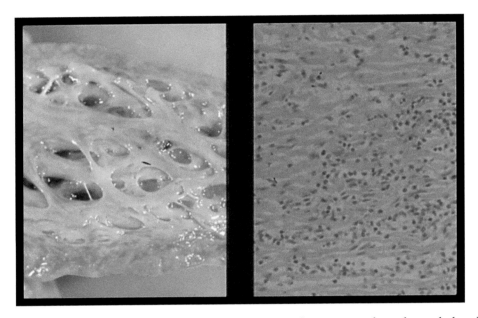

Figure 86   In this heart, the free wall of the left ventricle appears pale and mottled and was flabby. A representative microsection shows diffuse round cell infiltration with focal necrosis of cardiac myocytes. This heart is from a 5-year-old girl who had nausea, vomiting, and a slight fever for 3 days. She died in a car while being taken to the doctor. No virus was isolated by postmortem cultures of blood and myocardium. Serum collected at autopsy was negative for antibodies to several known cardiotrophic viruses.

myocarditis was the only pathology found at autopsy, 5 patients died suddenly, and 3 had no prodromal signs or symptoms (Smith et al., 1992).

Histologic diagnosis may be a problem because of sampling error and histologic interpretation. The disease has a patchy distribution. Thus, endomyocardial biopsy from the right ventricle may not sample an affected area. In contrast, autopsy allows for examination of representative sections from all heart chambers. The patchy distribution of myocarditis was demonstrated by a retrospective autopsy study of new histologic sections obtained from the hearts of supposed SIDS cases (Shatz et al., 1997). Rasten-Almquist and co-workers (2002) reported that retrospectively diagnosed myocarditis in supposed SIDS cases was located in the upper interventricular septum and adjacent right atrium. Also, a prospective autopsy study with detailed myocardial histology done on children and young adult SD cases recorded focal myocarditis in 28 of the 76 subjects with a macroscopically normal heart (Corrado et al., 2001b). Apart from sampling problems, lymphocytes are normally present in the myocardium (Tazelaar and Billingham, 1987). Also, myocyte necrosis may be difficult to define histologically, and immunohistochemical methods may be needed for demonstration (Kuhl et al., 1998; Dettmeyer et al., 1999). Myocarditis was noted as an incidental finding in 2 of 27 hearts of young infants who died violently (Rasten-Almquist et al., 2002). Also, it was reported in asymptomatic children who had biopsies because electrocardiogram changes were detected during mass screening (Nakagawa et al., 1999). Myocarditis was present in five childhood autopsy SD cases accompanying other diseases, such as bronchopneumonia or asphyxia (Smith et al., 1992).

A viral etiology for lymphocytic myocarditis is strongly suspected, because patients often have an antecedent influenza-like illness. Viral myocarditis is diagnosed clinically by a sharp rise in complement fixing antibodies for a specific virus. For example, a fourfold or greater rise in titer for Coxsackie enterovirus distinguished 42 viral myocarditis patients from another clinically similar group with a negative serum test (Levi et al., 1988). Those affected present with fever, tachycardia, hypotension, a reduced right ventricular function, increased creatine kinase, raised erythrocyte sedimentation rate, and a high white blood cell count, often with tachycardia and heart failure. A cardiotrophic virus may be isolated by blood or throat culture, but this or any other virus is rarely identified morphologically or by myocardial culture either at endomyocardial biopsy or autopsy. In the past, clinical diagnosis was based on an acute rise in antibody titer against strains of, for example, Coxsackie B3 enterovirus. Currently, polymerase chain reaction techniques are used to detect the viral genome (Grumbach et al., 1999; Hufnagel et al., 2000). Immunohistochemistry can also demonstrate enteroviral capsid protein in myocardium derived from autopsy or explanted hearts in cases of acute myocarditis or dilated cardiomyopathy. The results correlate well with polymerase chain reaction identification of enteroviral RNA in the same tissues (Li et al., 2000b).

Clinical and experimental evidence suggests that myocardial damage results from an immune-mediated mechanism rather than from any direct effect of the virus (Penninger and Bachmaier, 2000; Fairweather et al., 2001; Maisch et al., 2002). The good response to immunosuppressive therapy in some patients also points to this pathogenesis (Chan et al., 1991; Balaji et al., 1994; Ahdoot et al., 2000). Rheumatic heart disease is a prototypic immune-mediated heart disease although induced by a bacterial rather than a viral infection (see Chapter 6). Sudden death is reported in acute rheumatic myocarditis (Josselson et al., 1984).

Rarely, acute lymphocytic myocarditis results from a direct myocardial invasion by a virus. I have seen viral inclusion bodies, such as varicella and cytomegalovirus, within cardiac myocytes in many childhood cases. Lymphocytic myocarditis can also be caused by various drugs, poisons, and venoms, as well cardiotoxins produced by bacteria; in developed countries, a prime example from the past is diphtheria. These conditions are not discussed further except to mention the rare eosinophilic or granulomatous myocarditis sometimes associated with ventricular fibrillation and SD (Khoury et al., 1994; Aoki et al., 1996). We saw one florid case in a previously healthy 6-year-old girl who died suddenly after returning from a Caribbean holiday; an intense search for an infective, possibly parasitic, pathogen was negative (Figure 87). This case was referred elsewhere and reported by others (Sasano et al., 1989).

Subjects in our autopsy service who died due to acute lymphocytic myocarditis were most often ill for at least a few hours or days with, for example, a respiratory tract infection or nausea and vomiting, but some were found dead in bed and others dropped dead while under observation. In some, the heart was grossly flabby, and the ventricular myocardium was mottled (Figure 86). A minority showed mural thrombi in ventricles and the right atrium, the source of embolism to lungs or brain and, hence, an indirect cause of SD. Microscopically, a marked lymphohistiocytic infiltrate was present diffusely in all chamber walls, associated with focal myocyte necrosis. Postmortem virus cultures of heart and blood were negative. Serology was not diagnostic in virtually all instances. We had positive myocardial or blood cultures for different strains of Coxsackie virus in the past but only in newborn cases with generalized viral infection. We identified parvovirus B19 by polymerase chain reaction in the heart of an infant aged 1 year; she died suddenly of a stroke caused by thromboemboli from a left ventricular mural thrombosis. More recently,

Figure 87   Granulomatous myocarditis with many eosinophils as well as round cells and foreign body giant cell clusters are seen in this microsection from the heart of a 6-year-old girl who died suddenly after returning to Canada from a holiday in the Caribbean. Postmortem search for causative microorganisms including fungi was negative.

searches for myocardial enteroviruses and adenoviruses were done by the same technique in several autopsy cases with negative results. We believe the lack of positive cultures, serology, and positive polymerase chain reaction tests means that if a virus caused these cases of myocarditis, it would be by an immune-mediated mechanism. We were not able to link any case of myocarditis in children to therapeutic use of a drug (see Chapter 5).

Healing or Healed Myocarditis

A histologic review of 16 autopsy cases of young Swedish orienteers who died suddenly found active lymphocytic myocarditis together with varying degrees of reparative fibrosis and myocyte hypertrophy or degeneration (Larsson et al., 1999). Repeated biopsies in individual cases of acute myocarditis can demonstrate healing with fibrosis and cardiac myocyte hypertrophy (Billingham and Tazelaar, 1986; Hasumi et al., 1986). In our autopsy service, several cases of healing myocarditis diagnosed by one or more endomyocardial biopsies had a history of an acute influenza-like illness; some had a serological diagnosis of infection by a specific microorganism. We saw three cases of SD in children in whom autopsy revealed chronic active myocarditis in a dilated hypertrophied heart: two showing mild endocardial fibroelastosis in the left ventricle, and one was a 10-year-old boy who had a febrile illness 5 weeks prior to death. The heart was more than twice the expected weight, and its chambers contained extensive mural thrombi. Microscopically, a marked lymphocytic and plasma cell infiltrate was associated with extensive replacement fibrosis by mature collagen. Residual cardiac mycocytes were both hypertrophied and hyperplastic with large bizarre nuclei; no etiologic agent could be identified (Silver and Silver, 1992).

Dilated Cardiomyopathy

In sporadic dilated cardiomyopathy in childhood, SD is usually attributed to a cardiac dysrhythmia. Four of 28 (14%) young children followed with this disease died suddenly, and a ventricular dysrhythmia was documented in 3 (Muller et al., 1995). Sudden death occurred in 7 (13%) of 52 children with dilated cardiomyopathy whose enlarged left ventricles did not return to normal during echographic follow-up (Burch et al., 1994).

   Much clinical and experimental evidence suggest that lymphocytic, presumptively viral, myocarditis may evolve into a dilated cardiomyopathy. In children, the two entities are identical clinically, and both have a poor prognosis. They are diagnosed pathologically according to histologic findings on endomyocardial biopsy (Grogan et al., 1995). I believe a pathologist is unwise to make such a diagnosis. Rather, the histologic findings may be consistent with that diagnosis.

In infants, the prognosis is better if a biopsy shows acute myocarditis than if it shows endocardial fibroelastosis (Matitiau et al., 1994). Studies of dilated cardiomyopathy cases contain a proportion that show myocarditis in the biopsies (Dec et al., 1985; Kleinert et al., 1997; Olsen, 1992; Kasper et al., 1994), whereas others have active chronic myocarditis at autopsy (Morimoto et al., 1992). Investigation of family members of two new probands with dilated cardiomyopathy showed acute myocarditis in both them and their relatives (O'Connell et al., 1984). This may indicate that a familial anomaly in immune response is responsible for dilated cardiomyopathy in some families.

Among 42 patients with viral myocarditis and a fourfold or greater rise in antibody titer for Coxsackie complement-fixing antibodies, 10 subsequently died; none of 26 similar patients with a negative serum test had died after 15 years (Levi et al., 1988). The authors found that viral myocarditis resolved completely or evolved into chronic myocarditis (3 of 42 patients) or dilated cardiomyopathy (7 of 42 patients, 2 dying suddenly). Application of polymerase chain reaction techniques to endomyocardial biopsy samples detected enterovirus genome in nearly one third of patients with myocarditis or dilated cardiomyopathy; adenoviral genome was not found (Grumbach et al., 1999). Other studies using immunohistochemical techniques found that enterovirus persisting in the myocardium was important in the pathogenesis of this condition (Li et al., 2000b; Zhang et al., 2000a). A prospective randomized study using polymerase chain reaction and *in situ* hybridization demonstrated that viral persistence may contribute to the pathogenesis of inflammatory heart muscle disease, and that in chronic myocarditis, viral persistence (enterovirus, cytomegalovirus, adenovirus) occurred in 11.8% of nonselected patients (Hufnagel et al., 2000). In contrast, another investigation using serological and polymerase techniques to detect enterovirus and other cardiotrophic microorganisms did not find any evidence that microbial persistence in the heart was involved in the cases of 37 patients with end-stage idiopathic dilated cardiomyopathy (de Leeuw et al., 1999). A similar study on patients with early dilated cardiomyopathy failed to find enteroviral, adenoviral, or cytomegalovirus nucleic acids by polymerase chain reaction (Mahon et al., 2001). The latter study was conducted with relatives of proband cases of familial dilated cardiomyopathy. This case selection may explain the discrepant result.

Idiopathic (unexplained) dilated cardiomyopathy is familial in at least 35% of cases with various phenotypic subtypes, including associations with muscular dystrophies, cardiac conduction defects, and sinus node dysfunction and sensorineural hearing loss (Grunig et al., 1998; Fatkin et al., 1999; Becane et al., 2000). Honda and co-workers (1995) reported that most patients with the familial form died suddenly, whereas patients with the nonfamilial form often died of congestive heart failure. Certain kindreds of dilated cardiomyopathy associated cardiac conduction defects, such as atrioventricular block, have a high incidence of associated SD (Fatkin et al., 1999; Becane et al., 2000). A six-generation family with autosomal dominant progressive atrioventricular conduction block also had progressive dilated cardiomyopathy with degenerative myocardial changes and fibrosis, especially in the atrial myocardium (Graber et al., 1986).

The familial forms of the disease are associated with many different genetic defects found mainly in cytoskeletal and nuclear envelope proteins, all with particular transmission patterns if predominantly autosomal dominant (Arbustini et al., 2000; Towbin and Bowles, 2001), notably associated with lamins A and C gene mutations when atrioventricular block is present (Arbustini et al., 2002). Different mutations in the gene that governs the sarcomeric protein, troponin T, may cause either dilated or hypertrophic cardiomyopathy (Li et al., 2001). The overlap in genetic origin of these two dominantly transmitted diseases, both of which may cause SD, is discussed below. Desmin gene defects are associated with dilated cardiomyopathy and are also reported in restrictive cardiomyopathy and in desmin (myofibrillar) myopathies associated with atrioventricular conduction block (Park et al., 2000). Melberg and co-workers (1999) described a Swedish family with myofibrillar myopathy, atrioventricular conduction block, and sick sinus syndrome; some members also had right ventricular cardiomyopathy. In some probands and their relatives, myocarditis and dilated cardiomyopathy may represent the acute and chronic stages of an organ-specific autoimmune myocardial disease resulting from antibodies to certain isoforms of cardiac myosin (Caforio et al., 2002).

I have seen several cases of dilated cardiomyopathy in hydropic stillborn and newborn infants (Silver et al., 1996a). Some were presumed familial because a similarly affected sibling was found in our files. Also, we identified several cases upon reviewing autopsy cases diagnosed elsewhere as SIDS. However, familial dilated cardiomyopathy usually has a dominant mode of transmission, and familial cases occurring within a sibship, the offspring of normal parents, suggests an autosomal recessive transmission. When associated with endocardial fibroelastosis, it may be acquired during fetal life, caused, for example, by complete heart block, and this may also repeat within a sibship. One 5-year-old girl presented with dilated cardiomyopathy requiring cardiac transplantation. We reviewed the autopsy histology of her identical twin who was said to have died with SIDS and observed an acute myocarditis in heart sections. Presumably, the surviving twin, in early infancy, also had an acute myocarditis that healed and evolved into dilated cardiomyopathy (Silver, 1998). These instances emphasize the importance of obtaining a family and prenatal history in these cases occurring in infants and children.

The perinatal examples in our autopsy service had globally dilated hearts and biventricular endocardial fibroelastosis, and heart weight was not increased because hypertrophy would have started postnatally, after the left ventricle began supporting the systemic circulation. The presence of endocardial fibroelastosis in a newborn indicates that ventricular dilation occurred during fetal life. Several SIDS-like cases (SD between 6 weeks and 6 months of age) with dilated cardiomyopathy were normally developed and well grown for age: their hearts showed biventricular dilation, hypertrophy, and endocardial fibroelastosis, the latter worse in the left ventricle (Figure 74).

In our experience with many childhood cases of dilated cardiomyopathy, very few died suddenly, but most died with congestive heart failure. One 12-year-old male diagnosed with congestive heart failure only a week previously died in the hospital with documented ventricular fibrillation. The heart was twice its expected weight and globally dilated, with marked endocardial fibroelastosis in the left ventricle. Microscopically, cardiac myocytes showed hypertrophy and attenuation, with extensive interstitial as well as replacement fibrosis and no inflammation; viral etiology could not be proven (Silver and Silver, 1992). Subsequently, his 14-year-old brother had a heart transplant. The explanted heart was grossly and microscopically similar (Figure 88). Even if a cardiotrophic virus had been identified, such cases in siblings prompt a search for a genetic etiology. As well as advising

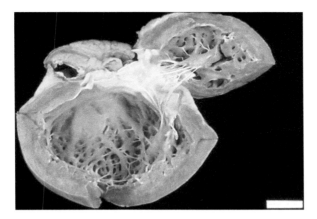

Figure 88   Left ventricular inflow tract showing a severely dilated left ventricle lined by a thick layer of endocardial fibroelastosis. This heart was excised from a 14-year-old boy who had a cardiac transplant for dilated cardiomyopathy. One year earlier, his 12-year-old brother developed heart failure 1 week prior to death. He died with documented ventricular fibrillation while in the hospital; his heart at autopsy closely resembled that shown here.

investigation of the family, the pathologist should preserve frozen blood and tissues for future molecular studies.

Syndromic Cardiomyopathies with Conduction Block or Muscle Disease

Familial dilated cardiomyopathy with conduction system disease is reported as an autosomal dominant gene defect in the rod domain of the lamin A/C gene (Arbustini et al., 2002). Lamins A and C are intermediate filament proteins in the nuclear envelope. Lamin A/C gene mutations may cause SD, although in adults rather than in children (Fatkin et al., 1999; Becane et al., 2000). Some affected families also show Emery-Dreifuss muscular dystrophy (Becane et al., 2000) caused by defects in a different domain of the same gene (Fatkin et al., 1999).

Myopathies and muscular dystrophies often have associated heart block, but only a few are associated with SD. Myotonic dystrophy, the most common, is related to an autosomal dominant gene on chromosome 19. Up to 30% of subjects, usually adult, die suddenly. Ventricular tachyarrhythmias may be more important than complete block in causing SD (Merino et al., 1998; Mammarella et al., 2000). The tachycardia is a result of bundle-branch reentry, and catheter ablation is used to abolish it (Merino et al., 1998). Among autopsy cases of myotonic dystrophy, the most frequently observed histopathologic lesions of the cardiac conduction system were fibrosis, fatty infiltration, and atrophy (Bharati et al., 1984; Nguyen et al., 1988).

Childhood onset autosomal dominant limb-girdle muscular dystrophy may be associated with progressive conduction block with syncope and SD (van der Kooi et al., 1996; Fang et al., 1997). An X-linked recessive gene causes Emery-Dreifuss muscular dystrophy. Sudden death was reported in a female carrier. Her ventricular myocardium showed interstitial fibrosis and scarring and the atrial myocardium fibro-fatty replacement, as is seen in men with the disease, but the conduction system appeared normal (Fishbein et al., 1993). Boys with Duchenne muscular dystrophies have characteristic scarring in the subepicardial infero-basal left ventricle. This causes the characteristic electrocardiographic

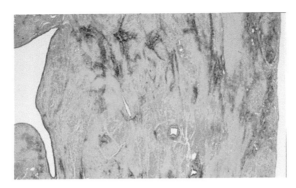

Figure 89 A 15-year-old boy with Becker muscular dystrophy received a heart transplant for dilated cardiomyopathy. Histologic sections from the native heart showed extensive interstitial and patchy replacement fibrosis in both ventricles, but the condition was worst in the left ventricular free wall, seen here. The fibroelastosis was worse in the subepicardial zone in a pattern clearly not due to myocardial ischemia.

pattern also seen in carriers (Sanyal and Johnson, 1982). Patients are liable to die suddenly. Chenard et al. (1993) described complex ventricular arrhythmias in this disease, but a later study linked mortality to left ventricular dysfunction rather than ventricular arrhythmias (Corrado et al., 2002). Vlay et al. (2001) described a family with progressive nonspecific cardiomyopathy presenting with atrial arrhythmias with later onset of life-threatening sustained ventricular tachycardia.

Becker muscular dystrophy is closely related to the Duchenne form and also depends on sex-linked transmission of a dystrophin gene defect (Finsterer et al., 1999; Falsaperla et al., 2001). We saw extensive interstitial scarring throughout the myocardium but particularly in the left ventricular free wall in the native heart of a boy with Becker muscular dystrophy who received a heart transplant (Figure 89).

Barth syndrome, an X-linked cardioskeletal myopathy with short stature, neutropenia, and abnormal mitochondria, may develop dilated cardiomyopathy with endocardial fibroelastosis (D'Adamo et al., 1997; Barth et al., 1999). Death occurs in early infancy before characteristic features of the disease are evident (Gedeon et al., 1995), but SD has not been reported. Another X-linked heart disease closely related to Barth syndrome is noncompaction of the left ventricular myocardium. It causes greatly enlarged hearts in male infants with early death and rhythm disturbances reported (Bleyl et al., 1997a), see also below.

Kearns-Sayre syndrome, a multisystem mitochondrial disorder that starts before age 20 and produces progressive external ophthalmoplegia, retinal pigmentary degeneration, and progressive impairment of cardiac conduction, is related to multiple mitochondrial DNA deletions in both cardiac and skeletal muscle. Subjects show complete or bundle branch block and may die suddenly (Katsanos et al., 2002).

Hypertrophic Cardiomyopathy (See also Chapters 5 and 9)

This disease is a familial autosomal dominant disease characterized by a ventricular hypertrophy mainly affecting the interventricular septum and associated with extensive myocardial fibers and myofibrillar disarray. Death from hypertrophic cardiomyopathy during infancy is usually caused by heart failure (Skinner et al., 1997), and SD is reported rather rarely in preadolescent children. However, as a cause of SD in young adults, it is second only to myocarditis (Basso et al., 1999). A clinical review done at this center from 1958 to

1997 of 99 patients younger than 18 years (71 male), followed 5 years, revealed that 18 died suddenly (Yetman et al., 1998). In a similar survey lasting 27 years, 6 of 31 preadolescent children died suddenly; ventricular dysrhythmia was present in 3 of them (Muller et al., 1995). Nevertheless, both studies were based on highly selected populations, and SD due to hypertrophic cardiomyopathy is relatively rare in childhood. Considering children up to age 20, Wren et al. (2000) suggested that a primary cardiac arrhythmia is probably about 10 times more common a cause of SD than unsuspected hypertrophic cardiomyopathy; the latter caused only one death per million person years.

In the natural history of this heritable disease, SD most likely occurs in asymptomatic adolescents and young adults, often during exercise or athletic competition (Maron and Fananapazir, 1992). Death was often the first presentation of disease, although individuals sometimes had a family history of SD and, frequently, exertional syncope. Often, they show marked left ventricular hypertrophy and myocardial ischemia as well as various dysrhythmias, including inducible ventricular tachycardia, clinically (Maron and Fananapazir, 1992). In contrast, a recent study showed that less than 15% of 44 cases of SD (age range 7 to 78 years) occurred after exercise, possibly because the subjects enrolled were informed they had hypertrophic cardiomyopathy (Maron, 2000).

Young subjects with extreme left ventricular hypertrophy are at high risk for SD; among 12 such patients less than 18 years old, 5 died suddenly (Spirito et al., 2000). In another group, however, most cases of SD occurred in patients with a septal wall thickness less than 30 mm (Elliott et al., 2001). A subset of young persons dying suddenly showed mild left ventricular hypertrophy. Their risk of SD may have been related to the mutant gene that governs the sarcomeric protein defect. Dominantly inherited hypertrophic cardiomyopathy is caused by many known mutations in the multigene families that encode at least eight sarcomeric proteins (3 myofilament proteins, 4 thin filament proteins, and 1 myosin-binding protein) within the contractile apparatus (Towbin and Bowles, 2001). Mutations in the gene encoding another sarcomeric protein, actin, cause both the dilated and the apical forms of hypertrophic cardiomyopathies (Olson et al., 2000). The latter is associated with lethal arrhythmias and SD (Okishige et al., 2001).

Mutations in genes encoding the sarcomeric thin filament protein troponin T cause either hypertrophic or dilated cardiomyopathy (Kamisago et al., 2000; Li et al., 2001). A particular mutation in this gene in one family was associated with partial transition of the hypertrophic form to the dilated form together with a high incidence of SD (Fujino et al., 2002). A specific mutation in troponin T in two kindreds produced mild or undetectable left ventricular hypertrophy but a high incidence of SD at a mean age of 17 (Moolman et al., 1997). In another study of subjects with a cardiac troponin T mutation, 8 of 9 patients between 6 and 39 years of age died suddenly. Histology of their myocardium showed severe myocardial disarray with only mild hypertrophy and fibrosis (Varnava et al., 2001). A family with a mutation in beta-myosin heavy chain had 4 deaths in children younger than 16 years among 32 family members in 4 generations (Hwang et al., 1998). Another family with a mutation in alpha-tropomyosin showed mild left hypertrophy, but 11 of 26 family members died suddenly (Karibe et al., 2001).

Maron (2000b) attributes SD in hypertrophic cardiomyopathy to ventricular tachyarrhythmias arising in an electrically unstable myocardial substrate characterized by disorganized cellular architecture, ischemia, cell death, and replacement scarring. A histologic study documented myocardial ischemic damage; as well as showing replacement fibrosis, the interstitial collagen was greatly increased (Basso et al., 2000). Another histologic study showed a significant increase of matrix collagen in the interventricular septum when compared with normal controls, as well as with hypertensive adults and infants with hypertrophic cardiomyopathy (Shirani et al., 2000). Similar changes were noted in myomectomy tissue from the left ventricular outflow tract (Factor et al., 1991).

Of more than a dozen hypertrophic cardiomyopathy cases in our autopsy service, only three children died suddenly. The first was a 7-year-old girl who died while walking to school. She was apparently healthy and well grown but had been followed in our cardiology clinic since the age of 6 months with a diagnosis of idiopathic endocardial fibrosis, confirmed by cardiac catheterization, and had taken digoxin since that age. The heart was enormously hypertrophied, weighing five times the expected weight. This was mainly due to hypertrophy of the left ventricle, which showed no endomyocardial fibrosis but showed mild endocardial fibroelastosis. The greatly thickened interventricular septum bulged into both outflow tracts. The second and third cases were brothers who died a year apart at age 12, each while riding a bicycle. Their hearts showed massive asymmetric left hypertrophy. We found three stored hearts from autopsies done between 1954 and 1957 on infantile cases diagnosed with idiopathic cardiomegaly or cardiac hypertrophy. All died of heart failure. Each heart weighed more than three times the expected weight and showed slit-like heart chambers. Additional histologic sections taken transversely from the septum in each case allowed for a retrospective diagnosis of hypertrophic cardiomyopathy, because we found extensive disarray in the hypertrophied mid-zone myocardium extending into the middle layer of the free walls of each ventricle (Figure 90). This morphologic finding is the *sine qua non* for diagnosis whether in infants, children, or adults. It requires viewing histologic transverse sections of the interventricular septum and appreciating that small amounts of disarray are seen focally in normal and hypertrophied hearts (Silver and Silver, 1992; Silver et al., 1996b).

Hypertrophic cardiomyopathy can mimic dilated cardiomyopathy both clinically and in gross appearance. Several of our cases showed mild left ventricular dilation and endocardial fibroelastosis. A clinical overlap with restrictive cardiomyopathy also occurs, but I was surprised to find that three cases of restrictive cardiomyopathy in our autopsy files were,

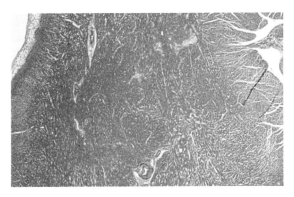

Figure 90 Section of the posteroseptal left ventricle from a 10-week-old female infant with hypertrophic cardiomyopathy. She presented with feeding difficulties, cough, and vomiting and was diagnosed by chest radiograph and electrocardiogram findings in the emergency room. On admission she had a cardiac arrest from which she was resuscitated, followed by persistent bradycardia and episodes of ventricular tachycardia and fibrillation leading to a final cardiac arrest. Note the wide zone of hypertrophied and disarrayed myocardium sparing a narrow layer deep to the epicardium (left) and the subendocardium (right).

on review, due to hypertrophic cardiomyopathy that presented in the terminal dilated phase with left ventricular endocardial fibroelastosis. All had congestive heart failure prior to death, and all died in the hospital after cardiac arrest. An infant sibling of one subsequently had a heart transplant. The explanted heart showed similar gross and microscopic findings. Possibly some cases of so-called familial restrictive cardiomyopathy in the literature are actually cases of familial hypertrophic cardiomyopathy.

Many conditions associated with hormones or drugs cause left ventricular hypertrophy with asymmetric septal hypertrophy, especially in the very young. We term this condition pseudohypertrophic cardiomyopathy (Silver and Silver, 1992). The first reported one followed increased transplacental passage of nutrients caused by maternal hyperinsulinemia promoting somatic overgrowth (Gutgesell et al., 1976). Subsequently, others incorrectly called this finding hypertrophic cardiomyopathy. No increased myocyte disarray is present in the interventricular septum in such infants of diabetic mothers. The phenomenon is also seen in infants with nesidioblastosis or islet cell adenoma of the pancreas, in Beckwith-Wiedermann syndrome, and after diazoxide therapy for hyperinsulinemic hypoglycemia. A similar phenomenon occurs in children with adrenocorticotropic hormone and corticosteroid therapy, with adrenal cortical hyperplasia, and with dexamethasone therapy for bronchopulmonary dysplasia in the preterm newborn. It is reported in newborns with rhesus hemolytic disease and in older children and adults with hypothyroidism. Reports of adult endocrine conditions such as myxedema, hyperthyroidism, and hyperparathyroidism associated with cardiac hypertrophy may have a similar basis.

We saw one case of SD in an infant with left ventricular hypertrophy due to adrenocorticotropic hormone therapy given to treat epilepsy. No more than the normal amount of cardiac myocyte disarray was present in transverse histologic sections of the interventricular septum. Also, we autopsied three overgrown infants, offspring of diabetic mothers, who died soon after birth with hypoglycemia, metabolic acidosis, and heart failure; each had asymmetric septal hypertrophy, and none had more than 5% disarray in transverse histologic sections of the interventricular septum.

Another group of conditions that may imitate hypertrophic cardiomyopathy clinically, especially if they show asymmetric septal hypertrophy, are genetic or chromosomal rather than acquired. They include Costello, Noonan's, LEOPARD (multiple lentiginosis), Friedreich's, Marfan, Senger, and Swyer syndromes, myotonic dystrophy, and neurofibromatosis. Sudden death is reported in many of these conditions. Most, with the exception of Noonan's syndrome, appear to be examples of pseudohypertrophic cardiomyopathy. Recorded cardiac findings in most reports are not sufficient to permit a histologic diagnosis of hypertrophic cardiomyopathy. When the gene defects that cause these familial syndromes are understood and sufficient cardiac histologic studies on them are available, we will learn whether or not they are actually causes of hypertrophic cardiomyopathy or another form of heart muscle disease. If these syndromes truly cause hypertrophic cardiomyopathy, then we will better understand how various gene defects influence heart muscle.

Concerning Noonan's syndrome, I saw myocardial disarray diagnostic of hypertrophic cardiomyopathy in three cases, including a 1-month-old male infant and a 9-year-old female, both of whom came to autopsy. Each heart showed more than 20% disarray in the interventricular septum. The third subject was a 15-year-old girl with hypertrophic cardiomyopathy diagnosed on two myomectomy specimens done 4 years apart and each resected from the obstructed left ventricular outflow tract. She had a history of a near-death episode from ventricular fibrillation following a cardiac catheterization at the age of 10 years. In a report on 33 patients with Noonan's syndrome, 17 had hypertrophic cardiomyopathy and 2 died suddenly with arrhythmia (Ishizawa et al., 1996). Some authors propose a gene linkage to hypertrophic cardiomyopathy (Burch et al., 1993). The latter is claimed to occur in LEOPARD syndrome, also related to Noonan's syndrome, and a case is reported in Turner's syndrome (see above), also closely related to Noonan's syndrome (Conte et al., 1995).

Inborn Errors of Metabolism Causing Generalized Cardiac Enlargement

The enlarged hypocontractile heart in an infant with Pompe's disease (glycogenosis Type II) represents glycogen storage in cardiac myocytes due to an inborn metabolic error. This storage may be either cytoplasmic or lysosomal depending on which enzyme is affected. The most common type of glycogenosis is late-onset pseudo-Pompe's disease. It is characterized by severe cardiomyopathy and mild myopathy appearing in the second or third decade, prominent arrhythmia, and Wolff-Parkinson-White syndrome (Verloes et al., 1997). Sudden death is reported in a case of lysosomal glycogen storage disease without acid maltase deficiency in cardiac myocytes (Tse et al., 1996). A phenotypically similar condition with vacuoles containing desmin-type intermediate filaments within cardiac and skeletal myocytes was reported in a mentally retarded boy with so-called hypertrophic cardiomyopathy; SD occurred in female relatives (Muntoni et al., 1994).

Genetic defects affecting respiratory chain enzymes may cause mitochondrial cardiomyopathy (Figure 91), not to be confused with oncocytic cardiomyopathy (see below). In the former, cardiac myocytes all show mitochondrial hyperplasia that causes generalized cardiac enlargement, although the heart, and skeletal muscle too, is hypotonic, and death is usually caused by heart failure (Bohles et al., 1987; Guenthard et al., 1995; Dipchand et al., 2001). The left ventricle shows concentric thickening without outflow tract obstruction (Guenthard et al., 1995). Nevertheless, cases with cardiac mitochondrial enzymopathies, for example, defects in cytochrome C oxidase, are often described, wrongly, as showing hypertrophic cardiomyopathy.

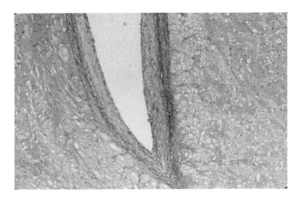

Figure 91 An 11-month-old female infant presented with a greatly enlarged heart initially diagnosed at 7 months as being caused by hypertrophic cardiomyopathy. However, skeletal muscle biopsy showed ragged-red fibers, and her heart was hypocontractile; the diagnosis was changed to mitochondrial cardiomyopathy after biochemical studies done on muscle. She died suddenly at home. This microsection from left ventricular myocardium shows slightly thickened elastotic endocardium and globular vacuolated cardiac myocytes. The latter change was generalized in all chamber walls. Electron microscopy confirmed mitochondrial hyperplasia.

Cardiomyopathy due to various mitochondrial DNA gene defects has a matrilineal inheritance (Arbustini et al., 2000; Towbin et al., 2001; Dipchand et al., 2001). The MELAS syndrome is a mitochondrial disease inherited from the mother and claimed to cause hypertrophic cardiomyopathy (Silvestri et al., 1997; Okajima et al., 1998). Sudden death is reported in families with maternally inherited mitochondrial diseases (Sweeney et al., 1993; Santorelli et al., 1996). Both Barth and Kearns-Sayre syndromes show abnormal mitochondria in skeletal muscle and myocardium (see above).

Restrictive Cardiomyopathies

This clinical type of cardiomyopathy is less common than either dilated or hypertrophic cardiomyopathy. It correlates histologically with findings in the myocardium that induce a functional stiffening, that is, clinical noncompliance. Thus, endocardial fibroelastosis or endomyocardial fibrosis can splint the ventricular myocardium and impede filling. In adults especially a specific infiltrate such as amyloid may surround individual myocytes (Silver and Silver, 1992). A family with autosomal dominant restrictive cardiomyopathy and progressive atrioventricular conduction block also developed myopathy after middle age; histology was nonspecific, but patchy endocardial and myocardial fibrosis was present in one autopsy case (Fitzpatrick et al., 1990). Another type of familial restrictive cardio-myopathy with distinctive ultrastructural morphology because of granulofilamentous deposits of desmin in cardiac and skeletal myocytes was associated with conduction block and mild myopathy (Arbustini et al., 1998).

In children, the usual finding in an endomyocardial biopsy is interstitial fibrosis. Based on a Texan series, SD is most likely in girls, especially those with chest pain and syncope. Prognosis is poor for all children. Rivenes et al. (2000) cite references to most pediatric cases of restrictive cardiomyopathy in the English literature. In an earlier paper from Texas, the authors noted that the etiology in 3 of their 12 patients was hypertrophic cardiomy-opathy, and the etiology in another 3 was cardiac hypertrophy with restrictive physiology (Denfield et al., 1997). Some case reports they cite also appear to be cases of hypertrophic cardiomyopathy, particularly a 6-year-old boy whose brother died of this form at age 3 (Nishikawa et al., 1992). Another case report described an endomyocardial biopsy finding of bizarre myocardial hypertrophy and disorganization (Maki et al., 1990).

Because histologic diagnosis will often be based on endomyocardial biopsy, a pathol-ogist reading that tissue should point out the nonspecificity of interstitial fibrosis, because autopsy may reveal an underlying hypertrophic cardiomyopathy in the late dilated phase (Figure 92A and Figure 92B), sometimes with endocardial fibroelastosis (see above). A puzzling fact in the three cases I examined (called restrictive clinically, but actually hyper-trophic cardiomyopathy) was that gross examination showed left ventricular hypertrophy in each instance with a heart weight two to three times than expected, whereas in life this finding had been absent or slight, as is required for that clinical diagnosis.

Right Ventricular Cardiomyopathy (see also Chapters 5 and 9)

Another familial autosomal dominant form of cardiomyopathy mainly affects the right ventricle, with partial or total atrophy of the myocardium and fatty or fibro-fatty replace-ment. A phase of active lymphocytic myocarditis apparently precedes myocyte necrosis or apoptosis (Pinamonti et al., 1996; Thiene et al., 2000). Pathologic studies note that about 50% of cases have left ventricular involvement (Gallo et al., 1992; Pinamonti et al., 1996;

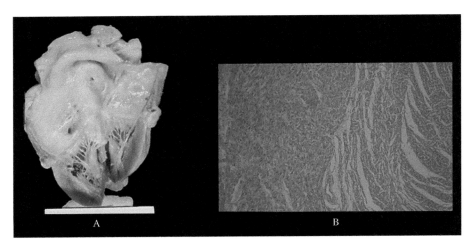

Figure 92 (A) Left ventricular inflow view of the heart from an 11-year-old boy who died in cardiogenic shock 3 days after admission with congestive heart failure. Restrictive cardiomyopathy was diagnosed. The heart weighed three times the amount expected for body size, but much of this was due to hugely dilated and hypertrophied atria. Both ventricles were mildly dilated and hypertrophied with slight endocardial fibroelastosis in the left. The upper interventricular septum was as thick as the left ventricular free wall. Thus, the gross appearance was in keeping with the clinical diagnosis. (B) However, microscopy of the upper interventricular septum revealed hypertrophied and disarrayed myocardium diagnostic of hypertrophic cardiomyopathy with normal appearing myocardium on the right. The zone of disarray is central in the septum and continuous with similar zones in the free walls of both ventricles. Pathologic diagnosis of hypertrophic cardiomyopathy depends on viewing histologic sections of one or more entire transversely oriented slices of the ventricles, the first one third of the distance from the atrioventricular rings to the apex.

Lobo et al., 1999; Thiene et al., 2000), with this present even in adolescents (Smith et al., 1999).

The prevalence of SD in this cardiomyopathy is related to dysrhythmias generated at the junctional regions between atrophic and normal myocardium; hence, the adjective "arrhythmogenic" is often added to the name. Sudden death is more likely in young adults than in children, and males are more susceptible to that event than females (Thiene et al., 1988), but pediatric SD cases are reported, the youngest aged 7 (Pawel et al., 1994) as well as sibling cases in adolescents (Smith et al., 1999).

Five gene loci on four different chromosomes were identified for this autosomal dominant familial disease (Thiene et al., 2000). The only gene so far identified to cause right ventricular cardiomyopathy (type 2) is the cardiac ryanodine receptor, the major calcium release channel in the sarcoplasmic reticulum, which also causes familial polymorphic ventricular tachycardia (see below; also see Marks et al., 2002; Keller et al., 2002). The latter condition was diagnosed together with arrhythmogenic right ventricular cardiomyopathy in living members of two families. Autopsies on juvenile family members who died suddenly confirmed fibro-fatty replacement of the apical segment of the right ventricle (Bauce et al., 2000). Although the diagnosis of the Brugada syndrome (ECG findings of right bundle branch block and persistent ST segment elevation in the right precordial leads; see below) requires a clinically normal heart, two of a 16-member family with this syndrome proved at autopsy to have right ventricular cardiomyopathy (Corrado et al., 1996). An autosomal recessive form for which the gene defect is known is associated with palmoplantar keratosis and wooly hair and occurs on the island of Naxos (Protonotarios et al., 2001).

I autopsied a 16-year-old boy who died while playing tennis; he was of Italian descent, and his uncle had died suddenly. The right ventricle was dilated, and its free wall was yellowish and partly translucent; this case taught me always to transilluminate the free wall of the right ventricle at autopsy. Microscopically it was replaced by fat except for hypertrophied subendocardial myocytes (Silver and Silver, 1992).

Cardiomyopathies in Inborn Errors of Fat Metabolism

Cardiac muscle derives most of its energy from beta-oxidation of fatty acids within mitochondria. Short- and medium-chain fatty acids enter mitochondria directly, but long-chain fatty acids are converted to acyl-coenzyme A forms in the cytoplasm and are then actively transported into mitochondria; this transport requires carnitine. Several genetic acyl-coenzyme A dehydrogenase deficiencies are known. Carnitine deficiency may be due to various defects in enzymes concerned with its absorption and transport (Figure 93). It tends to be deficient secondary to disorders of fatty acid metabolism (Stanley, 1995; Winter and Buist, 2000; Helton et al., 2000). A third type of mitochondrial fatty acid beta-oxidation defect occurs in the electron transport chain and is discussed above.

Beta-oxidation gene defects cause an abnormal response to fasting in those tissues (heart and skeletal muscle, liver) that use fatty acids for energy. Such inborn metabolic errors are heritable by autosomal recessive genes. Neurological findings are prominent in many of the more than 20 types identified so far and include hypotonia, myopathy (often with lipid storage), and peripheral neuropathy, episodic rhabdomyolysis, and hypoglycemia (Vockley and Whiteman, 2002). Arrhythmias and conduction system defects are common in children with such gene defects, and the accumulation of intermediary metabolites of fatty acids, such as long-chain acylcarnitines, may cause the arrhythmias (Bonnet et al., 1999).

Medium-chain acyl coenzyme A dehydrogenase beta-oxidation defects are the most common and can imitate SIDS or Reyes syndrome, leading to SD in infants and children

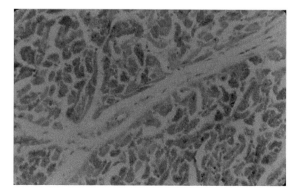

Figure 93 Following the death of a 19-month-old female with a Reye-like illness, part of the myocardium was snap-frozen at autopsy for frozen section. It revealed plentiful fat droplets in cardiac myocytes. The heart showed mild left ventricular hypertrophy and dilation with mild endocardial fibroelastosis. Dicarboxylic acid screening of urine collected at autopsy suggested long-chain acyl-coenzyme A dehydrogenase deficiency. The final diagnosis, after biochemical studies of the patient's cultured fibroblasts, was systemic carnitine deficiency.

(Iafolla et al., 1994; Vockley et al., 1994). The deficiency typically presents in the second year of life as hypoketotic hypoglycemia associated with fasting and may progress to liver failure, coma, and death. One novel mutation of the defect is linked to SD in infants (Brackett et al., 1995), and another caused SD in a 19-year-old female (Yang et al., 2000). Subjects usually do not have a cardiomyopathy, and the only histologic finding is fatty change in myocytes, hepatocytes, and renal tubular epithelium (Iafolla et al., 1994).

Long-chain 3-hydroxyacyl-CoA dehydrogenase deficiency, the defect associated with maternal acute fatty liver of pregnancy, was found in three families, children of which presented with sudden unexplained death or Reyes-like syndrome (Sims et al., 1995). This deficiency appears the most frequently diagnosed beta-oxidation defect in Finland, with dilated cardiomyopathy occurring in some cases (Tyni et al, 1999).

Very-long-chain acyl-CoA dehydrogenase deficiency causes dilated cardiomyopathy and SD in infancy and childhood (Strauss et al., 1995; Mathur et al., 1999). Sudden neonatal deaths are also reported due to enzymes concerned with carnitine transport (Chalmers et al., 1997; Rinaldo et al., 1997).

The known defects in fatty acid metabolism are often associated with a cardiomyopathy, and this may present clinically with a dilated, hypertrophic, or restrictive clinical pattern (Ino et al., 1988; Helton et al., 2000). Up to a quarter of such patients have cardiac arrhythmias, but the latter are absent in those with medium-chain acyl dehydrogenase deficiency (Bonnet et al., 1999). A documented acute arrhythmia was the cause of SD in 6 of 107 patients with fatty acid oxidation disorders (Bonnet et al., 1999).

Our autopsy service records include many SIDS-like cases caused by medium-chain acyl coenzyme A dehydrogenase deficiency. Gross autopsy findings that prompt a pathologist to collect a urine sample and cultured fibroblasts include a pale liver and a full bladder. Also, we collect frozen histologic sections of the myocardium (Figure 93) liver and kidney for fat stains in such cases. We have also observed one or two cases of many of the other beta-oxidation disorders mentioned here.

Sudden Death in Cardiac Tumors and Hamartomas

Malignant Cardiac Tumors

At any age, malignant tumors of the heart are usually metastatic. Primary tumors are rare, and most that occur in children are benign with less than 10% malignant (Takach et al., 1996). Sudden death occurred in a child with cardiac fibrosarcoma receiving chemotherapy (Bini et al., 1983). A 12-year-old boy who developed recurrent syncope and ventricular tachycardia had Hodgkin's lymphoma forming a mass in the right ventricular cavity with extension into the outflow tract (Manojkumar et al., 2001). A 16-year-old girl with acute lymphocytic leukemia and anthracycline cardiotoxicity who was also hypokalemic developed a life-threatening arrhythmia (Kishi et al., 2000). Sudden death in such cases could well be caused by rhythm disturbances induced by chemotherapeutic drugs (see below).

Benign Cardiac Tumors

Most benign cardiac tumors are not even true neoplasms but resemble hamartomas or congenital rests.

Congenital Tumor of the Atrioventricular Node

Congenital tumors of the atrioventricular node are usually microcystic and were previously called mesotheliomas but are now considered congenital entodermal rests (Burke et al., 1990). They cause SD at any age, especially in females (Lie et al., 1980) and in children, with evidence of heart block prior to death (Bharati et al., 1976; Ross, 1977; Thorgeirsson et al., 1983).

An interesting association with familial dilated cardiomyopathy was reported in a male who developed complete heart block at age 14, received a pacemaker at age 33, then developed dilated cardiomyopathy and died of heart failure at age 37 (Ford, 1999). At autopsy, the atrioventricular node was replaced by the cystic tumor under discussion, and the heart showed dilated cardiomyopathy; four close relatives had died suddenly with the latter disease but with no examination of their conduction systems at autopsy. Strom et al. (1993) described an atrioventricular node tumor in a young man with Emery-Dreifuss muscular dystrophy (discussed above) who died suddenly. These two reports reinforce the rule that the conduction system should be examined histologically in cases of documented chronic dysrhythmia, even if the subject has a disease known to be associated with SD.

Because the tumors are rarely visible grossly, a claim that one has caused SD usually means that, after negative gross and microscopic cardiac findings at autopsy, the conduction system was examined histologically. An atrioventricular node tumor, like cardiac rhabdomyoma, may also be an incidental finding at autopsy in cases where SD was clearly due to another cause. Also, both tumors are reported as incidental autopsy findings beyond the conduction system (Suarez-Mier et al., 1999).

Cardiac Rhabdomyoma, Rhabdomyomatosis

Cardiac rhabdomyoma, or congenital glycogen tumors, are the most common benign tumors. They are usually multiple and tend to regress during later infancy and childhood. Most prevalent in newborns, they are often associated with tuberous sclerosis (Bosi et al., 1996). The mechanism of their causing SD is likely a rhythm disturbance, particularly the Wolff-Parkinson-White ventricular preexcitation syndrome (Gotlieb et al., 1977; Case et al., 1991; Mehta, 1993; Bosi et al., 1996). Another possible presentation is with outflow tract obstruction. In an autopsy series, 2 of 17 childhood cases died suddenly (Burke et al., 1991b). Other case reports attributed SD in children to a solitary rhabdomyoma (Bohm et al., 1980; Violette et al., 1981; Burke et al., 1999a).

Multiple cardiac rhabdomyomas were found at autopsy in an infant thought clinically to have died of SIDS (Rigle et al., 1989). Here is another quandary for a pathologist — instead of two potentially lethal conditions being evident at autopsy, one is found, which if it were lacking would have allowed the negative diagnosis of SIDS. Multiple cardiac rhabdomyomas were interpreted as an incidental finding in a child who died of poisoning (Byard et al., 1991b).

It was demonstrated that accessory pathways connecting the right atrium to the ventricle contain rhabdomyomatous myocytes (Gotlieb et al., 1977; Mehta, 1993). I identified diffuse cardiac rhabdomyomatosis in the heart of an 8-year-old girl with a history of Wolff-Parkinson-White syndrome; rhabdomyomatous change was also demonstrated in multiple accessory pathways (Silver, 1998; see also Figure 94A and Figure 94B). I saw similar morphology in the native heart of a 6-year-old boy who had a cardiac transplant because

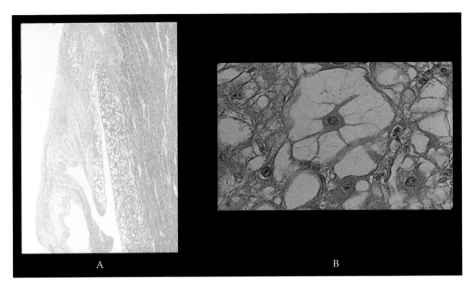

Figure 94 (A) An 8-year-old girl fell dead at school after having just returned from a skiing holiday. She had reentry supraventricular tachycardia and Wolf-Parkinson-White syndrome in infancy treated with propranolol. Thereafter, she developed normally. A full electrophysiological study at age 5 showed no anatomic or electrical abnormality. At autopsy, the heart was grossly normal. Microscopically, it contained numerous strands of rhabdomyomatosis seen here in the right ventricle deep to the tricuspid valve. They were also present within the AV node and His bundle and in two accessory atrioventricular connections. (B) High-power view of the typical "spider cells" of a cardiac rhabdomyoma. The nuclei are central, and the surrounding highly vacuolated cytoplasm appears empty due to glycogen being dissolved during histologic processing. Congenital glycogen tumor is an old name for this lesion, which appears to be hamartomatous.

of restrictive cardiomyopathy. Apart from these striking cases, I have seen many microscopic rhabdomyomas as incidental cardiac findings at autopsy in newborns, both those with normal hearts and some with congenital cardiac malformations.

Cardiac Fibroma

Cardiac fibroma is second in prevalence to rhabdomyoma in childhood. It is usually solitary and more often seen in the second decade of life, when it may present with SD (Amr et al., 1987; Burke et al., 1994). As with rhabdomyoma, another possible presentation is with outflow tract obstruction. Sudden death in infancy was reported with a cardiac fibroma (Mohammed and Murphy, 1997; Ottaviani et al., 1999). Cardiac fibromas occur in Gorlin's (nevoid basal cell carcinoma) and Sotos's (cerebral gigantism) tumor-associated syndromes (Vaughan et al., 2001).

Myocardial Hamartomas and Myocardial Noncompaction

The term "myocardial hamartoma" describes multiple discrete but nonencapsulated masses of markedly hypertrophied myocytes with structural disorganization, focal scarring, and thickened intramural arteries. Such nodules were found throughout atria and ventricles in three young patients, one of whom, a 9-year-old male, died suddenly (Burke et al., 1998b). A similar lesion caused SD in a 24-year-old man with a history of palpitations (Sturtz et al., 1998).

Hamartomatous malformation was the term used to describe two areas of deficient myocardium containing cavernous spaces surrounded by fibroelastic tissue in the left ventricle of a 17-year-old male who died suddenly during exertion (Koponen and Siegel, 1995). This description suggests a focal kind of noncompaction of the left ventricular myocardium, also called spongy myocardium in the newborn and left ventricular hyper-trabeculation in adults. Isolated diffuse noncompaction of the left ventricle may also affect the right ventricle and interventricular septum. It may be familial, occurs in both genders, and leads to dilated cardiomyopathy (Ichida et al., 1999; Guntheroth et al., 2002; Moura et al., 2002). The condition is closely related to Barth syndrome and, like the latter, it is sex linked and confined to males (Bleyl et al., 1997a). Rhythm disturbances were reported in two children with this lesion (Elshershari et al., 2001). Rhythm disturbances are also described with focal giant cardiac myocyte lesions described in infants. These also appear to be hamartomatous (Kapur et al., 1985; Drut and Drut, 1987).

Oncocytic Cardiomyopathy, or Cardiac Oncocytoma/Oncocytosis

Oncocytic cardiomyopathy is a multifocal, sometimes tumorous, oncocytic change in cardiac myocytes dating from early fetal life (Silver et al., 1980) (see also Figure 95A and Figure 95B). It must not be confused with mitochondrial cardiomyopathies discussed above, in which cardiac myocytes are affected diffusely with cardiac enlargement (Silver and Silver, 1992). Oncocytic cardiomyopathy has a high potential for SD before the age of two, although some have survived surgical removal of oncocytic tumors (Kearney et al., 1987; Tazelaar et al., 1992). Affected infants, usually female, die suddenly; lethal cardiac

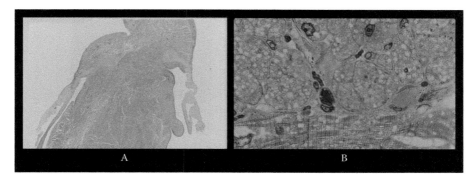

Figure 95 (A) Low-power view of the septal atrioventricular junction from the heart of an 11-week-old female infant who had been treated with digoxin since birth for supraventricular tachycardia. She died a few hours after admission with intractable tachycardia. Grossly, the heart showed numerous tiny yellowish nodules throughout the myocardium of all chamber walls as well as in the epicardium, endocardium, membranous septum, heart valves, and chordae tendiniae. Microscopically, the nodules were the result of cardiac myocytes transformed into oncocytes. In this section, the cells are rounded, pale, and foamy in contrast to the more deeply stained fascicles of normal myocardium seen centrally in the upper ventricular septum. Note the continuity between the atrial and ventricular myocardium through the defective central fibrous body. (B) One micron section epon-embedded and viewed by oil immersion shows cardiac oncocytes above and normal myocardium below. Electron microscopy confirmed the presence of sarcomeric remnants and intercalated discs in the transformed cells and showed that the foamy cytoplasmic appearance was due to accumulation of mitochondria. Compare with Figure 94A and B; like a rhabdomyoma, a cardiac oncocytoma appears to be a hamartomatous lesion.

dysrhythmias have been documented in many cases. Subjects commonly show Wolff-Parkinson-White syndrome (Keller et al., 1987). The multifocal oncocytic change in cardiac myocytes affects the conduction system (Silver et al., 1980; Malhotra et al., 1994; Koponen and Siegel, 1996) as well as the working myocardium and sites from which cardiac muscle generally disappears during fetal development, such as the epicardium, atrioventricular valves, and chordae tendineae (Silver et al., 1980). Other congenital anomalies and histologic features at autopsy suggest it is caused by a prenatal injury, such as a viral infection, but the female sex predominance fits nicely with an X-linked genetic defect described in one case (Bird et al., 1994).

There are many parallels between cardiac rhabdomyoma/rhabdomyomatosis and cardiac oncocytoma/oncocytosis. Both appear to be caused by a degenerative change of cardiac myocytes dating from fetal life; both affect the working myocardium as well as the conduction system (Figure 95A); both cause supraventricular tachycardia starting in fetal life and occur within accessory pathways and are associated with the preexcitation syndrome (Silver, 1998). Also, both may be associated with congenital heart disease (Franciosi and Singh, 1988; Burke et al., 1991b).

Cardiac Myxoma

Although it is the most common heart tumor in adults, cardiac myxoma is very rare in childhood. It may be familial. I have seen only two cases: one had a father with the same cardiac tumor. Such myxomas caused SD in infants due to massive pulmonary embolization (Parker and Embry, 1997) and to obstruction of the tricuspid valve orifice (Hals et al., 1990). A defect in the gene that regulates protein kinase A was found to cause myxoma (Vaughan et al., 2001).

Cardiac Lipomas

Cardiac lipomas are also thought to have a genetic origin (Vaughan et al., 2001).

Benign Cardiac Teratoma

Benign cardiac teratoma was reported in the interventricular septum of a young female child who died suddenly (Swalwell, 1993). We reported a similar tumor in a newborn with cardiac malformations but believe it was a combination of bronchogenic cysts and vascular hamartomas rather than a cystic teratoma (Davis et al., 1996).

Papillary Fibroelastomas

Papillary fibroelastomas are seen occasionally on heart valves in children. As in adults, they may give rise to emboli with lethal effects.

Sudden Death due to Cardiac Rhythm Disorders

The electrophysiologic mechanisms of cardiac arrest leading to death are either bradycardia leading to asystolic arrest or ventricular fibrillation (see Chapter 4). Both are commonly secondary to a major systemic or cardiac insult. Hemorrhagic shock, for example, produces increased extracellular potassium, with bradycardia and ultimately asystole. Regional myocardial ischemia due to coronary atherosclerosis is the most common cause of death from ventricular fibrillation (see Chapter 4). In that instance, local potassium efflux and calcium influx promote local premature impulses from the damaged myocardium. When a pulse is absent, an electrocardiographic (ECG) tracing may indicate whether the patient has asystolic arrest or ventricular fibrillation. The latter is less common in childhood than in adults. A study of 100 pediatric patients with cardiac arrest indicated that the most common terminal electrical activity in children, especially in newborns, was bradycardiac arrest, but the incidence of ventricular tachyarrhythmias was higher in patients who had congenital heart disease, with ventricular fibrillation being seen only in these patients (Walsh et al., 1983). In another study of 157 children, exclusive of diagnosis of SIDS, 19% had ventricular fibrillation and were resuscitated successfully, whereas most who presented with asystole died (Mogayzel et al., 1995). Among 79 consecutive pediatric prehospital cardiac arrest patients (SIDS, trauma, airway obstruction, near-drowning), asystole was the initial rhythm in nearly 80% and ventricular fibrillation in less than 4% (Kuisma et al., 1995). Bradycardia leading to asystole is the mechanism causing deaths directly due to anesthesia, with a threefold higher risk in children than in adults (Keenan et al., 1985).

Lethal Dysrhythmias in Children with a Structurally Normal Heart

In perhaps 10% of childhood SD cases, a lethal dysrhythmia is a primary event and occurs in a structurally normal heart not acutely exposed to either a global or a regional insult. Several different congenital (inherited) and acquired rhythm disturbances predispose adults and children with normal hearts to SD (see also Chapter 9). At all ages, ventricular dysrhythmias such as ventricular tachycardia (monomorphic) and polymorphic ventricular tachycardia (*torsades de pointes*) may degenerate into ventricular fibrillation and are considered so ominous that their onset justifies both medical (antidysrhythmic drugs) and surgical prophylaxis (implanted defibrillators). Silka et al. (1993) implanted cardioverter-defibrillators in 125 children who suffered an episode of SD or had drug-resistant tachyarrhythmias or syncopal episodes; the majority had either hypertrophic or dilated cardiomyopathy, but 26% had primary electrical diseases of the heart, a larger proportion than those with congenital heart disease (18%).

When a pathologist finds no anatomic cause of death at autopsy, and the heart is essentially normal, a primary cardiac dysrhythmic death will be suspected. For example, a previously well child who is directly observed to die during or soon after strenuous exertion or emotional distress has logically died because his or her heart stopped. If at autopsy it is anatomically normal grossly and microscopically, the pathologist will diagnose an unexplained death and suggest the likelihood of a primary cardiac dysrhythmic event. In cases with suspicion of abnormal cardiac conduction, usually related to known ECG abnormalities in the deceased or in his or her family, one may choose to examine the conduction system histologically or at least to preserve it for future examination. Preservation of blood and tissues snap-frozen in liquid nitrogen would also allow for genetic and molecular studies. Most importantly, a pathologist discerning the possibility of a sudden dysrhythmic death would propose clinical investigations of the family for a treatable familial electrophysiologic disorder.

Dysrhythmias associated with an increased risk of SD in children with a structurally normal heart include a prolonged Q–T interval, various ventricular tachyarrhythmias with normal Q–T interval, ventricular preexcitation, atrial tachyarrhythmias, and conditions of conduction blockage.

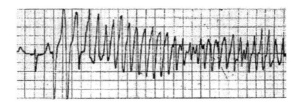

Figure 96 The pattern of polymorphic ventricular tachycardia known as "torsade de pointes." Note QRS complexes twisting around the isoelectric line with continuous changes in the QRS contour and amplitude. This type of dysrhythmia may degenerate into ventricular fibrillation.

Prolonged Q–T Interval

The Q–T interval determined from an ECG represents the period of myocardial repolarization following a systolic contraction. The upper limit of a normal Q–T interval corrected for heart rate (Q–Tc) is 0.46 (Keating, 1996). The risk of life-threatening arrhythmias increases exponentially as the Q–Tc increases. The long Q–T syndrome may be congenital and heritable (Romano-Ward syndrome; Jervall and Lange-Nielson syndrome) or acquired. Most cases are probably caused by a combination of genetic and environmental influences (Keating, 1996). Acquired prolonged Q–T interval may be the result of head trauma, severe bradycardia, various drugs, poisons, and electrolyte abnormalities.

Patients with prolonged Q–T intervals commonly develop ventricular tachyarrhythmias during periods of adrenergic stress, such as physical exercise or emotional upset. They may present with palpitations, syncope, supposed seizures, or SD, with the latter the first manifestation in some children (Garson et al., 1993). Syncope can follow onset of 2:1 block or ventricular tachycardia, characteristically a polymorphous tachycardia called *torsades de pointes*, because the peaks of the QRS complexes twist to one side of the isoelectric baseline and then to the other (Figure 96). There is a tendency for torsades de pointes to induce ventricular fibrillation, causing death unless heart rhythm is promptly restored.

The two heritable forms of long Q–T syndrome described below are both related to genetic defects in ion channel proteins that lead to delayed myocellular repolarization. The delay causes reactivation of secondary calcium channels, which favor development of the dangerous ventricular tachycardias described above (Keating, 1996). When physical or emotional stress promotes onset of torsades de pointes in congenital long Q–T syndrome, it is commonly called "adrenergic dependent." In acquired disease, its onset is always preceded by a long cycle length or pause, but this "pause/dependent" torsades de pointes may also occur in congenital long Q–T syndrome cases (Viskin et al., 1996).

Affected patients with the familial forms of the disease are diagnosed by protracted electrocardiographic studies, including stress testing and even an infusion of catecholamines. They are treated effectively with beta-adrenergic blocking agents or, if they survived an aborted SD, an implanted defibrillator. More specific drug therapy, for example, to block the sodium channel or open a potassium channel, is in the offing (Keating, 1996).

Congenital Long Q–T or Romano-Ward Syndrome

Congenital long Q–T syndrome, or Romano-Ward syndrome, has autosomal dominant transmission but can also be sporadic. Gene frequency is estimated at about 1 in 10,000 population (Chiang and Roden, 2000). The defective gene has a low penetrance in some families so that silent gene carriers would make this a gross underestimate (Priori et al.,

1999). So far, the six known subtypes are caused by many known mutations on several genes encoding cardiac ion channels, with five genes identified. These govern four proteins associated with potassium channels and one with the sodium channel (Keating, 1996; Chiang and Roden, 2000; Splawski et al., 2000). Subtypes 1 and 2 are the most common, and the others are quite rare (Splawski et al., 2000). Subtype 3 is related to defects in the sodium channel, and mutations here also cause the Brugada syndrome as well as familial conduction system disease. Our rapidly evolving understanding of cardiac electrophysiology is largely due to establishing how mutations in the cardiac ion channels, which produce the critical electrical currents in the heart, cause the long Q–T syndrome.

A clinical diagnosis of long Q–T syndrome is determined by the presence of a Q–Tc greater than 0.44 sec and a structurally normal heart (Marcus, 2000). A normal Q–Tc does not exclude a family member from being a genetic carrier. Some long Q–T genotypes tend to be associated with a distinctive electrocardiographic pattern (Zhang et al., 2000b), but genetic testing is still the only way to detect carriers (Kaufman et al., 2001). Among affected subjects, usually adolescent children or young adults, syncope, cardiac arrest, and SD are often triggered in a gene-specific fashion: exercise for subtype 1, emotion for subtype 2, and sleep for subtype 3 (Schwartz et al., 2001). Subtype 3 patients show lengthening of Q–Tc during sleep (Stramba-Badiale et al., 2000; Clancy et al., 2002). Apparently, subtype 1 and subtype 2 subjects have many life-threatening episodes, but LQ–T3 subjects are more prone to die during such events (Zareba et al., 1998). Death while swimming seems most associated with subtype 1; death associated with a loud noise seems most associated with subtype 2 (Moss et al., 1999). A 12-year-old boy with long Q–T syndrome was wearing an implanted cardioverter-defibrillator when he dived into cold water. It caused an irregular heart rhythm followed by further prolongation of the Q–T interval, then a premature ventricular complex followed by pulseless ventricular tachycardia, which was subsequently converted to sinus rhythm (Batra and Silka, 2002).

Among congenital long Q–T syndrome patients, Locati et al. (1998) found that the risk of cardiac events was higher in males until puberty and higher in females during adulthood. The same pattern was evident among subtype 1 gene carriers. The high rate of silent carriers is important, because they are also vulnerable to SD if exposed to certain drugs that block potassium channels (Keating, 1996; Priori et al., 1999).

Long Q–T syndrome presents in the fetus and newborn either as bradycardia due to 2:1 heart block or with ventricular tachycardia and intermittent bradycardia (Hofbek et al., 1997; Gorgels et al., 1998; Piippo et al., 2000). In one study, a prolonged Q–Tc in a newborn proved transient if less than 0.50, but four of eight cases with a Q–Tc greater than 0.60 died soon after birth (Villain et al., 1992). Piippo et al. (2000) reported a severe form of long Q–T in two siblings who had inherited the same potassium channel defect from both parents.

Among 287 children with long Q–T syndrome included in a multicenter study, 8% presented with SD. Those with a Q–Tc of more than 0.60 were considered at particularly high risk (Garson et al., 1993). In a prospective study, syncopal episodes in affected children were often misinterpreted as seizures, and 50% of probands of affected families had experienced either one syncopal episode or SD before they were 12 years old (Moss et al., 1991).

The entire conduction system and its blood supply were histologically normal in a 15-year-old male with congenital long Q–T syndrome (Moothart et al., 1976). James et al. (1978) studied the conduction system in eight patients who died suddenly and found focal neuritis and neural degeneration within the sinus node, atrioventricular node, His bundle, and ventricular myocardium in all cases. Subsequent similar studies in a few patients showed fibrosis and fatty infiltration in the conduction system (Bos et al., 1985; Bharati et al., 1985) as well as lymphocytic infiltration of cardiac sympathetic nerves (Bos et al., 1985) and of the ventricular myocardium (Bharati et al., 1985). With the new knowledge that the disease originates at the molecular level in cardiac myocyte ion channels, one must assume that the recorded conduction system abnormalities are secondary.

Cardioauditory Syndrome or Jervall and Lange-Nielson Syndrome

Jervall and Lange-Nielson syndrome includes deafness attributed to altered lymph stasis in the inner ear and a prolonged Q–T interval. The deafness is transmitted by an autosomal recessive gene common in Norway. It affects a potassium channel. The same gene defect responsible for the most common subtype of the autosomal dominant syndrome, subtype 1, produces the cardioauditory syndrome when both alleles carry the mutation (Neyroud et al., 1997; Tranebjaerg et al., 1999). Homozygous inheritance of alleles from a second closely related potassium channel gene, the one that causes autosomal dominant subtype 5, can also induce the cardioauditory syndrome. The occurrence of deafness requires two mutant alleles, but even a single one (as in the parents) appears to increase the risk of arrhythmia (Chiang and Roden, 2000).

Rett Syndrome

This severe neurodevelopmental disease is confined to females. Those with Rett syndrome are prone to SD, have significantly lower heart rate variability, a marker of autonomic disarray, and have prolonged Q–Tc (Guideri et al., 1999; Ellaway et al., 1999).

Familial Dysautonomia (Riley-Day Syndrome)

Riley-Day syndrome is a developmental disorder of the sensory and autonomic nervous system. Subjects are prone to SD during sleep (Axelrod et al., 2002). In them, a prolonged Q–T interval reflects sympathetic dysfunction, but they appear to be more prone to bradycardic asystole than polymorphic ventricular tachycardia leading via ventricular fibrillation to SD (Glickstein et al., 1999).

Andersen's Syndrome

Ventricular arrhythmias and a prolonged Q–T interval are described in patients with Andersen's syndrome, which features periodic paralysis, ventricular arrhythmias, and dysmorphic features. Subjects can develop syncope and die suddenly (Canun et al., 1999). The potassium-sensitive periodic paralysis is caused by a gene defect in potassium channels present in skeletal and cardiac muscle but is not related to those ion channel defects that cause congenital long Q–T syndrome. However, Q–T interval prolongation appears to be an integral feature of this syndrome (Sansone et al., 1997; Ai et al., 2002).

Corcos et al. (1989) described a Jerbian family with several cases of SD that had both congenital long Q–T syndrome and complete situs inversus. Marks and co-workers (1995) reported three unrelated children with both congenital long Q–T syndrome and bilateral cutaneous syndactyly, another autosomal dominant genetic disease. All also had congenital heart disease (patent ductus arteriosus in two), and two died suddenly.

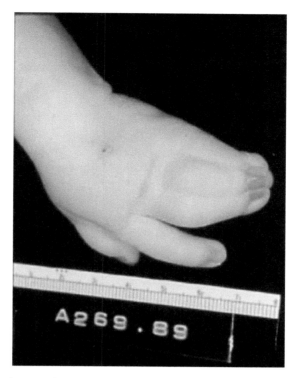

Figure 97 This newborn infant diagnosed with sporadic long Q–T syndrome died with 2:1 atrioventricular conduction block when 3 weeks old. He had bilateral symmetrical syndactyly, seen in this picture, as well as a ventricular septal defect. Syndactyly was inherited from his father. Others reported the association of long Q–T syndrome with (1) syndactyly or (2) ventricular septal defect. This case shows all three conditions.

Concerning associated congenital heart disease, two cases with congenital long Q–T syndrome presented during fetal life with ventricular tachycardia and atrioventricular block, and both had a ventricular septal defect. One died of intractable ventricular tachycardia at 4 days of age; despite having an implanted pacemaker, the second died suddenly at 10 months (Wu et al., 1999).

We saw a male infant who died when 3 weeks old with 2:1 heart block, after being diagnosed with congenital long Q–T syndrome that was apparently sporadic rather than familial. The infant also had cutaneous symmetrical syndactyly (Figure 97) inherited from the father, as well as a ventricular septal defect (Silver, 1998).

Acquired Prolongation of Q–T Interval

The Q–T interval is prolonged by various cardiac drugs, including the antiarrhythmics quinidine and flecainide; various extracardiac drugs, for example, tricyclic depressants such as thioridazine; as well as certain antibiotics, antimalarials, and antihistamines (Moss, 1999; De Ponti et al., 2002). Antiarrhythmic and pyschoactive drugs are the most common cause of acquired prolonged Q–T syndrome (Keating, 1996). Antipsychotic drugs are commonly prescribed in children and have caused SD (Riddle et al., 1993; Welch and Chue, 2000). Electrolyte imbalance, especially hypokalemia, hypomagnesemia, and hypocalcemia, impaired hepatic/renal function, and concomitant use of drugs (e.g., antibacterials with antiarrhythmics), also lengthen the Q–T interval, especially if a patient already has the congenital long Q–T syndrome, is a carrier, or has bradycardia (De Ponti et al., 2002). Inhaled volatile substances, for example, gasoline, butane, propane, and trichloroethane, sensitize the myocardium to catecholamines, so that a sudden alarm or exertion in a child sniffing gasoline may cause lethal dysrhythmia (Adgey et al., 1995). Cocaine also has this effect (Bauman and DiDomenico, 2002). Arsenic given to treat promyelocytic leukemia is also known to lengthen the Q–T interval and provoke torsade de pointes (Unnikrishnan et al., 2001). Anorexia nervosa cases show a significantly prolonged Q–T interval, with reversal to normal when the subject eats normally. Other electrocardiographic abnormalities are seen in these patients as well as reduced left ventricular mass and impaired myocardial performance, mitral valve prolapse, and SD (Cooke and Chambers, 1994).

Antiarrhythmic drugs may induce ventricular dysrhythmias. For example, two infants with cardiac malformation, atrial flutter, and congestive heart failure developed ventricular fibrillation a few days after starting amiodarone (Pohlgeers and Villafane, 1995). Parenthetically, the Q–T interval was normal in both, although amiodarone causes Q–T prolongation in adults (Torres et al., 1986). A newer antiarrhythmic drug, ibutilide, was given to four female patients to convert atrial fibrillation and flutter and caused torsade de pointes in each (Gowda et al., 2002).

Congenital Long Q–T Syndrome

Congenital long Q–T syndrome was proposed to cause SIDS (Schwartz et al., 1998b). One well-documented case of missed-SIDS was found to have a spontaneous mutation in the sodium channel gene (Schwartz et al., 2000b). The sodium channel defect seen in the LQ–T3 (autosomal dominant or sporadic) subtype of congenital long Q–T syndrome is related to SD during sleep rather than during exercise or stress (see above).

One study done between birth and day 3 showed no significant Q–T prolongation in its many subjects (Southall et al., 1986). However, an abnormally prolonged Q–T in newborns between 3 and 4 days old was subsequently suggested to mark infants who would die of SIDS (Sadeh et al., 1987; Schwartz et al., 1998b). Schwartz and co-workers (2000b) proposed two possible mechanisms: some have a developmental alteration in cardiac sympathetic innervation, which normally continues until 6 months of age, or some SIDS victims may either inherit or acquire long Q–T syndrome. The first mechanism is supported by the Q–T interval lengthening physiologically and temporarily in the first few months of life (Schwartz et al., 1982; Weinstein and Steinschneider, 1985). The second mechanism is supported by a single case report (Schwartz et al., 2000b). That infant was resuscitated successfully, which is very rare in SIDS.

The cardiac conduction system has been studied histologically in large groups of SIDS cases and smaller numbers of control hearts, the latter being hard to find in this age group. These studies, dating from 1968 to 2000, are recorded in the most recent report (Matturi et al., 2000), which is itself examined in an accompanying commentary (Anderson, 2000).

Ventricular Tachyarrhythmias with a Normal Q–T Interval

Polymorphic Ventricular Tachycardia

Mutations in the cardiac ryanodine receptor gene, the major calcium release channel in the sarcoplasmic reticulum, underlie this inherited arrhythmogenic disease that occurs in a structurally intact heart (Priori et al., 2001; Bauce et al., 2002). The mutations also cause malignant hyperthermia, responsible for many anesthetic deaths, and are associated with central core myopathy (Fortunato et al., 2000; Monnier et al., 2001).

The electrocardiographic pattern of polymorphic ventricular tachycardia closely resembles the arrhythmias associated with calcium overload and digitalis toxicity (Priori et al., 2001). The tachycardia is catecholamine sensitive and triggered by exercise, but the Q–T interval is normal. It presents during childhood, and children as well as young adults, especially males, are prone to syncope and SD during physical and emotional stress (Leenhardt et al., 1995; Priori et al., 2002).

A malignant type of polymorphic ventricular tachycardia was induced by exercise in two Finnish families. Members had an autosomal dominant gene located on the long arm of chromosome 1, later identified as the ryanodine receptor gene (Swan et al., 1999; Laitinen et al., 2000). A report on a clinically similar but autosomal recessive disease in two Bedouin families also localized the mutation to the same gene locus on chromosome 1 (Lahat et al., 2001).

Members of a family with a strong history of syncope and SD developed rapid polymorphic ventricular arrhythmias in response to emotional stress, isoproterenol infusion, or exercise. Autopsies in two young family members showed anatomically normal hearts (Fisher et al., 1999). Right ventricular cardiomyopathy, discussed above, was specifically excluded in that family. However, defects in the cardiac ryanodine receptor gene also cause type 2 right ventricular cardiomyopathy. Two families that experience five juvenile SDs during exercise had autopsy findings in three of them of fibro-fatty replacement of the apical segment of the right ventricle, and 16 living family members had polymorphic ventricular arrhythmias during effort; 15 of the latter were also diagnosed with right ventricular cardiomyopathy (Bauce et al., 2000).

Sporadic polymorphic ventricular tachycardia is similar to the familial disease and may be related to abnormal catecholamine sensitivity. The Q–T interval is normal. If present in children or adults, it may lead to SD (von Bernuth et al., 1982; Leenhardt et al., 1995; Eisenberg et al., 1995).

Brugada Syndrome

The heritable autosomal dominant rhythm disorder in Brugada syndrome is the result of mutations located in the cardiac sodium channel gene (Brugada and Brugada, 1992; Brugada et al., 2000). Patients, usually males, have a peculiar electrocardiographic pattern of right bundle branch block and an elevated ST segment in leads V1 to V3 (also known as "precordial injury pattern"). They can develop polymorphic ventricular tachycardia or ventricular fibrillation and have a strong tendency to succumb to SD, but that occurs in adults rather than in children. However, the syndrome can cause SD in children and even in very young infants, where it may be misdiagnosed as SIDS (Pinar Bermudez et al., 2000; Priori et al., 2000a; Suzuki et al., 2000). As in the congenital long Q–T syndrome, the heart is structurally normal, and Brugada subjects have a normal Q–T interval. They are treated with an implanted defibrillator, because death is not prevented by amiodarone or beta-blockers.

The electrocardiographic pattern that, in the presence of a normal heart and normal Q–T interval, allows a diagnosis of Brugada syndrome is not specific. It is seen occasionally in healthy controls (Viskin et al., 2000; Remme et al., 2001). In some patients, the pattern can be elicited by certain antiarrhythmic drugs (e.g., intravenous flecainide) that cause sodium channel blockade (Marcus, 2000). A similar precordial injury is seen in some anatomic heart diseases, for example, acute ischemia of the right ventricle and cardiomy-opathies, and with tricyclic drug overdose (Scheinman, 1997). In right ventricular cardi-omyopathy, the typical ECG pattern is similar (Corrado et al., 1996; Marcus, 2000). It was proven at autopsy that many supposed Brugada syndrome subjects had right ventricular cardiomyopathies (Corrado et al., 1996; Tada et al., 1998; Corrado et al., 2001a).

Patients with Brugada syndrome do not differ from those with idiopathic ventricular fibrillation and a normal ECG (see above) in their age group (young adults). Both groups show similar spontaneous and inducible arrhythmias. Patients with Brugada syndrome are, however, more often male, have a family history of SD, and develop ventricular fibrillation during sleep (Viskin et al., 2000). The latter might be due to increased vagal activity and less sympathetic activity at night (Matsuo et al., 1999). This is reminiscent of the tendency for death to occur during the sleep of subtype 3 congenital long Q–T syn-drome subjects (see above). An intravenous flecainide test can also induce ST segment elevation and Q–T interval shortening in subtype 3 patients. This demonstrates a pheno-typic overlap between subtype 3 and Brugada syndrome (Priori et al., 2000b). Bezzina and co-workers (1999) described a large eight-generation family with a high incidence of nocturnal SD, with Q–T-interval prolongation and a Brugada ECG pattern both present in the same subjects. Grant et al. (2002) described a four-generation family, including 17 gene carriers with long Q–T syndrome, Brugada syndrome, and conduction system disease, all linked to a single sodium channel mutation. Kyndt and co-workers (2001) reported on a French family whose members showed either the Brugada syndrome or isolated conduc-tion defect, due to the same mutation in the sodium channel gene. Subtype 3 and Brugada syndrome seem to have opposite molecular effects: the former induces a gain in cardiac sodium channel function, whereas Brugada syndrome mutations reduce that function (Deschenes et al., 2000).

Sudden Unexpected Nocturnal Death Syndrome

Sudden unexpected nocturnal death syndrome during sleep has long been known to be prevalent among young males from Southeast Asia and Japan. Such deaths continue in immigrants to North America and elsewhere. The mode of death is a polymorphic ven-tricular tachycardia followed by ventricular fibrillation. The testing of a group of 27 Thai men who survived cardiac arrest due to ventricular fibrillation while asleep revealed that 16 had a Brugada pattern ECG, whereas 11 had a normal ECG and a much lower incidence of both ventricular fibrillation and SD (Nademanee et al., 1997). Six of 11 patients in a Japanese series also had a Brugada ECG pattern (Kasanuki et al., 1997).

Mutations in the same sodium channel gene known to cause Brugada syndrome were identified in families of patients with SD during sleep. Vatta and co-workers (2002) found mutations in the sodium channel gene in affected families. This seemed to confirm that the syndrome of nocturnal death is genetically as well as phenotypically identical to the Brugada syndrome.

Autopsy studies on immigrants to North America who died in their sleep did not show any cases of right ventricular cardiomyopathy but suggested mild left ventricular hyper-trophy (Kirschner et al., 1986). The latter was denied by a Toronto report (Pollanen et al., 1996). Kirschner et al. (1986) described various abnormalities in the conduction system in 17 of 18 cases. A Japanese study that revealed fibrosis of the sinoatrial node with only minor anomalies of its artery claimed the changes were specific for this syndrome (Okada and Kawai, 1983). Again, the cause of sudden nocturnal death syndrome is now known to reside in a genetic defect — one may interpret such histologic changes in the conduction system as either nonspecific or secondary.

Miscellaneous Ventricular Tachyarrhythmias

Idiopathic Familial and Sporadic Ventricular Fibrillation

Young adults of both genders with a normal ECG may develop idiopathic ventricular fibrillation. In contrast to other polymorphic tachyarrhythmias, idiopathic ventricular fibrillation is not generally related to stress, and familial involvement is rare (Viskin and Belhassen, 1998). The "short-coupled variant of torsade de pointes" is a highly lethal dysrhythmia that resembles idiopathic ventricular fibrillation (Leenhardt et al., 1994; Viskin and Belhassen, 1998). In three families, three different mutations causing the latter were found in the cardiac sodium channel gene (Chen et al., 1998).

Commotio Cordis

This condition was named in the Victorian era but is still not well understood. In very rare cases, mainly in young athletes, death refractory to immediate resuscitation is caused by a low-energy, nonpenetrating, blow to the chest wall (Maron et al., 1995, 2002). The subject is often a male adolescent playing sports and wearing protective chest padding. Ventricular fibrillation was documented in subjects who were resuscitated. The SD is attributed to the blow occurring during a narrow window within the repolarization phase of the cardiac cycle, just before the peak of the T wave (Maron et al., 1999). See further discussion in Chapter 12.

Electocardiographic Patterns Associated with Sudden Death

Garg and co-workers (1998) described an 18-year-old male resuscitated following his SD due to ventricular fibrillation who had a terminal QRS abnormality on surface 12-lead ECG. They thought this represented a unique familial syndrome or possibly a variant of the Brugada syndrome. Idiopathic familial persistently short Q–T interval was described in a woman, her son and daughter, and an unrelated man, all of whom died suddenly (Gussak et al., 2000).

Chambers et al. (1995) reported on a family whose members died suddenly but whose hearts were normal on ECG. Hearts were also normal at autopsy in two cases. Two survivors had a clinical history of arrhythmic events; both had abnormal signal-averaged ECGs and inducible ventricular arrhythmias during electrophysiologic studies. The authors proposed that this ECG pattern is a harbinger for SD.

Ventricular Preexcitation Syndromes

Ventricular preexcitation is caused by accessory myocardial connections crossing from atria to ventricles. These allow electrical impulses traveling from atria to ventricles or in the reverse direction to reach the working myocardium faster than if they passed through the atrioventricular node and Hisian system. Conduction is usually retrograde, causing reentrant supraventricular tachycardia, but it may be antegrade with the danger of SD if a subject develops atrial fibrillation. The latter can be provoked by certain drugs used to treat supraventricular tachycardia (Janousek and Paul, 1998). During atrial fibrillation, transmission of more than 300 beats per minute through an accessory pathway can initiate ventricular fibrillation. Drugs such as digitalis that prolong the conduction time and AV node refractoriness are also dangerous because they may promote ventricular fibrillation. The patients at highest risk have atrial fibrillation and the shortest preexcited RR (ventricular pause) interval (Bromberg et al., 1996). Spontaneous onset of ventricular fibrillation is described in some cases (Timmermans et al., 1995; Sanatani et al., 2001). Those who survive cardiac arrest are treated by catheter radiofrequency ablation or surgical division of accessory atrioventricular pathways (Antz et al., 2002).

Wolff-Parkinson-White Syndrome

Wolff-Parkinson-White (WPW) syndrome describes patients who have symptoms (e.g., palpitations) and syncope. These are usually due to a reentrant supraventricular tachycardia but can be caused by atrial fibrillation or both dysrhythmias. Males are more likely to manifest the syndrome than females and develop it at an earlier age (Goudevenos et al., 2000). The condition commonly presents in early infancy (Deal et al., 1985). It is a fairly common and generally benign condition (Munger et al., 1993; Goudevenos et al., 2000) and is significantly associated with cardiac malformations — 30% of 60 children under 18 years with WPW syndrome (Bromberg et al., 1996). The most common association is with Ebstein anomaly (Deal et al., 1985). Accessory pathways are present around the anomalous Ebstein valve in about 30% of patients (Hebe, 2000).

Familial WPW syndrome occurring in association with conduction system disease and left ventricular hypertrophy (not hypertrophic cardiomyopathy), with autosomal dominant inheritance, is produced by mutation of a gene on the long arm of chromosome 7. It encodes a subunit of AMP-activated protein kinase regulating the glucose metabolic pathway in muscle (MacRae et al., 1995; Gollob et al., 2001). The syndrome also occurs in association with isolated noncompaction of the left ventricle, also called left ventricular hypertrabeculation, which may be familial (Ichida et al., 1999), see above. Also discussed above, but sporadic rather than familial, both cardiac rhabdomyomatosis and oncocytosis are associated with WPW syndrome.

A diagram of the conduction system is seen in Figure 98. The most common accessory pathway is a direct atrioventricular (AV) connection, or Kent bundle. It is often bilateral in the lateral AV ring and can be in right or left posterior septum or in the mid septum. Connections can be traced anatomically by serial section histology (Sugiura et al., 1989). In addition to AV connections consisting either of working or transitional myocardium, slow and fast atrial pathways exist that enter the AV node. They are composed of atrial myocardium. The names of the various connections have changed over the years, concomitant with the development of accurate electrophysiological mapping (Anderson and Ho, 1997). Conduction in the accessory connections is assessed clinically by stress and drug testing, but the propensity of the atria to develop fibrillation can only be tested electrophysiologically. Septally located accessory pathways are significantly more common in WPW patients with spontaneous ventricular fibrillation who survive a cardiac arrest (Timmermans et al., 1995). An autopsy series of 10 cases of SD in children and young adults with ventricular preexcitation demonstrated one or more subendocardial Kent bundles in all; most were left lateral in position (Basso et al., 2001b). Accessory AV connections can be multiple (Abbott et al., 1987) and familial (Vidaillet et al., 1987; Lu et al., 2000), the latter being more likely to have multiple connections (Vidaillet et al., 1987). However, multiple connections are not proven to carry a higher risk of SD (Timmermans et al., 1995; Bromberg et al., 1996).

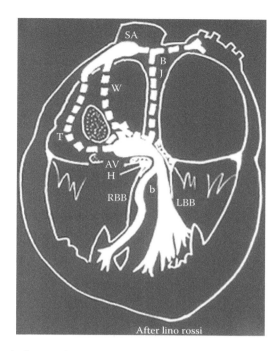

Figure 98 Diagram of the cardiac conduction system to show its normal anatomy. The preferential pathways between the sinoatrial node and the atrioventricular node are represented by dotted lines. They are composed of working myocardium. The bundle of His passes from the right atrium to the summit of the interventricular septum by penetrating the central fibrous body, where it bifurcates into right and left branches. It should be the only connection between atria and ventricles. Microscopically, cells composing the proximal Hisian system are larger and paler than working myocardium. The left main branch and distal right main branch are composed of even larger (Purkinje) cells.

I examined a young man who died suddenly with WPW syndrome and recurrent syncope. The septal leaflet of the tricuspid valve was fused to the septum, producing a mild form of Ebstein anomaly. The accessory connection was identified by electrophysiological mapping during life and demonstrated postmortem in the posterior septum, bypassing the AV node and joining the His bundle (Figure 99). Mild or micro-Ebstein anomaly together with WPW syndrome was reported to be associated with the SD of young athletes (Thiene et al., 1983).

Lown-Ganong-Levine Syndrome

Lown-Ganong-Levine syndrome is closely related to the WPW syndrome. It is caused by enhanced conduction through the AV node. Severe hypoplasia of the AV node and absence of accessory pathways was described in one young man who died suddenly with documented enhanced AV nodal conduction (Ometto et al., 1992).

Miscellaneous Atrial Tachyarrhythmias

Fetal supraventricular tachycardia (SVT) causes significant fetal and neonatal morbidity. It is highly associated with AV block in utero and with accessory AV connections that allow for AV reentrant tachycardia (Naheed et al., 1996; Etheridge and Judd, 1999). Since the

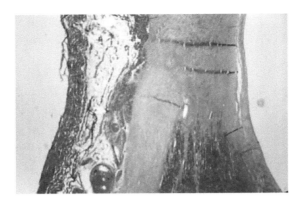

Figure 99   A 25-year-old man with Wolff-Parkinson-White syndrome died in his sleep. At autopsy, the heart showed a mild form of Ebstein anomaly, the septal leaflet of the tricuspid valve being fused to the underlying septum. This microsection shows the atrioventricular node just to the left of the central fibrous body. A large strand of atrial myocardium passed deep to the endocardium to the left of the bundle and entered the bundle of His at a lower level. We could not exclude other accessory connections, because the entire atrioventricular junction regions were not serially sectioned.

advent of fetal ultrasonography and echocardiography, it is clear that many cases of peri-natal hydrops are caused by SVT (Silver et al., 1996b). SVT occurs frequently in infancy and is controlled by drugs (Etheridge and Judd, 1999). Apart from reentrant tachycardia, other mechanisms of SVT include chaotic atrial and junctional ectopic tachycardias. Junc-tional ectopic tachycardia, an incessant tachycardia with normal QRS morphology and AV dissociation occurring in the newborn, can cause SD (Villain et al., 1990). Chaotic atrial tachycardia also happens in the newborn and can cause heart failure refractory to drug therapy. It has a high mortality, and such deaths are sometimes classed as sudden (Yeager et al., 1984). This arrhythmia in infants and children can be associated with cardiac malformation (Bisset et al., 1981). Among infants and children, incessant tachycardias such as SVT can induce secondary dilated cardiomyopathy. Sometimes radiofrequency catheter ablation treatment is needed to stop the tachycardia and reverse the ventricular dilation (Ciszewski et al., 1994; Lashus et al., 1997; Paul et al., 2000).

In children and adults, SVT is most often the result of retrograde conduction across accessory AV connections, as discussed above. It can cause disabling, potentially life-threatening symptoms (Wood et al., 1997). SVT caused cardiac arrest in approximately 5% of patients studied after an aborted SD (Wang et al., 1991). Certain drugs used to treat SVT are prone to promote arrhythmias. The best known is quinidine, but encainide or flecainide use was associated with three cardiac arrests or deaths among 15 young people who developed proarrhythmia in a group of 579 who received those drugs (Fish et al., 1991).

Atrial flutter can be dangerous in children. It is associated with late SD after surgical repair of tetralogy of Fallot (see above) but otherwise is rare in children with a normal heart (8% in a series of 380 cases) (Garson et al., 1985). Atrial fibrillation may be dangerous in WPW patients because it can initiate ventricular fibrillation (see above). Atrial fibrilla-tion, whether in a normal or an abnormal heart of an adult, is highly associated with stroke due to cerebral embolism and even from "normal heart," atrial biopsy consistently reveals histologic abnormalities, including myocarditis (Kobayashi et al., 1988; Frustaci et al.,

1997). Brugada and co-workers (1997) described atrial fibrillation in a family with the gene locus on the long arm of chromosome 10. The conduction system in a man with familial atrial fibrillation and congenital absence of sinus rhythm who died suddenly showed marked atrophy, degeneration, and isolation of the sinoatrial node (Bharati et al., 1992c).

Sudden Death in Conduction System Disorders

Many of these SDs appear to have an anatomic rather than functional cause. However, because a gene defect can cause both familial conduction system disease and primary "electrical" SD, one would suppose that familial conduction system diseases are functional at onset. Also, in many primary electrical disorders, bradycardias due to AV conduction block often initiate the ventricular tachyarrhythmias that cause SD. If anatomic accessory AV connections are present, they may allow an impulse to bypass the Hisian system, thereby exciting the ventricles prematurely; thus atrial fibrillation may initiate ventricular fibrillation (see above). As mentioned at the outset of this chapter, morphologic lesions in the conduction system occupy a gray area between form and function. They are included where pertinent.

The sinus node initiates the impulse for myocardial contraction (Figure 98). It is innervated by the autonomic system, as is the AV node. Exercise enhances sympathetic discharge and results in physiologic tachycardia. However, exercise in certain patients with a diseased conduction system can precipitate heart block. The AV node transmits the impulse to the ventricles via the bundle of His and its branches. The His-Purkinje system is relatively devoid of autonomic nerve supply and is not influenced by autonomic stimulation. Finally, the impulse reaches the ventricular myocardium and initiates systolic contraction. The majority of ventricular dysrhythmias result from disease at this level. The most common is coronary atherosclerosis, dealt with elsewhere. Myocardial ischemia is also the most likely cause of ventricular fibrillation in congenital heart disease and various coronary artery diseases (see above). Myocardial diseases have in common the damage and repair of cardiac myocytes with their loss, atrophy, hypertrophy, and disorganization as well as interstitial and replacement fibrosis. They share this morphology with myocardial ischemic damage, thus producing a substrate for dysrhythmia. Cardiac tumors cause dysrhythmias in different ways, mainly according to their sites and sizes in relation to the conduction system.

Sudden Death in Atrioventricular Conduction Block and Sick Sinus Syndrome

Compared to the dysrhythmic diseases previously discussed, AV conduction block and sinoatrial node dysfunction are less likely to lead to SD. Pacemakers are used to treat severe bradycardias but do not always prevent SD, especially in young adult males (Zehender et al., 1992). Sudden death occurred in 71 of 378 patients bearing an implanted pacemaker because of either AV conduction block or sick sinus syndrome, and SD was more likely in those with the former than with the latter (Mattioli et al., 1995). In children, sick sinus syndrome is usually a result of surgical damage to the sinoatrial node. For example, it was a problem after the Mustard procedure for transposition of the great vessels (see above) (Figure 77). However, the condition can occur in isolation. Sinus node dysfunction is usually a problem with aging but occasionally causes SD in middle years. In a clinical study of children with exercise-related syncope, one died suddenly because of sinoatrial node dysfunction (Noh et al., 1995).

Like congenital complete heart block, as discussed below, sinus node dysfunction and AV conduction block can also be caused by an autoimmune mechanism. When compared to age-matched controls, patients with anti-sinus node antibodies have a tenfold higher risk of developing sick sinus syndrome, whereas those patients with anti-AV node antibodies have a threefold increased risk of acquiring AV block (Ristic and Maisch, 2000).

Sugiura et al. (1976) reported histologic findings in sick sinus syndrome affecting elderly subjects. They noted marked reduction in sinoatrial nodal myocytes as well as moderate to marked fibrosis. Bharati et al. (1992a) reported marked fatty change in the sinoatrial node and its approaches in two elderly women bearing pacemakers.

Familial Atrioventricular Conduction Block and Sick Sinus Syndrome

Conduction system malfunction complicates several kinds of familial cardiomyopathies associated with SD. It is an integral part of that form of dilated cardiomyopathy caused by defects in the lamin A/C gene and is prevalent in many kinds of muscular dystrophy and myopathy discussed previously and below. It was a feature in two families with restrictive cardiomyopathy and mild desmin (myofibrillar) myopathy. Conduction system malfunction may also be important in SD occurring in hypertrophic cardiomyopathy. Electrophysiological studies in that condition show that most patients judged to be at increased risk for SD have sinoatrial or His-Purkinje conduction disease or inducible supraventricular or ventricular tachycardia (Maron and Fananapazir, 1992).

Conduction system malfunction is also associated with primary "electrical" heart disease. Mutations in the sodium channel gene that cause both the congenital long Q–T and the Brugada syndrome also cause progressive cardiac conduction disease. Other familial cases of conduction system dysfunction make those with the conditions prone to SD. Stéphan et al. (1997) reported a hereditary autosomal dominant conduction defect in a large Lebanese family that presented with right bundle branch block in infancy; some cases progressed to complete heart block. This change could be abrupt, leading to SD. In that family, the gene defect was mapped to the long arm of chromosome 19. Penetrance was low, and males were mainly affected. Brink et al. (1995) described a similarly affected South African family but with a higher penetrance and a higher incidence of SD. Again, the condition was more common in males. Those authors also localized the gene defect to the long arm of chromosome 19.

Hosoda and co-workers (1999) identified a gene defect that causes familial atrial septal defect and atrioventricular conduction disturbance, with some cases of SD. Two families studied did not show any symptoms in childhood, and one family also had anomalous pulmonary venous return (Gutierrez-Roelens et al., 2002). Barak et al. (1987) reported that adults in a family suffered syncope and Stokes-Adams attacks. They were diagnosed with both sinus node dysfunction and cardiac conduction disturbances after the proband, a third trimester fetus, was diagnosed with heart block by ultrasonography. Familial sinoatrial node dysfunction (sick sinus syndrome) may require pacemaker insertion (Isobe et al., 1998). Histology of the conduction system in a man with familial absence of sinus

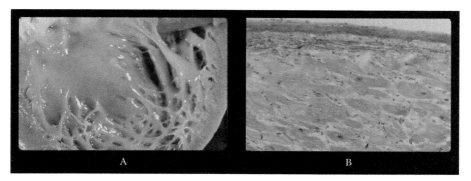

Figure 100   (A) This left ventricle shows hypertrophy, dilation, and slight endocardial milk-iness. It is from an 8-year-old boy with sensineuronal deafness since age 4. He developed heart failure 4 months before death and was diagnosed with dilated cardiomyopathy 1 month before death when progressive atrioventricular conduction block developed requiring a permanent pacemaker. The mode of death in the hospital was dysrhythmic with initial resuscitation after a cardiac arrest. This case was autopsied before the familial association between dilated cardio-myopathy, atrioventricular conduction block, and nerve deafness were recognized. (B) Micro-scopically, cardiac myocytes are hypertrophied and surrounded by interstitial fibrosis. The endocardium shows mild fibroelastosis. These nonspecific findings are usual in cases of dilated cardiomyopathy.

rhythm and atrial fibrillation who died suddenly showed marked atrophy, degeneration, and isolation of the sinoatrial node (Bharati et al., 1992c).

One of our childhood cases of dilated cardiomyopathy developed nerve deafness 4 years before presenting in heart failure and then AV conduction block requiring a pacemaker 1 month before death. He died with documented ventricular fibrillation in the hospital (Figure 100A and Figure 100B). Tissue from such a case would now be stored in liquid nitrogen for genetic studies.

Conduction Block Associated with Muscular Dystrophies and Myopathies

The gene locus for familial conduction system disorders on the long arm of chromosome 19 is very close to that for myotonic dystrophy, an autosomal dominant muscle disease associated with progressive AV conduction block and a high rate of SD (Melillo et al., 1996). Prolonged conduction detected electrophysiologically was the most frequent cardiac finding in 45 myotonic dystrophy patients (Rakocevic-Stojanovic et al., 2000), with the QRS interval increasing with increasing age (Merlevede et al., 2002). Histologic study of the conduction system in 12 such patients with a variety of ECG conduction defects revealed extensive fibrosis, fatty infiltration, and atrophy involving sinoatrial and AV nodes, AV bundle, bundle branches, and ventricular myocardium (Nguyen et al., 1988).

Proximal myotonic myopathy is another heritable autosomal dominant muscle disease linked to familial conduction block (von zur Muhlen et al., 1998). Autosomal dominant myofibrillar myopathy occurred in combination with right ventricular cardiomyopathy in a Swedish family with various tachyarrhythmias. Some members also had both AV block and sick sinus syndrome and required pacemaker insertion. The gene defect was mapped to the long arm of chromosome 10 (Melberg et al., 1999). In autosomal dominant limb girdle muscular dystrophy, patients often develop dysrhythmias and AV conduction disturbances presenting as bradycardia or syncopal attacks and require pacemaker implantation but are still prone to die suddenly (van der Kooi, 1996; Fang et al., 1997). Emery-Dreifuss muscular dystrophy patients develop a cardiomyopathy presenting most often as AV block. Some female carriers have AV conduction block (Emery, 1987). The conduction system appeared normal when examined histologically in one female carrier and two male patients (Fishbein et al., 1993).

So-called desmin cardiomyopathy is a familial condition that presents with dilated or restrictive cardiomyopathy. Patients also develop AV conduction block and mild myofibrillar myopathy.

Syndromic Atrioventricular Conduction Block

Sudden death from heart block is reported in Holt-Oram syndrome (Newbury-Ecob et al., 1996). Progressive heart block is also an integral part of Kearns-Sayre syndrome (Katsanos et al., 2002). An electrophysiologic study on two adolescents with Kearns-Sayre syndrome located the conduction delay to the Hisian tract beyond the AV node (Polak et al., 1989). Conduction defects and other dysrhythmias occur in inborn errors of fat metabolism (Bonnet et al., 1999). Intraventricular conduction defects and sick sinus syndrome were present in a three-generation family with brachydactyly (Ruiz de la Fuente and Prieto, 1980). Another three-generation family had brachydactyly and conduction defects (Hollister and Hollister, 1981).

Isolated Congenital Complete Heart Block due to Maternal Antiphospholipid Antibodies

Isolated congenital complete heart block, that is, heart block not associated with congenital heart disease, is diagnosed prenatally or at birth and is usually complete at onset. An affected fetus may become hydropic (fetal congestive heart failure), and a newborn is likely to show heart failure and be at risk for syncope and SD in postnatal life. Most newborn cases of complete heart block are caused by destruction of the conduction system during fetal life by maternal antiphospholipid autoantibodies, anti-Ro and anti-La, from a mother with a collagen disease, such as lupus erythematosus or Sjögren's syndrome. The disease may not be recognized clinically. Such IgG antibodies can cross the placenta and localize in the fetal conduction system and possibly in other cardiac tissues. Complete heart block due to antiphospholipid antibody damage is always congenital, even if it does not become evident for some weeks or months after birth. However, it is an acquired disease, even though it is acquired in utero. Some immune-mediated cases of heart block are not detected or are only partial at birth and progress later (Askanase et al., 2002). A child whose younger sibling had third degree block was diagnosed with first degree block when 10 years old; such cases suggest that conduction system damage in utero may be mild and capable of either resolution or progression (Askanase et al., 2002).

Congenital complete heart block has a risk of recurring in later-born siblings, particularly females and especially if the maternal disease is not recognized and treated (Julkunen et al., 1998). However, studies of pregnant women with connective tissue disease and antiphospholipid autoantibodies indicate that only a small percentage of their offspring are born with complete heart block (Brucato et al., 2002; Gladman et al., 2002). The antibodies that destroy fetal nodal and Purkinje cells at mid-gestation normally have no effect on the mother's conduction system (Gordon et al., 2001), but a case report of heart block in a woman with lupus erythematosus and another with Sjogren's syndrome suggest that autoantibodies do, sometimes, affect the AV node of adults (Mevorach et al., 1993; Lee et al., 1996). Infants born with complete heart block sometimes do well, but most require a pacemaker (Buyon et al., 1998). A small proportion develops dilated cardiomyopathy with endocardial fibroelastosis, which may occur even with a pacemaker inserted (Eronen et al., 2000; Moak et al., 2001), and is considered a reaction to chronic chamber dilation at this stage of life (Silver and Silver, 1992). Whether the dilated cardiomyopathy in congenital complete heart block is caused by the conduction defect or is a result of the original disease acting in utero is not clear. In fact, Nield and co-workers (2002) reported three perinatal cases of congenital endocardial fibroelastosis due to maternal antiphospholipids; the subjects did not have AV block.

Destruction of the conduction system at mid-gestation does not heal by scarring by the time of birth but results in the total disappearance of the conduction system, except for calcified strands in the expected sites of the AV node and Hisian system (Litsey et al., 1985; Silver and Silver, 1992; see Figure 101).

Atrial-axis discontinuity was seen in seven hearts from children with complete heart block born to mothers with anti-Ro antibodies. The anticipated site of the AV node was occupied by fibrous and adipose tissue (Ho et al., 1986). In the same study, an eighth child whose maternal serum was anti-Ro negative had nodoventricular discontinuity of the conduction system. Hackel (1988) examined the conduction systems from six cases of congenital complete heart block, four infants and two adults. All had an absence of fibers connecting the atrium with AV node and common bundle, as well as had partial or complete absence of the AV node. Histologic findings from three cases with complete heart block and SD included absence or fragmentation of the AV node and noninflammatory degeneration of the senoatrial node and both interatrial and internodal pathways, all attributed to apoptosis (James et al., 1996).

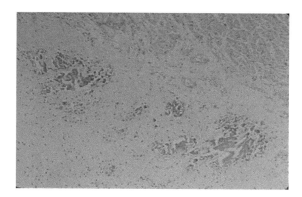

Figure 101 A woman with three pregnancy losses was at 26 weeks gestation when fetal bradycardia was detected. Cardiac ultrasonography showed normal heart structure and persistent complete heart block but no heart failure. Maternal serology was positive for both anti-Ro and anti-La autoantibodies, and she had Raynaud's syndrome clinically. A precipitous delivery of a stillborn male occurred at 33 weeks' gestation. The atrioventricular conduction system was examined histologically in step-serial section. The atrioventricular node was replaced by vascular young connective tissue. The common penetrating bundle and its proximal branches were not identified, but strands of calcified collagen were present in their location. This microsection shows calcific foci within young fibrous tissue in the position of the stem of the left bundle branch.

Congenital and Acquired Atrioventricular Conduction Block

Isolated AV conduction block not caused by immune-mediated conduction system destruction may also be congenital but usually starts weeks, months, or years after birth. It often presents as a partial block but progresses to a complete block with time (Frohn-Mulder et al., 1994). Because it is not associated with maternal autoimmunity, it has no risk for recurrence within a sibship unless the conduction disease happens to be familial. Familial kinds of autosomal dominant AV conduction block are discussed above. Isolated congenital AV conduction block in older children and adults has a tendency to cause SD during a Stokes-Adams attack. Among 102 patients with isolated complete permanent heart block followed many years, 27 had Stokes-Adams attacks and 8 died, 6 during the first attack (Michaëlsson et al., 1995).

Mediastinal radiation therapy causes acquired complete heart block that correlates with severe fibrosis in the conduction system and myocardium generally (Kaplan et al., 1997). Acute lymphocytic myocarditis, presumably immune mediated, can cause AV block that may recover or need pacemaker insertion (Heusch et al., 1996).

Conduction system histology in cases in adults with complete heart block clinically showed extremely severe destruction of the conduction tissues, including the sinoatrial node (Sugai et al., 1981). In another study of 14 cases, the authors correlated the sites of fibrosis in the Hisian system with the morphology of QRS complexes from those patients (Ohkawa et al., 1981).

Atrioventricular Conduction Block in Congenital Heart Disease

In the fetus, complete heart block associated with cardiac malformation is more common than isolated complete block. Among 6000 fetal patients examined by fetal echocardiography, 21 had heart block associated with congenital heart disease, and 15 had the condition in isolation (Machado, 1988). Only about 30% of newborn complete heart block cases have congenital heart disease, because hydropic fetuses with the latter are often stillborn (Gembruch et al., 1989).

The types of complex cardiac malformations encountered most often with AV conduction block include AV septal defect combined with left atrial isomerism (Machado, 1988; Gembruch et al., 1989); in right atrial isomerism, the conduction system is usually duplicated, and SD is attributed to arrhythmia (Cheung et al., 2002). The rare type of cardiac malformation, AV discordance (corrected transposition of the great vessels) is highly associated with AV conduction block (Gembruch et al., 1989). It develops spontaneously, sometimes with SD, especially if the interventricular septum is intact (Huhta et al., 1983). In Toronto, among 52 young adults with corrected transposition, 18 needed a permanent pacemaker — 9 following surgery and another 9 who had no operation (Connelly et al., 1996). Conduction system histology in corrected transposition cases showed two or three AV nodes, some blind ending as well as Kent bundles (Bharati et al., 1980).

A genetic defect was identified that causes both familial atrial septal defect and other forms of anomaly, as well as AV conduction defect (Gutierrez-Roelens et al., 2002).

Atrioventricular Conduction Block Following Cardiac Surgery

Surgery to repair many kinds of cardiac congenital malformations depends on accurate electrophysiological mapping of the conduction system prior to operation. In some cases, surgical damage cannot be avoided. It is treated with an implanted pacemaker. A retro-spective review of 6004 patients who had surgery for malformation showed 132 (2.2%) who needed permanent pacemakers. The most common defect was associated with a ventricular septal defect (Goldman et al., 1985). A pathologist would naturally tend to believe that acute (hemorrhage, edema) or chronic (scarring) damage in the sinoatrial or atrioventricular nodes or in conduction system fascicles would correlate with heart block or various dysrhythmias and, hence, with SD. Nevertheless, so few pathologists are expert in this type of pathological study that electrophysiologic studies of conduction system function are probably a better gold standard than morphology.

Acquired complete heart block in infants and children is sometimes caused during surgery. The approaches to both the conduction nodes were disrupted by scarring and old sutures in two cases of SD that followed 2 years after a patient underwent the Mustard operation for transposition of the great vessels (Bharati et al., 1979). Bharati and Lev (1988) also described suture granulomas in the sinoatrial node in late SD following surgical repair of atrial septal defects. Surgical damage to the conduction system was found histologically in two children who had atrial and ventricular septal defects repaired surgically (Ho et al., 1985). The conduction system is also at risk during repair of truncus arteriosus (Bharati et al., 1992b).

Direct conduction system trauma may cause SD. After radiofrequency catheter ablation of the atrioventricular node, done with pacemaker implantation to treat atrial fibrillation, the risk of SD was highest (1.2% of 334 consecutive patients) within 2 days, but similar deaths (0.9%) occurred 3 months later (Ozcan et al., 2002) and might be considered unexpected.

Histologic Examination of the Conduction System in Cases of Sudden Death

The step-serial transverse section method of examining the conduction system is best, with sections stained by hematoxylin-eosin, and with procurement of an elastic stain from each level, for example, on every tenth and eleventh section (Figure 102). Complete histologic study depends on the correct collection of tissue blocks and, hence, on an understanding of the anatomic locations of the sinoatrial and atrioventricular nodes (Lev and Bharati, 1981; Anderson et al., 1981b; Bharati, 2001). Song et al. (1997) proposed orienting the system longitudinally within four or five blocks. Taking random sections of the septal atrioventricular junction is useless and destructive.

An abbreviated study done by blocking the entire system in transverse section for histology and then viewing only one hematoxylin-eosin and one elastic stain from the face of each block has the advantage of preserving the entire system in case further examination by serial sections is required. In one large study of cardiac SD cases, Cohle and co-workers (2002) found lesions they considered lethal, by using this method. They found that the atrioventricular node artery was narrowed by fibromuscular dysplasia (seven cases) or tumors (four cases). Ford (1999) advocated a similar abbreviated method to demonstrate atrioventricular node tumors. Because elastic fibers are normally present in nodal tissues, an elastic stain (Movat pentachrome or elastic trichrome) helps a pathologist locate them. It also helps one assess the small coronary arteries supplying the nodes; their stenosis may help explain SD.

To avoid confusion and misinterpretation, an understanding of the development of the conduction system from fetal life onwards and its normal aging are fundamental for

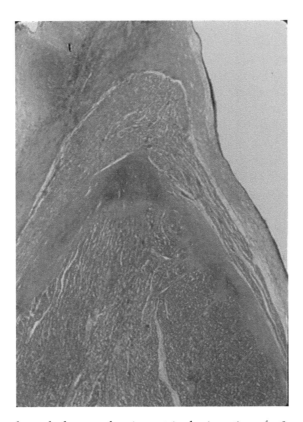

Figure 102   Section through the septal atrioventricular junction of a 3-month-old infant (control for the case depicted in Figure 23) shows the bifurcation of the His bundle. In this study, every tenth section was stained with H&E, and every eleventh, for elastic. This particular section was number 561.

examination. A lack of this understanding may explain the many confusing reports concerning conduction system morphology in SIDS. Studies done on subjects who died suddenly with the long Q–T syndrome are also confusing. Persistent fetal dispersion of the atrioventricular node and fragmentation of the bundle of His are lesions frequently described in the conduction system in cases of SD. They were a normal variation in a detailed study of 249 hearts from cardiac SD cases and 98 control hearts (Suarez-Mier et al., 1998). Studies on the conduction system of people of all ages agree that fat and fibrous tissue increase in these tissues with increasing age (Sugai et al., 1981; Bharati et al., 1992a; Suarez-Mier et al., 1995; Song et al., 2001). Song and co-workers (2001) also described patchy calcification. In a Japanese study, aging changes consisted of fatty infiltration, fibrosis and elastosis, disappearance of muscle fibers, and general atrophy that seemed related to increasing stenosis of small arteries supplying that tissue (Sugai et al., 1981). Another study of 33 cardiac SD cases described mild fibrointimal hyperplasia in the small arteries as a normal age change (Suarez-Mier et al., 1995).

An autopsy study of 197 young people with cardiac SD that included histology of the conduction system revealed that 24 of 76 cases with a macroscopically normal heart had conduction system abnormalities that the authors considered explained ventricular pre-excitation in 18 and heart block in 6; another 16 cases showed no microscopic abnormalities

(Corrado et al., 2001b). Parenthetically, only those AV connections closely related to the conduction system are found in serial sections of the system (Figure 102). To define any AV connections, one faces the prospect of embedding serial blocks of the entire AV rings and viewing step-serial sections of those blocks (Bharati, 2001). Fortunately, electrophysiological studies can map AV connection accurately, including defining their depth (subendocardial or subepicardial) as well as rate and direction of their conduction. They also help in defining location in anatomic studies.

My conclusion from personal experience with conduction system histology and from reading results of such studies is that it is wise to completely block the conduction system in cases of cardiac SD and to view face sections in certain selected cases. If this were done in every hospital and forensic autopsy, pathologists would soon become skilled at block selection and have a good understanding of conduction system histology, its normal variations and anomalies. I believe it unwise to proceed to cutting and viewing step-serial sections of each block unless a clear indication exists, such as a chronic dysrhythmia that was studied electrophysiologically during the life of the deceased or their families. Pathologists have a role in recognizing and defining changes in form that follow changes in function. Bear in mind the unfolding evidence of a molecular origin for familial electrical heart diseases. Histologic study of the cardiac conduction system will lead to useful morphologic findings if done on cases selected because clinical questions need answers.

Cardiac Alterations in Sudden Infant Death (SID)

8

THOMAS BAJANOWSKI AND BERND BRINKMANN

Contents

Definition

Sudden infant death is the most common cause of death during infancy in industrialized countries, with an incidence between 0.3 and 1.5 per 1000 live births (Fitzgerald, 2001). The term "sudden infant death syndrome" (SIDS) was defined in 1969 at the Second International Conference on Causes of Sudden Death in Infants, as "the sudden death of any infant or young child, which is unexplained by history, and in which a thorough post-mortem examination fails to demonstrate an adequate cause of death" (Beckwith, 1970).

This definition was criticized because the term is defined by exclusion, it links the previous history with the autopsy findings (which is often not done in practice), the standard of the autopsy is not defined, and last but not least this phenomenon cannot be characterized as a syndrome or an entity. The definition was later modified (Willinger, 1991; Valdes-Dapena, 1992) so that the upper age limit was defined as 1 yr at the time of death, and a extensive death scene investigation was also included.

Sudden infant death (SID) cases are characterized by a typical epidemiology, a typical death scene, and typical autopsy findings. Nevertheless, there are some difficulties for the pathologist in making this diagnosis. One is that a number of pathological changes can be observed, and the pathologist has to decide whether the findings can be judged to sufficiently explain the cause of death and exclude the case from SIDS or not.

Thus, pathological alterations can either exclude certain cases from SIDS or not. Because the severity of such alterations shows a continuous increase or decrease in a large collective, further subdivision into subgroups can be of assistance.

Macroscopic Findings in SID at Autopsy

SID victims are generally well nourished and show no evidence of injury. Lips and nails can be cyanotic (Jones and Weston, 1976). White or blood-stained pulmonary edema fluid may be found in the respiratory tract and sometimes also in the mouth and nares. The lungs are bulky with dystelectasis, edematous, and congested. Intrathoracic petechial hemorrhages are found in 68 to 95% of SID cases in the thymus, over the epicardium and on the pleural surface (Beckwith, 1989). The right ventricle of the heart is distended and contains liquid blood as does the left one and large blood vessels (Valdes-Dapena, 1983).

Pathologic Findings and Disturbances in the Heart

A number of reports can be found in the literature dealing with cardiac alterations in cases of sudden and unexplained infant death. While findings such as a previously unknown malformation, cardiomyopathy (Fried et al., 1979), or endocardial fibroelastosis (Valdes-Dapena, 1980; Williams and Emery, 1978) could be sufficient to cause death, other findings could be either coincidental (e.g., intimal thickening of nodal arteries; Kozakewich et al., 1982) to the cause of death or an expression of the physiological development during the first year of life (Maron and Fisher, 1977).

Myocarditis

In the earlier literature, inflammatory diseases of the heart were reported, especially interstitial myocarditis (Burgmeister, 1963; Mahnke, 1966; Müller, 1963; Seifert, 1961; Windorfer and Sitzmann, 1971), which may lead to the sudden unexpected death of an infant. Certainly, such cases have to be excluded from the SID group because of the significance of the pathological findings. Since modern diagnostic methods (e.g., immunohistochemistry and virus detection using PCR) are now available, a severe myocarditis is diagnosed in individual cases only (Dettmeyer et al., 1998). A mild infiltration of the myocardium by lymphocytes without signs of myolysis can be observed especially in cases with infectious diseases at other locations (Figure 103 and Figure 104) and has no significance for the cause of death. There is only one report where myocarditis was found more frequently (17%) in SID cases (Rambaud et al., 1994), but it would seem that this observation should be viewed with skepticism, because these authors also found other inflammatory diseases in a high number of SID cases (e.g., meningitis in 22.3%) (Damotte et al., 1994).

Hypoxia-Related Changes

Hypertrophy of the right ventricular wall was first described by Naeye et al. (1976), who suggested that this could be due to sleep apnea with repeated episodes of hypoxia. This

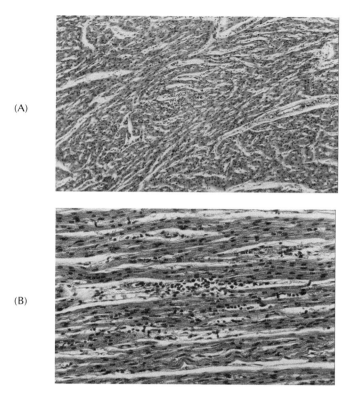

Figure 103   An 11-month-old female infant with interstitial pneumonia. (A) Mild focal infiltration of the myocardium by lymphocytes. (B) Perivascular edema and cellular infiltration of the interstitium. Damage of myocytes or myolysis could not be found.

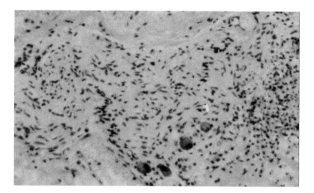

Figure 104   An 11-month-old boy with bronchitis and peribronchitis caused by *Haemophilus influenzae* infection. Focal infiltration by lymphocytes and granulocytes in the left ventricular wall around a ganglion — ganglionitis.

observation was evaluated by others using the same method of morphometry but could not be confirmed (Valdes-Dapena, 1980; Williams et al., 1979).

The quantitative analysis of myocardial mast cells in 25 SID and 15 non-SID cases (Riße and Weiler, 1997) resulted in an age-dependent increase during the first months of life up to the normal value of about 200 cells/cm² at the age of 6 months. Significant differences between both groups could not be observed, indicating no specific process of chronic inflammation, fibrosis, or repeated hypoxemia.

Even immunohistochemical investigations using an antibody to detect the $C5b-9_{(m)}$-complement complex, which is an early and sensitive marker for hypoxic cell damage, were completely negative in a series of 136 SID cases (Thomsen and Saternus, 1994).

Morphological Changes in the Cardiac Conduction System

Since 1968 (James, 1968), histological investigations of the cardiac conduction system have been performed in SID cases. Due to the time- and labor-intensive techniques involved, these investigations are not included in routine SID diagnostics. The dissection technique of the different morphological structures was described in detail by Zack and Wegener (1994). In addition to a number of case reports dealing with morphological variations of the conduction system, only very few systematic investigations exist (Table 46) where the results in SID cases are compared to those obtained in a much smaller number of controls (cases of explained death). The results obtained are not consistent, and the interpretation of the results is partially contradictory. Authors (James, 1968; Ferris, 1972), for example, described a uniform process of resorptive degeneration of His bundle and AV node with cellular necrosis. This process should be responsible, in general, for fatal cardiac arrhythmia. Other authors (Valdes-Dapena et al., 1973; Dudorkinova and Bouska, 1993) discussed the findings of James and Ferris to be age-related physiologically and denied the existence of an active process of necrosis in the myocardium of infants. Bharati et al. (1985) described a left-sided His bundle and postulated that this morphological variant of the cardiac conduction system could be more affected by left ventricular pressure than the normal variant, leading to a higher vulnerability for cardiac arrhythmia. This interpretation is in contrast to others (Suarez-Mier and Aguilera, 1998), who described His bundle dispersion (Figure 105) and left-sided His bundle as normal anatomical variants of the cardiac conduction system in infants with no functional significance. Suarez-Mier and Aguilera (1998) observed accessory fasciculoventricular tracts in only 7 of 55 SID cases (Figure 106) and discussed a possible pathological significance for preexcitation syndromes.

Immunohistochemical and morphometrical investigations of the cardiac conduction system are also rare. Fu et al. (1994) demonstrated a relative lack of nerve fibers in the AV node and His bundle using S100-antibody. Ho and Anderson (1988) reported on three cases of infants showing cardiac arrhythmia prior to sudden unexpected death, where hypoplasia of the SA node or AV node could be found as a possible cause for the disturbances of cardiac rhythmogenic function. Because normal values for the size of the various structures of the cardiac conduction system are not available in this age group at present, these results have no objective background.

Disturbances of Rhythmogenic Function

Changes in cardiac control centers of the brain stem and autonomic imbalance leading to cardiac arrhythmia were also proposed as possible causes of sudden and unexpected infant death (Schwartz, 1976, 1987, 1989). Therefore, some investigators (Rossi, 1995, 1999; Hunt, 1995) included the brain stem in the morphological investigations and indicated the possibility of a fatal reflexogenic mechanism (Filiano and Kinney, 1992; Rossi and Matturri, 1995, 1997), including an inappropriate activation of the diving reflex producing severe bradycardia with apnea (Lobban, 1991; Kelly et al., 1991).

Table 46 Overview of Important Histological Findings in the Cardiac Conduction System in SID Cases

Author(s)	Year	Number of SID Cases	Number of Controls	Important Findings
James	1968	40	16	Resorptive degeneration of the His bundle and the atrioventricular node without inflammation
Anderson et al.	1970	18	12	Focal hyperplasia of the atrioventricular node artery in 35% of the SID cases and in 10% of the controls
Ferris	1972	47	0	Resorptive degeneration (James) confirmed
Ferris	1973	50	0	11 SID cases showing hemorrhages in the SA node and in internodal tracts
Valdes-Dapena et al.	1973	31	16	Findings like those of James (1968), but another interpretation: normal anatomic variant
Anderson et al.	1974	15	15	Accessory atrioventricular tract in one SID case, hemorrhages in specific myocytes in both groups
Kendeel and Ferris	1974	38	28	Increased amount of connective tissue in the atrioventricular complex in SID cases
Lie et al.	1976	26	24	Resorptive degeneration in all cases, hemorrhages in 27% of the SID cases and in 29% of the controls
Anderson and Hill	1982	40	0	Fibromuscular hyperplasia of the atrioventricular node artery ($5\times$) and of the SA node artery ($1\times$)
Marino and Kane	1985	7	0	Accessory tracts ($2\times$) and dispersion of the atrioventricular node and His bundle ($4\times$)
Bharati et al.	1985	15	8	Left-sided His bundle in eight SID cases and in two controls
Ho and Anderson	1988	30	19	Wide variety of pathomorphologic changes in both groups (e.g., hyperplasia of the atrioventricular node artery, dispersion of the atrioventricular node and the His bundle, accessory tracts, hemorrhages)
Dudorkinova and Bouska	1993	21	6	No signs of resorptive degeneration
Suarez-Mier and Aguilera	1998	55	15	Fasciculoventricular tracts in 7 of 55 SID cases and in none the controls

In the 1970s, it was proposed that the long Q–T syndrome (LQ–TS) could be responsible for some cases of sudden infant death (Maron et al., 1976; Southall et al., 1979) due to ventricular tachycardia leading to ventricular fibrillation (Schwartz, 1988; Jervell and Nielsen, 1957; Romano et al., 1963; Ward, 1964). As the underlying mechanism is inhomogeneity of repolarization, relevant genetic mutations leading to changes in protein structure seem to affect proteins controlling the myocardial ion channels (Rossi, 1995; Wang et al., 1995, 1996).

Schwartz (1982) reported three cases of SIDS that occurred among 4205 prospectively investigated infants and showed a significant prolongation of Q–Tc. The authors speculated that LQ–TS could be the main cause of SID. However, these results could not be confirmed by other investigators (Southall et al., 1983a, 1986; Weinstein and Steinschneider, 1985) (Table 47); therefore, considerable doubt about the validity of these conclusions arose. In

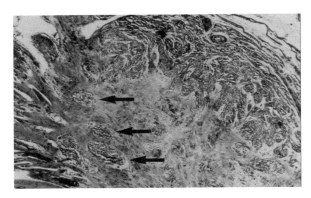

Figure 105   A 10-month-old female infant, showing mild inflammation of the upper respiratory tract caused by *Pseudomonas aeruginosa* infection: cardiac conduction system. Atrioventricular node consisted of archipelagoes of specialized fibers (arrows).

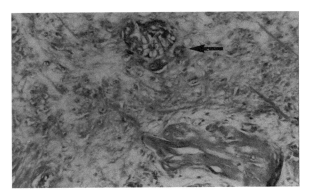

Figure 106   A 14-week-old boy, showing an accessory fasciculoventricular tract (arrow) in the central fibrous body near the atrioventricular node.

Table 47   Review of the Literature on Studies on Q–Tc Measurements of Infants Who Subsequently Died from SIDS

Authors	Year	Number of Infants Investigated	Number of SIDS Cases	Number of SIDS Cases with Significant Q–Tc Prolongation
Schwartz et al.	1982	4205	3	3
Southall et al.	1983	6914 full-term and 2337 pre-term	29	0
Weinstein and Steinschneider	1985	1000	8	0
Southall et al.	1986	7254	15	1
Schwartz et al.	1998	34,442	24	12

1998, the working group of Schwartz published a new investigation of more than 34,000 infants in whom an electrocardiogram (ECG) was performed during the first week of life. A prolonged Q–Tc was found in 50% of 24 SIDS cases that occurred in the investigation group. The authors evaluated this result as being confirmation of their previous hypothesis with regard to the causes of SIDS, but the assumption that 50% of all SIDS cases could be caused by LQ–TS does not seem justifiable.

New molecular genetic investigations could bring further light to this field because mutations on five different genes (KCNQ1, HERG, SCN5A, KCNE1, KCNE2) on chromosomes 3, 4, 7, and 11 were described in the 1990s, and an association with defects in cardiac sodium and potassium channels leading to LQ–TS (Curran et al., 1995; Jiang et al., 1994; Keating et al., 1991; Schott et al., 1995; Wang et al., 1995) could be shown.

To the present, no systematic molecular genetic analyses in SID cases were reported, and only one case report dealing with this theme was published (Bajanowski et al., 2001). In this report, two SID victims who were suspected of having died due to LQ–TS were investigated using specific PCR techniques combined with SSCP analysis and sequencing. None of the known mutations responsible for LQ–TS could be detected, but a number of polymorphisms (frequency in the population >1%) could be found that either do not lead to amino acid substitution or the amino acid substitution is of no detectable significance. It needs to be stressed that results of these molecular genetic investigations do not totally exclude the existence of this syndrome. The mutation analysis using fluorescent SSCP is a screening method, and the sensitivity is about 90% (Orita et al., 1989). Furthermore, LQ–TS are multigenetic diseases. Other, still unknown loci could be involved in the pathogenesis, and the significance of some of the molecular genetic changes defined to be "polymorphisms" is still unknown. Theoretically, it has to be taken into consideration that the presence of two or more polymorphic sites can lead to functional disturbances if they are combined.

The first systematic investigations of a greater number of SID cases are in progress, and new results will be generated.

Brugada syndrome (Brugada et al., 1992), which was first described in 1992, is characterized by a right bundle branch block and ST elevation on electrocardiogram that may lead to idiopathic ventricular fibrillation and is associated with sudden cardiac death. As a molecular genetic basis, mutations of the SCN5A gene were reported (Bezzina et al., 1999; Baoroudi et al., 2000; Veldkamp et al., 2000) to influence the function of the cardiac sodium channel. Priori et al. (2000) reported five children from the same family who died after cardiac arrest, and a mutation in the cardiac sodium channel could be detected that confirmed the diagnosis. It can be assumed that this disease, which should be more frequent in the population than LQ–TS (Brugada et al., 1999), could also be responsible for some SID cases (Priori et al., 2000).

Other Pathologic Findings and Functional Disturbances

Unknown cardiac malformations may cause single cases of unexpected sudden death, but, of course, these cases have to be excluded from SIDS. Other diseases, for example, tumors of the myocardium (Figure 107), were reported (Bajanowski et al., 1993). Their significance in the cause of death depends on the kind of tumor, its localization, and size. Hemorrhages are often the result of resuscitation. In cases where resuscitative attempts were not carried out (Table 48) (Karch and Billingham, 1984; Matsuda et al., 1997), hemorrhages may contribute to death if they affect the main structures of the cardiac conduction system (Figure 108) (Bajanowski et al., 1993).

An elevation of the mean heart rate (Schechtmann et al., 1988; Southall et al., 1988) and a reduced heart rate variability during the waking state (Schechtmann et al., 1989) were also reported in SID cases, as well as isolated cases of Wolff-Parkinson-White syndrome in infants (Lipsitt et al., 1979).

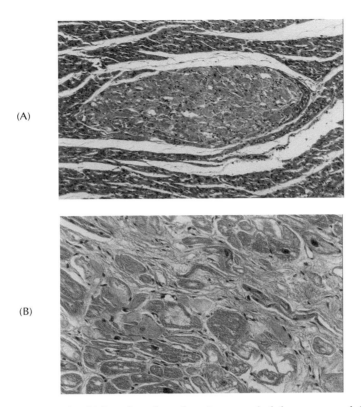

(A)

(B)

Figure 107 An 18-week-old female infant showing two rhabdomyomas of the left ventricular wall. (A) The tumors are circumscribed but not encapsulated. HE, ×40. (B) An irregular vacuolization of the cytoplasm can be seen. The vacuoles vary in size and are separated by strands of cytoplasm. Typical "spider cells" are not present in this case.

Table 48   Changes Due to Resuscitation

Changes induced by countershock
- Myocardial hemorrhages
- Hypereosinophilia on the epicardial surface or just below
- Epicardial damage of myocytes
- Disruption of intercalated discs
- Subjacent contraction bands
- Coagulative necrosis

Catecholamine or pressure-induced changes
- Focal microinfarcts with mild acute inflammatory infiltrates

Conclusions

Although the cardiac findings in SIDS are sometimes contradictory, disturbances of cardiac function could be one pathophysiological mechanism leading to sudden death in a subgroup of SIDS victims. Problems in diagnosing these disturbances are caused by different factors, such as the following:

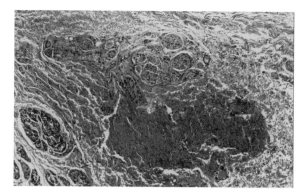

Figure 108 An 18-week-old male infant with a large hemorrhage in the connective tissue near the SA node. Resuscitation was not carried out in this case.

- Morphologic variations of unknown significance (e.g., in the cardiac conduction system) that are not demonstrable by routine autopsy
- A lack of standardized investigation techniques
- Loss of a suitable definition of control cases
- The need to use time- and cost-intensive, highly specialized molecular genetic techniques
- Problems in the classification of SIDS cases

Sudden Cardiac Death and Channelopathies

9

EMANUELA TURILLAZZI, KATHRYN A. GLATTER, AND
MARGHERITA NERI

Contents

Introduction

In Europe, the United States, and much of developed society, sudden cardiac death remains the biggest killer (Escobedo and Zack, 1996; Virmani et al., 2001; Zheng et al., 2001; Goraya et al., 2003; Chugh et al., 2004). Although many of these cases are due to known causes, such as atherosclerosis and myocardial infarction, some cases will not have clear findings at necropsy. Unexplained sudden death with an unremarkable autopsy is not rare (Davies, 1999; Chole and Sampson, 2001), and the coroner or pathologist should consider genetically associated causes of disease, as discussed below (Chugh et al., 2004; Priori and Napolitano, 2004; Ji et al., 2004; Sarkozy and Brugada, 2005a,b; Vincent and Zhang, 2005; Yoshioka and Lee, 2003; Roberts, 2006). We will discuss the clinical presentation; genetic, molecular, and cellular abnormalities; and diagnostic evaluation of these causes, including ion channelopathies and cardiomyopathies (see list below).

These diseases should be considered when there is no clear cause for death. Because they have a genetic basis, other family members (who are still living and could seek treatment) could also be affected.

Disorders of the Ion Channel:

1. Long QT syndromes (LQTS)
2. Brugada Syndrome (BS)
3. Catecholaminergic Polymorphic Ventricular Tachycardia (Ryanodine receptor defect) (CPMVT)
4. Short QT Syndrome (SQTS)

Disorders of the Heart Muscle:

1. Hypertrophic Cardiomyopathy (HCM)
2. Arrhythmogenic Right Ventricular Dysplasia/Cardiomyopathy (ARVD/C)

The Forensic Approach

Molecular Diagnosis

When an autopsy is perfomed in cases of sudden death, samples for full toxicological analysis should be obtained. Blood, urine, liver, bile, ocular fluid, gastric contents, and so forth, should be retained for potential analysis (Ackerman et al., 2001; Di Paolo et al., 2004).

In addition, samples of peripheral blood and tissue should be retained for genetic workup to rule out a genetic cause of cardiac sudden death (SD). Systematic postmortem genetic analysis of archived tissue may identify etiologies of SD.

So, when coming across a negative autopsy, the medical examiner and the pathologist should carefully investigate the cause of sudden death, the history of disease, and the family history, and then rule out the possibility of the above disorders.

With marked advances in molecular techniques, DNA testing of peripheral blood and tissue has revolutionized the diagnosis of genetic causes of sudden death. We will describe the basic methods of tissue preparation and DNA analysis as a useful overview for the clinical pathologist and coroner.

Collection of DNA from Blood Samples

It is easiest to amplify DNA that will be used for genetic testing when it is taken from blood samples (Higuchi, 1989; Bajanowski et al., 2001). Ideally, the coroner or pathologist would collect at the time of autopsy 15 ml of blood in several tubes that contain ethylene-diaminetetraacetic (EDTA) acid, to prevent coagulation and degradation of the DNA. The tubes are then stored at 4°C until the DNA is extracted for analysis, which should be within 1 week, although we have sometimes extracted DNA 4 months after collection. If the blood samples are collected in tubes that do not contain an anticoagulant, the DNA should be extracted promptly (within days of the initial collection).

Collection of DNA from Tissue Samples

The extraction of high-quality DNA from tissue that can be used for polymerase chain reaction (PCR) amplification is much more problematic than using blood samples. It is often difficult to amplify long fragments of DNA from formalin-fixed and paraffin-embedded tissue because of the fixation time in the formalin, the often long storage time in the tissue blocks prior to analysis, and the formation of formic acid in the sample (Sato et al., 2001; Cao et al., 2003). Formic acid hydrolyzes DNA and creates single-strand nicks in it. In postmortem tissues fixed in nonbuffered formalin (usually in tissue preserved more than 20 years ago), DNA fragments longer than 90 base pairs cannot be amplified.

There are a variety of published methods for extracting DNA from preserved tissue (Konomi et al., 2002; Mygind et al., 2003). Many involve a phenol-chloroform digestion and washing step. Commercial kits are also available that may simplify the methodology. One paper described a "pre-PCR restoration process" in which single-stranded DNA nicks are repaired with Taq polymerase prior to PCR amplification, which greatly improved the length of DNA pieces that could be amplified (Bonin et al., 2003).

An alternative method with which to obtain usable DNA from tissue is to snap-freeze fresh myocardial tissue collected at autopsy in liquid nitrogen and store at 80°C until DNA extraction is performed. These tissues can be used many months later.

Clearly, collection and preservation of tissue or blood samples for future DNA analysis are cumbersome, time-intensive, and costly. However, it is very helpful for the pathologist or coroner to carefully preserve such biologic material for future DNA testing in those cases where a genetic cause of sudden death is suspected, as is outlined below.

Disorders of the Ion Channel

Long QT Syndromes

Long QT syndrome (LQTS) is one of the more common and well-known of the ion channelopathies. It can be inherited as a dominant gene or can be seen in cases of acquired LQTS after taking common drugs, including antipsychotics, antiarrhythmic drugs, or allergy medications (Vincent, 1998; Wehrens et al., 2002; Zeltser et al., 2003; Al-Khatib et al., 2003).

Epidemiology

Currently, it is estimated that 1 in 5000 people carry a long QT syndrome genetic mutation (Vincent, 1998; Wehrens et al., 2002). It is one of the more common genetic causes of sudden death and has been diagnosed with increasing frequency as more coroners are educated regarding LQTS (Roden, 2001; Yang et al., 2002).

Clinical Features

At least 10% of affected LQTS patients may present with sudden death as their first (and last) symptom (Moss et al., 1991; Zareba et al., 1998; Schwartz et al., 2001). However, most patients with an LQTS mutation will never experience any symptoms. The majority of LQTS families are discovered when a young person tragically dies suddenly with a normal autopsy, and other family members are found to have a prolonged QT interval on electrocardiogram (ECG). Each genetic subtype (described below) has its own trigger for events (Table 49) (Moss et al., 1991; Locati et al., 1998; Zareba et al., 1998; Schwartz et al., 2001). LQ–T1 patients usually experience symptoms (syncope, cardiac arrest, or sudden death) during adrenaline-driven types of activities, such as exercise or running or with strong emotion (e.g., during an argument). An unexplained drowning in a person who is a good

Table 49   Genes Associated with Sudden Death

Disease	Chromosome Locus	Gene	Gene Product
LQ–T1	11p15.5	KVLQT1	I_{Ks}, α subunit
LQ–T2	7q35–36	HERG	I_{Kr}, α subunit
LQ–T3	3p21–23	SCN5A	Na channel
LQ–T4	4q25–27	*ankyrin 2*	ankyrin-B
LQ–T5	21p22.1	*mink (KCNE1)*	I_{Ks}, β subunit
LQ–T6	21p22.1	*MiRP1 (KCNE2)*	I_{Kr}, β subunit
LQ–T7	Chromosome 17	KCNJ2	I_{Kr}, α subunit

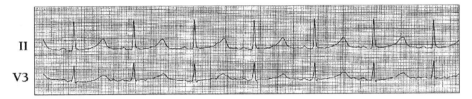

Figure 109   Electrocardiogram of a 10-year-old, asymptomatic girl with incidentally discovered long QT syndrome. She was genotyped as having LQ–T3, the rare sodium channel subtype. Ten members of the family also have LQ–T3, with no sudden deaths or syncopal events. Her Q–Tc interval is extremely long at >550 ms (paper speed = 50 mm/sec).

swimmer could be due to an LQ–T1 mutation (Zareba et al., 1998; Moss et al., 1999; Ali et al., 2000; Schwartz et al., 2001). LQ–T2 (HERG) mutations may cause sudden death due to auditory triggers; for example, an alarm clock or the telephone ringing. The rare LQ–T3 (sodium channel) subtype may occur during sleep or during periods of slow heart rate (Figure 109) (Moss et al., 1985; Schwartz et al., 2001).

Although LQTS is an autosomal disease, females are far more likely to experience symptoms than males (Locati et al., 1998; Schwartz et al., 2001). LQTS can be diagnosed in some cases by noting a prolonged QT interval (>450 ms) on the ECG. However, up to 30% of gene-positive patients may have a normal or only borderline prolonged QT interval, making diagnosis difficult in some cases (Schwartz et al., 2001; Moss et al., 1985, 1999; Zareba et al., 1998; Priori et al., 1999).

There is a small association in the literature between SIDS (sudden infant death syndrome) and LQTS, although probably fewer than 5% of all SIDS cases are due to ion channel mutations (Schwartz et al., 2000; Ackerman et al., 2001; Bajanowski et al., 2001; Ackerman, 2005; Gunteroth and Spiers, 2005). Other causes of SIDS are likely far more common, such as placing the infant prone, co-sleeping with adults, or inborn errors of metabolism.

Pathophysiology

The fundamental defect in LQTS is prolonged ventricular repolarization and a tendency toward torsades de pointes (polymorphic ventricular tachycardia) and ventricular fibrillation. Beta-blocker medications (described below) do not shorten the QT interval; they are believed to act, in part, by blocking early after-depolarizations (EADs) that initiate the ventricular arrhythmias.

Genetics

To date, a total of seven genes were identified as causing long QT syndrome (Keating et al., 1991; Jiang et al., 1994; Wang et al., 1996; Abbott et al., 1999). The mutant ion channel that causes clinical LQTS is inherited in an autosomal dominant fashion with incomplete penetrance and was originally known as the Romano-Ward syndrome. With the advent of genetic testing, it has become clear that each LQTS genetic subtype represents a unique disease, with different triggers to arrhythmias. The genes that encode the potassium channels KVLQT1 (on chromosome 11) and minK (on chromosome 21) interact to form the cardiac I_{Ks} (inward slow potassium) current; mutations in each cause LQ–T1 and LQ–T5, respectively (Keating et al., 1991; Jiang et al., 1994; Vincent, 1998). The potassium channels HERG (on chromosome 7) and MiRP1 (on chromosome 21) interact to form the I_{Kr} (inward rapid potassium) current, and defects in each cause LQ–T2 and LQ–T6, respectively (Abbott et al., 1999). Mutations in the sodium cardiac channel SCN5A cause LQ–T3 (on chromosome 3) (Wang et al., 1996). The gene responsible for LQ–T4 on chromosome 4 was identified as ankyrin 2. LQ–T7 or Andersen's syndrome is due to a defect in the α-subunit of the I_{Kr} channel (gene product KCNJ2) (Fodstad et al., 2004).

The potassium channel mutations cause a loss of function in the channel (or a dominant-negative effect, in the case of the HERG mutation), whereas defects in the sodium channel cause a gain of function.

In the unlikely event that a mutant copy of the I_{Ks} channel is inherited from each parent (mutations in the KVLQT1 and minK genes), the child will suffer from a clinically severe form of autosomal dominant LQTS and from autosomal recessive congenital deafness. This condition is known as the Jervell and Lange-Nielsen syndrome (JLNS) (Splawski et al., 1997; Chen et al., 1999). It is actually quite rare, with an estimated incidence of 1.6 to 6 cases per million (Splawski et al., 1997).

Treatment

There is no consensus on how to treat patients with LQTS (Zareba et al., 1998; Chiang and Roden, 2000; Wehrens et al., 2002). Many physicians would advocate an implantable defibrillator (ICD) for those patients who survived a cardiac arrest, or possibly even in those with syncopal events. Most physicians would advocate beta-blocker therapy in asymptomatic LQTS patients. The exact dose or type of beta-blocker medication to be used is unclear. Restriction from heavy physical activity is also suggested in affected patients.

Autopsy Findings

As with many ion channelopathies that cause unexplained sudden death, autopsy findings are unremarkable. The presence of LQTS should be considered in all cases of SCD where autopsy is negative for anatomic and histopathological findings. In these cases, after an accurate anamnesis, a genetic screening should always be performed (Lunetta et al., 2003). DNA can be isolated from venous EDTA/blood and tested at a research laboratory for the presence of LQTS ion channel mutations. EDTA/blood or 5 to 10 g of heart, liver, or spleen tissue flash-frozen and stored at 80°C would provide the best source of genetic material. In addition, archived, paraffin-embedded myocardial tissue blocks obtained at postmortem examination can be used as a source of deoxyribonucleic acid (DNA) for genetic analysis. In these cases, the possibility of degradation of DNA and subsequent effects on genetic analysis must be considered. However, this approach has been used successfully.

Brugada Syndrome

Brugada syndrome (BS) is another inherited ion channelopathy that causes unexplained sudden death, particularly in middle-aged males (Gussak et al., 2000; Antzelevitch, 2001; Antzelevitch et al., 2002; Glatter et al., 2005). It is relatively common in southeast Asia and should particularly be considered in the autopsy of subjects with this ethnicity (Nademanee et al., 1997).

Epidemiology

A Brugada syndrome consensus report published in 2002 estimated the incidence of the disease worldwide to be up to 66 cases per 10,000 people (Wilde et al., 2002). In contrast to LQTS, it affects males more commonly than females, in an 8:1 male-to-female ratio, although it is also an autosomal dominant gene. However, the gene is much more prevalent in southeast Asia than in the United States. Brugada syndrome is thought to cause the entity known as Lai Tai ("death during sleep") in Thailand, a relatively common cause of sudden unexplained death among young healthy men (Nademanee et al., 1997).

Clinical Features

Brugada syndrome (BS) was first described in 1992 in patients with right bundle branch block patterns on the ECG who suffered unexplained cardiac arrests (Priori et al., 2002). Since then, more has been learned about BS, although much about the disease remains unknown (Brugada and Brugada, 1992; Alings and Wilde, 1999).

It is not known why some patients with BS become symptomatic and others do not. However, once patients with BS experience a symptom (syncope or aborted cardiac arrest), it becomes a lethal disease with a high clinical penetrance. Most arrhythmic events occur for the first time when the patient is in his or her early forties, but episodes have occured in a wide age range (2 to 77 years). Patients with symptomatic BS experience polymorphic ventricular tachycardia degenerating into ventricular fibrillation, leading to syncope or even death. The episodes occur most commonly during sleep but may also happen with exercise or at rest.

The ECG of a patient with BS is frequently abnormal and represents the best way to diagnose BS. A right bundle branch block type of pattern is often noted in the right precordial leads V1 to V3 with ST segment elevation. In many patients with BS, the ECG abnormalities can normalize or be unmasked by pharmacologic challenge with a sodium-channel-blocking drug like procainamide, flecainide, or ajmaline.

Many patients with BS will have abnormal test results during invasive electrophysiology (EP) study. Inducibility of malignant ventricular arrhythmias is not rare and portends a worse clinical prognosis than for those patients who have normal EP studies. The usual cardiac tests in determining BS include echocardiogram, cardiac MRI, and biopsy (Brugada et al., 1998, 2002, 2003; Chen et al., 1998; Priori et al., 2000; Kanda et al., 2002; Smits et al., 2002; Juntila et al., 2004).

Pathophysiology

The mutation in the SCN5A gene results in either a reduced sodium channel current or failure of the sodium channel to express. The disease is caused by a defect in the α-subunit of the cardiac sodium channel gene (SCN5A) (Alings and Wilde, 1999; Balser, 2001; Antzelevitch, 2001; Kurita et al., 2002). Numerous SCN5A mutations have been described which produce BS, but most lead to a loss of function in the cardiac sodium channel. Interestingly, LQ–T3 (a completely different disease) is also due to mutations in the SCN5A gene but leads to a gain of function in the sodium channel (Yan et al., 1999; Balser, 2001; Kurita et al., 2002; Clancy and Rudy, 2002; Glatter et al., 2004).

The mutant sodium channel demonstrates more abnormal function at higher temperatures. There are numerous reports in the literature of patients with BS who experience symptoms during febrile illnesses (Gussak et al., 2000; Antzelevitch, 2001, 2002).

Genetics

Brugada syndrome is an ion channelopathy inherited in an autosomal dominant fashion. To date, only 20% of Brugada cases have been linked to the SCN5A gene; the precise ion channel mutations causing the remaining 80% of cases are unknown (Priori et al., 2002; Alings and Wilde, 1999; Brugada and Brugada, 1992). The SCN5A gene is one of the largest ion channel genes known, with at least 28 exons identified thus far (Balser, 2001; Kurita et al., 2002; Glatter et al., 2004).

Treatment

Medications are largely ineffective at treating BS (Wilde et al., 2002). The recommended treatment for symptomatic patients with BS is ICD implantation, particularly as the recurrence rate for such subjects is high. Patients who have not yet experienced an arrhythmic event but spontaneously exhibit the abnormal ECG findings are at intermediate risk for an episode and may benefit from prophylactic ICD therapy (Chen et al., 1998; Kakishita et al., 2000; Laitinen et al., 2001; Kanda et al., 2002).

Autopsy Findings

Autopsy findings in patients with BS are also unremarkable. As in LQTS, in order to conduct the postmortem molecular analyses, proper collection and storage of autopsy tissue are critical.

Catecholaminergic Polymorphic Ventricular Tachycardia

Catecholaminergic polymorphic ventricular tachycardia (CPMVT) is a newly described inherited disorder of cardiac calcium channels. It is another arrhythmogenic disorder characterized by sudden unexplained death associated with exercise.

Epidemiology

CPMVT has thus far been characterized in several Finnish and Italian families (Leenhardt et al., 1995; Swan et al., 1999; Priori et al., 2001, 2002). The epidemiology of this disorder has not yet been fully described and is, so far, limited to small case series. Its true incidence is likely much higher than is currently appreciated because most cases are undiagnosed (Massin et al., 2003).

Clinical Features

CPMVT was first described in 1995 in 21 children (Lahat et al., 2001). This disorder is characterized by syncopal spells in childhood and adolescence that are often triggered by exercise or stress (catecholamines). The disease has a mortality of 30 to 50% by the age of 30 in affected individuals (Swan et al., 1999). Due to its autosomal dominant nature, there is often a family history of unexplained sudden death.

The resting ECG of a patient with this disorder is usually unremarkable, as are cardiac imaging studies (echocardiogram, angiogram, cardiac MRI, etc.) (Leenhardt et al., 1995; Swan et al., 1999; Priori et al., 2001, 2002). Patients with CPMVT may experience bidirectional or polymorphic ventricular tachycardia with exercise stress testing, with emotional stress, or during infusion of adrenaline (isoproterenol) (Tunwell et al., 1996; Fisher et al., 1999). Up to 30% of such patients were initially misdiagnosed as having LQTS in one study (Priori et al., 2001).

Pathophysiology

Defective calcium channels formed as a result of the mutations in the ryanodine receptor gene RyR2 lead to abnormal conduction, which predisposes the heart to ventricular tachycardia and sudden death (Leenhardt et al., 1995; Swan et al., 1999; Priori et al., 2001, 2002; George et al., 2003; Lehnart et al., 2004). Sudden death is hypothesized to occur as the result of torsades de pointes or ventricular fibrillation due to the abnormal calcium channel handling.

Genetics

CPMVT is due to a defect in the cardiac ryanodine receptor (RyR2) gene, which is inherited in an autosomal dominant fashion. Ryanodine receptors are intracellular calcium channels that regulate the release of calcium from different cell sites. They are the largest ion channels yet described. RyR2 (encoded by 105 exons) is characteristically found in the heart, while RyR1 is found in skeletal muscle. Because this entity is newly described and the genes encoding the mutant calcium channel are so large, no commercial genetic screening is currently available for CPMVT.

Treatment

Beta-blockers form the mainstay of therapy in this condition. In patients who have survived cardiac arrest or are felt to be at particularly high risk for sudden death, an ICD is offered (Tunwell et al., 1996; Priori et al., 2001).

Autopsy Findings

Autopsy findings in CPMVT subjects are generally normal.

Short QT Syndrome

A new ion channelopathy, which has been associated with sudden cardiac death and premature atrial fibrillation, is the short QT syndrome (SQTS). It was first described in 2000 by Gussak et al., and the molecular defect has been defined. Several families have been identified with family histories of unexplained sudden death, syncope, and palpitations. Uniformly, they demonstrate very short QT intervals, with measured QT_C intervals <300 msec (normal QT_C interval = 400–440 msec). These entities are autosomal dominant, like LQTS, and appear to be due to a "gain of function" mutation in the potassium channel (as opposed to a "loss of function" mutation for LQTS). Mutations in the *KCNH2* and *KCNQ1* genes have been found. Several cases of early onset atrial fibrillation in teenagers have also been found in these patients. Quinidine has been proposed as a possible treatment, although defibrillator therapy is probably the treatment of choice for symptomatic patients (Bjerregaard and Gussak, 2005; Perez Riera et al., 2005; Brugada et al., 2005).

Disorders of the Heart Muscle

Arrhythmogenic Right Ventricular Dysplasia/Cardiomyopathy (ARVD/C)

Arrhythmogenic right ventricular dysplasia/cardiomyopathy (ARVD/C) is a newly recognized disorder that is a cause of unexplained sudden death in otherwise healthy young adults, particularly young athletic men (Fontaine et al., 1999; Corrado et al., 2000; Thiene et al., 2001; Marcus et al., 2003). Especially in the early stages, affected patients may have grossly normal heart function. (See also Chapter 5 and Chapter 7.)

Epidemiology

The true incidence of ARVD/C is unknown (Dalal et al., 2005). In a prospective, autopsy-based study in the Veneto region of northern Italy, 20% of unexplained sudden deaths in subjects under age 35 were found to have ARVD/C, including 22% of young athletic men who died suddenly in the region (Thiene et al., 1988). It is unclear if northern Italy simply has an abnormally high incidence of the disease or if this reflects its true incidence. However, it is likely that ARVD/C is a much more common entity than initially appreciated, as most cases go undetected.

Clinical Features

Unfortunately, the initial presentation of ARVD/C clinically is often unexplained sudden death in a healthy, athletic male. However, ARVD/C manifests by means of a wide spectrum of clinical presentations, from symptom patients to asymptomatic relatives of patients with this cardiomyopathy (Hulot et al., 2004). Patients experience ventricular arrhythmias from the diseased right ventricle ranging from benign premature complexes (PVCs) to ventricular tachycardia or even ventricular fibrillation and cardiac arrest (McKenna et al., 1994; Corrado et al., 1997; Nava et al., 2000; Obata et al., 2001). The Study Group on ARVD/C defined specific criteria to aid in the diagnosis of ARVD/C (Table 50) (Corrado et al., 2001).

ECG findings include a complete or incomplete right bundle branch block during normal sinus rhythm with T wave inversion in leads V_1 to V_3. An epsilon wave, a terminal notch in the QRS, may also be present (McKenna et al., 1994; Corrado et al., 1997, 2000; Fontaine et al., 1999; Nava et al., 2000; Thiene et al., 2001). A signal-averaged ECG (SAECG) is also characteristically abnormal (Corrado et al., 2000).

Echocardiographic findings may be normal or may reveal a variety of abnormalities in the right ventricle, including wall thinning, dilatation, or dysfunction (Mckenna et al., 1994; Basso et al., 1996; Corrado et al., 1997, 2000; Fontaine et al., 1999; Nava et al., 2000;

Table 50 Arrhythmogenic Right Ventricular Dysplasia/Cardiomyopathy (ARVD/C) Diagnostic Criteria

Family history
   Confirmed at autopsy
Structural findings
   Right ventricular global hypokinesis with preserved left ventricular function
Arrhythmias
   Right ventricular tachycardia or premature beats
Electrocardiogram findings
   Epsilon wave of QRS in leads V1 to V3
   Late potentials on signal-averaged electrocardiogram

Thiene et al., 2001). Cardiac MRI can sometimes be useful, as it may reveal the fibro-fatty infiltration of the right ventricular free wall. Biopsy of the right ventricular septum (done in the septum and not in the free wall, due to free wall thinning) is often not helpful, because involvement of the septum in ARVD/C is sporadic.

Pathophysiology

The pathophysiology of ARVD/C is unclear. It likely represents a complex interplay between genetic predisposition, cellular mechanisms, and unknown environmental factors (Fontaine et al., 1999; Corrado et al., 2000; Thiene et al., 2001). Several consistent features of ARVD/C can be noted: apoptosis (programmed cell death) (Valente et al., 1998; Nishikawa et al., 1999; Fox et al., 2000; Basso et al., 2004; Yamaji et al., 2005), a component of inflammatory heart disease (Bowles et al., 2002; Calabrese et al., 2006), and myocardial dystrophy. The disease is progressive over decades in some patients, whereas it is relatively quiescent, for unknown reasons, in others.

Genetics

At least seven distinct chromosomal loci have so far been located in association with ARVD/C (Rampazzo et al., 1994; Ahmad et al., 1998; Melberg et al., 1999; Li et al., 2000; Danieli et al., 2002). These loci include two on chromosome 10, two on chromosome 14, and one each on chromosomes 1, 2, and 3. There is no commercial genetic testing currently available to diagnose ARVD/C. For most cases of ARVD/C, the genetic linkage is unclear. Up to 30 to 50% of cases will have an associated family history consistent with ARVD/C (including sudden death).

Treatment

There is no consensus for how to treat ARVD/C. In those patients who survived cardiac arrest, implantation of a defibrillator (ICD) is generally recommended to avoid sudden death (Calkins, 2006). Pharmacologic therapy with beta-blocker or antiarrhythmic medications was also suggested. Radiofrequency ablation during electrophysiology study of ventricular arrhythmias was also attempted.

Autopsy Findings

ARVD/C is characterized by lack of myocardium in the RV free wall, which is replaced by fatty or fibrofatty tissue (Marcus et al., 1982; Thiene et al., 1988; Basso et al., 1996; Corrado et al., 1997b, 2000). The disease may be segmental or involve the right ventricle diffusely; when focal, it is most frequently located at the angles of "the triangle of dysplasia": the pulmonary infundibulum, the right ventricular apex, and the inferior wall of the right ventricle (Gallo and d'Amati, 2001). At macroscopic examination, the heart discloses no or only slight right ventricular dilatation, and transmural fibro-fatty replacement of the right ventricular musculature (Figure 110). Cases of ARVD/C in which left ventricular myocardium was involved were described; the relatively high prevalence of left ventricle involvement described by some authors suggests that this disease should no longer be considered as limited to the right ventricle.

Finally, in ARVD/C, the spectrum of morphological appearances may range from concealed right ventricular myopathic changes to biventricular cardiomyopathy; even in

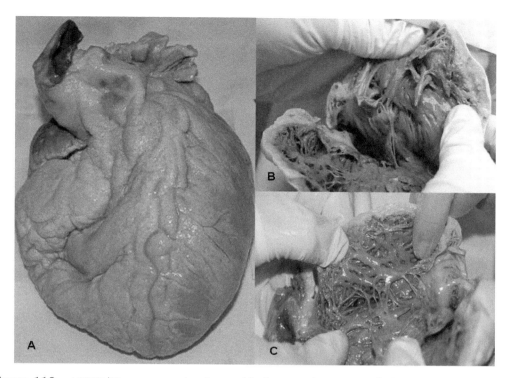

Figure 110 ARVD/C macroscopic view with fatty replacement of the right ventricle in a 32-year-old woman who died suddenly with no previous symptoms (A). Transillumination of the posterior right ventricular wall (B). Section of the right section of the heart: fatty replacement of the entire ventricular wall (C).

autopsies, in fact, the disease is often overlooked. The right ventricle should be extensively sampled for histopathological analysis in all cases of sudden death, especially those associated with strenuous exercise and young age.

Two pathological types were proposed: a typical form with fat infiltration and scarring (fibro-fatty) and a form characterized solely by fat replacement (fatty form). In the first, myocardial loss is replaced by both fibrous and fatty tissue, which may occur in virtually any area of the right ventricle and will generally result in myocardial thinning. In the fatty variety, myocardial loss is replaced only by fatty tissue with predominant involvement of the apex and infundibulum, in absence of fibrosis and inflammatory infiltraes (Thiene et al., 1988; Marcus et al., 1995; Basso et al., 1996; Nava et al., 1997; Corrado et al., 1997b; Burke et al., 1998a). The individuation of the fatty form of ARVD/C is problematic because it is known that significant fat infiltration of the right ventricle occurs in >50% of normal hearts in elderly patients.

Other authors are not in agreement with these definitions and consider that the presence of fibrous tissue is necessary for the diagnosis of ARVD/C. Others prefer a distinction based on histological evaluation of both the nature of the replacing tissue and the myocellular features, so distinguishing an infiltrative and a cardiomyopathic pattern. In the infiltrative pattern, normal or slight myocytes are replaced by mature adipocytes in a lacelike way. This form of ARVD/C is frequently associated with a right ventricular localization, with infrequent involvement of all the sites of triangle of dysplasia. An adipose

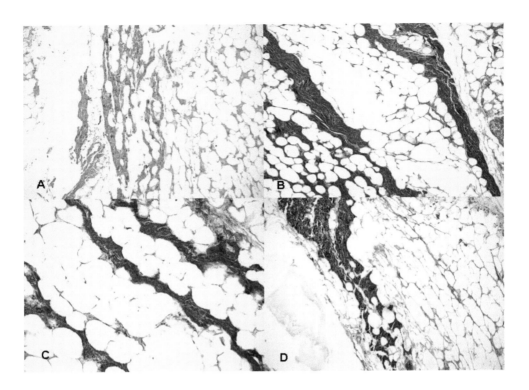

Figure 110 bis   Fibro-fatty ARVC in a 42-year-old woman who died suddenly. View of the anterior right ventricular free wall, with transmural marked fibro-fatty replacement. (A) Van Gieson (40×); (B) Mallory stain (60×); (C) Azan stain (100×); (D) PTAH (10×).

pattern is maintained and is usually limited to the outer ventricular layers. In the cardio-myopathic pattern, there is a massive myocardial replacement by fibro-fatty tissue. It is associated with extensive right ventricular involvement and an almost constant extension to the left ventricle.

In suspected cases of ARVD/C, histological myocardial sections can be stained with hematoxylin and eosin and Masson's tricrome for collagen network; other elective stains, such as azan techniques or hematoxylin-phloxine-safranin O stain can be useful.

Histologically, ARVD/C is characterized by substitution of the myocardium with adipose or fibro-adipose tissue (Figure 110 bis); inflammatory infiltrates with focal myocyte necrosis are often present. The following pathological features must be assessed and graded in histological specimens: myocardial atrophy and fatty replacement, myocardial fibrosis, myocyte degeneration or necrosis, and interstitial cell infiltrates (Marcus et al., 1982; Maron 1988; Lobo et al., 1992; Schionning et al., 1997; Nava et al., 1997; Fornes et al., 1998; Burke et al., 1998; D'Amati et al., 2001; Gallo and D'Amati, 2001; Michalodimitrakis et al., 2001; Chimenti et al., 2004; Basso and Thiene, 2005).

Recently, the myocardial loss in ARVD/C was related to apoptosis, which is observed in the myocardium of patients with ARVD/C. However, it is still unclear whether apoptosis is the cause for the myocyte loss, or whether it occurs as a secondary phenomenon induced by myocytes stress or ischemia followed by myocardial fibro-fatty replacement.

Hypertrophic Cardiomyopathy (See Chapters 5 and 7)

Hypertrophic cardiomyopathy (HCM) is one of the oldest known causes of sudden death. It was first described in 1958 by Teare. It has been called hypertrophic obstructive cardiomyopathy and also idiopathic hypertrophic subaortic stenosis, despite the fact that 75% of affected patients do not have a sizable resting outflow gradient (Braunwald et al., 1964; Maron and Epstein, 1979). It is a polygenic, relatively common genetic cause of sudden death, particularly in young athletes.

Epidemiology

Hypertrophic cardiomyopathy is actually the most common genetically associated form of sudden cardiac death. It is estimated that 1 in 500 people (0.2% of the general population) carry an HCM genetic mutation (Maron et al., 1995; Ommen and Nishimura, 2004). However, the phenotypic presentation or clinical penetrance of the disease is much lower. Most patients with an HCM mutation will not show signs of the disease during life.

Clinical Features

The clinical diagnosis of HCM during life is made most reliably by echocardiography. Severe ventricular wall thickening can be seen. A normal left ventricular wall thickness is generally <12 mm, with thicknesses >30 mm not unusual in severe cases of HCM (Louie and Maron, 1986; Spirito et al., 1997, 2000; Elliot et al., 2001; Nishimura and Holmes, 2004). This marked septal hypertrophy is often an age-dependent effect and may not be seen initially in young patients. In most cases, the left ventricle may be affected diffusely or may demonstrate asymmetric septal hypertrophy (ASH). In contrast, in the Japanese variant of HCM, the apical left ventricle is primary affected and shows abnormal thickening (Maron and Roberts, 1979; Louie and Maron, 1986; Spirito et al., 1997, 2000; Elliot et al., 2001; Nishimura et al., 2004).

Pathophysiology

Syncope in these subjects may occur due to arrhythmias or from obstruction due to ventricular hypertrophy and cavitary obliteration (Spirito et al., 1997; Maron, 2002; Nishimura et al., 2004). Dehydration can trigger a syncopal event in such patients. Sudden death is thought to occur in HCM due to a primary electrical abnormality by ventricular arrhythmias (McKenna et al., 1981; Spirito et al., 1997; Elliot et al., 2000; Maron et al., 2000; Watkins, 2000; Maron, 2002). In support of this view, one large study of HCM patients in whom defibrillators were implanted demonstrated that nearly 25% had documented ventricular arrhythmias over a 3-year follow-up period (Maron et al., 2000).

The disease may be progressive in some patients. The myocyte hypertrophy continues over years in a clinically silent manner and may ultimately lead to an end-stage, dilated cardiomyopathic picture. Depending upon the time frame during which the patient is evaluated, the HCM-affected heart could appear grossly normal, markedly hypertrophied, or even dilated, making the diagnosis difficult.

Genetics

The polygenic and multicellular nature of HCM makes it a frustratingly complicated disease to diagnose unless gross histopathologic abnormalities are found on echocardiogram

Table 51   High-Risk Features in Hypertrophic Cardiomyopathy

Family history of sudden death
High-risk genotype (e.g., Arg719Gln)
History of sustained ventricular arrhythmias
Previous cardiac arrest
Exertional syncope
Massive left ventricular wall hypertrophy (≥30 mm)

or at autopsy. At least 10 different genes encoding the cardiac sarcomere have been implicated in HCM (Spirito et al., 1997; Seidman and Seidman, 2001; Maron, 2002; Nishimura et al., 2004). Over 150 unique mutations have been reported to date since the first genetic cause for HCM was identified in 1990 (Geisterfer-Lowrance et al., 1990). Most are missense mutations found in the proteins of the cardiac sarcomere and are located in the β-myosin heavy chain, cardiac troponin T, or myosin binding protein-C. Although the disease is autosomal dominant, a family history of syncope or sudden death may be lacking, and the disease has widely variable clinical penetrance.

Within the β-myosin heavy chain gene (MYH7), numerous mutations have been described as malignant mutations associated with a poor clinical prognosis. These particular mutations seem to be associated with a severe clinical phenotype including progression to end-stage heart failure or sudden death, a relatively high penetrance of the disease, and extreme left ventricular wall thickness (Table 51) (Geisterfer-Lowrance et al., 1990; Watkins et al., 1992, 1995; Moolman et al., 1997; Tesson et al., 1998; Marian, 2000; Enjuto et al., 2000; Seidman and Seidman, 2001).

Treatment

There are no formal guidelines for treating asymptomatic patients with hypertrophic cardiomyopathy. In those patients with symptoms of shortness of breath, medical therapy with medications that reduce the outflow gradient remain the mainstay of therapy. Such medications include beta-blockers or calcium channel blockers. In symptomatic patients with a large (>50 mm) gradient, the outflow gradient can be reduced by surgical myomectomy or by catheter-based alcohol ablation. The latter is a relatively new technique that causes a controlled myocardial infarction and thus reduces the outflow gradient. In those patients deemed high risk for an arrhythmic event (see Table 51), an ICD may be implanted to avert sudden death (Spicer et al., 1984; Gilligan et al., 1993; Lakkis et al., 2000; Qin et al., 2001).

Autopsy Findings

The classical anatomic form of HCM described by Teare involved thickening of the basal anterior septum, which bulges beneath the aortic valve and causes narrowing of the left ventricular outflow tract (Teare, 1958).

Gross heart anatomy is marked by myocardial hypertrophy with thickened left ventricular wall and structural derangement. The characteristic gross morphologic feature is a hypertrophied and nondilated left ventricle It is rarely "symmetric," a diffuse concentric hypertrophy of the left ventricle; the symmetrical form of HCM accounts for about 42% of cases and is characterized by concentric thickening of the left ventricle with a small cavity dimension. The majority of cases present with "asymmetric" hypertrophy that may involve any ventricular portion. Generally, the maximal increase is in the interventricular septum (interventricular septum-to-free wall ratio equals 3:2) which, as a result, protrudes into the left ventricular cavity, drastically reducing it. This obstruction generally involves the median-superior portion of the interventricular septum. The whole septum or its apical region are rarely involved. Davies and McKenna (1995) reported the following distribution of left ventricular hypertrophy: anteroseptal in 23 of 43 asymmetric cases (53%); anteroseptal plus lateral in 8; anterior, septal, or posteroseptal in 2 cases each; and anterolateral or posterolateral in 1 case each. Also, the right ventricle is rarely affected — 18% in the study by Davies and McKenna. Frequently, the pattern of wall thickening is heterogeneous, and contiguous segments of the left ventricle may differ greatly in thickness.

The morphological spectrum of HCM is further complicated by the observation that HCM may progress to a dilated or "burned-out" phase, and left ventricular cavity dilatation may develop.

Other findings may include endocardial fibrosis marked by a white thickening of endocardium, where the anterior mitral leaflet beats against the septal wall, and the ventricular aspect/chordae tendineae of anterior mitral valve (75%), atrial dilatation (100%), and a prominent septal band connected with the superventricular crista in the right ventricle. In cases with congestive heart failure, the heart has a gross aspect typical for dilated cardiomyopathy (Maron et al., 1980, 1982, 1987; Gallo and d'Amati, 2001; Hughes, 2004).

Histology remains the cornerstone for the diagnosis of HCM (Hughes, 2004). The characteristic change is disarray, which is defined as star-like disposition of adjacent myocytes, aligned obliquely or perpendicular to each other, and joined together by short, generally hypertrophic myobridges, with interconnecting myofibrils and interstitial fibrosis. The myocytes are disposed in any direction with myofibrillar connections. The structural derangement results in bundles of hypertrophied myocytes often arranged in a disorganized pattern with adjacent myocells forming oblique or perpendicular angles to each other, forming a herringbone pattern. The myocytes can assume bizarre forms; they are characterized by bizarre nuclei that exhibit nuclear enlargement, pleomorphism, and hyperchromasia (Figure 111).

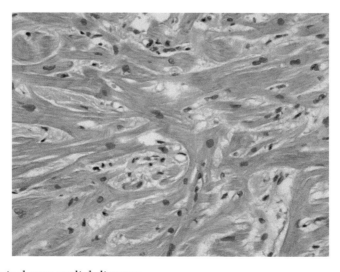

Figure 111   Typical myocardial disarray.

In cases of hypertrophic cardiomyopathy, disarray may be either symmetric or asymmetric, and it may be extensive. Between the disarrayed myocytes is an increased interstitial matrix and a myofibrous hyperplasia of arterioles. From a practical standpoint, pathologic disarray can be diagnosed when it involves more than 20% of the myocardium in at least two tissue blocks. In this case, the detection of myocardial disarray is a highly sensitive and specific marker of HCM.

Patients with HCM may present various degrees and distributions of fibrous-tissue formation in the left ventricular myocardium; this may range from patchy interstitial fibrosis to extensive and grossly visible scars that may even be transmural.

The real significance of myocardial disarray is not yet completely defined (Fineschi et al., 2005); recently, experimental studies suggested the need to reconsider the significance of myocardial disarray. Small foci of disorganized myocardial cells can be found in the normal human interventricular septum. Myocardial disarray is found around scars, in congenitally malformed hearts, with lentiginosis (Carney's syndrome), in Friedreich's ataxia, in Turner's syndrome, and in hyperthyroidism. The real incidence and the significance of myocardial disarray in different human conditions must be explored further.

Interpretation of Pathological Changes in Sudden Death

10

GIORGIO BAROLDI

Contents

Meaning of Pathological Findings in Sudden Death

Most reports on sudden death relate its cause to chronic lesions or congenital anomalies that preexisted for years without symptoms and signs. In particular, coronary atherosclerosis is defined as the most frequent cause of sudden cardiac arrest through ischemia caused by luminal stenosis. Previously, we questioned the ischemic effect of a stenosing plaque that we believe is nullified by an adequate "compensatory" collateral flow. If so, the atherosclerotic plaque as a chronic obstructive lesion does not cause the sudden death. Similarly, all other chronic conditions, including congenital ones, as for instance the dislocation of coronary ostia (Figure 112) or the "mural" tracts of coronary arteries, coexist with a perfectly normal life. At the most, such chronic "compensated" conditions can be considered predisposing factors for an acute and sudden trigger mechanism.

A sudden cardiac arrest must have an acute etiopathogenetic mechanism, and the pathologist should seek acute morphological changes. In all coronary and noncoronary cases we studied (Baroldi et al., 2004), sudden death was associated with catecholamine

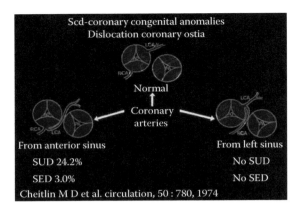

Figure 112   Dislocation of coronary ostia and sudden death.

myocardial necrosis, which, in general, showed early and late stages in the same heart with an extent significantly greater than that found in the hearts of healthy controls without cardiopulmonary resuscitation. In monitored subjects and experimentally, we demonstrated that this necrosis was linked with ventricular fibrillation and experimentally prevented by beta-blocking agents. This suggests what the mechanism of sudden death in all can be.

In a recent review of the cardiovascular causes of sudden death, the latter is interpreted as a symptom and not a disease. This is a correct definition if one believes that all associated chronic lesions cause the sudden demise (Thiene et al., 2001). However, I believe that any time a cardiac arrest results from a malignant arrhythmia/ventricular fibrillation, an acute pathognomonic adrenergic mechanism causes the sudden death. Similarly, when the cardiac arrest is a result of bradycardia/asystole generally secondary to a chronic or occasionally to an acute–subacute congestive heart failure, the histological hallmark is colliquative myocytolysis, that is, a progressive edematous vacuolization following a loss of myofibrils (Baroldi et al., 2004). These two patterns are not related to ischemia, and each becomes a physiopathologic entity and not only a symptom.

A last point should be made related to acute pathologic findings. A ruptured aorta with cardiac tamponade is a classic example of a true cause of sudden death; other acute findings should be carefully evaluated. From experiments, we know that an acute occlusion of a normal coronary artery, in a short time and before an infarct is formed, may lead (40%) to ventricular fibrillation, always combined with catecholamine necrosis, both being prevented by beta-blockers. Therefore, an acute coronary occlusion by an embolus or other does not mean that the embolus caused sudden death. Again, it is a predisposing ischemic/asynergic factor that can start a preventable chain of adrenergic events that may end in ventricular fibrillation.

Review of Specific Conditions

Reflow or Reperfusion Necrosis

The definitions of "reflow necrosis" and "reperfusion necrosis," often associated with ventricular fibrillation, were provided above, and experimental reperfusion was distinguished from infarct necrosis. Clinically, emergency revascularization or reperfusion is indicated for any acute coronary syndrome, both as the first episode of ischemic disease or as an acute event in chronic patients. If these procedures supplied more blood to an ischemic myocardium, we might expect extensive reflow necrosis and ventricular fibrillation as occurs experimentally, a situation apparently contradicted by innumerable "revascularizing" procedures performed everywhere in the world. The specific enzymes found after these invasive techniques are likely due to catecholamine necrosis following traumatic manipulation and not to persisting ischemia as proposed (Schoen, 2001). Sometimes, therapy acts differently from what we believe. On this argument, a recent new method of revascularization (Saririan and Eisenberg, 2003) is based on the concept of reestablishing a lacunar type of vascularization as in early ontogenesis. By laser, several transmural "channels" are formed in the wall of the ischemic left ventricle during open chest surgery. I had the opportunity to examine the heart of a patient revascularized by this method who died after 6 months. There were no signs of new vessel formation, and the channels formed by laser were replaced by dense fibrous avascular scarring.

Coronary Artery Stenosis

At present, efforts are directed at preventing reocclusion after angioplasty or bypass by stents, a mechanical structure to maintain the lumen patent as well as specific endothelizing substances to prevent thrombosis. In our experience, the slowing of blood flow leading to stasis consequent to the increased intramyocardial resistance is the main possible cause of accelerated atherosclerosis and hemorrhage within a plaque, its rupture, and thrombus formation. Obviously, a reopened lumen restores the native flow, with the disappearance of preexisting collaterals that immediately reappear if occlusion is reestablished (Gregg, 1950). What is the meaning of these invasive revascularizing procedures? This is a pertinent question related to surgical bypass because an intimal thickening of the implanted vein occurs because of increased hemodynamic forces (Figure 113). On this background in predisposed patients, an atherosclerotic process may develop after time. However, atheroma must be distinguished by atheromasia of an old coagulum. Moreover, the frequent occlusion of a bypassed stenosis (Aldridge and Trimble, 1971; Griffith et al., 1973; Figure 114) is a good example of occlusion following stasis at the plaque level because bypass flow is equal to a collateral distal flow.

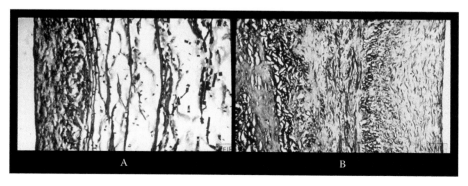

Figure 113 (A) Normal saphenous vein. (B) Media-intimal thickening in a saphenous vein in aorto-coronary bypass graft.

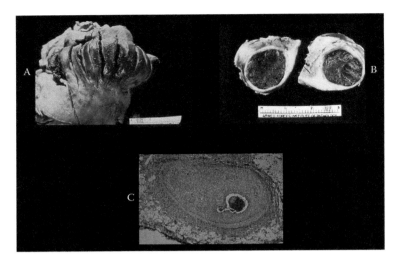

Figure 114 (A) Severe stenosis of left anterior descending (LAD) branch. (B) Venous bypass of the stenosis. (C) Occlusion of the stenosis following bypass. The latter induces a blockage of blood flow within the stenosis, which favors occlusion by laminar thrombi. The bypass flow is equivalent to a collateral compensatory flow. (Courtesy of Dr. Griffith.)

Myocardial Disarray

When found in a specific zone of the cardiac muscle, the focal disarray is located at the site of the directional change of muscle bundles. This suggests that nodal junctions are capable of anchoring different bundles to help in contraction. In contrast, when there is widespread disarray, the myocardium becomes asynergic, that is, it loses pumping action. In turn, in proportion to the extent of the disarrayed zone, the myocytes normally aligned in parallel will hypertrophy to compensate for the latter. Therefore, in this cardiomyopathy, we have two concurrent types of hypertrophy. One is consequent to a useless increasing contraction of the disarrayed myocardium, and the other is the "normal" compensatory hypertrophy with tridimensional neomyofibrillar mitochondrial-sarcomerogenesis and marked irregular enlargement and branching of the nucleus for increased surface exchange (Figure 115 and Figure 116). Furthermore, the frequency and extent of pathologic disarray in human cardiology are not known (see below). It is of interest that diffuse disarray in an otherwise normal heart was found in a 28-year-old woman with hyperthyroidism and clinical diagnosis of restrictive cardiomyopathy (Baroldi et al., 1998c). Other similar cases were reported (McKenna et al., 1990; see Chapter 7). Therefore, myocardial disarray may occur in a spectrum of diseases, with the need for a better semantic interpretation of hypertrophic cardiomyopathy, a term that should be substituted by "disarray cardiomyopathy," inclusive of different clinical patterns, from asymmetric to restrictive cardiomyopathy.

Pacemaker Implantation

Mention should be made of myocardial changes secondary to different types of pacemakers (Wilson, 1991), including infection at the site of implantation, dislocation or rotation of the device, or rupture of an electrode. Around the latter, a layer of contraction bands may be visible. Experimentally (Reichenbach and Benditt, 1969) and in human heart defibrillated by electrical shock, extensive subepicardial contraction band necrosis forms (Karch,

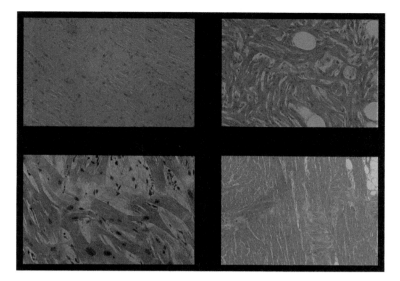

Figure 115 Myocardial disarray. Only this type of architectonic disorganization of the myocardial cells (type I according to Maron et al., 1978) was considered in our study as the structural background of myocardial asynergy.

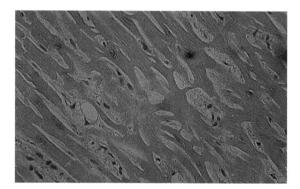

Figure 116 Myocardial disarray in a transplanted heart.

1987). Its peculiar location differentiates it from catecholamine necrosis. Atrioventricular node ablation for refractory atrial fibrillation in 344 patients resulted in sudden death in 9, 3 within 48 hours, 1 after 4 days, 3 after 3 months, and 2 unrelated to the procedure (Ozcan et al., 2002).

Stem Cell Therapy

There are two possible therapies now being considered in cardiology: neomyocytogenesis to rebuild a destroyed myocardium and neovasogenesis to improve the myocardial blood supply. It was previously emphasized that in the natural history of many heart diseases studied, I could not demonstrate any morphologic hallmark of either neomyocytogenesis or new vessel formation as might be expected in many different conditions with functional and structural damage. Nevertheless, apart from the previous references cited, several recent contributions apparently show that stem or progenitor cells can be implanted within the myocardium with their further differentiation and proliferation in myocardial cells or endothelial elements. So far, neomyocytes were recognized in transplanted human hearts (Quaini et al., 2002); in adult rat hearts in which human mesenchymal cells were implanted (Toma et al., 2002); fetal and neonatal but not adult cardiac myocytes in cryoinjured or infarcted animals (Reinecke et al., 1999; Leor et al., 1996; Orlic et al., 2001); neovascularization following implantation of smooth muscle cells of rats, resulting in thickened scarring, preventing dilatation (Reinecke et al., 1999); and mobilization of endothelial progenitor cells and their putative CD-34 positive, mononuclear cell precursor in the blood in cases with an acute myocardial infarction (Shintani et al., 2001).

Some recent reviews of this therapeutic approach are very critical: "The completion of this paradigm (regenerative capacity of myocytes) from the dogma of nonregenerating myocardium to that of a self-repairing heart requires unquestionable proof since other authors do not confirm such extraordinary regenerative capacity of the myocardium. Although we recognize the importance of challanging dogmas for the progress of science and the welfare of human kind, we also urge all involved in the field to rigorously evaluate the data since false-positive imaging of new myocyte occurs (Taylor et al., 2002) with the need of randomized trials to assess the balance between likely benefit and theoretical risk of stem cell therapy" (Forrester et al., 2003). In conclusion, there is no proof of a spontaneous regeneration to repair or support an altered myocardium. The demonstration, per se, of survival and differentiation of stem cells is not enough to prove a physiologic repair because it is difficult to understand how the new elements can integrate with the still functioning structures of the tissue. May new endothelial cells, myocytes, or neurons with their complex connections restitute normal function?

Cardiovascular Genomics

The other point is what today is defined as cardiovascular genomics. The existence of monogenic inherited diseases was recognized long ago. The relatively recent identification of the draft sequence of human genome and the ability to exploit innumerable gene sequences opened a new era of genomic medicine (Seidman and Sampson, 2001; Nabel, 2003). Gene mapping and the hetereogenity of gene defects comprise an essential step in helping us to understand the different mechanisms of onset and various complications in the natural history of both monogenic and nonmonogenic diseases. The variations in gene expression help us to recognize the biomarkers of any single symptom or sign. This is an extremely promising but complex field, with the need to carefully discriminate between achievements and pitfalls (Yoshioka and Lee, 2003).

Adrenergic Stress and Related Morphopathology in Cardiology

All of our morphofunctional data indicate that the sympathetic nervous system may have an important role in explaining the pathogenesis of several cardiovascular conditions in general, and ischemic heart disease and congestive heart failure in particular. An adrenergic stress theory (Table 52) appears to be a viable alternative to the hydraulic, occlusive, unifying theory. In fact, the natural history of contraction band necrosis may help interpret complications and causes of cardiac arrest in acute coronary syndromes, no matter whether

Table 52 Adrenergic Stress in Coronary Heart Disease, Atherosclerosis, and Congestive Heart Failure

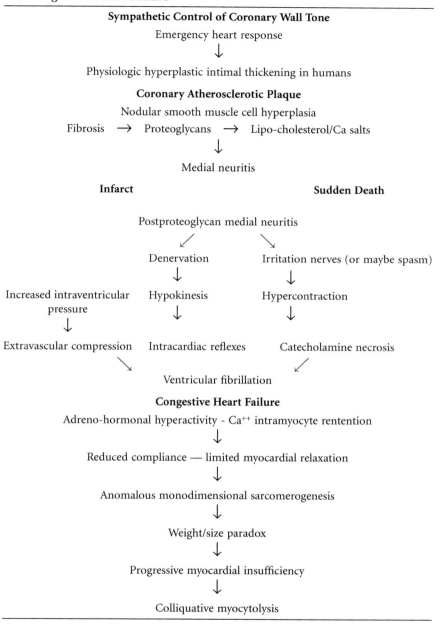

it is a first episode in an apparently healthy subject or it occurs in a patient with a chronic "ischemic" condition. The term "ischemic" needs to be in quotation marks, because the role of ischemia appears to be restricted to specific conditions and is often absent in human pathology. In contrast, adrenergic stress with correlated catecholamine myocardial necrosis is an ongoing process, even in the absence of an acute cardiac infarct. This process (*catecholamine cardiopathy*, if you will) may explain progressive myocardial fibrosis — a concept supported by a lack of correlation between infarct size (measured by Technetium 99m sestamibi/end-systolic and end-diastolic volume/ejection fraction) at discharge from the hospital and severe dilatation of the left ventricle 1 year later (Chareconthaitawee et al., 1995).

Adrenergic overstimulation is also a possible factor in explaining the genesis of human myohyperplastic atherosclerotic plaques. Catecholamines promote smooth muscle cell hyperplasia (Blaes and Boissel, 1983; Schwartz et al., 1986; Yamori et al., 1987; Velican and Velican, 1989; Leenen, 1999), explaining the physiologic intimal thickening (Wolkoff, 1929) found only in humans and not, for instance, in the dog, despite an identical extramural disposition of the main coronary vessels and an equal type of diphasic blood flow (Gregg, 1950). The adrenergic control of coronary arterial tone (Rushmer, 1963), more frequently stimulated in humans (see below), could explain physiologic intimal thickening and a predisposition to nodular smooth muscle cell hyperplasia (monoclonal hyperplasia after Benditt, 1974) as the starting point of a myohyperplastic atherosclerotic coronary plaque and endomyoelastofibrosis.

Another relevant finding is the lympho-plasma-cellular infiltrates around nerves of the media (medial neuritis; Figure 117) at the time of proteoglycan accumulation. Statistically more frequent and extensive in patients displaying any pattern of coronary heart disease than in normal controls with the same degree of obstructive coronary atherosclerosis, this medial neuritis was present in all coronary plaques of the same hearts, independent of the degree of lumen reduction, and in aortic plaques of hearts displaying coronary heart disease in contrast to hearts of controls (Baroldi et al., 1988). This finding may explain that the prognosis in this disease is related to the number of all plaques and not to the number of plaques with severe stenosis (Bigi et al., 1999). This may be a possible pathogenic factor in determining regional asynergy of the distal myocardium with extravascular intramural compression and stasis in the main artery followed by secondary plaque hemorrhage/rupture and thrombosis. What needs to be stressed again, in contrast to other opinions (Libby, 1995; Vink et al., 2001; Heistad, 2003), is that coronary atherosclerosis does not begin as an inflammatory process, and medial neuritis — the unique flogistic process found — is a relatively late complication after proteoglycan storage in the deep sclerotic intima (see

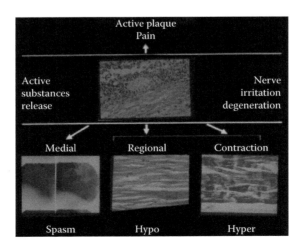

Figure 117   Possible interaction between medial neuritis and different forms of morphofunctional myocardial disorder.

above). Myohyperplastic coronary plaque, as defined previously, is not an inflammatory process, and its relation to several (*Chlamydia pneumoniae, Helicobacter pylori*, Herpes simple virus, Cytomegalovirus, etc.) inflammatory agents (Ridker, 2002; Espinola-Klein et al., 2002) as well as to coronary thrombosis (Topol et al., 2003) needs to be carefully reconsidered, keeping in mind that C-reactive protein, a sign of inflammation, is produced by smooth muscle cells of the coronary artery (Calabrò et al., 2003).

Another fact to be considered is the previously discussed regional denervation of the myocardium demonstrated by scintigraphy in various conditions. Such denervation could be caused by different mechanisms, such as nerve fragmentation in local asynergic myocardium, extensive interstitial myocarditis, or, as previously mentioned, neuritis of medial nerves, etc. Further investigation is needed in this field, because denervated myocardial cells become more sensitive to intramyocardially released or blood-borne catecholamines (Donald, 1974), with an increased concentration of norepinephrine in the synaptic cleft at the myocyte membrane level because of reduced uptake (Bohn et al., 1995) or loss of autonomic innervation (Young et al., 1993). On the other hand, some sympathetic reinnervation is obvious in one third of patients who underwent heart transplants and survived more than 3 years (Halpert et al., 1996) or in 48% of these hearts after 1 year (De Marco et al., 1995), with reinnervation a regional heterogeneous phenomenon (Wilson et al., 1993). It is apparently restored after 15 years (Bengel et al., 1999). This limited information applies to the global denervation in transplanted hearts, without any knowledge of possible regrowth of adrenergic nerves in local denervation. On the other hand, hyperinnervation (maybe compensatory) around a denervated zone could explain myocardial sensitization following sympathetic denervation in diabetes mellitus (Stevensen et al., 1998) and an increased risk of malignant arrhythmia in congestive heart failure (Cao et al., 2000) and dilated cardiomyopathy with T-wave alteration (Harada et al., 2003); as demonstrated experimentally by an excess mortality rate in swine with hibernated myocardium following coronary stenosis (Luisi et al., 2002). Of some interest would be a study that would discriminate whether or not angioplastic balloon inflation with severe disruption of atherosclerotic plaques may result in regional denervation by breaking coronary nerves at the plaque site. Finally, an equivalent dysfunction without nerve damage may result from impaired turnover of catecholamines in the interstitium; for instance, a decreased noradrenaline uptake was shown in the hypertrophied zone but not in the normal zone of patients with hypertrophic cardiomyopathy (Li et al., 2000) or at the border of a cardiac aneurysm (Bevilacqua et al., 1986).

The asynergic and arrhythmogenic effects of myocardial disarray were discussed. Several experiments substantiate a link between this structural disorganization and adrenergic system/endocrine disorders as shown by administration of nerve growth factors (Witzler and Kaye, 1976) or acetic analogue of tri-iodothyronin (triac) (Hawkey, 1981) or sub-hypertensive doses of norepinephrine with an increase of tissue norepinephrine (Laks, 1979; Blautass, 1975). The cardiac direct overexpression of human beta 1-adrenergic receptors in transgenic mice leads to myofibrillar disarray, marked cardiac hypertrophy, interstitial fibrosis, cardiac dysfunction, and sudden death (Bisognano et al., 2000). This finding suggested a review of the frequency and extent of myocardial disarray (Figure 55, Figure 56, Figure 115, and Figure 116) in a series of conditions including those in which an adrenergic stress is admitted (Table 36) (Baroldi and Silver, 2004). In these groups, the myocardial disarray was significantly more frequent and diffuse in areas where it was not normally found. An interesting note is that it was absent in transplanted hearts with a survival less than 1 week (Fineschi et al., 2005). This invites a reconsideration of myocardial disarray as an arrhythmogenic factor (Slodes et al., 1993).

The histologic evidence of myocardial adrenergic toxicity, the effect of the sympathetic system on many phenomena, such as platelet aggregation, arterial wall tone, hyperlipidemia, myocardial asynergy (Figure 109), and so forth (Eisenberg, 1966; Levine and Welles, 1964; Kaplan, 1987, 1988; Yamori et al., 1987; Cruickshank et al., 1987; Ablad, 1988; Wikstrand et al., 1988; Wikstrand and Kendall, 1992; Ross, 1993; Leenen, 1999), and the relationship between immunologic response and adrenergic stress (Benshop et al., 1994; Maisel, 1994), all require further investigations of their roles in the natural history of cardiac pathologies, with particular attention paid to ischemic heart disease and congestive heart failure (see above).

Coronary Heart Disease: Deficiences in Current Causal Hypotheses and Proposal for New Ones

Despite preventive measures aimed to reduce risk factors and all of the therapeutic approaches instituted during the past 50 years, coronary heart disease remains a prime killer and the first cause of morbidity in consumer-technologic societies. In some countries, mortality from acute coronary syndromes was lowered, but morbidity is unchanged, and lowering the death rate from acute episodes results in older patients with chronic disease, a human and social calamity. This is not a satisfactory outcome after more than 50 years of procedures aimed at reperfusing an "ischemic" myocardium. It is time to critically review current dogma and include in our theoretical approach other lines of investigation.

The tenet that ischemia is the cause of acute coronary syndromes becomes increasingly debatable. Coronary syndromes have a cloudy pathogenesis. Apart from inconsistencies discovered by our investigations, we cite other paradoxes:

1. Atrial pacing to increase metabolic demand in coronary patients produces vaso-constriction (or maybe myocardial asynergy with vascular compression) rather than maximal vasodilatation (Sambuceti et al., 1997; Marzilli et al., 2000).
2. Global flow reduction in patients with acute infarct is 45% less than normal in both nonculprit and reopened culprit artery (Gibson et al., 1999).
3. A similar outcome in results was shown at 1 year, independent of the application of more aggressive therapy in the U.S. than in Canada (Mark et al., 1994).
4. No difference in results was shown in patients with acute infarct treated by intravenous fibrinolytic therapy versus angioplasty (Dauchin et al., 1999), keeping in mind that the latter reestablishes a "normal" lumen, while reopening a vessel lumen after fibrinolysis does not change severe preexisting stenosis.
5. No blood-flow-related recovery of contraction by dobutamine was observed in dysfunctioning myocardium supplied by a nearly occluded vessel (Barilli et al., 1999).
6. Contradictory recovery of function was shown in hibernating and stunned segments after surgical revascularization (Bax et al., 2001).
7. There was an absence of reperfusion injury following angioplasty within 6 h in patients with an acute myocardial infarct (Lepper et al., 2002).

8. Despite significantly less restenosis following coronary stent insertion, the rate of patient survival did not change (Kelley et al., 1999).
9. Patients with heart failure associated with single vessel disease and no history of myocardial infarction should be classified as nonischemic subjects for prognostic purposes (Felker et al., 2002).

The epidemic of coronary heart disease seems to start any time there is a change from an agricultural to an industrial-technologic lifestyle leading to an affluent society. This means inexhaustible competition starting early in childhood and a stressful effort to achieve increasingly more, stimulated by mass media. "Society has need for such men, produces them and destroys them" (Morris and Gardner, 1969, p. 676); and destruction means mainly "ischemic" heart/brain disease that develops often during the period of maximal productivity.

Two documented epidemics of sudden death — one recorded in 1705 at Rome and the present one — question whether or not they have any common denominator. The precise, limited autopsy notes of Lancisi (1745) do not permit a comparison of postmortem findings. Did the apparently opposite societies affected, 1700 Rome at its maximum decadence and poverty, and our rich, technological society, share a similar mental depression without hope? For example, an epidemic of sudden death occurred among highly skilled technology employees of the U.S. National Aeronautics and Space Administration following a budget cut in 1968 with difficulty in staff relocating (Eliot, 1994; Eliot and Buell, 1985), producing depression related to socioeconomic factors (Salomon et al., 2000). We need to understand the link between psychological factors, acting in "ischemic" patterns (Rozanski et al., 1999; Freasure-Smith et al., 2000) and heart/brain interaction (Skinner, 1985; Lown, 1979; Lown et al., 1977). An impaired balance of the vegetative nervous system may explain cardiac arrest and coronary syndromes. Acute brain lesions produce focal myocardial contraction band necrosis (Connor, 1969; Baroldi et al., 1997a) that is prevented by betablocking agents (Hunt et al., 1972). Also, arrhythmias following a sudden increase of intracranial pressure were inhibited by sectioning in an experimental model the vagal and sympathetic nerves with an intact spinal cord (Estanol et al., 1977). In dogs, coronary occlusion plus psychological stress by electrical shock provokes major arrhythmias (Corbalan et al., 1974). In subjects with preclinical coronary heart disease, a mental stress with physical exercise can induce myocardial ischemia (Kral et al., 1997).

The concept of adrenergic stress is based on several facts. The adrenergic system provides an alarm mechanism that alerts the body to react quickly to any acute emergency (Cannon, 1942; Raab, 1970; Kubler, 1992; Kubler and Strasser, 1994; Meerson, 1993; Meerson et al., 1982). In permanently stressful conditions, exacerbated by infinite stimuli from the enviroment with an intense psychological response (Rozanski et al., 1999), adrenergic excitement, particularly in susceptible individuals may become a damaging element. This is a fact apparently proven by the effectiveness of beta-blocker therapy (Wikstrand and Kendall, 1992; Airaksinen et al., 1998); sympathetic overactivity in unstable angina (Schwartz et al., 1992; Neri Serneri et al., 1993; McCance et al., 1993); a higher frequency of sudden death with a heart rate of 65 beats per minute (Algra et al., 1993); premature termination of cardiac arrhythmia suppression trials because of a higher mortality in patients not treated with beta-blockers (Peters et al., 1994); the deleterious effects of cigarette smoking that increase sympathetic outflow (Narkiewicz et al., 1998); reduction of the ventricular fibrillation threshold by use of beta-blockers (Baroldi et al., 1977; Anderson et al., 1983; Clusin et al., 1982); stimulation of the adrenergic system by cocaine (Vong-patanasin et al., 1999) and its role in coronary heart disease and congestive heart failure (Du et al., 1999); effects shown in brain injury (White et al., 1995) and in many other conditions, such as sympathetic stimulation or sympatholysis impaired blood-flow-mediated dilatation (Hijmering et al., 2002); reduced left ventricular filling and mechanoreceptor activation despite an increased contractility (Hesayen et al., 2002); different alpha and beta receptors increase in hypo-akinetic segments with respect to normal ones (Shan et al., 2000); and improved survival after surgical revascularization in patients showing myocardial viability with low-dose dobutamine ecocardiography (Sicari et al., 2003).

Histological documentation is needed in patients of different behavioral types (Kahn et al., 1987) who died suddenly after emotional stress, as, for instance, following an earthquake (Leor et al., 1996). Myofibrillar degeneration (or contraction band necrosis) has been documented in victims of homicidal assaults without internal injuries (Cebelin and Hirsch, 1980), particularly in ischemic patients in whom the prominent feeling was hopelessness or helplessness (Engel, 1971). The question is whether intense emotion, per se, is sufficient to alter neurovegetative control (Lynch et al., 1977; Eliot and Buell, 1985) or whether it needs to be associated with a morphofunctional derangement as described previously for coronary cardiopathy, which, in turn, may also be an expression of repetitive adrenergic stimulation. Passive emotion or fear seem to be insufficient, per se, to explain sudden cardiac death, as proven in World War II, in which no epidemics of sudden death were reported in either the civilian or the military population despite stressful conditions in so many countries.

On this subject, recall the still mysterious "voodoo deaths" of aborigines who may predict the date of their demise (Cannon, 1942), and the cases, one with a cerebral infarct and the other with a primary brain tumor, both of whom had a typical clinical pattern of a transmural anteroseptal cardiac infarct yet at postmortem showed only focal myocytolysis and severe coronary atherosclerosis (Duren and Becker, 1976). These are two examples that emphasize a heart/brain relationship and the lack of electrocardiogram (ECG) specificity in discriminating between different myocardial lesions. We have seen cases with typical ECG and serum enzymes change, which allowed for a clinical diagnosis of acute myocardial infarction that at autopsy presented only extensive contraction band necrosis without infarct necrosis in the affected area of the myocardium. Infarcts related to cocaine abuse seem to have the same pathophysiologic background (Fineschi et al., 1997). In experimental intravenous catecholamine infusion, in the absence of blood flow reduction, ischemic ST depression was the unique ECG sign of catecholamine myonecrosis (Todd et al., 1985b) — an ECG sign that was present at exercise tests done in patients who were resuscitated from out-of-hospital cardiac arrest (Sharma et al., 1987). In contrast, severe malignant arrhythmias/ventricular fibrillation plus contraction band necrosis were elicited by monocoronary noradrenaline infusion in the dog (unpublished data). This demonstrated that only a local neuromyotoxic impairment may unbalance the cardiac pump. In our institute, a 33-year-old man with unstable angina during angioplasty had an irreversible cardiac arrest following ventricular fibrillation. The left anterior descending branch and the right coronary artery presented 80% lumen reduction due to atherosclerotic plaques, while the main left coronary artery and the left circumflex branch had a 50% stenosis. The right coronary artery showed a recent (>3 days) occlusive thrombus at plaque level without evidence of myocardial infarction in the dependent territory. All atherosclerotic plaques showed perimedial nerve lympho-plasma-cellular infiltration. The numbers of foci and myocells with nonhemorrhagic contraction band necrosis were, respectively, 219 and 272 and 594 and 461 × 100 mm² in the anterior and posterior left ventricle, 19 and 36 in the interventricular septum, and 6 and 22 in the right ventricle. Myocardial fibrosis was 40% in posterior and 10% in anterior left ventricle, 30% and 15% in right ventricle, and absent in the interventricular septum. The numbers of fibrin-platelet thrombi were 9 in left, absent in right ventricle, and 16 in the interventricular septum. We interpreted the findings as an adrenergic overstimulation leading to contraction band necrosis and ventricular fibrillation in a patient with chronic coronary heart disease with a silent and ineffective occlusive thrombosis of the right coronary artery, extensive myocardial fibrosis (maybe silent infarct or progressive catecholamine damage), lympho-plasma-cellular medial neuritis of atherosclerotic coronary plaques, and fibrin-platelet intramural thrombi consequent to an adrenergic storm because no rupture or fibrin-platelet mural thrombi were found in coronary arteries. Such storms seem to be frequent in various conditions, with overexpression of β-adrenergic receptors in myocardial fibrosis and heart failure (Liggett et al., 2000), induced by free radicals and prevented by beta-blocking (Flesh et al., 1999; Fineschi et al., 2001a). Regional ischemia by angioplastic balloon inflation elicits a profound sympathetic response (Joho et al., 1999) and possibly related oxidation stress (Buffon et al., 2000).

Conclusions

When introducing coronary heart disease (see above), the unifying theory of etiopathogenesis was reported. Recently it was updated:

> The incidence of plaque rupture appears to be reduced in patients receiving cholesterol-lowering theraphy, beta-adrenergic blocking agents and possibly angiotensin-converting enzyme inhibitors and antioxidants. Not all ruptured coronary plaques produce an acute coronary syndrome. The consequence of plaque rupture depends on the extent of thrombus formation over the fissured plaque. This is determined by flow characteristics within the vessel as well as the activity of the thrombotic and fibrinolytic systems. Recent advances in cardiovascular molecular biology, coronary diagnostic techniques and cardiac therapy have opened windows of opportunity to study and modify the factors leading to plaque rupture. The local modification of gene expression to alter plaque composition and to elucidate and subsequently inhibit the prothrombotic and fibrinolytic defects that promote coronary thrombosis may in future prevent plaque rupture and its consequence. (MacIsaac et al., 1993, p. 1362)

This is optimistically presumed to occur in the first 20 years of the new millenium (Braunwald, 1997).

A synopsis of arguments in favor of and against vascular factors in the natural history of ischemic heart disease is given in Table 53. We note that in the current clinical interpretation of coronary heart disease, the different forms of myocardial necrosis we described are not often differentiated (Thygesen and Alpert, 2000). Their possible interactions were defined and discussed in relation to complications and death to show the complex relationships between regional loss of contraction; mechanoreceptor stimulation (Malliani et al., 1979); and increased contractility (Goldstein et al., 1972) and blood flow (Hood,

Table 53   Synopsis on Significance of Main Morphofunctional Factors in Favor of or Against Cause–Effect Relation between Coronary Heart Disease

Coronary	Pros	Cons
ATS Stenosis Occlusion	Ischemia	Adequate collateral function
Experimental normal artery	Infarct/sudden death Infarct size stable/ventricular fibrillation	No valid model with respect to human coronary heart disease; infarct size from <10 to >50%
Thrombus postmortem	Large transmural infarction	Multivariant/secondary in severe stenosis + collaterals already functioning
Coeval infarct	Not considered	Often noncoeval
Vascular territory	Not estimated	Infarct smaller, often invading nonischemic territory
Angiographic *in vivo*	87% within 4 h Never demonstrated before infarct onset	Not shown in infarct case cinenagio monitored- pseudocclusion
Rupture/fissuration ATS ↓	Subintimal intracoronary injected radiopaque material through fissure	Through adventia-intima collateral plexuses; fissuration often artifact; lack of dissecting aneurysm in CHD
Microembolization ↓	Few emboli reported	Atheromatous exceptional in myocardium Fibrino-platelet thrombi secodary to infarct necrosis
"Microinfarcts"	"Sudden death in unstable angina"	Postinfarct nonhemorrhage contraction band necrosis
Rupture/thrombosis of small ATS not seen angio	Inconsistent hypothesis	Never demonstrated postmortem
Spasm endothelial-derived resistance vessels	Unproved hypothesis	What are the triggers? Too often, a small local necrosis in respect to billions of endothelial cells
	Myocardium	
Different forms of necrosis	Never considered	Different morphology, dysfunction biochemistry, pathogenesis
Infarct necrosis/size	?	Size not related to death, number of stenoses, extent occluded vascular territory
Catecholamine necrosis	Ischemia	Always associated with infarct/sudden death, nonhemorrhagic and nonischemic
Colliquative myocytolysis	Ischemia	Secondary CHF
Reflow necrosis	Spasm (never demonstrated)	Contract band necrosis + massive interstitial hemorrhage not seen in natural history of CHD

Note: ATS = atherosclerotic plaque; CHF = congestive heart failure.

1970) of the normal myocardium jeopardized by catecholamine release with more favorable conditions for calcium influx and peroxidation (Mak and Weiglicki, 1988; Ferrari et al., 1990; Hori et al., 1991; Buffon et al., 2000). A complex ionic and molecular disorder occurs in progressive sequence from early damage to the repair process. Little do we know of other parameters and, in particular, of the role of the intramyocardial and coronary nerves, lymphatics, collagen matrix, and myocellular cytoskeleton (Weber, 1989; Gonate et al., 1993; Heine et al., 1995, 1996), alterations of which could be primary or secondary events. The structural adaptive changes of the heart are the end result of many concurrent factors, triggering other etiopathogenetic mechanisms in the course of the same disease (Figure 109). These factors, accepted in current dogma, may be epiphenomena.

We are living in an era of specialization in which one becomes an expert in one discipline or technique to be used practically and to be defended theoretically. The results are technological windows or breakthroughs that cannot substitute for an overall scientific vision that includes an entire problem. The damaging effect of specialization — "a means of institutionalizing, justifying and paying highly for the disintegration of the various functions of character: workmanship, care, conscience and responsibility" (Berry, 1977, p. 13) — is to reduce to fragments the complex "ischemic universe" that cannot be resolved simply by preventing plaque rupture; and in which ischemia needs to be redefined.

What is alarming is the loss of intellect, time, and money in searching — generally using expensive, sophisticated, and often uncontrolled techniques that will never resolve the problem and, worse, will delay its solution. The comment of Hurst (1967), "some words prevent learning," can be paraphrased to "some morphologic findings prevent learning." The "attractive" plaque rupture concept focuses preventive attention on innumerable substances pouring from each cell. A way to inhibit or stop discussion is to state that it "is hardly credible that there should be continuing debate about what is ostensibly so simple a morphologic problem; the relationship of coronary thrombosis to acute myocardial infarction" (Davies et al., 1976, p. 663), while the application of the unifying theory for acute coronary syndromes to cases where a coronary artery thrombus is not present flies counter to fact and does not consider other variables or recognize different forms of myonecrosis associated with myocardial infarction. Not all of them may be caused by ischemia.

The time for a reassessment of cause of death (Laner et al., 1999) should be a collaborative clinical-pathological effort, to understand where the truth lies.

Clinical Techniques for Anatomical and Functional Evaluation of the Heart

11

ANTONIO L'ABBATE

Contents

Parameters of Interest

Until a few decades ago, the only "instrumental" information available to the cardiologist was the arterial blood pressure measured by Riva-Rocci sphygmomanometer, the profiles of the heart and the great and pulmonary vessels obtained by chest x-ray, and the electrical function of the heart presented by the electrocardiogram. Later, technological progress allowed an increasingly detailed visualization of the heart and vessels, including coronary arteries to the point that imaging of cardiac anatomy, mechanical function, and metabolism has become the basis for the diagnosis and the treatment of cardiovascular diseases.

In brief, the following clinical information can be obtained by different techniques:

Anatomy of the heart and related large vessels, including coronary arteries

Flow velocity (cm/sec) and *flow rate* (ml/min) in selected arterial and venous vascular segments, including distal pulmonary veins, coronary arterial segments, coronary sinus, and great cardiac vein

Regional myocardial perfusion, the flow reaching the various regions of the heart (left ventricle) expressed either in relative terms (regional flow heterogeneity) or in absolute terms (ml/min per unit mass or unit volume of myocardial tissue)

Right and left *ventricular function*, including global and regional mechanical function, systolic and diastolic function, cardiac output, and right and left pressures

Intracardiac flows, that is, blood flow velocity through the cardiac valves or intracardiac abnormal communications

Myocardial metabolism, the myocardial uptake or production of a variety of metabolites (arterial-venous difference) or imaging of specific metabolic pathways

Myocardial tissue typing, the texture of the myocardium or the vascular wall reflecting the biophysical proprieties of the tissue

Electrical activity, the formation, conduction, and propagation of the electrical impulse

Myocardial ischemia, the presence, location, and extent of ischemia and its pathogenesis

Myocardial necrosis and myocardial residual viability, the presence, location, and extent of necrosis; the presence of residual viable myocardium in dysfunctioning regions of the left ventricle

Available Techniques

The multiparametric approach to the diagnosis of cardiovascular diseases still requires the use of multiple techniques, due to the fact that no one technique can provide all required information. The so-called "one-stop shop cardiac imaging" is still a slogan more than a reality, although this goal is likely not far from being accomplished.

Imaging techniques are based on the use of different kinds of energies able to cross the body. An energy source (*transmitter*) can be either outside the body or introduced into the body, as with radioactive tracers. Energy passing across the body or its reflection (ultrasound) is acquired by special detectors (*receiver*). The use of many transmitters and receivers, lying on a single plane and aligned around the body, together with the use of sophisticated computerized data processing permit the reconstruction of a tomographic image (slice) of the portion of the body explored. Sequential or contemporary acquisition of multiple contiguous slices allows for the imaging of the entire organ of interest up to its three-dimensional reconstruction (3-D imaging).

Spatial resolution (minimal distance between resolved neighbor points) varies among techniques, being high for ultrasound, magnetic resonance (MRI), and x-ray, and much lower for nuclear techniques. Temporal resolution (minimal time between successive images) also varies among techniques, being highest for ultrasound, intermediate for MRI and x-rays, and lowest for nuclear imaging. Among others, spatial and temporal resolutions define the peculiarity of the information and thus the clinical indication for each technique. As an example, while the assessment of macroscopic anatomy gains advantage from high spatial resolution, the investigation of cardiac function requires high temporal resolution

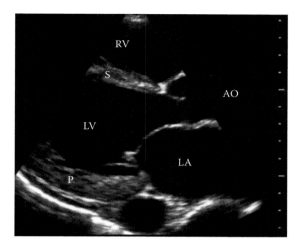

Figure 118 Parasternal view of the middle and basal portions of the left ventricle (LV), the left atrium (LA), and the aortic root (Ao) by two-dimensional (2D) echocardiogram. Note that S: interventricular septal wall; P: posterior wall; RV: right ventricle, hardly visible in this projection. An instant of the mid-systole of a cardiac cycle is "frozen" in this frame, as documented by the closed mitral valve as opposed to the open aortic valve.

in order to image the rapid movement of the heart more than its detailed structure. In addition, due to heart movement (related to both cardiac and respiratory activities), any need for increased spatial resolution contrasts with longer times of acquisition and, thus, lower temporal resolution. To this purpose, it should be considered that hundreds of frames are acquired in a single cardiac cycle by ultrasound techniques (Figure 118); conversely, the ECG-gated acquisition of hundreds of cycles is needed to reconstruct a few frames of a single "virtual" cardiac cycle using nuclear techniques (Figure 119). Thus, the anatomical and functional imaging of the heart has forced the development of "fast" and "very fast" machines specifically dedicated to cardiology and the adoption of the ECG-gated acquisition of the signal as a common means to image the heart at predefined time intervals during the cardiac cycle.

Clinically established imaging and nonimaging techniques are summarized in Table 54, together with the information they provide and their invasiveness.

Cardiac Anatomy

Macroscopic anatomy of the heart is no more the prerogative of surgeons and pathologists. Detailed knowledge about cardiac cavities and wall thickness can be obtained noninvasively by all the imaging techniques with the exception of nuclear ones (Table 55). Ventricular volumes and left ventricular mass can be computed with an approximation that is inversely correlated with the number of tomograms and thus of planes explored. Cardiac and paracardiac masses can generally be assessed and their nature established in some instances. Fibrotic wall lesions, such as scars secondary to myocardial infarction, can be directly visualized by nonnuclear techniques, due to the high density of the fibrous tissue as compared with normal myocardium. Nowadays, echocardiography is by far the most used technique for cardiac imaging.

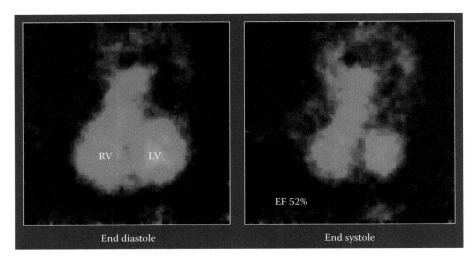

End diastole   End systole

Figure 119 Radionuclide ventricular angiography (RNA). End-diastolic and end-systolic frames of a "virtual" cardiac cycle obtained by the ECG-gated acquisition of hundreds of cycles. The cavities of the two ventricles and the vascular peduncle are visualized by labeling the blood pool with a radioactive gamma-emitting tracer. Note the much lower temporal and spatial resolution as compared to the single frame of the echocardiogram shown in Figure 118. Note that RV: right ventricular cavity; LV: left ventricular cavity; EF: left ejection fraction.

Table 54   Imaging and Nonimaging Techniques

Method	Information	Mode
	Imaging Techniques	
Chest Rx	Anatomical profiles	Noninvasive
Echo 2D	Anatomy and function	Noninvasive
Echo 2D Doppler	Anatomy and function and flow velocity	Noninvasive
Echo 2D Contrast	Anatomy and function, flow velocity, and myocardial perfusion	Noninvasive
Intravascular Echo	Coronary wall	Invasive
Angioscopy	Coronary lumen	Invasive
RNA	Right and left ventricular volumes; right and left ventricular function	Noninvasive
ECG-gated SPECT	Myocardial perfusion; left ventricular function	Noninvasive
PET	Myocardial perfusion; myocardial metabolism	Noninvasive
MRI	Anatomy and function; flows and myocardial perfusion; myocardial metabolism	Noninvasive
Fast CT	Anatomy and function	Noninvasive
Rx Ventriculography	Right and left ventricular cavities; left ventricular function	Invasive
Coronary angiography	Coronary lumen	Invasive
	Nonimaging Techniques	
ECG	Electrical function	Noninvasive
Echo M-mode	Dimensions and function	Noninvasive
Doppler	Flow velocity	Noninvasive
Doppler catheter	Flow velocity	Invasive
Tracer wash-out	Myocardial perfusion	Invasive
Thermodilution	Cardiac output	Invasive
Coronary sinus (CS) thermodilution	CS flow	Invasive

Table 55   Cardiac and Coronary Anatomy

Method	Tracer/Contrast
Cardiac Anatomy	
Echo M-mode	
Echo 2D	
Chest Rx	
MRI	(Contrast)
CT	Contrast
Coronary Anatomy	
Angiography	Rx Contrast
Intravascular echo	
MRI	(Contrast)
EBCT, spiral CT (calcifications)	

By MRI, peculiar additional information can be obtained about such things as myocardial edema or myocardial "infiltration," which can modify the biophysical characteristics of the cardiac tissue. Postprocessing of the image, commonly addressed as "tissue typing," allows, in some instances, for the identification of a pathological process even beyond its macroscopic morphology.

As compared to the cardiac walls, imaging of cardiac valves still remains a technological challenge due to their thin structure and rapid movement, unless they are thickened and impeded in their movements by pathologic processes. To overcome these limitations, functional rather than anatomical features, such as transvalvular flow velocity and direction, are searched for in order to diagnose valvular abnormalities.

Coronary Anatomy

Available techniques for coronary imaging are summarized in Table 55. Information on the anatomy of the coronary circulation is invasively obtained by the selective injection of a radiopaque contrast medium into the right and left coronary arteries (coronary angiography). Injections are repeated in different projections for the right (usually three) and the left (usually five) coronary artery. A sequence of frames is acquired for each injection by analogic (film) or digital cine-mode. In this way, the arterial tree can be visualized down to vessels 0.5 to 1 mm in diameter. Large veins can also be visualized, provided the acquisition is maintained during the washout of the contrast medium and its concentration is still sufficiently high. Obviously, also within the limits of spatial resolution of the instrumentation used and the limits imposed by the bidimensional representation of a tridimensional structure, an angiographic picture does not entirely reproduce the anatomic one but rather a functional one. This consideration is particularly true for smaller vessels, collaterals included, that are remarkably underrepresented compared to anatomical casts.

Angiographic information refers to the lumen profile of visualized vessels; thus, no information is available on the vascular wall (Figure 120). Absolute measurement of segmental lumen can be obtained, even automatically, by single-frame image processing (computerized quantitative angiography), using the known size of the catheter tip as reference. Dynamic changes in the morphology of the vascular profile, due to changes in tone or in lumen content, can be appreciated by repeated injections during the course of

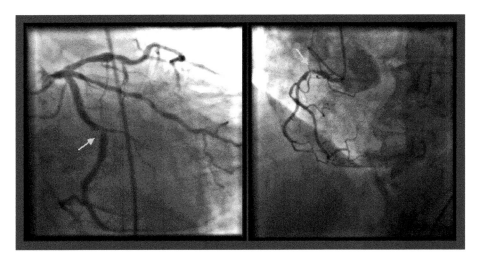

Figure 120 Coronary angiography. Following the retrograde catheterization of the aortic root, the selective injection of the radiopaque contrast medium into the right and left coronary arteries allows visualization of the arterial coronary tree. Here only one frame for each selective injection is shown. Repeated injections are needed to visualize the coronaries by multiple projections.

a coronary event, such as an ischemic attack or an acute myocardial infarction, or following pharmacological or mechanical interventions.

The lumen of the proximal portion of coronary arteries can also be visualized angioscopically. By the introduction of the angioscope directly into the coronary vessel, and following the transient removal of the blood by saline injection, the inner surface of the vascular wall can be visualized as can any obstructive material adherent to the wall. The push–pull movement of the catheter along the vessel allows for its detailed exploration. The clinical use of this technique is limited to a few centers.

Information on both arterial coronary lumen and wall can be provided by intravascular echo technique (IVE). By this means, a miniaturized ultrasonic transducer mounted on the tip of a thin catheter emits and receives signals over 360°, encompassing the entire vascular wall. Due to the different ultrasound reflecting properties of different tissues, the intima can be distinguished from the media as well as the different components of an atherosclerotic plaque, that is, lipid deposits, fibrous tissue, and calcification. Clinical use of IVE is limited.

New approaches to the noninvasive imaging of the coronary arteries use nuclear magnetic resonance imaging (MRI). However, at present, only the proximal portions of the main coronary arteries are inconsistently visualized by this technique, and its clinical applicability is limited (Figure 121). Noninvasive detection of coronary calcifications can be produced by electron beam computed tomography (EBCT) or spiral computed tomography (CT).

Coronary Vascular Flow

Phasic blood flow velocity can be obtained in a well-defined segment of a large coronary vessel by Doppler technique using ultrasound as the energy source (Table 56). The signal

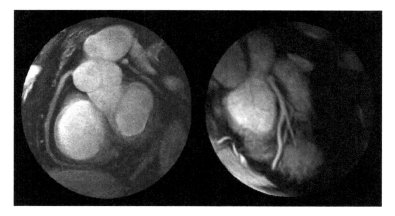

Figure 121   Imaging of the right (left panel) and left (right panel) coronary arteries by nuclear Magnetic Resonance. Note the incompleteness of information as compared to Figure 120.

Table 56   Coronary Vascular Flow and Myocardial Perfusion

Method	Tracer/Contrast
Coronary Vascular Flow	
Echo 2D Doppler	(Echo contrast for enhancement)
Doppler catheter	
CS Thermodilution	Cold saline
MRI (coronary arteries and CS)	
Myocardial Perfusion	
Tracer wash-out	Diffusible indicators
PET	Positron isotopes
SPECT	Gamma isotopes
Echo 2D	Echo contrast
MRI	Contrast
Fast CT	Contrast

can be obtained invasively using either a Doppler catheter or a thin Doppler wire advanced into a coronary vessel. Velocity values (cm/sec) can be transformed into absolute flow values (ml/min), provided that the measurement of the vascular section at the side of the Doppler transducer by quantitative coronary angiography is available. Analogous information can be obtained semi-invasively by transesophageal echo Doppler (TEE). Using TEE, only the proximal left anterior descending coronary artery can be explored. Thus, information is limited to this vessel. More recently, progress in echo technology has made it possible to obtain the same information more simply by trans-thoracic echo-Doppler. By both approaches, simultaneous visualization of the vessel by echo allows for the conversion of velocity values into absolute flow values.

Absolute flow values in the coronary sinus can be obtained by thermodilution. In this case, a special catheter, instrumented with a couple of thermistors positioned a few centimeters apart along the distal portion of the catheter, is introduced into the coronary sinus or advanced into the great cardiac vein. Flow is measured by infusing cold water upstream of the catheter tip and measuring the difference in blood temperature between the two thermistors, which is inversely related to flow. Finally, recent attempts to measure flow in the proximal segments of the coronary arteries or in the coronary sinus by MRI are worth mentioning.

Myocardial Perfusion

Different from vascular flow, measurement of myocardial perfusion provides information on the amount of flow perfusing the myocardium, independent of its vascular source (Table 56). Some techniques provide information on quantitative flow per unit mass of tissue (specific flow, ml/min/g); others provide only relative differences in flow among different regions of the heart.

Specific flow values in different regions of the left ventricle can be obtained noninvasively only by positron emission tomography (PET; Figure 122). The high number of slices provided by modern PET allows for the exploration of the entire heart; however, the low diffusion and the high costs of PET perfusion studies greatly limit their clinical use. Specific flow can also be assessed by techniques based on the inhalation or selective coronary injection of diffusible tracers and the recording of their dilution curves on the precordium (for radioactive tracers) or from the coronary sinus. These techniques were largely employed in the past for research purposes and are now practically abandoned.

The established clinical techniques for perfusion studies do not provide absolute measurement of flow but rather images of flow distribution. In this way, it is possible to identify and localize "perfusion defects," that is, left ventricular regions that receive less flow than others. Imaging of flow distribution is accomplished by nuclear techniques using gamma-emitting flow tracers, such as Thallium 201 or Tc-99m MIBI, that deposit in the heart according to flow, and planar or tomographic (SPECT) gamma-cameras to acquire tracer cardiac distribution (Figure 123). More recently, new techniques of noninvasive myocardial

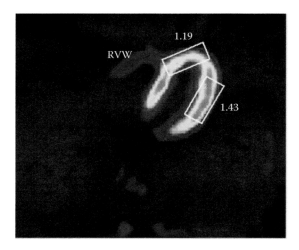

Figure 122 Images of myocardial perfusion obtained through use of positron emission tomography (PET). The uniqueness of PET is that myocardial blood flow values in ml/min/g of tissue can be obtained. In the tomogram shown here, the flow values in two separate regions of interest (rectangles) are reported.

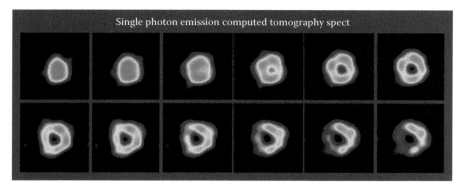

Figure 123 Tomographic perfusion images of the left ventricle obtained with SPECT. Transverse "slices" go from the apex (upper left) to the base (lower right) of the ventricle. A relatively lower perfusion is apparent in the septal wall toward the basis (green, blue colors).

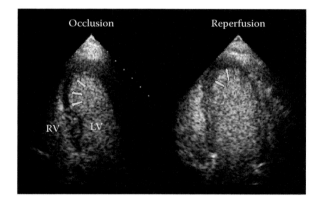

Figure 124 Echo-contrast imaging of myocardial perfusion in a patient with acute anterior myocardial infarction, before and after primary percutaneous coronary angioplasty (PTCA). The two frames were obtained following two successive intravenous injections of echo-contrast, at the time the microbubbles have contrasted in turn the right ventricle (RV), the left ventricle (LV), and then the myocardium. Perfused myocardium is whitened by the contrast, as are the ventricular cavities, while the nonperfused area (arrows) remains black. In spite of an angiographically successful PTCA, only a partial refilling of the perfusion defect is evident in the right frame.

perfusion imaging are in the field of ultrasound (contrast echocardiography, Figure 124), fast CT, and MRI by the assessment of intramyocardial distribution of suitable intravenously injected indicators.

Ventricular Function

The evaluation of ventricular function is a major challenge in cardiology due to both theoretical and technical complexities. Schematically, ventricular function is generally distinguished into global and regional, systolic and diastolic.

Table 57   Ventricular Function

Method	Tracer/Contrast
Echo 2D	
RNA	Gamma isotopes
ECG-gated SPECT	Gamma isotopes
MRI	Contrast
Rx ventriculography	Rx contrast

Global Function

In order to evaluate global systolic and diastolic function, knowledge of ventricular volumes in the various phases of the cardiac cycle and the corresponding values of intraventricular pressure are required. Due to difficulties of obtaining such measurements, simplified indices of global ventricular function are commonly used in the clinical setting.

Stroke volume (SV) and cardiac output (CO; normal value approximately 5 l/min in the adult) are the most obvious indices of the integrated cardiac function. They can be noninvasively assessed by tracer dilution curves, sampling of blood flow velocity in the aortic root by echo-Doppler, or, alternatively, by measuring the ventricular volume excursion from end-diastole to end-systole. Invasive approaches include thermodilution in the pulmonary artery and left ventriculography.

Ventricular ejection fraction (EF) is the most commonly used index of global systolic function. It measures the percentage of the end-diastolic blood content pumped by the ventricle at each beat (normal value >50%). Its calculation requires knowledge of relative changes in ventricular volumes that can be noninvasively obtained by all cardiac imaging techniques (Table 57), including ECG-gated nuclear ones, such as radionuclide angiocardiography (RNA) and ECG-gated SPECT.

As far as intraventricular pressure measurements, they can be assessed only by cardiac catheterization. However, systolic left ventricular pressure can be assimilated to the systolic arterial blood pressure (as long as aortic valve stenosis can be excluded), and left diastolic pressure can be indirectly estimated by transmitral flow velocity (which is directly correlated with the pressure gradient across the valve). According to this principle, noninvasive clinical indices of left ventricular diastolic function are commonly derived by use of the Doppler echocardiogram.

Regional Function

Regional function can be assessed by imaging techniques only, either by measuring the dislocation of different regions of the ventricular wall, typically from the periphery toward the center, or their thickening during systole. Figure 125 shows a typical way of assessing regional mechanics from comparison of end-diastolic and end-systolic internal profiles of the left cavity, manually or automatically derived from a two-dimensional image of the left ventricle. A four-scale score is generally used to quantify regional abnormalities: 0 equals normal, 1 equals hypokinetic, 2 equals akinetic, and 3 equals dyskinetic.

A three-dimensional dynamic reconstruction of the left ventricle is routinely obtained by myocardial GATED-SPECT scintigraphy (Figure 126). Note that, differently from RNA, perfusion as well as function (both global and regional) can be simultaneously studied by this technique. While with RNA ventricular *cavities* are visualized by labeling the blood, with SPECT the left ventricle *walls* are labeled by the flow tracer, which deposits into the

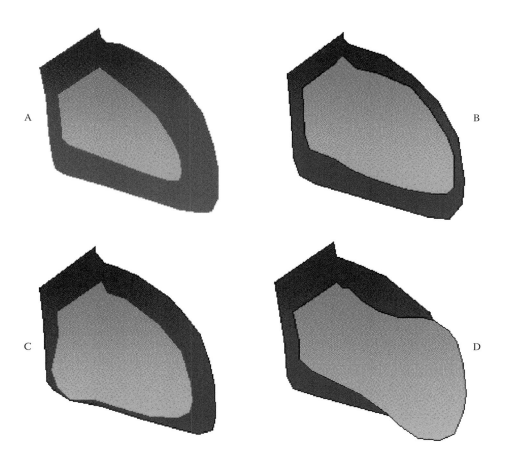

Figure 125   Way of classifying regional ventricular wall motion starting from the endocavitary profiles obtained at end-diastole and end-systole. (A) Normal; (B) antero-lateral hypokinesis; (C) inferior akinesis; and (D) apical dyskinesis.

myocardium. As compared to SPECT, RNA provides the opportunity to study the global and regional functions of both left and right ventricles (the latter being poorly visualized by SPECT due to the relatively low myocardial mass and flow).

Intracardiac Flows

The assessment of intraventricular blood flow falls within the domain of Doppler echocardiography. Flow velocity and its direction can be sampled at definite spatial points of the heart chambers and visualized in color-coded tomographic images (color Doppler echocardiography). Thus, the degree of valvular narrowing and regurgitation as well as intracardiac shunts are currently assessed by using this technique (Figure 127). In addition, as transvalvular flow velocities are proportional to pressure gradients, blood pressures in the heart chambers and in pulmonary circulation can be indirectly estimated, avoiding the need for cardiac catheterization. Possibilities for assessing intracardiac flow by MRI are currently

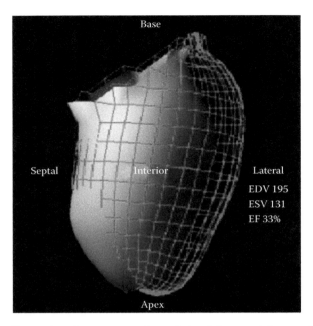

Figure 126 Three-dimensional reconstruction of the left ventricle starting from ECG-gated SPECT acquisition of the myocardial uptake of a flow tracer. Standstill frame of a dynamic sequence. The ECG-gated acquisition allows us to follow the ventricular wall excursion during a reconstructed "virtual" cardiac cycle. Movement of the wall toward the center is represented by dislocation of the inner colored solid from the external net. Thus, akinesis keeps the solid close to the net, and dyskinesis pushes it outside the net. In the specific case shown here, the end-systolic frame documents an akinetic–dyskinetic septal wall. Note that EDV: end-diastolic volume; EDS: end-systolic volume; EF: ejection fraction.

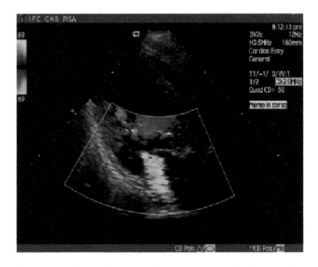

Figure 127 Color Doppler echocardiogram from a patient affected by mitral valve insufficiency. Blood flowing back into the atrium is colored blue according to its direction opposite to the transducer (located at the apex of the triangular image).

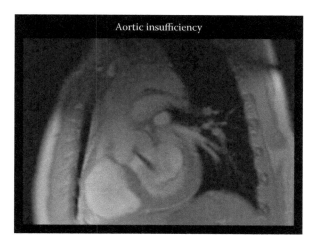

Figure 128   Magnetic resonance imaging documentation of aortic valve regurgitation. The regurgitant black jet contrasts with the white blood content of the cavity.

under evaluation; the preliminary results are promising, especially in the detection of valvular insufficiency (Figure 128).

Myocardial Metabolism

The noninvasive approach to the study of myocardial metabolism is based on the use of PET and MRI spectroscopy. Invasively, cardiac metabolism can be assessed by measuring coronary blood flow and arterial-venous concentration gradients of different substrates across the myocardium by coronary sinus catheterization. The study of human cardiac metabolism is still confined to research protocols, with its diagnostic clinical use being limited to the imaging of residual viable myocardium in dysfunctioning regions of the left ventricle (see the section below on cardiac viability tests).

Myocardial Tissue Typing

The possibility of studying the texture of the myocardium, a feature reflecting its biophysical properties, not necessarily or not exclusively related to its histological characteristics, was explored by ultrasound backscatter analysis and postprocessing analysis of echo as well as MRI images. Despite some success in differentiating different types of myocardial hypertrophy or in characterizing arterial plaque composition, until now, texture analysis has not found large clinical application.

Electrical Alterations

The continuing use of the electrocardiogram in cardiac diagnosis is based on its capability to detect most cardiac alterations, such as atrial and ventricular enlargement, ventricular hypertrophy, myocardial ischemia, myocardial infarction, and arrhythmias.

Electric cardiac activity is conventionally obtained from the chest surface and the extremities using 12 standardized leads, generally sufficient to explore the entire heart. The number of derivations is reduced from 12 to 2 or 3 during continuous monitoring of the electrocardiogram (known as the Holter technique). In this case, the number of beats registered (about 100,000 in a 24 h recording) and the peculiarities of the information regarding rhythm, and the presence and type of arrhythmia make up for the use of fewer leads and the consequent reduction of sensitivity in diagnosing ischemia. More recently, continuous monitoring has been used as the only means of study for sympatho-vagal balance, through the analysis of spontaneous variability of heart rate. The two branches of the autonomous system modulate heart rate with an opposite effect on the intrinsic rhythm of the sinus node that determines oscillations with characteristic periods depending on the prevalence of one or the other branches.

Through the modern developments in electrocardiography, electric signals can be recorded directly from either the internal surface of the heart (intracavitary ECG) or from the inside of the esophagus (intraesophageal ECG). Moreover, sophisticated computer analysis of multiple-lead ECG acquisition has made it possible to image the spreading of the electric potentials over the surface of the trunk by means of color-coded maps. Until now, however, this effort has not found wide clinical use. Conversely, the recent application of this approach to invasive multilead intracavitary ECG has rapidly become a valid tool for both the diagnosis and interventional treatment of arrhythmias (Figure 129).

Direct comparison of the electrocardiogram with modern imaging techniques has clearly documented its low sensitivity and specificity in the diagnosis of hypertrophy, ischemia, and necrosis. Nevertheless, the ECG is still the leading diagnostic technique used in the detection of cardiac disorders, such as arrhythmias, intraventricular blocks, preexcitation, and long Q–T syndrome. In all these conditions, the ECG is essential for clinical diagnosis, and the type of electrocardiographic alteration constitutes the fingerprint of the disease.

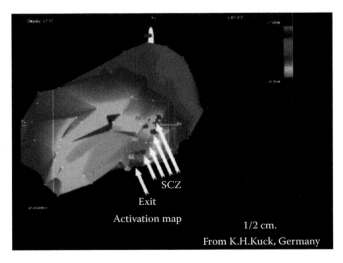

Figure 129 Electrocardiogram activation color-coded intracavitary map in a case of post-infarction ischemic ventricular tachycardia showing the entrance at the border of the scar (SCZ) and the exit (EXIT) into the normal tissue. (From K.H. Kuck.)

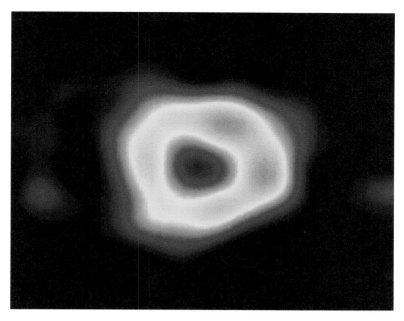

Figure 129A   Scintigraphic perfusion defect in the septal and inferior walls of the left ventricle during exercise stress testing. The absence of the defect in the scintigram obtained in resting conditions is indicative of exercise induced ischemia.

Provocative Tests for Ischemia

The classical way to provoke ischemia is effort stress testing, which mimics in the laboratory the physiologic way of inducing ischemia by the increase of oxygen myocardial demand in the presence of limited coronary flow supply. Alternatively, the administration of scalar doses of either inotropic or chronotropic drugs such as dobutamine and atropine or vasodilator drugs such as dipyridamole or adenosine are used. Vasodilator drugs redistribute coronary flow in favor of better perfused regions and induce ischemia by the so-called "blood flow steal". Coronary spasm and ischemia can be provoked in patients with vasospastic angina by hyperventilation or, if negative, by the administration of scalar doses of ergonovine.

The single ECG or preferably its combination with an imaging technique such as echocardiography or scintigraphy and more recently MRI can be used to detect ischemia. At echocardiography and RNA, the diagnosis is provided by the occurrence of a transient impairment in regional contraction while with perfusion scintigraphy the occurrence of a transient perfusion defect is required (Figure 129A). When mechanical or perfusional regional alterations are permanent rather than transient, the provisional diagnosis of necrosis is made with the reserve of the possible coexistence of residual viable myocardium.

Tests for Residual Myocardial Viability

In past years, the diagnosis of myocardial viability has become a relevant issue in the management of patients with IHD. This is due to the documentation that dysfunctioning

Flow-Metabolism Mismatch

NH3
Resting

NH3
Dipyridamole

FDG

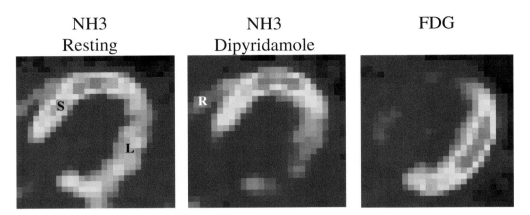

Figure 129B   Study of perfusion and metabolism in a patient with post-infarction severe dysfunction of the antero-lateral wall. Study of perfusion at rest (left panel) shows a marked reduction in flow in the lateral wall (L), which further decreases following the administration of a vasodilating drug such as dipyridamole; however, a high uptake of ^{18}F-deoxyglucose is present in the same area indicating the persistence of a significant portion of still alive myocardium. S: septal wall; R: free right ventricular wall.

ventricular regions following myocardial infarction may actually contain a variable amount of viable myocardium and that functional recovery by coronary revascularization is proportional to the residual viable tissue. The approaches to the diagnosis of myocardial viability are of two types. The first is based on the assessment of wall motility and on the assumption that a region containing a sufficient amount of viable myocardium can transiently improve its function if adequately stimulated. In this setting the most frequently used tests are the low-dose echo-dobutamine and the echo-dipyridamole stress tests. The second approach is based on the imaging of myocardial uptake of radioactive metabolic tracers. To this purpose, the same tracers used for imaging myocardial perfusion are commonly employed because their uptake requires the myocardial cell integrity. Thus, by the use of ECG-gated SPECT scintigraphy, global and regional left ventricular function as well as regional perfusion and viability can be assessed in a single test. The use of PET for the investigation of residual viability is required only for selected cases. In this instance ^{18}F-deoxyglucose, which traces the first step of the cellular glucose pathway, is used generally in association with the study of regional perfusion (Figure 129B).

Cardiovascular Traumatic Injuries $\quad$ 12

EMANUELA TURILLAZZI AND CRISTOFORO POMARA

Contents

Introduction

Thoracic traumas are of great importance, second only to craniocerebral traumas as a cause of morbidity and mortality. Actually, it is estimated that about one fourth of deaths due to trauma are caused by thoracic injuries, with a 50 to 75% incidence of cardiovascular traumas.

Because of its peculiar anatomical position, placed between the sternum and the thoracic vertebrae, the heart is, in fact, exposed to sudden and abrupt forces applied to the thoracic wall (Sanbar, 1989; Lasky, 1972) (Figure 130 and Figure 131).

The severity of cardiovascular traumas greatly varies in relation to the kind of cardiovascular injury. There is a wide range of postraumatic cardiovascular injuries, from cardiac concussion to heart and great vessels rupture with acute cardiac tamponade (Rashid et al., 2000; Tenzer, 1985; Dick et al., 2000).

The different clinical expression and severity of the different kinds of cardiovascular trauma (heart injury scale) are expressed in Table 58 (Asensio et al., 2001).

In the international literature, it is common to divide and describe traumatic injuries to the cardiovascular system in separate chapters pertaining to penetrating and nonpenetrating trauma. We will follow this pattern in this chapter.

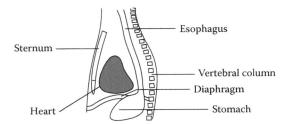

Figure 130   Thorax: medial sagittal section.

Anatomic cardio-thoracic deformation

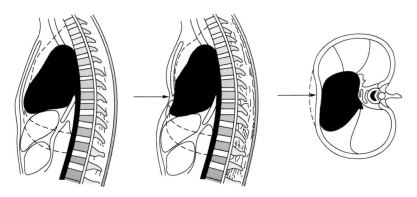

Figure 131   Anatomic cardiothoracic deformation.

Nonpenetrating Trauma

Mechanism of Injury

Blunt thoracic traumas are more frequently involved in the pathogenesis of cardiovascular injuries. Generally, they are caused by motor vehicle accidents, industrial accidents, sport activities, and falls from great heights.

Pathogenetic mechanisms of cardiovascular injuries in blunt thoracic traumas are summarized as follows and are represented in Figure 132:

1. Direct impact to the precordial region which causes an abrupt compression of the heart placed between the sternum and the vertebral column
2. Sudden deceleration that generally determines great vessels injuries
3. Abrupt blood pressure increase in cardiac chambers following chest or abdominal compression or crushing (blast injuries) (Pollak and Stellwag-Carion, 1991)

Of particular interest, also from a medico-legal point of view, is the so-called "collision triad" (Lasky, 1972), suggested to describe the physical stress to which the body and the cardiac region in particular are subjected in motor vehicle accidents (Figure 133 and Figure 134).

Table 58   American Association for the Surgery of Trauma (AAST) Organ Injury Scaling (OIS): Heart Injury Scale

Grade[a]	Injury Description
I	Blunt cardiac injury with minor electrocardiographic abnormality (nonspecific ST- or T-wave changes, premature atrial or ventricular contraction, or persistent sinus tachycardia)
	Blunt or penetrating pericardial wound without cardiac injury, cardiac tamponade, or cardiac herniation
II	Blunt cardiac injury with heart block (right or left bundle branch, left anterior fascicular, or atrioventricular) or ischemic changes (ST-depression or T-wave inversion) without cardiac failure
	Penetrating tangential myocardial wound up to, but not extending through, endocardium, without tamponade
III	Blunt cardiac injury with sustained (>5 beats/min) or multifocal ventricular contractions
	Blunt or penetrating cardiac injury with septal rupture, pulmonary or tricuspid valvular incompetence, papillary muscle dysfunction, or distal coronary arterial occlusion without cardiac failure
	Blunt pericardial laceration with cardiac herniation
	Blunt cardiac injury with cardiac failure
	Penetrating tangential myocardial wound up to, but not extending through, endocardium, with tamponade
IV	Blunt or penetrating cardiac injury with septal rupture, pulmonary or tricuspid valvular incompetence, papillary muscle dysfunction or distal coronary arterial occlusion producing cardiac failure
	Blunt or penetrating cardiac injury with aortic or mitral valve incompetence
	Blunt or penetrating cardiac injury of the right ventricle, right atrium, or left atrium
V	Blunt or penetrating cardiac injury with proximal coronary arterial occlusion
	Blunt or penetrating left ventricular perforation
	Stellate wound with <50% tissue loss of the right ventricle, right atrium, or left atrium
VI	Blunt avulsion of the heart; penetrating wound producing >50% tissue loss of a chamber

[a] Advance one grade for multiple penetrating wounds to a single chamber or multiple chamber involvement.

From www.aast.org.

Pathology

The main pathological changes that may follow blunt thoracic trauma are shown in Table 59 (Lobo and Heggtviet, 2001).

Pericardial Injuries

Pericardial contusions are the most frequently detected pathological manifestation following blunt, nonpenetrating thoracic trauma. In less serious trauma, they are simple contusions consisting of localized blood effusions or superficial lacerations of the pericardial serous membrane, without concomitant involvement of cardiac muscle. Hemopericardium, also described in blunt trauma, occurs more frequently in penetrating injuries.

Pericardial rupture is a relatively rare occurrence. Generally, it is associated with other cardiac and great vessels injuries (May et al., 1999), and it is often diagnosed only postmortem. It is associated with a high mortality rate (Janson et al., 2003; Mattila et al., 1975), although in the last 30 to 40 years, cases of survival following traumatic pericardial rupture were reported in the literature (Fulda et al., 1991).

In patients with blunt thoracic trauma, pericardial ruptures are diagnosed with even more difficulty than cardiac ruptures. In an exhaustive study on this subject (Fulda et al., 1991), 22 patients with pericardial rupture following blunt thoracic trauma are described. Their distribution is shown in Table 60.

Five sources of cardiac injury

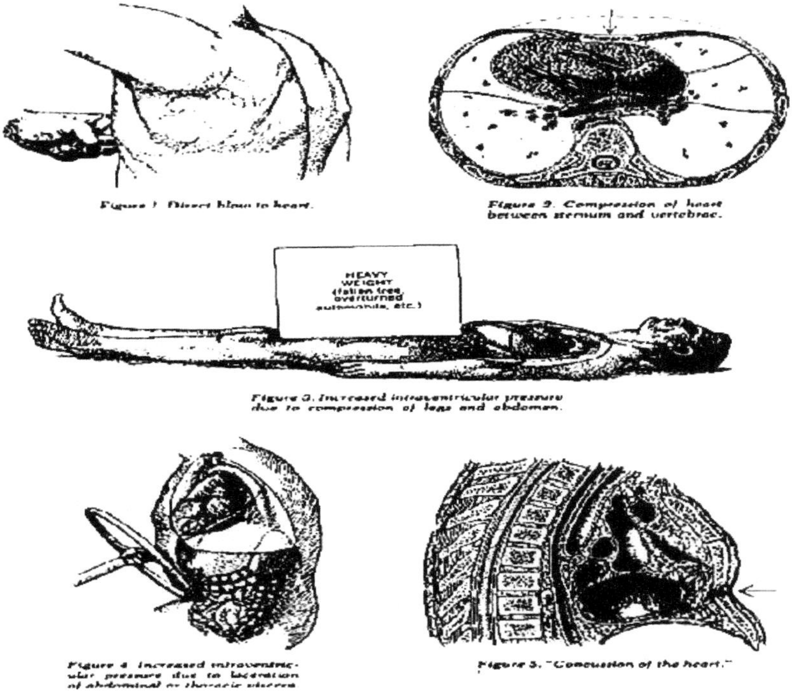

Figure 132 Five sources of cardiac injury. (From Lasky and Davis, Management of heart disease caused by nonpenetrating chest trauma, *Hosp. Med.*, April, 1965. Courtesy of *Hospital Medicine*.)

Collision triad

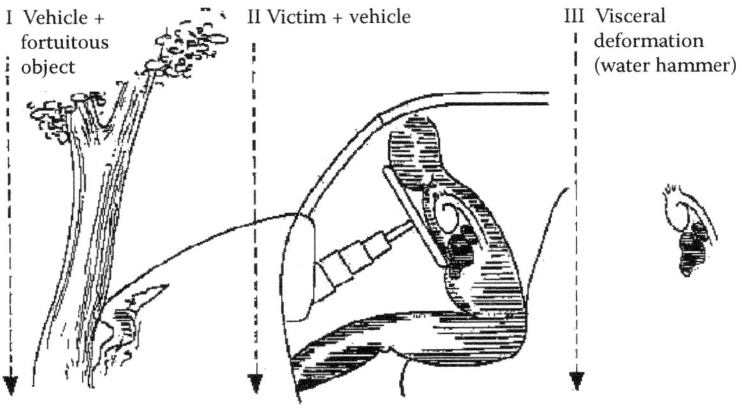

I Vehicle + fortuitous object

II Victim + vehicle

III Visceral deformation (water hammer)

$$F = M \times \frac{V_2 - V_1}{T} \quad < \quad F = M \times \frac{V_2 - V_1}{T} \quad < \quad F = M \times \frac{V_2 - V_1}{T}$$

Figure 133 Collision triad.

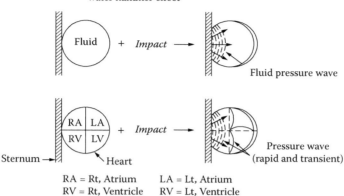

Cardiac "water hammer effect"

Enclosed fluid + kinetic energy (impact)→ water hammer effect

Fluid pressure wave

Pressure wave (rapid and transient)

RA = Rt, Atrium   LA = Lt, Atrium
RV = Rt, Ventricle   RV = Lt, Ventricle

Impact may produce distension, shearing or rupture of the heart
1. The walls of the heart have variable thickness and variable Youngs modulus
2. The water hammer effect is a correlate of magnitude, speed and direction of impact. (Second derivative of velocity– "Heart Jerk")

Figure 134 "Water hammer effect." (From Lasky et al., Automotive cardiothoracic injuries: a medical-engineering analysis. Report 680052, January 1968. Courtesy of Society of Automotive Engineers, Inc., New York.)

Table 59   Blunt Cardiovascular Traumas

Pericardial injuries
  • Contusions
  • Lacerations or ruptures
  • Hemopericardium
  • Serofibrinous or suppurative pericarditis
  • Recurrent pericarditis
  • Chronic constrictive pericarditis
Cardiac injuries
  • Concussions
  • Contusions
  • Lacerations or ruptures
Valvular injuries
  • Contusions
  • Lacerations or ruptures
  • Lacerations or ruptures of papillary muscles or chordae tendinae
Coronary arteries injuries
Cardiac rhythm disturbances
Great vessels injuries
  • Lacerations or ruptures
  • Aneurysms
  • Dissections
  • Thrombosis

Table 60   Location of Pericardial Tears in 22 Patients

	Number of Patients	Percent (%)
Left side	14	64
Diaphragmatic	4	18
Right side	2	9
Mediastinal	2	9

The overall mortality in this series was high (14/22). However, in relation to the presence of concomitant injuries, the reported survival percentage was 47% when pericardial laceration was the only injury, whereas it was 33% when cardiac herniation was associated with pericardial laceration.

Pericardial lacerations can occur basically at two levels (Borrie and Lichter, 1979): diaphragmatic pericardium or pleural pericardium. When lacerations of diaphragmatic pericardium occur, abdominal organs can herniate into the pericardial sac and cause abrupt cardiac compression with consequent cardiogenic shock (de Rooij and Haarman, 1993); on the other hand, if the laceration involves the pleural surface, cardiac herniation into a pleural cavity may ensue (Poletti et al., 2005).

Posttraumatic pericarditis is often associated with blunt cardiac trauma and generally consists of simple serous or hematic effusions and fibrinous exudate. Chronic constrictive pericarditis with calcifications can also represent a late sequel of recurrent posttraumatic pericarditis or can result from the organization of hematic effusions into the pericardium sac.

Cardiac Concussion (Commotio Cordis)

Cardiac concussion, also known as commotion cordis, is a relatively uncommon and frequently mortal condition (only 15% of patients survive). It has well-defined characteristics that clearly distinguish it from cardiac contusion. This last condition occurs far more frequently following blunt thoracic traumas.

Commotio cordis results from a sudden and blunt impact to the precordial area that alters the normal cardiac electrical activity and causes cardiac arrest or sudden death, without evidence of anatomical and structural cardiac injuries. The impact causing commotio cordis is generally of low energy and low speed. Death is caused by a cardiac arrhythmia, and there is evidence that ventricular fibrillation is the most common in fatal cases. In 60% of victims of cardiac concussion, death occurs immediately, presumably due to ventricular fibrillation; the remaining 40% of victims show a brief period of survival before terminal cardiac arrest. These descriptions of cases in which death does not occur immediately suggest the hypothesis of a brief period of arrhythmia (as ventricular tachycardia) rapidly evolving into terminal ventricular fibrillation (Curfman, 1998; Maron et al., 2002).

The relatively rare occurrence of cardiac concussion is partly explained by its pathogenetic mechanism, which implies the convergence of some requirements: a topographical one, namely, the traumatic impact localized to the precordial area (Link et al., 2001), and a chronological one, with reference to the coincidence of the traumatic impact with the vulnerable period of ventricular repolarization, right before the T wave peak. Experimental studies on animals (Link et al., 1998) identified the exact timing of the vulnerable period of the cardiac electrical cycle in a time window of 15 msec before the T wave peak. Recently, it has been hypothesized that activation of stretch-sensitive channels could be crucial for this electrophysical phenomenon (Garan et al., 2005).

According to the traditional interpretation, cardiac concussion is characterized by the absence of cardiac injuries and of any kind of myocardial cellular damage, studied either with histopathological or immunohistochemical methods, in the absence of elevation of MB isoenzyme of creatine phosphokinase (CPK). However, recent experimental studies on animals (Guan et al., 1999) showed ultrastructural alterations of myocardial cells with the presence of relaxed myofibers with a widening of the I band and myofibrillar hypercontraction with formation of contraction bands. Also shown (Lindsey and Navin, 1978; Frazee et al., 1986) was an increase in myocardial cells membrane permeability, which could cause the elevation of MB-CPK found in commotio cordis by clinical experience.

Another relevant element in the diagnosis of cardiac concussion is the absence of any form of injury of traumatic origin in all other organs and systems, with the exclusion of small contusive lesions on the left thoracic side in the shape of generally circular or oval excoriations or ecchimosis, related to the object that caused the commotio cordis (Maron et al., 1995).

Cardiac concussion typically occurs during sports activities, such as baseball, hockey, and karate, and children or adolescents can be involved (Abrunzo, 1991; Viano et al., 1992; Kaplan et al., 1993; Maron et al., 1999, 2005; Pearce, 2005; Hamilton et al., 2005). Cases of sudden death due to cardiac concussion were reported as being a consequence of fists to the chest during fighting (Frazer and Mirchandani, 1984) and of low-speed car accidents without evidence of other, different, blunt thoracic injuries (Michalodimitrakis and Tsatsakis, 1997). Cases of commotio cordis occurring in the murder of children were also described in the literature (Baker et al., 2003; Boglioli et al., 1998; Denton and Kalelkar, 2000).

It should be noted that the particular thoracic conformation of children and adolescents makes them particularly vulnerable to cardiac concussion.

Cardiac concussion may be an underestimated event, as diagnosis is sometimes not recognized, and not all cases are correctly described and reported.

The U.S. Commotio Cordis Registry in Minneapolis, Minnesota (Maron et al., 2002), recognizes cases of cardiac concussion on the basis of well-defined selection criteria:

1. Blunt thoracic trauma testified for by the immediateness of cardiovascular collapse
2. Documentation about the traumatic event
3. The absence of structural alterations to sternum and ribs, such as, for example, fractures, as well as of the heart
4. The absence of preexistent cardiac anomalies

Cardiac Contusion

Cardiac contusion is the most frequent injury in blunt chest traumas, with an incidence rate that varies between 7 and 27% in some surveys (Darok et al., 2001) and between 7 and 71% in other reports (Sakka et al., 2000).

The great variability in incidence rates reported is due to the fact that cardiac contusion is not often diagnosed because of the difficulty of making this clinical diagnosis (Sybrandy et al., 2003). It is an event associated with a low mortality rate, less than 15%.

Cardiac contusion can occur following a rapid deceleration that causes an impact of the heart against the sternum and a "crush injury" in which the heart is compressed between the sternum and the thoracic vertebrae.

Clinically, the diagnosis of cardiac contusion is very uncertain, so it was defined a "capricious syndrome" (Jones et al., 1975). Sometimes, external thoracic injuries are present. The thoracic injuries most frequently associated with cardiac contusion are rib and clavicular fractures, pulmonary contusions, pneumothorax, hemothorax, flail chest, sternal fractures, and great vessels injuries.

Symptoms of cardiac contusion vary greatly and sometimes are not recognized or are underestimated in polytraumatized patients, in whom cardiac symptoms can be masked due to the clinical importance of concomitant musculoskeletal lesions. Generally, patients present with precordial pain, often of anginal or infarct kind. Electrocardiogram (ECG) does not show specific patterns, and rhythm disturbances are the ECG abnormalities most frequently reported, in particular, atrial arrhythmias, ventricular extrasystoles, bundle branch block, and atrioventricular (AV) conduction disorders. Repolarization abnormalities were also described. In every case, ECG abnormalities are frequent and can occur either immediately after the traumatic event or after a short period of time. In more than 70% of cases, they occur within 3 days after hospitalization for the traumatic event. Clinically significant is the almost constant increase in isoenzyme CK-MB serum concentration that occurs immediately after the traumatic event and returns to the "normal" range within 3 to 4 days. LDH1 and LDH2 isoenzymes are increased in all cases for about 2 weeks. With immunochemical determination of cardiac troponin (cTnI and cTnT), it is possible to reach a 100% diagnostic specificity (Fulda et al., 1997; Adams et al., 1996; Ferjani et al., 1997; Meier et al., 2003; Peter et al., 2005). In conclusion, the difficulty in clinically diagnosing cardiac contusion and the lack of univocal criteria, varying from criteria of simple suspicion to diagnostic criteria based on ECG findings, on determination of enzymes indicators of cardiac necrosis and on transthoracic and transesophageal echocardiography, explain the great variability of incidence rates reported in the literature. In autopsy studies, these rates vary between 15 and 20% and, as already mentioned, between 7 and 9% and more than 70% in clinical studies (Cachecho et al., 1992; McLean et al., 1992; Lindstaedt et al., 2002).

However, cardiac contusion is a well-defined pathological entity that can be identified and diagnosed either during macroscopical examination of the heart at autopsy or during microscopical examination. Subepicardial, intramyocardial, or subendocardial hemorrhagic effusions can be detected that are associated with tiny lacerations of cardiac myofibers. It is histologically well characterized with a sharp delimitation of the contused area from the surrounding undamaged myocardium. Histological abnormalities consist of foci of necrosis with contraction bands, localized and limited in their extension, and interstitial hemorrhages (Figure 135 and Figure 136); reparative processes begin rapidly and are characterized by the presence of leukocyte infiltrates, blood and myocardial necrotic tissue reabsorption, and subsequent formation of scars with fibrous substitutions of myocardium (Darok et al., 2001; Yoshida et al., 1992).

Formation of ventricular wall aneurysms can occur as a late sequela in cases of extensive transmural myocardial contusions.

Cardiac Rupture

Cardiac ruptures occur in about 0.5% of all chest traumas (Darok et al., 2001); in 10 to 15% of cases, they are a consequence of motor vehicle accidents.

Cardiac rupture carries a high mortality rate and is frequently diagnosed postmortem at autopsy due to the difficulty of clinical diagnosis (Brathwaite et al., 1990). Traumatic

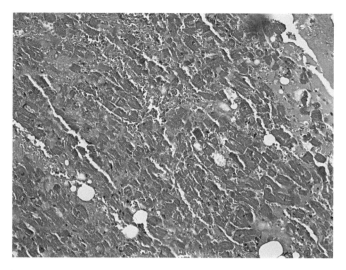

Figure 135 Cardiac contusion: interstitial hemorrhages and foci of contraction band necrosis.

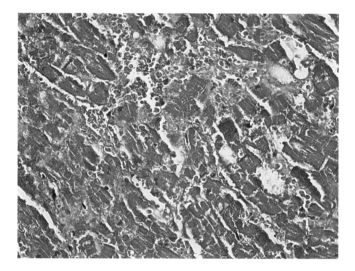

Figure 136 Hemorrhagic effusions associated with fragmentations and lacerations of cardiac myofibers.

cardiac ruptures are not always associated with thoracic wall injuries or with rib fractures (Durak, 2000).

The pathogenetic mechanisms that lead to cardiac rupture are basically the same as those involved in determining cardiac contusion: the direct mechanism is a consequence of a blunt impact against the anterior surface of the chest — this seems to be the main mechanism involved in ventricular ruptures — and the most vulnerable time for it to occur is during late diastole and early systole, when ventricular chambers are completely extended. An indirect mechanism is an increase in intrathoracic pressure caused by an increase in abdominal pressure or pressure in the lower extremities. A bidirectional mechanism is the compression of the heart between the sternum and the vertebral column, and acceleration/deceleration.

Cardiac ruptures can partially involve the myocardial wall (incomplete) or they can extend through the whole wall thickness (complete) and can be single or multiple. The right ventricle is most frequently involved in cardiac ruptures, followed, in decreasing order, by the left ventricle, the right atrium and the left atrium, and the interventricular and interatrial septum. Multiple ruptures of cardiac chambers are also described.

Septal ruptures following blunt chest trauma deserve particular attention because of a relatively high survival rate when they represent the main cardiac lesion. There is no area of the interatrial septum that seems to be particularly vulnerable to rupture, even if lacerations involve the borders of foramen ovale. Interventricular septum ruptures are often associated with ventricular rupture; the most frequently involved area seems to be the muscular part of the septum followed, in incidence, in combined ruptures of the muscular and membranous part and, finally, by the less frequent isolated lesions of the membranous part (Parmeley et al., 1958; Symbas, 1989; Liedtke and DeMuth, 1973).

Injuries may vary from partial and incomplete endomyocardial lacerations to lacerations involving the entire thickness of the cardiac wall. Generally, injuries have a vertical orientation in ventricles and a horizontal orientation in atria and tend to extend more toward the endomyocardial surface than toward the epicardial surface (Lobo and Heggtviet, 2001).

In cases of traumatic forces of great severity (e.g., massive compression of the chest), complex and extensive ruptures may occur that involve all cardiac chambers.

Cardiac ruptures are generally immediate, but cases of cardiac rupture occurring 2 weeks after the traumatic event are reported in the literature, probably following vacuolization and necrosis in the contused parts, before cicatrization takes place (Lobo and Heggtviet, 2001).

The histologic picture in initial lesions is characterized by alterations in myocardial cells, contraction bands, coagulative necrosis areas, sarcoplasmatic vacuolization, platelets aggregation, and fibrin deposition (Pollak and Stellwag-Carion, 1991).

Death is generally caused by cardiac tamponade following heart rupture or by hemorrhagic shock.

Clinical diagnosis of cardiac rupture is usually based on the early recognition of symptoms and clinical signs of cardiac tamponade.

Aortic Injuries

The incidence of traumatic aortic injuries following blunt chest trauma (blunt traumatic aortic rupture [BTAR]) greatly increased in recent years as a consequence of the ever-increasing incidence of motor vehicle accidents, which represent with no doubt one of the most frequent causes. Recent studies (Fabian et al., 1997) show that in the U.S. and Canada, approximately 7500 to 8000 deaths each year are caused by BTAR — the great majority of these occur as a consequence of motor vehicle accidents.

However, on the whole, BTAR is a rare occurrence. In 1998, a study (Gammie et al., 1998) reported that 4.4% of 1660 yearly hospital admissions for chest trauma were diagnosed as BTAR.

Aortic rupture is associated with high mortality. In his fundamental study, Parmeley et al. (1958) observed an almost immediate mortality at the scene of the traumatic event due to massive hemothorax in 80% of cases. Almost 50 years later, Richens et al. (2003) reported 130 cases of death by BTAR out of a total of 132 observed cases, with a total survival percentage of 1.5%.

Aortic lacerations following blunt chest traumas occur in most cases (>90%) at the aortic isthmus immediately before (approximately 2 to 3 cm) the origin of the left subclavian artery, where the extremely mobile part of the aorta is located between two more fixed aortic segments. The aortic arch is fixed by the great vessels of the neck; the descending thoracic aorta is fixed to the chest by the arterial ligament (ligamentum arteriosum) and by the intercostal arteries. The most mobile segment of aorta (distal portion of aortic arch and proximal portion of descending aorta) is fixed only loosely to the thorax by parietal pleura (Figure 137, Figure 138, and Figure 139).

Aortic ruptures generally occur at the border between these aortic segments due to the different responses to forces applied to the chest (Pasic et al., 2000).

A study by Kim and Busuttil (1996) of 66 cases of BTAR affirms the possibility of aortic lacerations occurring at other sites.

Sites of Traumatic Aortic Rupture

Aortic root	1
Ascending aorta	5
Junction between aortic arch and thoracic descending aorta	44
Thoracic aorta distally to aortic arch	10
Inferior segment of thoracic descending aorta	9
Abdominal aorta	5

In the series of 275 cases of traumatic aortic ruptures reported by Parmeley et al. (1958), the authors reported 171 cases of isolated aortic rupture and 104 cases of rupture associated with other cardiac injuries. In the 171 isolated injuries, the most frequently involved site was the isthmus (56%), while in circumstances of aortic lacerations associated with cardiac rupture, the aortic isthmus was involved in only 28% of cases, with the aortic arch being the preferred site (45%). This suggests a different pathogenetic mechanism of aortic rupture in the two different circumstances. Generally, the aortic laceration was single, although cases of multiple aortic lacerations in the same patient were reported.

There are four basic pathogenetic mechanisms involved in traumatic aortic injuries, as described by Symbas (1989) (Figure 140).

Because of the greater resistance of the aortic wall to longitudinal stress than to horizontal stress, aortic lacerations usually appear as transverse tears that may partially or totally involve the entire wall thickness. It is possible to find intimal tears that may vary in depth to transmural extension and complete lacerations of the aortic wall. In less severe traumas, the injury may appear as a circumferential intimal tear more or less extending into the aortic media. In these cases, the adventitia is intact and may contain blood within the vessel, allowing for possible survival of the patient. Alternatively, blood pressure may force blood into the layers of the aortic wall, causing the formation of a pseudoaneurysm. In more severe trauma, the laceration may extend to the adventitia to a complete transection of the aortic wall. In these cases, there is a massive mediastinal and pleural hemorrhage that is generally fatal, but cases of survival due to prompt diagnosis and reparative treatment are described in the literature (Richens et al., 2003).

Aortic lacerations may be associated with various injuries: rib fractures, sternum fractures, thoracic vertebral fractures, pulmonary lacerations and contusions, hemopericardium, hemothorax, cranio-cerebral injuries, hepatic and splenic injuries, and so forth (Kim and Busuttil, 1996).

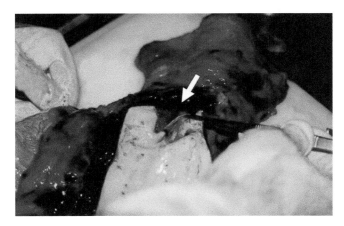

Figure 137   Aortic laceration: a transverse tear partially involves the entire wall thickness.

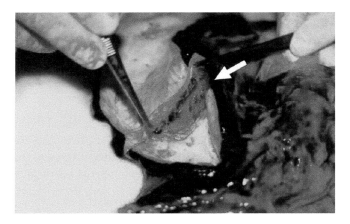

Figure 138   Intimal tear.

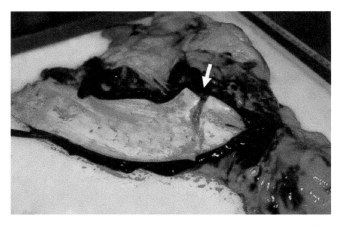

Figure 139   Transverse tear that involves the entire wall thickness.

Of particular interest for its obvious medico-legal implications in the reconstruction of injury dynamics in motor vehicle accidents, which, as already mentioned, are the most frequent causes of BTAR, is a recent study (Richens et al., 2003) that examines 132 cases

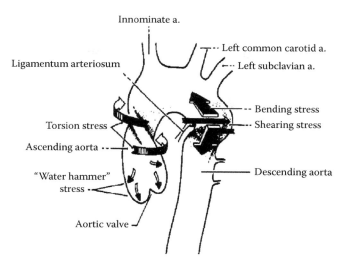

Figure 140   Diagram of the stresses contributing to rupture of the aorta from blunt trauma. (From Symbas, P.N., *Cardiothoracic Trauma*, Philadelphia, W.B. Saunders, 1989.)

of BTAR following motor vehicle accidents, with particular reference to the modality of the accident, the victim's position in the car, and the use of safety measures (seat belts, airbags, etc.). BTAR was found to occur in the occupants either of front seats or of rear seats of the cars involved in the accident. Use of seat belts and the presence of airbag devices do not seem to eliminate the risk of BTAR: in 74 out of 132 cases, seat belts were securely fastened at the time of the impact.

A study by Arajärvi et al. (1998) shows, proportionally, a higher incidence of ascending aortic ruptures in subjects who at the time of the car accident were not protected by seat belts, while occupants of vehicles who wore safety belts showed a higher number of descending aortic injuries. However, no incidence difference was detected in regard to the typical site of aortic injuries, namely, the aortic isthmus. Apparently no correlation emerged between the modality of impact and aortic rupture occurrence, which are described in all hypotheses about the dynamics of motor vehicle accidents. The higher number of cases occurring as a consequence of frontal or lateral impacts is correlated with the more frequent occurrence of this kind of impact in motor vehicle accidents.

The severity of the impact, expressed as ETS (equivalent test speed) seemingly does not correlate with the incidence of BTAR, as cases of traumatic aortic rupture were reported in the literature even in accidents where the ETS was 30 km/h[1] (Richens et al., 2003).

Even if aortic lacerations are injuries that typically occur in motor vehicle accidents, a high incidence of BTAR is also reported in pedestrians who are victims of traffic accidents (Brundage et al., 1998) — 12.7% of 220 pedestrians involved in accidents over a period of about 6 yr. The study showed that aortic lacerations occured in sites different from that more frequently observed in victims of motor vehicle accidents. Left subclavian artery, aortic arch, and ascending and descending aorta lacerations are reported rather than lacerations in the most common site, the ligamentum arteriosum. This particular topography of aortic lacerations in pedestrians who are victims of traffic accidents suggests a pathogenetic mechanism in which compressive forces rather than deceleration mechanisms predominate. The high incidence reported in this retrospective analysis of multiple rib fractures and thoracic vertebral fractures suggests that the mechanism prevalently operating in causing aortic laceration is one of compression or flexion-distraction.

Mason (1978) reported a significant incidence of BTAR in plane crash victims. Moreover, BTARs are also described in victims of motorcycle accidents and of falls from great heights.

Aortic dissection following thoracic blunt traumas is not common because in these cases, rupture is the most frequently encountered injury involving the aorta. Trauma is rarely described as a causal factor of aortic dissections. However, cases of traumatic aortic dissections are reported in the literature (Rice and Wittstruck, 1951; Potter and Hopkins, 1968; David and Blumberg, 1970; Faraci and Westcott, 1977; Wilson and Hutchins, 1982; Gates et al., 1994; Gammie et al., 1996; Goverde et al., 1996; Rogers et al., 1996; Ono et al., 1998; Mimasaka et al., 2003; Anakwe, 2005).

The new classification of aortic dissections reported by the Task Force on Aortic Dissection and by the European Society of Cardiology (2001) basically consists of a subdivision of the classic classifications by Standford and De Bakey (Virmani and Burke, 2001) and distinguishes five classes of aortic dissections: class 5 includes traumatic and iatrogenic aortic dissections.

We note that intimal tears from trauma may cause aortic aneurysms years after an accident, with their rupture causing sudden death.

Valvular Apparatus Injuries

Posttraumatic sequelae involving cardiac valvular apparatus consist of contusions, lacerations, and ruptures; however, these are rare occurrences. Moreover, it is difficult to define either their clinical presentation or the corresponding morphological pictures because of the remarkable anatomical variability of the valvular apparatus as well as of the entity and the kind of traumatic action applied to the chest wall in blunt trauma.

The injury may consist of rupture of a papillary muscle (most common), rupture of chordae tendineae, or laceration of both (less common). A rupture of a papillary muscle may be complete or incomplete.

The aortic valve is the most commonly injured, followed by the mitral and tricuspid valves.

Isolated injuries of the tricuspid valve are less frequent and have less severe hemodynamic consequences than isolated injuries of the mitral valve. Mitral valve rupture occurs more frequently when a sudden compressive force is applied to the chest during diastole or early systole, in the brief time between the closing of the mitral valve and the opening of the aortic valve (Cuadros et al., 1984; McDonald et al., 1996; Smedira et al., 1996; Stahl et al., 1997; Bruschi et al., 2001).

However, these injuries are not common; Parmeley (1958) does not report any isolated injury of the mitral valve and only eight cases of laceration of the mitral valve apparatus associated with other cardiac injuries.

The clinical picture of mitral valve injuries may vary greatly: from sudden cardiogenic shock to cases in which the injury remains asymptomatic for many years.

Diagnosis is currently based on transthoracic or transesophageal echocardiography (TTE).

Penetrating Thoracic Traumas

Penetrating thoracic traumas may involve all the deep structures (pericardium, heart, great vessels, and lungs). The most frequent causes are wounds produced by cutting or sharp instruments (knives, blades, etc.) that produce particular cardiac and great vessels injuries.

Different kind of injuries may occur following penetrating wounds produced by firearms with various morphological features, depending on the kind of weapon (single or multiple shot), type of munition, and the distance between the weapon and the body.

Cardiac and great vessels penetrating injuries may be determined by a direct or indirect mechanism. In the first case, there is a direct penetration through the thoracic wall with involvement of pericardium and heart. In the second case, a strong and violent contusive impact to the thoracic area can cause sternum and rib fractures that, in turn, are responsible for cardiac and vascular injuries. This is an occurrence frequently described in motor vehicle accidents.

Pericardium, all cardiac chambers, and great vessels may be involved in cases of penetrating trauma. The resulting lesions may vary greatly, from lacerations limited to the pericardium without cardiac and vessel involvement, to lacerations involving and even perforating the entire thickness of cardiac walls and, especially in cases of injuries produced by firearms with great potential, to real interruptions of the cardiac wall (Figure 141 through Figure 144).

Cutting instruments generally produce limited and localized injuries that extend to varying extent into the transmural thickness and sometimes may completely perforate the cardiac wall (Rich and Spencer, 1978; Asfaw and Arbulu, 1977; Parmeley et al., 1978).

Histologically, along the myocardial injury border and in the surrounding area, there is evidence of contraction bands with hypercontracted sarcomeres and thickened Z lines, without myofibrillar rupture (Yoshida et al., 1992)(Figure 145 and Figure 146). This aspect is similar to "cutting hedge" hypercontraction, an artifact observed at the borders of myocardium samples taken during biopsies and in hearts resected for transplantation (Todd et al., 1985; Baroldi et al., 2001).

Patients with cardiovascular injuries produced by firearms generally have a threefold higher mortality rate than those with injuries produced by steel weapons, and hospitalization time and morbidity are higher in cases of injuries produced by firearms than in those produced by steel weapons (Pate and Richardson, 1969; Madiba et al., 2001; Thourani et al., 1999).

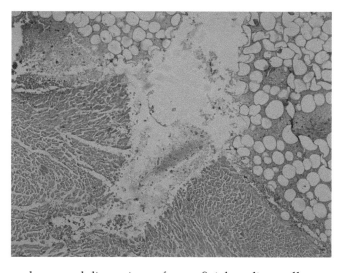

Figure 141   Hemorrhages and disruptions of superficial cardiac wall.

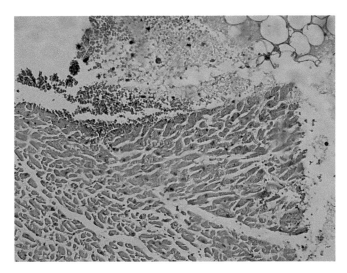

Figure 142   Subepicardial hemorrhages.

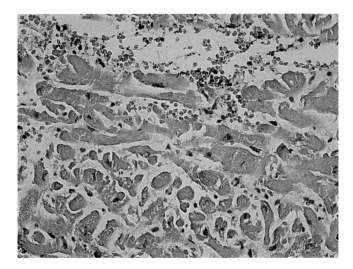

Figure 143   Interstitial hemorrhages.

The right ventricle is the anatomical site most frequently involved in penetrating thoracic trauma (35%), followed by right atrium and left ventricle (25%) (Moore et al., 1991); the left atrium is rarely involved because of its position. The left ventricle muscular wall can close a small injury more easily than the thinner right ventricle wall; the atria seem to completely lack this capacity (Karmy-Jones et al., 1997). Perforating injuries can involve the interatrial and interventricular septum, the valvular apparatus, the papillary muscles, the chordae tendinae, and the great vessels originating from the heart (Figure 147 through Figure 154).

Coronary arteries, in particular the anterior descending artery, may be involved in cases of penetrating injuries (3 to 9%) (Reissman et al., 1992), even if the real incidence of these injuries is difficult to determine because they are often not recognized.

Recently, we described a case in which death was attributed to cardiac tamponade due to a selective penetrating injury of the right coronary artery (Turillazzi et al., 2005). Penetrating

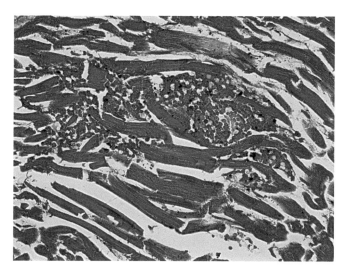

Figure 144   Interstitial hemorrhages and fragmentations of myocytes.

heart wounds are associated with coronary artery injuries in up to 9% of cases, although the documented incidence of such wounds is low, because they often remain undetected. The incidence of traumatic coronary vessel rupture is not accurately established, because the clinical diagnosis is difficult, and the damage is often unrecognized before death. Frequently, traumatic coronary damage is revealed only after postmortem heart examination. Because of their small size and protected location, selective penetrating injuries to the coronary arteries are uncommon. The left anterior descending artery is the most frequently injured vessel because of its location and relative size, followed by the right and the circumflex coronary arteries, respectively.

Hemopericardium, cardiac tamponade, and ischemic injury in the territories perfused by the injured coronary vessel may ensue, as well as myocardial infarction that, in some cases (injuries of proximal segments of coronary arteries associated with wide areas of myocardial ischemia), may require coronary bypass surgery (Karin et al., 2001). Thoracic great vessels injuries are diagnosed with increased frequency because of the escalating use of automatic weapons. The overall mortality of thoracic aortic injuries following penetrating thoracic trauma is higher than 90%, and in subclavian vascular injuries it is higher than 65%. Most of these patients reach the hospital dead or in severe shock (Demetriades, 1997). The clinical presentation of penetrating cardiovascular injuries ranges from a condition of complete hemodynamic stability to acute cardiovascular collapse and immediate cardiac arrest. Symptoms may be related to several factors, including the wounding mechanism; the length of time that elapsed between the trauma and arrival at an equipped hospital; the extent of the injury, which, if sufficiently large, causes copious hemorrhage in the thoracic cavity; and, finally, the presence or absence of cardiac tamponade (Mittal et al., 1999).

Cardiovascular injuries generally carry a high mortality rate, as they may be conditioned by several factors (Rhee et al., 1998; von Oppell et al., 2000; Campbell et al., 1997). Significant factors affecting survival are the number of injuries, the presence of cardiac tamponade, right ventricle involvement, the number of cardiac chambers affected by the injury, and the presence of pleural lacerations (Mandal and Sanusi, 2001). In particular, mortality is strongly affected by the presence of cardiac injuries: a study examining 3049 patients with penetrating

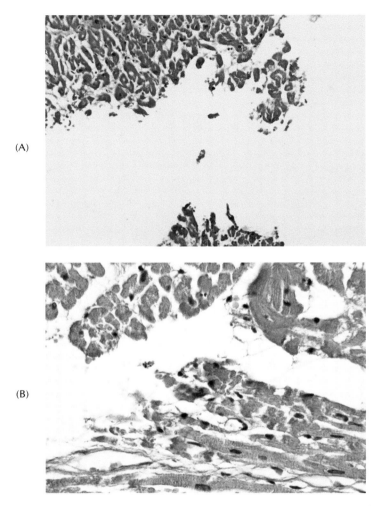

(A)

(B)

Figure 145 (A) Cardiac injury produced by firearm. (B) Along the myocardial injury border and in the surrounding area, there is evidence of contraction bands with hypercontracted sarcomeres and thickened Z lines.

thoracic injuries showed that when cardiac injuries are present, the mortality rate reaches 21.9%, while in patients without cardiac injuries, the overall mortality rate is 1.5%.

Death is caused by hemopericardium with cardiac tamponade, massive hemothorax in the presence of pericardial lacerations communicating with the pleural cavity, cardiac arrhythmia, and conduction disturbances (Figure 155).

In the presence of penetrating injuries, even of slight extent (such as lacerations limited to the pericardial membrane), there is a possibility of superimposed infective processes to set in (pericarditis, abscesses, myocarditis, and endocarditis).

Iatrogenic Trauma

Iatrogenic traumas are a polymorphous category including different kinds of cardiac and vascular injuries that may occur following cardiac massage, surgical procedures, and invasive diagnostic and therapeutic procedures (such as catheterization, pacemaker implantation, etc.).

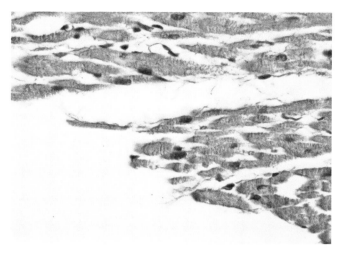

Figure 146   Hypercontracted sarcomeres and thickened Z lines, without myofibrillar rupture.

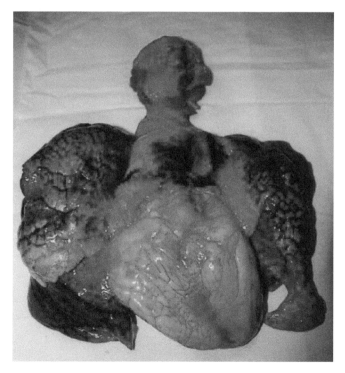

Figure 147   Heart–lung "en bloc" technique: hemorrhage in the corresponding ascending aorta.

Resuscitative Measures

Cardiopulmonary resuscitation (CPR) is a procedure that has been widely used since its introduction in 1960 (Kouwenhoven et al., 1960). Its possible complications are well described in the literature (Krischer et al., 1987; Bedell and Fulton, 1986; Powner et al.,

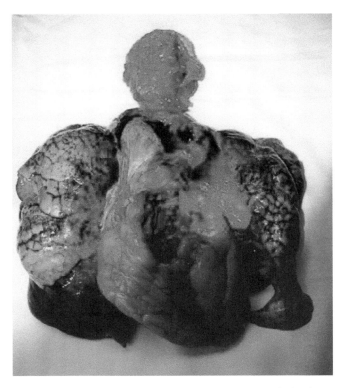

Figure 148 Heart–lung "en bloc" technique: posterior view of the heart with large atrial laceration.

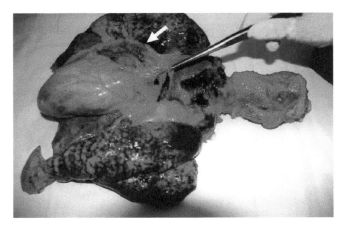

Figure 149 Gross appearance of Figure 148.

1984), and among them, rib and sternal fractures are particularly frequent. In particular, the reported incidence of sternal fractures following CPR with cardiac massage varies from 0 to 43%, while the incidence of rib fractures after CPR is 19 to 80%. The most serious complications, such as pericardial ruptures or lacerations, myocardial ruptures or lacerations with cardiac tamponade, often caused by the penetrating action of bony stumps from a fractured sternum, are generally rare occurrences (Bodily and Fischer, 1979; Noffsinger

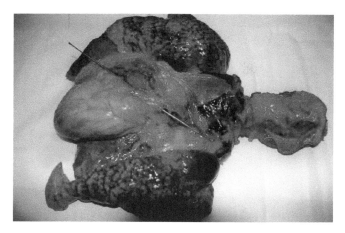

Figure 150   Injuries produced by firearm.

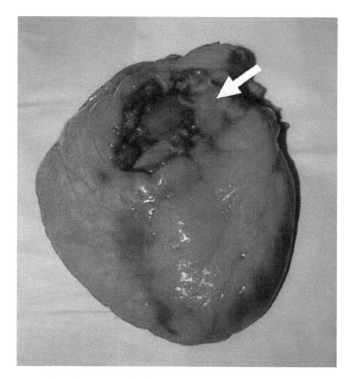

Figure 151   Large atrial and ventricular laceration produced by firearm.

et al., 1991; Fosse and Lindberg, 1996; Kempen and Allgod, 1999; Machii et al., 2000; Sokolove et al., 2002). Cases of right ventricular rupture were also described in the absence of sternal and rib fractures (Baldwin and Edwards, 1976). Injury severity depends on several factors, such as the patient's age, his or her physical constitution, the duration of CPR, and the experience and skill of the reanimation team: in aged subjects, because of bone fragility due to osteoporosis, sternal and rib fractures are frequent even when cardiac massage is performed correctly.

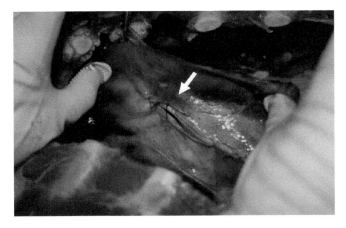

Figure 152   Pericardium: surgical repair of penetrating injury.

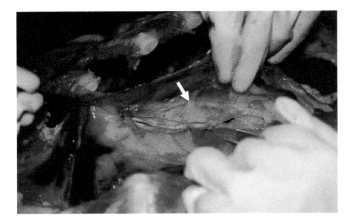

Figure 153   Surgical repair of pericardium lesion produced by knife injury.

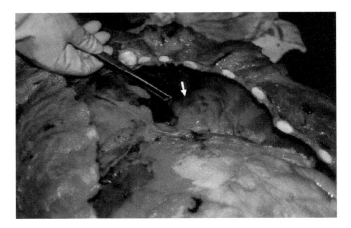

Figure 154   Hemopericardium.

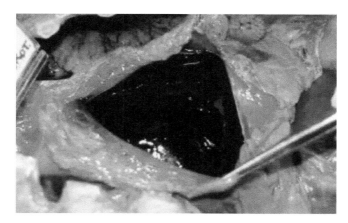

Figure 155   Cardiac tamponade due to massive hemopericardium.

Active chest compression/decompression (ACD), a cardiopulmonary resuscitation procedure introduced in 1990 (Lurie et al., 1990) that uses a drain suction pump, is burdened by complications. Cases of skin abrasions at the site where the pump is applied, of rib and sternal fractures, as well as of pneumothorax were described (Rabl et al., 1996). Following ACD, cardiovascular injuries may occur, such as cardiac and pericardial lacerations or ruptures (Rabl et al., 1997; Baubin et al., 1999a,b).

Klintschar et al. (1998) reported multiple rib fractures, sternal fractures, rupture of pericardial sac, rupture of right ventricle, rupture of right and left atrium, and rupture of ascending aorta in a patient who received CPR followed by ACD for 15 min.

Catheterization

Catheterization procedures via different sites of vascular access are widely used for diagnostic and therapeutic purposes. They produce a penetrating trauma to the vessel where the catheter is inserted as well as the possibility of further injuries as the catheter proceeds.

Central venous catheters are widely used when the need to administer long-term endovenous medication eliminates the possibility of using a peripheral venous access to give fluids, blood products, medications, and nutrients. The use of central venous catheters is associated with several adverse events that are well known and described in the literature. They include pneumothorax, hemothorax, hydrothorax, air embolism, detachment and embolism of parts of the catheter, injury and perforation of the venous wall or of the heart, sepsis, thrombophlebitis, hemorrhagic diathesis, and cardiac arrhythmias (Conces and Holden, 1984; Agarwal et al., 1984; Cobb et al., 1992).

Cardiac tamponade is a serious complication of a central venous catheter (Forauer et al., 2003; Shields et al., 2003). It was reported for the first time by Turner and Sommers (1954), who described right atrial perforation following right median cubital vein catheterization. Since then, more than 100 cases of cardiac tamponade following central venous catheter use were reported in the literature (Collier and Goodman, 1995), associated with a 65% mortality rate (Collier et al., 1984). It is particularly significant that about 52% of complications associated with central venous catheter insertion seem to be related to errors in insertion (Scott, 1988).

The site of the injury caused by venous catheter insertion is generally at the vessel used for venous access, the right atrium, and with less incidence, the right ventricle.

The different pathogenetic mechanisms of cardiac perforation are widely described. Cardiac perforation can occur immediately after catheter insertion through direct injury to the myocardial wall or some time after the invasive procedure through a mechanism of progressive microtraumatization of the myocardial wall, because the catheter may get entangled in intracavitary structures. Brandt et al. (1970) described the mechanism of cardiac and vessel perforation in catheterization injuries: the initial endothelial injury produced by the catheter tip would cause thrombus formation with consequent adhesion of the tip to the adjoining cardiac wall. Cardiac muscle contraction together with the blood turbulence would contribute to subsequent perforation of the myocardial wall.

Cardiac catheterization is a procedure widely used for diagnostic and therapeutic purposes. The most common site of vascular access is the femoral artery. The most common complications are injuries of the vessel wall that may cause localized hemorrhages, contusions of the vessel wall, and lacerations. Endothelial injury may facilitate intraluminal thrombus formation with subsequent embolization. Arteriovenous fistulae, pseudoaneurysms, hemorrhages, and infections are complications also described.

These are complications that despite the ever-increasing improvement of the procedure, still represent a significant problem. A recent review of 3723 cardiac catheterizations showed a 0.9% incidence rate of vascular injuries (Manuel-Rimbau et al., 1998). In particular, arteriovenous fistulae formation is always possible following vascular catheterization: about 1% of patients who underwent transfemoral cardiac catheterization will develop arteriovenous fistulae (Kelm et al., 2002).

Iatrogenic injuries were described during aortography and angiography, coronarography, pacemaker insertion (Howell and Bergen, 2005), valvuloplastic surgery, as well as during orthopedic surgery, vascular surgery, oncologic-related surgery, and general surgery (Natali and Benhamou, 1979; Lazarides et al., 1991; Friedrich et al., 1994; Ricci et al., 1994; Devlin et al., 1997; Joseph et al., 1997; Heuser, 1998; Nashed et al., 1999; Elsner and Zeiher, 1998; Lazarides et al., 1998; Lewis et al., 1999; Dunning et al., 2000; Von Sohsten et al., 2000; Chandrasekar et al., 2000).

Cardiac Surgery and Forensic Medicine

13

ALBERTO REPOSSINI

Contents

Introduction

Heart surgery is probably one of the most complex surgical disciplines because of multi-organ involvement and both technical and physiopathologic perspectives related to extra-corporeal circulation.

During the last 10 years, many new cardiac surgical centers were created throughout the world, raising the number of surgical operations per year to 1000 new cases per million population in Europe and the U.S. Among adults, coronary bypass grafting has the highest (60 to 80%) incidence of operation in Western countries, followed by valvular surgery. The decrease of rheumatic disease in the latter countries is balanced by an increasing frequency of degenerative aortic and mitral valvular diseases due to aging, which need surgical repair. In childhood, congenital cardiac malformations prevail.

At experienced centers, operative and postoperative mortality are very low, ranging from 1 to 5%. However, severe or lethal complications may occur. In general, heart operations have a good outcome, and we may distinguish complete and palliative care. Complete care means recovery to normal function and life, while palliative care improves function or prolongs life or both. However, this distinction does not consider complications during and after surgery. Failure is the correct definition when no positive results are obtained, because postsurgical complications may result in an early death, or permanent damage due to such events as thromboembolism, or paraprosthesic leak following valvular correction or vessel graft. Most, if not all, of these complications are the consequence of human error due to a lack of technical-scientific information, or technical support of the surgical unit, resulting in, for example, use of an incorrect surgical approach, incomplete discrimination of risk factors, or incorrect postsurgical treatment. Such errors can be neutralized by timely corrective action. On the other hand, hospitalization or premature death may follow untreatable cardiac shock, respiratory or kidney failure, neurologic disorders, irreversible coma, etc. (multiorgan insufficiency), mirrored pathologically by different forms of myocardial necrosis (Baroldi et al., 2004), acute respiratory distress syndrome findings, renal tubular necrosis, hepatic cholestasis or necrosis, adrenal cortical necrosis, neuronal eosinophilia, and, in some cases, by necrosis of the gut lining.

The correct identification and understanding of these potentially lethal complications are essential for evaluating risk factors.

By retrospective analysis, significant risk factors were included in a score system to predict perioperatory mortality. The more popular is the following Euroscore (Nashef et al., 1999), for which the sum of the total score gives a percentage in terms of risk of death:

	Score
Extracardiac Factors	
Age >60 (1 for 5 more years)	1
Female gender	1
Chronic bronchopneumonia	1
Peripheral arteriopathy	2
Neurological history	2
Cardiac reintervention	3
Creatinine >200 m micromol/l	2
Active infective endocarditis	3
Critical preoperative state	3
Cardiac Factors	
Unstable angina pectoris	2
Left ventricular dysfunction (ejection fraction <30%)	3
Recent myocardial infarction (<90 days)	2
Pulmonary hypertension (>60 mm Hg)	2
Surgical Factors	
Emergency	2
Associated procedures	2
Thoraco-aortic procedure	3

Extracorporeal Circulation

Most cardiac surgical procedures are performed with extracorporeal circulation and using cardioplegic cardiac arrest.

Extracorporeal circulation consists of diverting the systemic venous blood through a cannula placed in the right atrium, vena cava, or femoral vein, into a membrane oxygenator, allowing for gas exchange, and raising the PO_2, with a heat exchanger controlling temperature and a pump (roller or centrifuge) to return blood to the aorta or another great artery. Surgery is performed at normal body temperature, namely, a rectal temperature of 35° to 36°C; some centers prefer a moderate 28° to 30°C hypothermia able to protect against metabolic and coagulation imbalance. The pump ensures a systemic pulsating pressure ranging from 50 to 80 mm Hg and corrects any significant hypotensive or hypertensive variations of vascular tone. The extracorporeal blood transit needs heparin to prevent thrombotic events, maintaining a time of coagulation higher than 400. At the end of a procedure, after decannulation of venous sites, protamine sulfate is given to restore normal coagulation (100/120).

The major complications encountered with extracorporeal circulation are as follows:

1. *Massive air embolism* consequent to faulty technique during arterial cannulation or circuit preparation, or level below the safety limit of reservoir and perfusion by air bubbles
2. *Thrombotic embolism* due to insufficient anticoagulant use (heparin)
3. *Atheromatous embolism* from ruptured atherosclerotic aortic plaques
4. *Aortic dissection* at cannulation or clamping sites, generally due to preexisting cystic medionecrosis or less commonly to faulty technique (Figure 156)

Extracorporeal circulation in healthy young adults may last 2 to 3 h without major systemic complications. After 4 h, signs of hypoperfusion, such as hemolytic crisis, kidney failure, and pulmonary and brain edema with cerebral vascular and splanchnic insufficiency, may ensue. Older subjects and those affected by multiorgan vascular pathology are at risk of such complications after a shorter period of extracorporeal circulation.

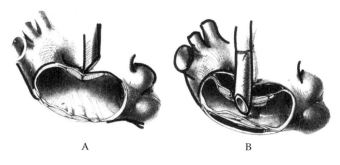

A   B

Figure 156 Aortic dissection at cannulation. A calcific atherosclerotic plaque may favor this complication.

Other minor complications include the following:

1. Activation of a systemic inflammatory response, especially dangerous in patients with neoplasia or who are immunodepressed
2. Systemic microembolization, especially affecting the brain, with loss of memory, concentration, and altered behavior

Surgery of the aortic arch may require circulatory arrest and deep hypothermia (20°C) with or without selective perfusion of the carotid artery to assure cerebral protection. In these cases, the risk of cerebral/systemic hypoperfusion increases proportionally to the length of circulatory arrest; if less than 30 to 40 min, complications are generally rare.

The heart, excluded from the circulation by clamping the ascending aorta, stops contracting and is perfused with a cardioplegic solution able to protect it from the lack of nutritional blood. Cardiac perfusion may be antegrade through the ascending aorta or coronary ostia or retrograde by cannulation of the coronary sinus.

The adopted solutions are as follows:

1. A crystalloid solution (physiological solution plus KCl) at 4°C with perfusion repeated every 20 to 30 min and eventually associated with ice-cooling of the myocardium with the risk of paresis of the left phrenic nerve
2. A blood solution with KCl added directly from the pump (30 to 35°C) repeated every 10 to 15 min (Calafiore technique)
3. A crystalloid solution added with antioxidant (Breitschneider); a single perfusion protects for 3 h

A surgeon selects the best protective technique for the myocardium depending upon the type of surgery and personal experience. Errors in timing and method may result in complications at declamping. At that time, transient disorders (arrhythmias, ischemia, contractile deficits) may occur or severe and irreversible myocardial damage may become obvious. The latter may impair recovery of normal global cardiac function after extracorporeal circulation.

New surgical techniques for myocardial revascularization without extracorporeal circulation and cardiac arrest have been introduced (see below).

Myocardial Revascularization

From the earliest and now abandoned attempts at revascularization by Beck using pericardiopexia or oxygenated arterial blood flow into the coronary sinus or the insertion of a mammary artery into the myocardium (after Vineberg), myocardial revascularization is currently accomplished through aortocoronary grafts adopting autologous saphenous vein (Favaloro, 1969) or the internal mammary artery, or less often, the radial, gastroepiploic, or superficial epigastric arteries.

Venous Grafts

Venous grafts are harvested from the leg or thigh either after direct exposure of the entire segment of a saphenous vein or by small cutaneous incisions along its course and removal of the whole vein with care to avoid endothelial damage and acute thrombosis.

Sampling in the perimalleolar zone, near the saphenous nerve, may induce paresthesia or zonal dysesthesia. The venous segments are reversed and anastomosed, that is, the anatomic distal end is connected proximally to the aorta, to prevent venous valves from blocking flow. The aortocoronary bypass is grafted into the ascending aorta and by terminal or lateral anastomosis into a section of an intact tract of the coronary artery distal to the atherosclerotic stenosis. The procedure can be perfomed by sequential laterolateral anastomoses between different coronary arteries, especially for vessels of the posterolateral heart. By this method, perfusion is more physiological compared to a single bypass (Lemma et al., 2003), but the procedure is more complicated with the need for a perfect orientation and exact graft length to avoid its kinking or angulation (Figure 157A and B) or flattening because of insufficient length, with stretching and consequent occlusion of the anastomosis (Figure 158 and Figure 159).

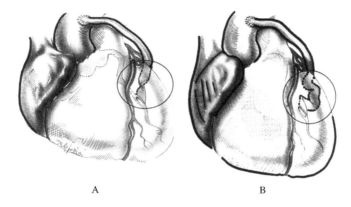

A   B

Figure 157 Aortocoronary venous graft at the margin of the left circumflex (LCX) artery. Erroneous orientation (A) and excessive graft length (B) are causes of occlusion.

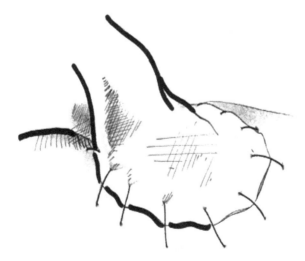

Figure 158 Coronary surgical anastomosis with stretching and flattening of the latter causing flow reduction and risk of occlusion.

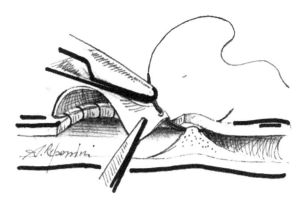

Figure 159 Coronary surgical grafting: posterior arterial wall erroneously included within the suture and consequent graft occlusion.

The more frequent technical errors are as follows:

1. No inversion of the venous graft with blockage of blood flow by its valves
2. Distal anastomosis made proximal to the coronary stenosis or at the level of an atherosclerotic coronary segment with mobilization of atheromatous material and vessel dissection and subsequent thrombosis
3. Proximal anastomosis at the level of an atherosclerotic ascending aorta with atheromatous embolization into the venous graft and resultant ischemic-necrotic damage of the dependent myocardium

Internal Thoracic or Mammary Artery Graft

The left mammary artery arising from the subclavian artery is generally grafted (Dion, 1996), in a "skeletal" way, namely, the artery alone, excluding by hemostatic clips the intercostal secondary branches or "pedicled" (i.e., together with the two satellite veins and with the internal thoracic band). The distal end of the artery is anastomized to the coronary artery beyond its stenosis. The left mammary artery is particularly employed for grafting the left anterior descending branch and its diagonal branches, while the right mammary artery is employed for the left circumflex or marginal branch. Because of its length, it is not always possible to insert it into the posterior interventricular branch. The mammary artery is usually free from atherosclerosis. However, one must check it to assure the absence of plaques or intimal thickening.

Frequently, in females and diabetic patients, increased tissue fragility may lead to the risk of a hematoma or dissection at implantation with a possible early occlusion. Only a macroscopically intact artery showing a normal blood flow should be used. Occlusion of all collateral branches must be carefully controlled to avoid the perioperative risk of acute hemorrhage.

Radial Artery Graft

Through an incision in the anterior forearm, segments of the radial artery before its bifurcation into the ulnar branch are isolated. The integrity and validity of the collateral ulnar circulation to the palmar arch must be checked (Allen test). The radial artery is generally used to complete revascularization with mammary artery grafts without saphenous vein grafts. The arterial radial segment is used for an aortocoronary bypass. Because a direct anastomosis with the aorta is prone to early occlusion, implantation of the anastomosis is done to a saphenous vein patch on the aortic wall. Other possibilities are a lateral or end-to-end anastomosis with the mammary artery.

Gastroepiploic Artery Graft

Almost abandoned in use, this vessel was anastomosed to the posterior-lateral arteries of the heart. After sternotomy and laparatomy, the artery was taken from the curve of the stomach with all collateral branches being occluded (Suma, 1996). Then, through a stomach and hepatic edge route, a passage was created in the diaphragm, and the distal end of the artery was introduced into the pericardial cavity and grafted to a coronary artery.

Incomplete Revascularization

This definition is adopted when surgery does not accomplish revascularization of a whole myocardial territory distal to a critically stenosed coronary artery (anatomical incompleteness). Often, if the distal coronary segment has a thin caliber (less than 1 mm) or is severely calcified, the anastomosis cannot be performed. Generally, when the myocardial protection is good, no immediate postoperative complication exists; however, risk of ischemia with perisurgery heart attack may occur. An anatomically incomplete but functionally effective revascularization can be obtained by participation of a compensatory collateral flow.

Beating Heart Coronary Surgery

This revascularization is achieved using current techniques but without extracorporeal circulation and cardiac arrest. The myocardium is exposed, and the coronary section to be anastomosed is kept steady by using a "stabilizer" instrument. Such revascularization can be performed in all myocardial districts. However, the lateral wall of the left ventricle is the most difficult part to reach, and particular skill is needed to expose and stabilize the cardiac wall. In cases of instability or other technical problems, use of extracorporeal circulation is preferred. In general, the anastomosis is done to a proximal coronary stenosis using suturing thread and intracoronary shunt to guarantee the blood flow during the operation. When well performed, no difference exists between such a procedure done on a beating heart or one cardioplegically arrested. At present, most surgeons prefer to use the beating heart technique in patients (for example, the elderly or those with severe aortic atherosclerosis, brain vasculopathy, or severe bronchopneumonia; Calafiore et al., 2003; Ascione et al., 2003) at high risk if extracorporeal circulation is used. Worth mentioning is the minimally invasive surgery by left *minithoracotomy*, namely, the anastomosis of a mammary artery to the anterior descending branch of the left coronary artery (Repossini et al., 2000).

Postsurgical Complications of Myocardial Revascularizations

Postsurgical complications of myocardial revascularizations include the following:

1. Transient or persistent ischemia, with the need for cineangiographic examination
2. Heart attack due to possible occlusion of the graft or poor blood flow in the coronary branch, with the need for aortic counterpulsation balloon therapy
3. Heart attack due to ventricular arrhythmia

Ventricular Aneurysmectomy

The indications for the surgical excision of a ventricular aneurysm are cardiac failure, embolization, or ventricular arrhythmia. Surgery is performed by opening the apex of the left ventricle laterally to the course of the left anterior descending branch; removing the contained thrombus material; excising the ventricular dyskinetic ventricular wall; and repairing with a Dacron patch sutured at the endocardium along the border between fibrotic and normal myocardium. The ventricular apex is rebuilt by joining it to the patch. When there is ventricular arrhythmia, the arrhythmogenic centers along the transitional zone between normal myocardium and fibrous scar tissue are eliminated by cryoablation or radiofrequency.

Possible technical errors are as follows:

1. Injury of a coronary artery during ventricular wall opening or its occlusion by suture after excision of the aneurysm and subsequent myocardial necrosis
2. Excessive reduction of the residual ventricular cavity leading to a low cardiac output
3. Tearing of suture stitches with heart rupture, massive hemorrhage, and cardiac tamponade

Valvular Surgery

Valvular surgery can be reparative or substitutive. Reparative means that the pathological valve is remodeled, while substitutive surgery uses either mechanical or biological valvular prostheses to substitute for the natural one. Implantable devices must be marked and registered by Health and Welfare authorities.

Mechanical Prostheses

Mechanical prostheses are composed of carbon, titanium, and steel to form mobile elements (leaflets or disks) fixed with hinge pins or zippers to a supporting ring. Around this is a band of synthetic material to allow fixation to the native valvular ring. A prosthesis can be damaged during implantation by an incorrect placement, and its function must be monitored by echocardiogram. Furthermore, prostheses may have a structural fault not detectable during surgery, and they may break or become occluded later on, often as a consequence of an incorrect anticoagulant therapy (Figure 160). At surgery, prosthesis malfunction is not predictable, but when it shows an unusual frequency, manufacturers must revise the faulty series.

Biological Prostheses

Porcine aortic valves fixed on a metallic or plastic support (stent) coated by a synthetic material that permits suturing to the native ring or biological tissue valves (bovine pericardium usually) form stented biological prostheses. Most are fixed and preserved in glutaraldehyde solution that must be totally removed before implantation, by many prolonged washings in physiologic solution. Incomplete washing may result in dysfunction because of premature tissue hardening.

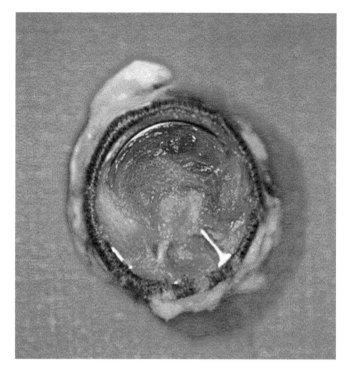

Figure 160   Prosthetic valve associated with thrombosis.

Damage may occur during implantation, and the leaflets may be perforated or torn, as shown by a postsurgical echocardiogram. Such prostheses may last 10 to 15 years without certainty as to limiting factors (i.e., structural failure, fatigue rupture, tearing, calcification). Their major advantage is that patients do not need anticoagulant therapy.

In the aortic area, porcine or pericardial bovine valves (xenografts) or human valves (homografts) can be fixed directly to the aortic ring or aortic wall in a supra-annular/sub-coronary position without any structural support. Implantation of these valves requires a more sophisticated surgical technique, but they guarantee more favorable hemodynamics.

A prosthesis is anchored to the native valvular ring by use of simple stitches, U stitches with or without reinforcement by Teflon material, or a continuous suture.

Holes that develop between the native valve ring and prosthetic support are defined as "paravalvular leaks" and are demonstrated by echocardiogram. Particularly at risk are heavily calcified or fragile rings (as seen in patients with endocarditis and the elderly) or a prosthesis with a diameter smaller than the native annulus.

Aortic Valve Surgery

In Western countries, the most frequent indication is a calcific degeneration of the aortic valve, in contrast to rheumatic disease in other countries. Degenerative aortic pathology occurs between the seventh and eighth decade of life and is often associated with athero-sclerosis of vessels, such as the carotid, renal, and coronary arteries. This association raises risk for surgery. The presence of atheromatous or calcific plaques in the ascending aorta

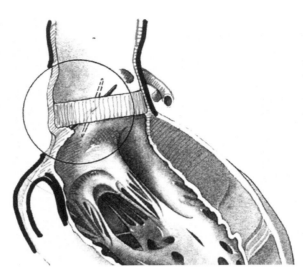

Figure 161   Prosthetic aortic valve larger than the natural annulus, resulting in its dysfunction.

may cause embolism. The replacement of this aortic segment is imperative when an implant or aortotomy is needed. The native valve has to be excised carefully, removing calcification to allow for correct prosthesis implantation. Calcium fragments may fall into the ventricular cavity and embolize into systemic arteries. On the other hand, excessive decalcification may result in extracardiac ruptures or left atrial communications, or in hematoma formation that weakens the ring at the aortic origin with subsequent perioperative rupture. Prevention or repair of these complications can be achieved by replacing the ascending aorta and aortic valve with a prosthesic valve tube graft in addition to reimplanting the coronary arteries (Bentall procedure).

Correct measurement of the valve ring permits selection of a prosthesis with an adequate caliber (Figure 161). Whenever the native ring does not fit perfectly with the prosthetic one, paravalvular leaks are possible. When oversized, the prosthesis may cause coronary ostial occlusion, the function of mobile components may be limited, or a difficult aortotomic suture may result. The maximal valve caliber with respect to the native ring (better transprosthesic gradients) has to be used. When a native ring is fragile or weakened by decalcification, U stitch implantation on pledgetted felt is recommended. Annular suture stitches passing deeply in noncoronary and right coronary positions of the ring can damage the conduction system with permanent atrioventricular block. Similarly, an endocarditis with annular abscess or massive calcification of the right coronary ostium may result in an atrioventricular block with need for pacemaker insertion. Furthermore, deep stitches near the left coronary ostium can cause severe lesions of the left coronary main trunk with lethal complications. In the case of an extremely small aortic annulus, we proceed to an enlargement toward mitral anterior leaflet by a pericardial patch.

Sutures have to be correctly knotted without excessive tension to avoid the tearing of annular tissue with paravalvular detachment and must be cut close to the knot, because if they are too long, they may damage and perforate biological prosthetic cusps or interfere with normal excursion of mechanic elements. The sutured aortotomy must be hemostatic. If aortic tissue is weak, as in the presence of cystic medionecrosis, or a fragile dilated aorta is encountered, reinforcement in the form of a felt strip may prevent postoperative aortic dissection. If the ascending aorta dilatation is greater than 5 cm in diameter, replacement by a straight Dacron prosthesis is mandatory.

David and Feindel (1992) developed a new technique for aortic valve repair in cases with aortic incompetence due to annular swelling and loss of cusp coaptation, where the native valve was repaired by suturing the cusps inside a Dacron tube fixed at the origin of the aorta with coronary arteries reimplantation.

Mitral Valve Surgery

Myxomatous degeneration of the valve is the most common pathology leading to surgery in Western countries, while in others, rheumatic pathology predominates. In ischemic heart disease, mitral incompetence is frequently caused by postinfarct posterior papillary muscle dysfunction with dyskinesis of the left posterolateral ventricular segment plus annular dilatation, the mitral leaflets generally being free from primary pathological alteration.

Rheumatic mitral stenosis or insufficiency or both, associated with calcification, commissural and subvalvular fusion, and retraction, are corrected by biological or mechanic prosthesis replacement. However, degenerative or postischemic mitral incompetence is repaired by annuloplasty or valvuloplasty (Deloche, 1990). All interventions are done using extracorporal circulation and cardioplegic cardiac arrest by left atriotomy in the interatrial groove or transeptal, by right atriotomy at fossa ovalis level, when the left atrium is small. In our institute, we proceed as follows:

1. *Posterior leaflet prolapse with chordal rupture or lengthening*: quadrangular resection of prolapsing portion, direct suturing and posterior reinforce annuloplasty
2. *Anterior leaflet prolapse with chordal rupture or lengthening*: transposition of chordae from posterior to anterior edge; artificial chordae (Goretex) from the free anterior margin to the apex of papillary muscle; shortening chordae with plicature; suture of anterior leaflet free margin to posterior one (edge-to-edge technique) creating a double mitral orifice
3. *Both leaflets prolapse*: using all the techniques described above

After edge rebuilding, in order to stabilize coaptation, an annuloplasty is performed using a prosthesis with a flexible or rigid ring and single U stitches. The ring size is set by measuring the anterior leaflet length. A wide leaflet resection may result in insufficient residual tissue with tissue tension when rebuilding by suture. On the contrary, a limited resection may produce a persistent prolapse and valve incompetence. A prosthetic ring that is too small may cause a relative stenosis, whereas large ones lead to persistent loss of leaflet coaptation. Myxomatous valves with a tissue surplus and small ring correction may cause an anomalous postoperative mobility of the anterior leaflet that in systole fills the left ventricular outflow tract with subaortic valvular stenosis, severe mitral incompetence, and low cardiac output syndrome.

Some studies highlighted the importance of the subvalvular system in maintaining the normal geometry and function of the left ventricle. Therefore, during mitral valve replacement, we try to preserve as much of the subvalvular integrity as possible and the native chordae tendineae, or we replace them with artificial Goretex chordae between papillary muscle apex and annulus.

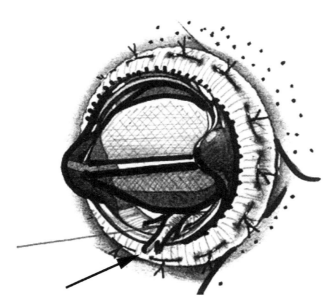

Figure 162 Prosthetic mitral valve. Exceedingly long chordae tendineae or calcification may result in dysfunction of the prosthesis.

Mitral leaflet edges that are seriously degenerated and calcified must be removed and replaced by a prosthesis (Figure 162). Excision of the native valve must be precise in order to avoid damage to the ring or endocardium, especially posteriorly, with resultant subendocardial hematoma leading to heart rupure in the postoperative period, an event that occurs in 3% of cases with valve replacement. The pathogenetic mechanism is a progressive dissection of blood from the left ventricular cavity into the myocardial wall, during systole, leading to heart rupture because of atrioventricular disruption. It is more frequent in cases with heavy calcification of the posterior ring and after biological prosthesis implantation.

When a hematoma develops, intervention must be quick, with immediate removal of the prosthesis and fixation of an atrioventricular pericardial patch, fixed to the endocardium, and infiltration of the area with a biological glue (Resorcinol). The prosthesis will be reimplanted subsequently on a pericardial patch (Repossini, 1993).

The technique used to implant a prosthesis depends on the anatomy and the surgeon's preference: simple U stitches (with or without pledgettes) or a continuous suture. In all cases, sutures have to be passed perpendicular to the ring at a proper depth. If too superficial, they may not fix the prosthesis in position, with resultant paravalvular leaks due to ring laceration; if placed too deep, damage to structures close to the ring may occur with the following possibilities:

1. *Occlusion of left circumflex coronary artery (rear ring)*: ischemia or perioperative infarct
2. *Tethering of aortic valve cusp (anterior commissure)*: postoperative aortic insufficiency
3. *Involvement of atrioventricular node (anterior ring)*: conduction disorders
4. *Extension into the coronary sinus (posteromedial commissure)*: fissure formation

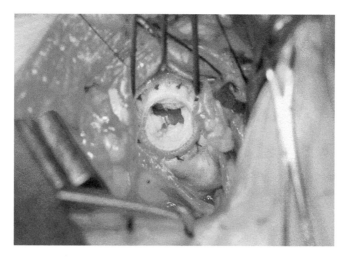

Figure 163 Biological prosthetic mitral valve with degeneration and rupture of leaflets.

An oversized prosthesis may be partially supra-annular with the obstruction of its mobile components or compression of the circumflex coronary artery. Once implanted, a mechanical prosthesis has to be checked by echocardiography to assure that its mobile elements have the proper function (Figure 163).

Drugs of Abuse and Pathology of the Heart

14

EMANUELA TURILLAZZI AND IRENE RIEZZO

Contents

Introduction

Investigating drug-related deaths is an important task of forensic pathologists. Although toxicological analyses are indispensable for the final clarification of the cause of death, postmortem (external and internal) examination remains a fundamental challenge in forensic investigation. Because heart disease is responsible for most abusers' deaths, the study of macro and micro cardiac alterations is an essential step.

First, we want to report a classification of the causes of death in drug abusers according to a recent paper (Webb et al., 2003):

Overdose: death directly related to an episode of toxicity; for example, where cause of death is given as the physiological effects of the substance implicated or its direct consequences, that is, respiratory depression, pneumonia, anoxia, renal failure, and multiple organ failure

Drug-related medical condition: toxicity superimposed on a preexisting medical condition related to substance misuse, that is, hepatitis B, hepatitis C, chronic endocarditis, and cirrhosis of the liver; cases were placed in this category where such a condition was mentioned as a contributing cause of death

Acute infection: death caused by acute infection related to an index episode of drug abuse, that is, septicemia, acute endocarditis, and necrotizing fasciitis

Acute physical event: nontraumatic physical/physiopathological event related to an index episode of intoxication/drug misuse, that is, aspiration of gastric contents, postural asphyxia, and hypothermia; deaths by acute nontraumatic physical event were deemed drug-related where intoxication is likely to have contributed to an inability to protect oneself from harm

Next we will discuss the adverse effects of drugs of abuse.

Cocaine

Cocaine is one of the most commonly used illicit drugs; it is perhaps the most frequent cause of drug-related death (Lange and Hillis, 2001). In the U.S., the number of habitual users of cocaine is estimated to be 1.5 million; it is reported that 25 million people between the ages of 26 and 34 years have used cocaine at least once.

Cocaine is prepared from the leaves of the plant *Erythroxylon coca* and is available as cocaine hydrochloride (a water-soluble powder or granule that can be taken orally, intravenously, or intranasally) and as "freebase" or "crack" cocaine (heat stable, melting at high temperatures, thus allowing it to be smoked) (Karch, 2005).

The pharmacokinetics of cocaine are well known: cocaine is rapidly degraded by plasma and liver cholinesterases to metabolites benzoylecgonine and ecgonine methylestere. Unlike cocaine, benzoylecgonine and ecgonine methylestere are slowly excreted by the body (Karch, 1996).

The local effects of cocaine are well known: cocaine acts as a local anesthetic because of its property of inhibiting membrane permeability to sodium during depolarization and blocking transmission of electrical impulses.

The systemic effects of cocaine are mediated through alterations in synaptic transmission because of its ability to block the presynaptic reuptake of norepinephrine and dopamine and to produce an excess of these neurotransmitters at receptor sites on the postganglionic neuron (Lange et al., 2001). Cocaine acts as a powerful sympathomimetic agent.

The physiologic responses to cocaine may cause severe pathologic effects that were reported in nearly every organ and system including: brain, heart, lung, kidneys (Di Paolo et al., 1997), gastrointestinal tract, musculature, etc.

Among the many complications induced by cocaine use, cardiovascular toxicities are very prominent. Short-term or long-term abuse of cocaine can cause or exacerbate hypertension, atherosclerosis (Noris et al., 2001), coronary artery spasm, myocardial ischemia, cardiac infarction, myocarditis, arrhythmias including ventricular fibrillation (Kloner et al., 1992), cerebral infarction, nontraumatic intracranial hemorrhage, cerebrovasospasm, stroke, and sudden death (Li et al., 2004).

The effects of cocaine on the cardiovascular system were extensively documented both in animal models and in humans. However, the relationship between cardiac morphological alterations and cardiac disorders in cocaine abusers is still controversial. Many reports postulated a relationship between cocaine abuse and cardiac abnormalities, particularly of an ischemic nature. Some of the toxic effects due to cocaine use are due to a blockage of catecholamine reuptake, the accumulation of which leads to numerous alterations. Animal and clinical studies also suggested that cocaine use was associated with increased concentrations of catecholamine in the circulation and that chronic exposure to high levels of adrenaline and noradrenaline can damage the heart. The oxidative metabolism of catecholamines may have damaging effects due to the generation of reactive oxygen species (ROS) and the formation of oxidation products (Fineschi et al., 2001a; Dietrich et al., 2005).

However, the exact mechanisms leading to cardiac injury and irreversible myocardial cellular changes remain elusive. Central and peripheral stimulation of the sympathetic and renin–angiotensin system by cocaine appears to form a major component of the acute toxicity of this alkaloid, which seems to result in the uncontrolled loss of physiologic mechanisms used for normal regulation. Contrary to the general opinion that excess catecholamines produce cardiotoxicity mainly through binding to adrenoceptors, there is increasing evidence that catecholamine-induced deleterious effects may also occur through oxidation of catecholamines, resulting in the formation of highly toxic substances such as aminochromes (e.g., adenochrome) and free radicals, and by virtue of the latter's actions on different types of heart membranes, they cause intracellular Ca^{2+} overload and myocardial cell damage.

Recently, animal studies (Moritz et al., 2003) confirmed the central role of ROS in the development and progression of cardiomyopathy after cocaine abuse, demonstrating that cocaine administration induces early ROS production that precedes the sustained left ventricular (LV) dysfunction seen after repeated administration of the drug, and that cocaine-induced cardiac dysfunction is prevented by antioxidant treatment.

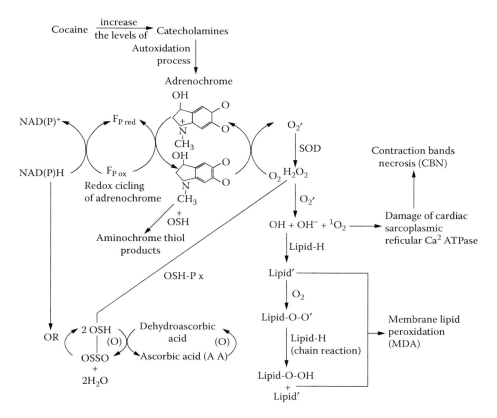

Scheme 1

Cardiovascular Complications

Many cardiovascular complications of cocaine use were reported in the medical literature. They include: endocarditis, myocardial infarction, aortic dissection, left ventricular hypertrophy, arrhythmias, sudden death, and cardiomyopathy (Frishman et al., 2003).

Endocarditis

Endocarditis due to illicit drug use is an ever-increasing problem; a greater incidence of endocarditis in cocaine abusers than in other illicit drugs addicts was reported (Lange and Hillis, 2001). Drug use is a well-known predisposing factor. Endocarditis may present with a multitude of signs and symptoms of varying severity. Patients may complain of only vague symptoms consistent with a viral syndrome, or they may present with a neurologic or cardiovascular catastrophe. When endocarditis occurs, it mainly affects the aortic and mitral valves, as it is the left-sided heart valves that are subjected to the maximum degree of hypertension-related stress and damage. Endocarditis is often associated with unusual organisms such as candida, pseudomonas, or klebsiella, and frequently has an aggressive clinical course with severe valvalar destruction, abscess formation, and a requirement for surgical intervention (Ghuran and Nolan, 2000). Multiple valve involvement was reported to be more common in intravenous drug users than in nonaddicts.

Myocarditis

Clinical evidence suggests that cocaine abuse increases the incidence of myocarditis; both lymphocytic and eosinophilic myocarditis were reported in cocaine abusers (Wang et al., 2002; Jentzen, 1989; Virmani et al., 1988). Animal studies provided direct evidence of the following issues:

1. Exposure to cocaine significantly increased the mortality of viral myocarditis in mice and exacerbated the severity and time course of viral myocarditis
2. Acute or chronic preexposure to cocaine did not affect the mortality or degree of myocardial damage of viral myocarditis
3. Cocaine treatment significantly increased myocardial NE concentration
4. Adrenalectomy abolished the enhancing effects of cocaine on viral myocarditis
5. Treatment with a beta-blocker decreased the effects of cocaine on mice with viral myocarditis (Wang et al., 2002)

Aortic Dissection

Aortic dissection is a rare but recognized complication of crack cocaine inhalation. Single cases or a summary of individual case reports were previously documented (Rashid et al., 1996; Perron and Gibbs, 1997; Famularo et al., 2001; Hsue et al., 2002). Data from the International Registry for Aortic Dissection (IRAD; Eagle et al., 2002) show that only 0.5% of aortic dissections were associated with cocaine use, suggesting that cocaine use would be responsible for no more than 1% of aortic dissections. The location of the dissection appears to be approximately equally divided between types A and B (Famularo et al., 2001). Several pathophysiological mechanisms were proposed to explain the occurrence of aortic dissection in cocaine addicts.

First, the increase in heart rate, blood pressure, and myocardial contractility due to the cocaine-induced sustained adrenergic stimulation is thought to cause a shear effect on the aortic walls (Perron and Gibbs, 1997), especially in the presence of an underlying process that weakens the elastic media of the aorta.

Another hypothesized mechanism relates to the premature atherosclerosis induced by chronic cocaine use. It was postulated that recurrent cocaine exposure makes the endothelium more permeable to atherogenic low-density lipoprotein and may accelerate the migration of leukocytes to the aortic wall (see below).

Finally, animal studies in rats showed that the percentage of apoptotic aortic vascular smooth muscle cells (VSMCs) is increased after cocaine administration, so demonstrating that aortic VSMCs can undergo rapid apoptosis in response to cocaine in a concentration-dependent manner. Cocaine-induced apoptosis may thus play a major role in cocaine-abuse-induced aortic dissection (Su et al., 2004).

In addition to aortic dissection, acute coronary dissection was also reported in the setting of cocaine use (Jaffe et al., 1994; Eskander et al., 2001) as well as intramural aortic hematoma of the ascending aorta (Neri et al., 2001). Furthermore, an association between cocaine use and coronary artery aneurysms (CAAs) was reported. Satran et al. (2005) found a high prevalence of CAAs among cocaine users (30.4%).

Myocardial Infarction

Myocardial infarction (MI) associated with cocaine use was originally reported in 1982 (Coleman et al., 1982). Since then, several reports documented the association between cocaine use and myocardial infarction, with over 200 cases reported (Erwin et al., 2004). Cocaine was implicated as a trigger of acute myocardial infarction in patients free of coronary artery disease and, more frequently, in patients with underlying coronary atherosclerosis (Mittleman et al., 1999).

Cocaine-related myocardial ischemia or infarction reveal themselves in the hour immediately after cocaine use; data from the Determinants of Myocardial Infarction Onset Study showed that users of cocaine sustained a transient 24-fold increase in risk of myocardial infarction in the hour immediately after cocaine use, and that the elevated risk rapidly decreased thereafter (Mittleman et al., 1993).

Three pharmacological mechanisms likely acting in combination are involved in the pathogenesis of cocaine-related myocardial ischemia. First, cocaine, due to its adrenergic stimulation, causes an increase in heart rate, blood pressure, and left ventricular contractility. Second, documented in several animal studies (Hale et al., 1989; Kuhn et al., 1990, 1992; Hayes et al., 1991; Zimring et al., 1994; Shannon et al., 1995; Egashira et al., 1991) is coronary vasoconstriction associated with cocaine use, due to the stimulation of coronary arterial α-adrenergic receptors (Benzaquen et al., 2001). These studies demonstrated that cocaine administered intravenously caused a decrease in coronary artery caliber and decreased coronary blood flow. Evidence also suggests that in diseased coronaries there is increased vasoreactivity to cocaine. A number of human studies were performed on the effects of cocaine on coronary artery caliber (Lange et al., 1989; Flores et al., 1990; Moliterno et al., 1994; Brogan et al., 1992). However, it appears from these animal and human studies that cocaine has a dual effect on the coronary vasculature consisting of an early vasodilation followed by a more sustained vasoconstriction. These effects seem to be dose dependent and probably stem from different pharmacologic properties. The vasodilatation may be mediated by the anesthetic properties of cocaine, whereas the vasoconstriction appears to be mediated by the potentiation of the α-adrenergic response of the coronary arteries (Benzaquen et al., 2001).

Third, it was documented that cocaine causes an increase in platelet aggregability in *in vivo* and *in vitro* testing (Togna et al., 1985; Kugelmass et al., 1995); its use was associated with intravascular thrombosis at a number of different sites: coronary and pulmonary circulation, peripheral venous circulation, and skin and renal vasculature (Stenberg et al., 1989; Delaney and Hoffman, 1991; Zamora-Quezada et al., 1988; Sharff, 1984; Lisse et al., 1989).

Finally, mismatch between myocardial oxygen supply and demand from cocaine-induced vasoconstriction and increased myocardial workload are often invoked as the major postulated mechanism by which cocaine induces myocardial ischemia.

Coronary Atherosclerosis

Evidence of cocaine-induced accelerated atherosclerosis was documented in angiographic (Dressler et al., 1990; Om et al., 1992) and autopsy studies (Kolodgie et al., 1991; Wilson, 1998). In cocaine users, Karch et al. (1995) found larger hearts than in the controls, and more severe coronary atherosclerosis; the degree of myocardial hypertrophy documented in this study was highly significant but because the increase is modest (around 10%), it is likely to go unrecognized at autopsy.

The exact mechanisms underlying this accelerated atherosclerotic process are not completely understood (Erwin et al., 2004; Roberts et al., 1989). Experimental studies suggest that the proatherogenic effects on the vascular endothelium are multifactorial. First, cocaine decreases endothelial nitrous oxide production, thereby enhancing leukocyte migration and adhesion molecule expression. In addition, cocaine compromises the endothelial barrier, allowing permeation of low-density lipoproteins into the subintimal space. Finally, Kolodgie et al. (1991) demonstrated a significantly increased number of adventitial mast cells in stenotic coronaries of deceased subjects with evidence of cocaine intoxication. These results suggested that cocaine may be implicated in the proliferation of coronary mast cells and that mediators released by these cells could result in accelerated atherosclerosis and the promotion of thrombosis.

Dysrhythmias

Cocaine is considered likely to exacerbate cardiac arrhythmias. Lange et al. (2001) reported several cardiac dysrhythmias and conduction disturbances associated with cocaine use. These include: sinus tachycardia, sinus bradycardia, supraventricular tachycardia, bundle-branch block, complete heart block, accelerated idioventricular rhythm, ventricular tachycardia, ventricular fibrillation, asystole, torsade de pointes, and Brugada pattern.

Several different mechanisms are thought to be involved in cocaine-induced cardiac dysrhythmias. First, cocaine is a well-known sympathomimetic drug that may increase ventricular irritability and lower the threshold for ventricular fibrillation. In addition, it has sodium-channel blocking effects that may inhibit the generation and conduction of the action potential, acting similarly to an I class antiarrhythmic agent. Third, cocaine increases the intracellular calcium concentration, which may result in afterdepolarizations and triggered ventricular arrhythmias. Fourth, animal studies advanced the hypothesis that cocaine also exerts a major vagolytic action that contributes importantly to its positive chronotropic effects.

Experimental studies showed that cocaine depresses sinus node automaticity and blocks conduction at the atrioventricular (AV) node, prolonging AH and HV intervals. Localized spasm involving the sinoatrial artery, inferior myocardial infarction, vagal stimulation, overdrive suppression after supraventricular tachycardia, and a direct toxic effect may all contribute in the etiology of cocaine-induced bradyarrhythmias (Castro and Nacht, 2000).

Prolongation of the Q–T interval due to cocaine's use was described in 1997 (Perera et al., 1997); Gamouras et al. (2000) found an increased Q–T interval in patients admitted to the hospital after cocaine use.

Recently, the effects of cocaine on the sodium-potassium channel were stressed: cocaine is a slow on–off sodium blocker and a fast on–off potassium blocker (Baumann and DiDomenico, 2002). Effects on repolarization are biphasic: at low concentrations, cocaine delays ventricular recovery, whereas at higher levels, cocaine hastens it.

Cardiac Findings

The pathological alterations in the myocardium related to cocaine use were extensively documented in both animal models and in humans.

The typical morphologic lesion in cocaine-related death is an expression of catecholamine myotoxicity and is represented by contraction band necrosis (CBN), also referred to as coagulative myocytolysis, which is a necrosis of the myocardial cells in a hypercontracted state (tetanic death) characterized by rhexis of the myofibrillar apparatus, an anomalous hypereosinophilic crossband formed by segments of hypercontracted sarcomeres with extremely thickened Z lines (Figure 164 and Figure 165).

In a previous study on 26 cocaine-associated deaths in chronic cocaine abusers, no histological evidence of ischemic myocardial necrosis was found; the only lesion identified was a microfocal myocardial necrosis, pathognomonic of catecholamine myotoxicity (Fineschi et al., 1997). A following experimental study in a rat model (Fineschi et al., 2001a) demonstrated that after chronic cocaine exposure, rats showed CBN and patchy myocardial fibrosis (Figure 166); platelet aggregation in coronary vessels, infarct necrosis, or other pathological changes were not detected in the acute model.

Furthermore, a reported fatal case of cocaine "body stuffer" syndrome showed typical cardiac alterations, consisting of rhexis of the myofibrillar apparatus and anomalous,

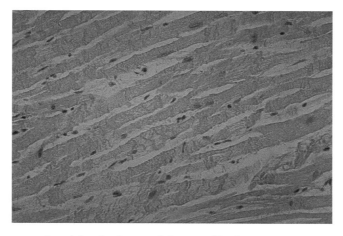

Figure 164 Cocaine-related death: rhexis of the myofibrillar apparatus and anomalous, hypereosinophilic crossbands formed by segments of hypercontracted sarcomeres with extremely thickened Z lines.

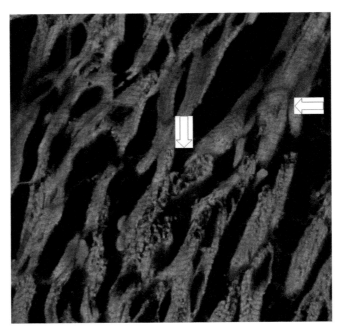

Figure 165   Contraction band necrosis (arrows) with fragmentation of the myofibers: three-dimensional view produced by a high-resolution confocal microscopy system.

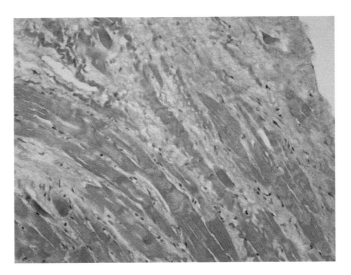

Figure 166   Area of old myocardial fibrosis in a chronic cocaine abuser.

hypereosinophilic crossbands formed by segments of hypercontracted sarcomeres with extremely thickened Z lines (Fineschi et al., 2002).

We found that this form of myocardial cell necrosis, both in experimental and human studies, is secondary to oxidative stress and is not related to ischemia, representing the histological hallmark of an acute adrenergic stress linked with malignant arrhythmia. TUNEL assay was positive in the heart with clear evidence of apoptosis (Figure 167).

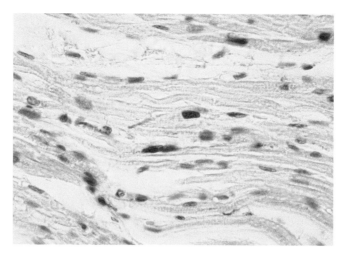

Figure 167 Apoptotic cells detected in histologic sections by specific techniques (TUNEL assay).

MDMA and MDEA

MDMA, or Ecstasy, and 3,4-methylenedioxyethylamphetamine (MDEA, or Eve) have emerged as popular recreational drugs of abuse over the last decade (Koesters et al., 2002). In the U.S., an estimated 6.4 million individuals have used MDMA (Landry, 2002). Ecstasy is the popular or "street" name for a substance identified chemically as N-methyl-3,4-methylenedioxy-amphetamine or 3,4-methylenedioxymethamphetamine. MDEA differs from MDMA only in having a 2-carbon ethyl group, rather than a 1-carbon methyl group, attached to the nitrogen atom of the amphetamine structure.

MDMA and MDEA can exist either as freebases or as salts of various acids; they are, generally, prepared as single-dose tablets for oral assumption. MDMA is absorbed in the intestinal tract. MDMA metabolism is regulated by the levels of CYP2D6 (debrisoquine hydroylase), a member of the cytochrome P450 superfamily of enzymes, and COMT (catechol-O-methyltransferase): both these enzymes exhibit some genetic polymorphism. It is excreted by the kidney.

Pharmacological studies indicate that these substances produce a mixture of central stimulant and psychedelic effects, many of which appear to be mediated by brain monoamines, particularly serotonine and dopamine. MDMA is a potent releaser and reuptake inhibitor of presynaptic serotonin (5-HT), dopamine (DA), and norepinephrine (NE). These actions result from the interaction of MDMA with the membrane transporters involved in neurotransmitter reuptake and vesicular storage systems. The effects of Ecstasy begin approximately 20 min after ingestion and generally last up to 6 h, although larger doses can have an effect for up to 48 h (de la Torre et al., 2004).

Pathological Findings

Acute toxic effects are related to its pharmacologic actions; in a review of adverse reactions to MDMA, Henry et al. (1992) concluded that the toxicity was not attributable to "overdose."

Some individuals can use the drug without immediate harm, while others, at the same dosage, have experienced severe toxicity, including death (Burgess et al., 2000). The range of activity of the enzymes involved in MDMA metabolism may account for some interindividual differences in terms of toxic responses to the drug.

These drugs have serious toxic effects, both acute and chronic, due to the excess of sympathomimetic actions. Recent reports of deaths resulting from MDMA/MDEA abuse led to an increased understanding of the pathology of their misuse. Toxic effects and the occasional deaths following ring-substituted amphetamine misuse were extensively reported (Schifano et al., 2000, 2003a,b; Schifano, 2004). Recent data show that the risk of using Ecstasy varies between one death in 2000 first-time users to one death in 50,000 first-time users (Gore, 1999).

The clinical manifestations of MDMA intoxication are related to the so-called "sympathomimetic toxidrome" (Goldfrank et al., 1998), due to the stimulation of the central and peripheral nervous systems with tachydysrhythmias, hypertension, hyperthermia, diaphoresis, delirium, agitation, and acute psychosis, even leading to life-threatening hyperthermia, rhabdomyolysis, acute renal failure, DIC, and death. The serotonin syndrome consists of altered mental status, altered muscle tone or activity, autonomic instability, hyperthermia, and diarrhea (Gill et al., 2002).

A recent review of the literature revealed over 87 Ecstasy-related fatalities caused by hyperpyrexia, rhabdomyolysis, intravascular coagulopathy, hepatic necrosis, cardiac arrhythmias, cerebrovascular accidents, and drug-related accidents or suicide (Kalant, 2001). After the onset of deaths related to MDMA ingestion, reports were compiled about fatal arrhythmias or cases of hyperthermia followed by DIC. Clinical proof of hyperthermia, rhabdomyolysis, and DIC are also evident in deaths caused by MDEA intoxications. Similar pathological findings are described in the cases of death due to combined intoxication of MDMA and MDEA (Milroy et al., 1996; Byard et al., 1998; Dowling et al., 1987; Suarez and Reimersma, 1988; Henry et al., 1992; Rella and Murano, 2004; Zorec et al., 2005).

MDMA and other ring amphetamine substitutes lead to an increase of several neurotransmitters (serotonin, noradrenaline, and dopamine). It is well known that the major toxic effects on the cardiovascular system are noradrenaline mediated. These effects result in hypertension and hypertension-related diseases and tachycardia with a consequently increased cardiac workload and a resulting risk of heart failure (Kalant, 2001).

Acute myocardial infarction associated with MDMA abuse is described in the literature (Qasim et al., 2001; Lai et al., 2003). Potential explanations include coronary vasospasm, excessive catecholamine discharge resulting in ischemic myocardial necrosis, and catecholamine-mediated platelet aggregation with subsequent thrombus formation. The syndrome closely resembles acute myocardial infarction due to cocaine abuse (Fineschi et al., 1999a). Aortic dissection was reported by Duflou and Mark (2000), who described a case of aortic dissection and cardiac tamponade after ingestion of Ecstasy at a "rave" party in a man with no history of hypertension or other risk factors for aortic dissection. The likely mechanism is paroxysmal increases in blood pressure that can lead to aortic dissection.

Macroscopic myocardial alterations found in MDMA-related deaths are mainly represented by alterations induced by disseminated intravascular coagulation (Figure 168).

In a previous report, we refer to three cases of death by myocardial coagulative necrosis following ingestion of MDMA/MDEA (Fineschi et al., 1999a). We observed purifocal foci of myocells with hypercontraction of the whole myocell and myofibrillar rhexis with an

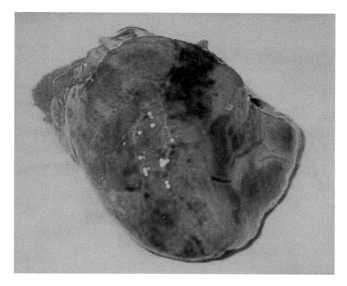

Figure 168   Subepicardial petechiae in MDMA-related death.

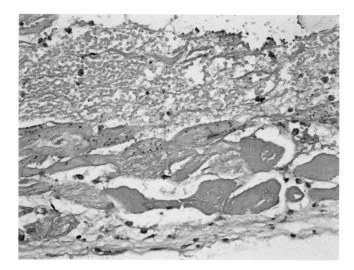

Figure 169   Subendocardial hemorrhages.

anomalous deep eosinophilic crossband formed by hypercontracted sarcomeres. More advanced stages of coagulative myocytolysis (alveolar or healing pattern) and old myocardial fibrosis were absent or minimal. Subendocardial hemorrhages are described (Figure 169). No histological signs of infarct necrosis were found. Coagulative myocytolysis, even if confined to few myocells, can be interpreted as a histological sign of adrenergic overdrive.

Narcotic Agents

The most commonly misused narcotic analgesics are heroin and morphine. Heroin (diacetylmorphine) is a semisynthetic analogue of morphine that is slowly metabolized to the parent compound. Heroin is more lipid soluble than morphine and therefore acts more rapidly. Morphine has a plasma half-life of 2 to 3 h. It undergoes rapid hepatic metabolism, and the metabolites are excreted in the urine. The duration of renal excretion is highly variable and is affected by the dose, the precise chemical composition of street preparations, the user's previous drug habits, and individual variations in renal and hepatic function. In general, metabolites can be detected for up to 48 h in occasional users and for several days in chronic misusers.

Narcotic analgesics are commonly injected, smoked, or ingested orally. In the last few years, routes of drug administration have been changing. Intranasal administration (snorting) and pulmonary inhalation (smoking) of heroin are now more common. An increasing proportion of mortalities resulting from an overdose of heroin that involve routes of administration other than injection were reported (Thiblin et al., 2004).

The major pharmacological effects of heroin can be traced to some structural properties of the morphine molecule. The analgesic effects of heroin derive from the two active metabolites, 6-O-acetylmorphine and morphine, which bind specifically to the μ-opioid receptors of the central nervous system (CNS). The mu-receptors also mediate other pharmacological actions of heroin, that is, respiratory depression, euphoria, and physical dependence. Heroin is more potent and faster acting than morphine as an analgesic drug. Chronic administration of heroin results in the development of tolerance. It is characterized by a shortened duration and decreased intensity of the analgesic, euphoric, sedative, and other CNS-depressant effects. Tolerance to opioids is due to increased adaptation of the cells, which changes their receptor sites after chronic exposure to the drug.

Pathological Findings

Respiratory arrest is the most common cause of death among heroin addicts. Narcotic analgesics act directly on the vasomotor center to increase parasympathetic activity, reduce sympathetic activity, and release histamine from mast cells. These effects combine to produce bradycardia and hypotension. Drug-induced bradycardia along with enhanced automaticity can precipitate an increase in ectopic activity, atrial fibrillation, idioventricular rhythm, or potentially lethal ventricular tachyarrhythmias.

Moreover, several implications for research arise from the literature on deaths attributed to heroin overdose. In a certain proportion of cases, blood morphine levels cannot account for the fatal outcome of a heroin overdose. The wide range of concentrations found in postmortem blood samples confirms that the term "overdose" is relative and does not sufficiently characterize death associated with heroin addiction. In the cases presenting with low drug concentrations, the mechanism of death is not quite clear. Some of these deaths are undoubtedly due to a loss of tolerance. Anaphylactic or allergic reaction may be responsible in other cases.

Recent studies suggest that an alternative explanation for these acute heroin-associated deaths may be that they are the result of a rapid drop in blood pressure and respiratory and cardiac failure due to acute, nonspecific mast-cell degranulation and the release of histamine, tryptase, and other mediators caused by heroin and morphine.

Previous studies found statistically significant levels of tryptase in the drug addicts, indicating that many of these deaths were probably preceded by mast-cell degranulation. Opioids such as morphine are potent histamine- and tryptase-releasing agents, and anaphylactic reactions to these drugs were attributed to this property. A study showing that commonly abused drugs (morphine, cocaine) release histamine from mast cells when in the presence of oxidative enzymes (Di Bello et al., 1998) suggests that the massive release of mast-cell histamine is an additional risk factor in heroin and cocaine overdose. Our findings (Fineschi et al., 2001c) in 48 heroin-related deaths suggested that some cases of heroin-related death might be caused by anaphylactoid shock.

The risk of contracting bacterial or viral infection in heroin addicts is well known (Spijkerman et al., 1996). A recent review (Passarino et al., 2005) of histopathological findings in 851 autopsies of drug addicts (90% heroin related) showed a high incidence of inflammatory heart diseases (3.7%), mainly myocarditis with predominant lymphocytic infiltration. Endocarditis was also reported in high incidence in the population of active intravenous drug users (Jones et al., 2002).

The development of arterial mycotic pseudoaneurysm (AMP) as a sequela of injection drug use is much less frequently reported (McLean et al., 1998; Tsao et al., 2002).

Androgenic Anabolic Steroid

Anabolic androgenic steroids (AASs) are synthetic derivatives of testosterone. These substances are used worldwide to help athletes gain muscle mass and strength, although the prohormones of testosterone and nandrolone are on the list of forbidden substances of the International Olympic Committee (IOC) and of the International Sports Federations (IOC 2001) (Hartgens and Kuipers, 2004).

AASs show a wide range of clinical effects. The mechanism of anabolic action is represented in Figure 170 (Peters et al., 2002). Several diseases, including cardiac and liver diseases and tumors as well as other disorders occur in athletes who self-administer high doses of androgens.

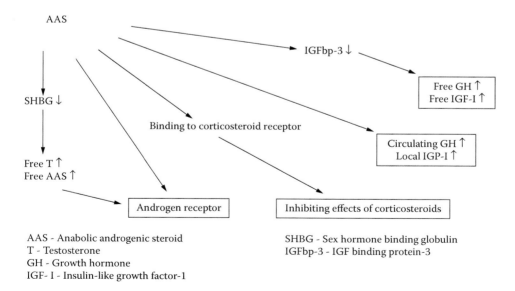

Figure 170 Suggested mechanisms of anabolic action of supraphysiological doses of anabolic androgenic steroids. (From Peters et al., 2000.)

Cardiovascular Complications

The heart is one of the organs most frequently affected by administration of anabolic steroids, as shown in Figure 171. The different targets for anabolic steroid effects in the heart can be summarized as shown in Figure 172 and 173.

Melchert and Welder (1995) suggested that there are at least four hypothetical models of AAS-induced adverse cardiovascular effects: an atherogenic model involving the effects of AAS on lipoprotein concentrations; a thrombosis model involving the effects of AAS on clotting factors and platelets; a vasospasm model involving the effects of AAS on the vascular nitric oxide system; and a direct myocardial injury model involving the effects of AAS on individual myocardial cells (Figure 174).

Evidence shows a proatherogenic effect. Anabolic steroids were shown to elevate serum levels of low-density lipoprotein (LDL), cholesterol, and triglycerides by 40 to 50% and to reduce high-density lipoprotein (HDL) by 50 to 60% (Friedl et al., 1993). It is well known that reduced HDL levels as well as other lipoprotein disorders are associated with athero-sclerosis. Moreover, it was suggested that the increase in total cholesterol by the use of

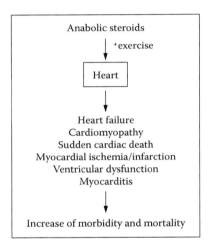

Figure 171   Effects of anabolic steroids on the heart.

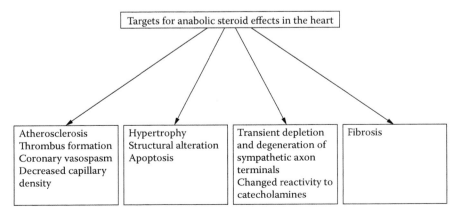

Figure 172   Different targets in the heart can be affected by anabolic steroids.

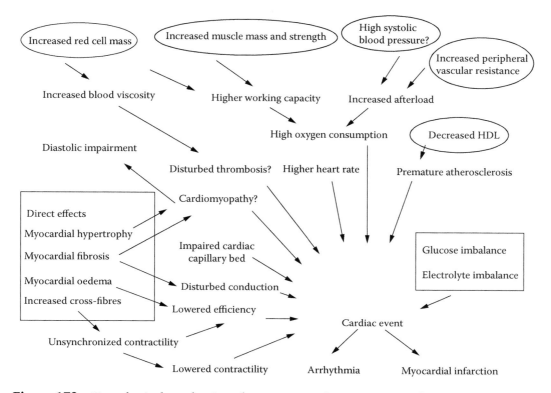

Figure 173   Hypothetical mechanism that may contribute to AAS cardiotoxicity.

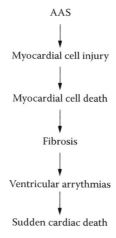

Figure 174   Hypothetical model of direct myocardial cell injury induced by androgenic anabolic steroids (AAS).

anabolics enhanced the coronary artery response to catecholamines (Kennedy and Lawrence, 1993; Sader et al., 2001).

Nieminem et al. (1996) demonstrated a possible role of AAS in increasing procoagulant factors, platelet aggregation, endothelium release, and protein C and S, and also in decreasing fibrinolytic activity and synthesis of prostacyclin.

Another hypothesized mechanism that may contribute to AAS cardiotoxicity is coronary artery spasm due to androgens' action by receptor-dependent and independent mechanisms in the cardiovascular system (Appleby et al., 1994).

Finally, a direct myocardial injury is supposed on the basis of the demonstration that anabolic steroids are associated with marked hypertrophy in myocardial cells, extensive regional fibrosis and necrosis, increased fibrous, as collagen and decreased elastic proteins in the coronary arterial wall (Behrendt and Boffin, 1977; Luke et al., 1990; Hausmann et al., 1998). Cellular modifications include changes in the contractile apparatus (i.e., the sarcomere), as shown by disintegration and Z-band distortions or dissolution, and disturbances in the energy unit of the cell (i.e., the mitochondria), as shown by swelling damage and lipid storage (Behrendt and Boffin, 1977).

Among the numerous documented toxic and hormonal effects of AAS, attention has been focused especially on the cardiovascular effects during recent years (Madea and Grellner, 1998). The intake of these steroids has been associated with vascular complications, cardiomyopathy, coronary atherosclerosis, and cardiac hypertrophy; in several clinical reports, AAS intake was associated with an acute myocardial infarct. Increases in blood pressure and peripheral arterial resistance are known from experimental studies, but there are also effects on the heart muscle, primarily left ventricular hypertrophy with restricted diastolic function. Severe cardiac complications, such as cardiac insufficiency, ventricular fibrillation, ventricular thromboses, myocardial infarction, or sudden cardiac death, in individual strength athletes with acute AAS abuse were also reported (McNutt et al., 1988; Luke et al., 1990; Faenchick and Adelman, 1992; Kennedy and Lawrence, 1993; Huie, 1994; Hausmann et al., 1998; Sullivan et al., 1999; Fineschi et al., 2001b; Urhausen et al., 2004).

Cardiac Findings

The cardiac pathological findings in AAS death were previously analyzed. We reported four cases of sudden death in healthy bodybuilders who were using androgenic anabolic steroids (Fineschi et al., 2001b, 2005). In one case, we were able to demonstrate the presence of typical infarct necrosis (Figure 175 and Figure 176). The lesion was characterized by

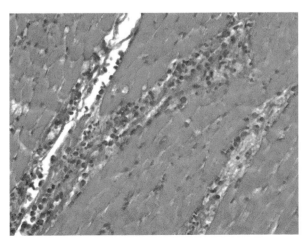

Figure 175   Infarct necrosis with marked polymorphonuclear leukocytic infiltration.

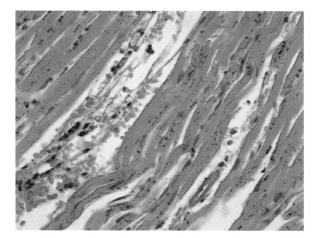

Figure 176   Central area of the large infarct where mildly eosinophilic, stretched, dead myofibers persist.

dead, hyperdistended myocardial cells with sarcomeres in registered order, circumscribed at its periphery by young collagen tissue containing numerous macrophages. External to the infarct necrosis was a large layer of contraction band necrosis formed by hypercontracted, deeply eosinophilic myocardial cells with rupture of the myofibrillar apparatus in anomalous bands, with no macrophagic reaction. Occasional foci of contraction band necrosis and few fibrotic microfoci in the internal portion of the posterior left ventricle and interventricular septum were also seen.

The other cases showed only occasional, isolated myocardial cells with contraction bands and segmentation of the myocardial cells, spotty areas of fibrosis (Figure 177).

In our cases, and in most of those reported in the medical literature, lesions at any level of the coronary system were absent, even in the presence of a myocardial infarction.

The presence of extensive early contraction band necrosis strongly supports the hypothesis that sudden death should be related to adrenergic stress (Fineschi et al., 2005).

Figure 177   Spotty areas of fibrosis.

Cannabis and Cannabinoids

Different preparations of the plant *Cannabis sativa* are widely used around the world for their euphoric effects, making cannabis one of the most frequent drugs of abuse (Gledhill-Hoyt et al., 2000; Miller and Plant, 2002). The most commonly used preparations are marijuana and hashish. Cannabinoids also include synthetic agents and endogenous substances termed endocannabinoids, which include anandamide (2-arachidonoylethanolamide) and 2-arachidonoylglycerol.

Delta(9)-tetrahydrocannabinol (THC) is the main source of the pharmacological effects caused by the consumption of cannabis, both the marijuana-like action and the medicinal benefits of the plant. Natural cannabis products and single cannabinoids are usually inhaled or taken orally; the rectal route, sublingual administration, transdermal delivery, eye drops, and aerosols have been used in only a few studies and are of little relevance in practice today. The pharmacokinetics of THC vary as a function of its route of administration. Pulmonary assimilation of inhaled THC causes a maximum plasma concentration within minutes, psychotropic effects start within seconds to a few minutes, reach a maximum after 15 to 30 minutes, and taper off within 2 to 3 hours. Following oral ingestion, psychotropic effects set in with a delay of 30 to 90 minutes, reach their maximum after 2 to 3 hours, and last for about 4 to 12 hours, depending on the dose and specific effect (Grotenhermen, 2003).

The biological effects of marijuana and its main psychoactive ingredient are mediated by specific G protein-coupled cannabinoid (CB) receptors. To date, two different receptors have been identified: the CB1 receptor, which is highly expressed in the brain but is also present in heart and vascular tissues, and the CB2 receptor, which is expressed primarily by haematopoietic and immune cells (Pacher et al., 2005). Cannabinoids effects on the cardiovascular system have been widely interpreted as being mediated by CB1 receptors, although there are a growing number of observations, particularly in isolated heart and blood vessel preparations, that suggest that other cannabinoids receptors may exist (Hiley and Ford, 2004).

The increase in use of cannabis and cannabinoids is of concern because of their adverse effects on health (Kalant, 2004). Cannabis may cause mental illnesses, most notably schizophrenia and depression (Rey and Tennant, 2002), but it is also worth examining its potential to cause other illnesses, especially those of the heart and respiratory system (Henry et al., 2003). Neurologic dysfunction occurring in relation to cannabis use has been described and corresponding cerebral imaging studies have documented focal ischemic changes and vessel abnormalities (Moussouttas, 2004).

Cannabinoids also elicit potent cardiovascular effects that are well known and have recently been highlighted (Hillard, 2000; Kunos et al., 2000; Jones, 2002; Ralevic et al., 2002; Randall et al., 2002, 2004; Frishman et al., 2003; Fisher et al., 2005).

The acute physiological effects of marijuana include a substantial dose-dependent increase in heart rate generally associated with a mild increase in blood pressure (Jones, 2002; Sidney, 2002); at low or moderate doses, the drug leads to an increase in sympathetic activity and a reduction in parasympathetic activity, producing tachycardia and increase in cardiac output. At high doses, sympathetic activity is inhibited and parasympathetic activity increased, leading to bradycardia and hypotension (Ghuran and Nolan, 2000). Orthostatic hypotension may occur acutely as a result of decreased vascular resistance (Sidney, 2002). An increase in myocardial oxygen demand with a decrease in oxygen supply has been described, which is due in part to an increase in carboxyhemoglobin; this results in a lower anginal threshold in patients with chronic stable angina (Aronow and Cassidy, 1974, 1975). Reversible ECG abnormalities affecting the P and T waves and the ST segment have been reported, although it is not clear if these changes are related to drug ingestion independently of effects on heart rate (Kochar and Hosko, 1973). Cases of serious cardio-vascular events precipitated by repeated heavy cannabis use were described (Lindsay et al., 2005). Smoking marijuana has been reported as a trigger of acutemyocardial infarction in otherwise low-risk individuals (Charles et al., 1979; Collins et al., 1985; Pearl and Choi, 1992). Mittleman et al. (2001) studied 3882 patients with recent myocardial infarctions, 124 of whom were marijuana users. Compared to the patients who were not marijuana users, the users were more likely to be males, cigarette smokers, and overweight. The risk of myocardial infarction during the first hour after smoking marijuana was calculated to be increased 4.8-fold over the baseline in non-users. Myocardial infarction and cardiac arrhythmias have been reported due to the combined action of cannabis and other drugs (Wilens et al., 1997; McLeod et al., 2002).

Non-fatal cases of paroxysmal atrial fibrillation after smoking cannabis have been described (Kosior et al., 2000, 2001). Moreover, Gupta et al. (2001) reported the death of a 25-year-old man with a known history of rheumatic heart disease, who died within 24 hours of drinking a small dose of bhang, a traditional Indian form of cannabis.

Sudden deaths have been attributed to smoking cannabis by Bachs and Morland (2001), who reported six cases of possible acute cardiovascular death in young adults, where very recent cannabis ingestion was documented by the presence of tetrahydrocan-nabinol (THC) in postmortem blood samples. At postmortem examination, cardiac find-ings consisted of widespread atheromatosis in the coronary arteries and aorta, hypertrophied heart, and signs compatible with an older and a recent myocardial infarction (case 1); atheromatosis in the coronary arteries and insinuated narrowing in the left coronary artery next to the aorta (case 2); widespread atheromatosis, enlarged heart, and signs of an earlier infarction (case 3); widespread atheromatosis, narrowing of coronary arteries, and signs of pulmonary emphysema (case 4); slightly enlarged heart without signs of atheromatosis (case 5); slight atheromatosis in the coronary blood vessels and narrowing of a coronary artery (case 6).

Recently, a case of unexpected death of a cannabis bodypacker was described (Barnett and Codd, 2002). A 40-year-old man was found dead in his apartment. At postmortem examination, 55 packages of cannabis resin wrapped in cellophane were found in the large intestine.

Appendix

The forensic autopsy aims are characterized by some fundamental requirements:

1. Using uniform forms and clear orientation in the execution
2. Applying particular techniques to conditions and situations to single out and localize reports or to define pathological tissues or samples (for instance, a different dissection for heart and great vessels opening when an embolism is suspected)
3. Using national and international uniform guidelines to record findings
4. Using fixed and standard samples to perform histological, biochemical, toxicological, ballistic exams, and so forth
5. Keeping parts of samples and reports in order to guarantee that the investigations will be repeatable

These and other recommendations are useful to collect and file necessary preliminary evidence for forensic judgment. It presumes an absolutely clear objectivity (supported by photographic, microfilm documents) and a trusted inclination to check.

On March 23, 1999, The European Council finally promulgated an additional protocol that contains a recommendation (R-99-3) about harmonization of forensic autopsy rules. It assert that, "it is an essential initiative" not only "in order to [meet] epidemiologic and statistic aims" but also to list data and make them available so that others have knowledge of the cause of death and identification, making it easier to avoid conducting another autopsy on a subject who died in a foreign state (there is also a project underway to establish a unique scheme for the autopsy protocol). In addition to the general rules, this recommendation underlines particular ways of checking and drawing conclusions concerning the cardiovascular apparatus and, specifically, sudden death.

In Italy, there is a national protocol for the sectioning of the heart (Melfi–Foggia Consensus Conference, 2005).

Sectioning of the Heart

Before removing the heart, examine it, noting its volume; the state of filling of single cavities (atrium and ventricles); the general and particular shape of the apex and the roles the two ventricles play; the course of the coronary vessels and their state; and both the color and consistency (rigor mortis) of the organ.

In the case of death suspected from gaseous embolism, before touching the heart, incise the right ventricle by a shortcut made under water, observing if gas bubbles emerge.

In ordinary cases, while the heart maintains its normal function, open every cavity and determine the quantity of blood and its state of coagulation and its general features in each. Furthermore, explore the dimension of the auricular-ventricular orifices, found by delicately pushing with two fingers from the side of the chest.

If you are examining a sudden death case, always open the pulmonary artery (*in situ*) to look for the possible presence of embolism. After you excised the inferior vena cava emerging from the diaphragm and the aorta and pulmonary arteries (as high as possible), extract the heart, and record the quantity and the appearance of the subepicardial fat and the epicardium's features; then weigh the organ after you empty the cavities. Later, examine the efficiency of the aortic valve, putting some water into it, taking care not to let any coagulum pass through the left ventricle into the aorta, least it prevent the semilunar valves from flattening under the water pressure: this could mimic an insufficiency of the valve. Then keep cutting the origin of the big arteries in order to explore the intima and the valves. At this point, finish cutting the atrium and ventriculas, so you can explore the atrioventricular valves, their papillary muscles, and the state of the ventricular endocardium. The following step is to describe the myocardium in regard to its thickness (do not include the trabecular stratum in your measurement), the color, and their other features. In any case, keep fragments of the organ in liquids such as formalin and Muller liquid to verify the supposed pathologies. If you suspect a drowning, before removing the heart, wash the vessels, and separately collect blood from the left side of the heart and from the right side (ventricle) for the cryoscopic exam and for other research required to diagnose this cause of death. In cases of unexpected death, remove the heart without previously opening the cavities *in situ*, in order to better evaluate the quantity of blood in each cavity. In this case, in sudden death, keep the heart completely in 10% formalin solution for possible research in addition to the state of the myocardium (Figure 178, Figure 179, and Figure 180).

HEART MORPHOLOGY STUDY

Ord. No. _____ Source _____ Autopsy No. _____

Death–autopsy interval (hours) _____

Last name _____ First name _____

Sex   1. Male   2. Female

Age (yr) _____

Body weight (kg) _____

Height (cm) _____

Interval first   minutes _____   hours _____
episode–death

Ischemic heart disease

           1. No   2. Angina   3. Infarct   4. Unknown

Cardiac failure

           1. No   2. Yes   3. Unknown

Cardiac arrest

           1. Ventricular fibrillation   2. Asystole   3. Failure

           4. Electromechanical dissociation   5. Unknown

Other _____ Data _____

_____

_____

_____

HEART GROSS

Weight (gm) _____ Body weight (kg) _____ % _____

Diameter (mm)   Longitudinal _____

                    Transverse _____

                    Anteroposterior _____

Wall thickness (mm)   ANT/SUP_____

                    POST/SUP _____

                    LV_____

                    RV _____

                    SPT _____

Other data _____

_____

_____

_____

HEART HISTOLOGY							
CORONARY ARTERIES	LM	LAD	LCX sup	RCA ant	RCA marg	RCA post	RCA
Stenosis (%)	___	___	___	___	___	___	___
Stenosis type	___	___	___	___	___	___	___
1. Nodular **2.** Semilunar **3.** Concentric							
Plaque	___	___	___	___	___	___	___
1. Fibrous **2.** +s.m.c **4.** Basophilia **8.** Atheroma		**16.** Calcify **32.** Hemorrhage **64.** Lymph-plasm					
Thrombus mural	___	___	___	___	___	___	___
1. No 2. Yes							
Thrombus occlusive	___	___	___	___	___	___	___
1. No 2. Acute	3. Recent 4. Organized						

Other findings _____

MYOCARDIUM	LV ant	LV post	RV ant	RV post	SPT
Area mm²	___	___	___	___	___
Infarct necrosis %	___	___	___	___	___
Histologic pattern	___	___	___	___	___
1. Eosinophils + PMN 2. PMN exudate 3. Macrophages		4. Early fibrosis 5. Fibrous + necrotic tissue			
Wall location	___	___	___	___	___
1. Subendocardial 2. Internal 4. Subepicardial					
Base-apex location	___	___	___	___	___
1. Superior 2. Middle		4. Inferior 8. Apex			

Coagulative myocytolysis					
No. foci	————	————	————	————	————
No. of myocells	————	————	————	————	————
Wall location	————	————	————	————	————
1. Subend					
2. Intern.					
4. Subep					

Type	————	————	————	————	————
1. Monofocal 3. Confluent					
2. Plurifocal 4. Massive					

Base-apex location	————	————	————	————	————
1. Superior 4. Inferior					
2. Middle 8. Apex					

Histological pattern	————	————	————	————	————
1. Hypercontraction + rhexis 8. Alveolar					
2. Holocytic 16. Organizing					
4. Paradiscal					

Associated monocytes	————	————	————	————	————
1. No					
2. Micro					
3. Extensive					

Myofiber breakup/VF	————	————	————	————	————
1. No					
2. Yes					

	LV ant	LVP ant	LV post	RV ant	RV post	SPT
Colliquative myocytolysis						
Grade 0, 1, 2, 3	____	____	____	____	____	____
Base-apex location						
1. Superior 4. Inferior						
2. Middle 8. Apex						
Wall location	____	____	____	____	____	____
1. Subend						
2. Intern						
4. Subep						
Histologic pattern	____	____	____	____	____	____
1. Lysis						
2. Vacuolar						
4. Alveolar						
Fibrosis %	____	____	____	____	____	____
Age	____	____	____	____	____	____
1. Old						
2. Recent						
Type	____	____	____	____	____	____
1. Monofocal 8. Perivascular/interfascicular						
2. Plurifocal 16. Intermyocellular						
4. Confluent						
Wall location	____	____	____	____	____	____
1. Subendocardial 4. Subepicardial						
2. Internal						
Base-apex location	____	____	____	____	____	____
1. Superior 4. Inferior						
2. Middle						

Fibrosis endocardial						
1. No	8. +monocytes					
2. Yes	16. s.m.c. sine end; fibrosis					
4. +s.m.c.						

Type						
1. Focal light	3. Diffuse light					
2. Focal severe	4. Diffuse severe					

Fibrosis epicardial						
1. No	4. + Monocytes					
2. Yes	8. + Fibrin					

Type						
1. Focal light	3. Diffuse light					
2. Focal severe	4. Diffuse severe					

Constrictive Pericarditis						
1. No						
2. Yes						

	LV ant	LV post	RV ant	RV post	SPT
Lymphocyte Infiltrates					
Number of foci	___	___	___	___	___
Perivascular	___	___	___	___	___
Intramyocardial	___	___	___	___	___
Intramyocardial + myocardial necrosis	___	___	___	___	___
Other Infiltrates	___	___	___	___	___
1. No 4. Eosinophils					
2. PMN 5. Eosinophils + necrosis					
3. PMN + necrosis					
Extension	___	___	___	___	___
1. Focal light 3. Diffuse mild					
2. Focal severe 4. Diffuse severe					
Hypertrophy	___	___	___	___	___
1. No					
2. Yes					
Disarray	___	___	___	___	___
1. No					
2. Focal					
3. Diffuse					

Other findings _____

_____

_____

_____

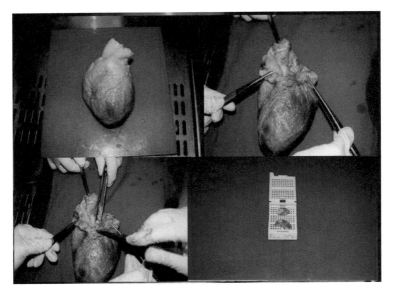

Figure 178   Section of the heart and collection of coronary arteries.

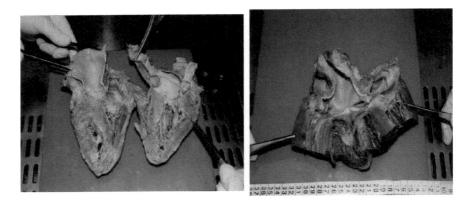

Figure 179   Different methods in the sectioning of the heart.

Figure 180   Macroscopic examination of the ventricles.

References

Abad C, Coronary artery bypass surgery for patients with left main coronary lesions due to Takayasu's arteritis, *Eur J Cardiothorac Surg* 9, 661, 1995.

Abbott GW, Sesti F, Splawski I, et al., *MiRP1* forms IKr potassium channels with *HERG* and is associated with cardiac arrhythmia, *Cell* 97, 175, 1999.

Abbott JA, Scheinman MM, Morady F, et al., Coexistent Mahaim and Kent accessory connections: diagnostic and therapeutic implications, *J Am Coll Cardiol* 10, 364, 1987.

Ablad B, Beta adrenergic mechanisms and atherogenesis, in *Hypertension — The Tip of an Iceberg*, Adis Press, Manchester, 1988, p. 23.

Abrunzo TJ, Commotio cordis. The single, most common cause of traumatic death in youth baseball, *Am J Dis Child*, 145, 1279, 1991.

Ackerman MJ, Cardiac causes of sudden unexpected death in children and their relationship to seizures and syncope: genetic testing for cardiac electropathies, *Semin Pediatr Neurol* 12, 52, 2005.

Ackerman MJ, Siu BL, Sturner WQ, et al., Postmortem molecular analysis of *SCN5A* defects in sudden infant death syndrome, *JAMA* 286, 2264, 2001.

Ackerman MJ, Tester DJ, and Driscoll DJ, Molecular autopsy of sudden unexplained death in the young, *Am J Forensic Med Pathol* 22, 105, 2001.

Adams JE, Davila–Roman VG, Bessey PQ, et al., Improved detection of cardiac contusion with cardiac troponin I, *Am Heart J*, 131, 308, 1996.

Adgey AA, Johnston PW, and McMechan S, Sudden cardiac death and substance abuse, *Resuscitation* 29, 219, 1995.

Agarwal KC, Khan MAA, Falla A, et al., Cardiac perforation from central venous catheters: survival after cardiac tamponade in an infant, *Pediatrics*, 73, 333, 1984.

Ahdoot J, Galindo A, Alejos JC, et al., Use of OKT3 for acute myocarditis in infants and children, *J Heart Lung Transplant* 19, 1118, 2000.

Ahmad F, Li D, Karibe A, et al., Localization of a gene responsible for arrhythmogenic right ventricular dysplasia to chromosome 3p23, *Circulation* 98, 2791, 1998.

Ahmed M, Age and sex differences in the structure of the tunica media of the coronary arteries in Chinese subjects, *Acta Anat* 73, 431, 1969.

Aho A, On the venous network of the human heart and its arterio-venous anastomoses, *Ann Med Exp Fenn* 28 (Suppl 1), 1, 1950.

Ahrens PJ, Sheehan FH, Dahl J, et al., Extension of hypokinesia into angiographically perfused myocardium in patients with acute infarction, *J Am Coll Cardiol* 22, 1010, 1993.

Ahronheim JH, Isolated coronary periarteritis: report of a case of unexpected death in a young pregnant woman, *Am J Cardiol* 40, 287, 1977.

Ai T, Fujiwara Y, Tsuji K, et al., Novel KCNJ2 mutation in familial periodic paralysis with ventricular dysrhythmia, *Circulation* 105, 2592, 2002.

Aintablian A, Hamby RI, Hoffman I, et al., Coronary ectasia: incidence and results of coronary bypass surgery, *Am Heart J* 96, 309, 1978.

Airaksinen KEJ, Tahvanain KUD, Ecberg DE, et al., Arterial baroreflex impairment in patients during acute coronary occlusion, *J Am Coll Cardiol* 32, 1641, 1998.

Akhtar S, Meek KM, and James V, Ultrastructure abnormalities in proteoglycans, collagen, fibrils and elastic fibers in normal and myxomatous mitral valve chordae tendineae, *Cardiovasc Pathol* 8, 191, 1999.

Aldridge HE and Trimble AS, Progression of proximal coronary artery lesions to total occlusion after aorta-coronary saphenous vein by pass grafting, *J Thorac Cardiovasc Surg* 62, 7, 1971.

Algra A, Tijssen JGP, Roelandt JRTC, et al., Heart rate variability from 24-hour electrocardiography and the 2-year risk for sudden death, *Circulation* 88, 180, 1993.

Ali RH, Zareba W, Moss AJ, et al., Clinical and genetic variables associated with acute arousal and nonarousal-related cardiac events among subjects with long QT syndrome, *Am J Cardiol* 85, 457, 2000.

Alings M and Wilde A, "Brugada" syndrome: clinical data and suggested pathophysiological mechanism, *Circulation* 99, 666, 1999.

Al-Khatib SM, LaPointe NM, Kramer JM, et al., What clinicians should know about the QT interval, *JAMA* 289, 2120, 2003.

Allen JM, Thompson GR, Myant NB, et al., Cardiovascular complications of homozygous familial hypercholesterolaemia, *Br Heart J* 44, 361, 1980.

Alpert JS, Thygesen K, Antman E, et al., Myocardial infarction redefined — a consensus document of the joint European Society of Cardiology/American College of Cardiology, Committee for the redefinition of myocardial infarction, *J Am Coll Cardiol* 36, 959, 2000.

Alter BR, Wheeling JR, Martin HA, et al., Traumatic right coronary artery–right ventricular fistula with restained intramyocardial bullet, *Am J Cardiol* 40, 815, 1977.

Amaral F, Tanamati C, Granzotti JA, et al., Congenital atresia of the ostium of the left coronary artery; Diagnostic difficulty and successful surgical revascularization in two patients, *Arq Bras Cardiol* 74, 339, 2000.

Amati F, Mari A, Digilio MC, et al., 22q11 deletions in isolated and syndromic patients with tetralogy of Fallot, *Hum Genet* 95, 479, 1995.

Amato MC, Moffa PJ, Werner KE, and Ramires JA, Treatment decision in asymptomatic aortic valve stenosis: role of exercise testing. *Heart* 86, 381, 2001.

Ambrose JA, Tannenbaum MA, Alexopoulos D, et al., Angiographic progression of coronary artery disease and the development of myocardial infarction, *J Am Coll Cardiol* 12, 56, 1988.

Ambrose JA, Winters SL, Arora RR, et al., Coronary angiographic morphology in myocardial infarction: a link between the pathogenesis of unstable angina and myocardial infarction, *J Am Coll Cardiol* 6, 1233, 1985a.

Ambrose JA, Winters SL, Stern A, et al., Angiographic morphology and pathogenesis of unstable angina pectoris, *J Am Coll Cardiol* 5, 609, 1985b.

Amr SS and Abu al Ragheb SY, Sudden unexpected death due to papillary fibroma of the aortic valve: report of a case and review of the literature, *Am J Forensic Med Pathol* 12, 143, 1991.

Amr SS, Abu Al-Ragheb SY, Soleiman NA, Al-Debs NR, Sudden, unexpected death due to cardiac fibroma: report of a case and review of the literature, *Am J Forensic Med Pathol* 8, 142, 1987.

Amsterdam EA, Pan H, Rendig SV, et al., Limitation of myocardial infarct size in pigs with a dual lipoxygenase-cyclooxygenase blocking agent by inhibition of neutrophil activity without reduction of neutrophil migration, *J Am Coll Cardiol* 22, 1738, 1993.

Anakwe RE, Traumatic aortic transection, *Eur J Emerg Med*, 12, 133, 2005.

Anand R and Mehta AV, Progressive congenital valvar aortic stenosis during infancy: five cases, *Pediatr Cardiol* 18, 35, 1997.

Anderson JL, Rodier HE, and Green LS, Comparative effects of beta-adrenergic blocking drugs on experimental ventricular fibrillation threshold, *Am J Cardiol* 51, 1196, 1983.

Anderson KR, Bowie J, Dempster AG, and Gwynne JF, Sudden death from occlusive disease of the atrioventricular node artery, *Pathology* 13, 417, 1981a.

Anderson KR and Hill RW, Occlusive lesions of cardiac conducting tissue arteries in sudden infant death syndrome, *Pediatrics* 69, 50–52, 1982.

Anderson RA, Sudden and unexpected death in infancy and the conduction system of the heart, *Cardiovasc Pathol* 9, 147, 2000.

Anderson RH, Bouton J, Burrow CT, and Smith A, Sudden death in infancy: a study of cardiac specialized tissue, *BMJ* 2, 135–139, 1974.

Anderson RH, Ho SY, Smith A, et al., Study of the cardiac conduction tissues in the paediatric age group, *Diagn Histopathol* 4, 3, 1981b.

Anderson RH and Ho SY, Anatomy of the atrioventricular junctions with regard to ventricular preexcitation, *Pacing Clin Electrophysiol* 20, 2072, 1997.

Anderson T, Assessment and treatment of endothelial dysfunction in humans, *J Am Coll Cardiol* 34, 631, 1999.

Anderson WR, Edland JF, and Schenk EA, Conducting system changes in the sudden infant death syndrome, *Am J Pathol* 59, 35a, 1970.

Angelini A, Benciolini P, and Thiene G, Radiation-induced coronary obstructive atherosclerosis and sudden death in a teenager, *Int J Cardiol* 9, 371, 1985.

Angelini A, Ho SY, Anderson RH, et al., The morphology of the normal aortic valve as compared with the aortic valve having two leaflets, *J Thorac Cardiovasc Surg* 98, 362, 1989.

Angelini A, Thiene G, Frescura G, and Baroldi G, Coronary arterial wall and atherosclerosis in youth (1–20 years): a histologic study in northern Italian population, *Intern J Cardiol* 28, 361, 1990.

Angelini P, Villason S, Chan AV, et al., Normal and anomalous coronary arteries in humans, in *Coronary Arteries Anomalies: A Comprehensive Approach*, Angelini P, Ed., Lippincott, Wilkins, Philadelphia, 1999.

Anguera I, Quaglio G, Ferrer B, et al., Sudden death in *Staphylococcal aureus*–associated infective endocarditis due to perforation of a free-wall myocardial abscess, *Scand J Infect Dis* 33, 622, 2001.

Antman EM, Tanasijevic MJ, Thompson B, et al., Cardiac-specific troponin I levels to predict the risk of mortality in patients with acute coronary syndromes, *N Engl J Med* 335, 1342, 1996.

Antoci B, Baroldi G, Costanzi G, et al., Atherosclerotic plaque in heart and brain arteries, in *Atherosclerosis: first international workshop on ischemiac and brain disease*, Baroldi G, Bonomo L, Fieschi C, and Maseri A, Eds., Pensiero Scinentifico, Rome, 1980, p. 39.

Antz M, Weiss C, Volkmer M, et al., Risk of sudden death after successful accessory atrioventricular pathway ablation in resuscitated patients with Wolff-Parkinson-White syndrome, *J Cardiovasc Electrophysiol* 13, 231, 2002.

Antzelevitch C, The Brugada syndrome: ionic basis and arrhythmia mechanisms, *J Cardiovasc Electrophysiol* 12, 268, 2001.

Antzelevitch C, Brugada P, Brugada J, et al., Brugada syndrome: a decade of progress, *Circ Res* 91, 1114, 2002.

Aoki Y, Nata M, Hashiyada M, and Sagisaka K, Sudden unexpected death in childhood due to eosinophilic myocarditis, *Int J Legal Med* 108, 221, 1996.

Appleby M, Fisher M, and Martin M, Myocardial infarction, hyperkalemia, and ventricular tachycardia in a young male bodybuilder, *Int J Cardiol* 44, 171, 1994.

Applefeld MM and Wiernik PH, Cardiac disease after radiation therapy for Hodgkin's disease: analysis of 48 patients, *Am J Cardiol* 51, 1679, 1983.

Araie E, Fujita M, Ejiri M, et al., Relation between the preexistent coronary collateral circulation and recanalization rate of intracoronary thrombolysis for acute myocardial infarction, *J Am Coll Cardiol* 15, 218A, 1990.

Arajärvi E, Santavirta S, and Tolonen J, Aortic ruptures in seat belt wear – injury, *J Thorac Cardiovasc Surg*, 98, 355, 1998.

Arbustini E, De Servi S, Bramucci E, et al., Comparison of coronary lesions obtained by directional coronary atherectomy in unstable angina, stable angina, and restenosis after either atherectomy or angioplasty, *Am J Cardiol* 75, 675, 1995.

Arbustini E, Morbini P, Grasso M, et al., Restrictive cardiomyopathy, atrioventricular block and mild to subclinical myopathy in patients with desmin-immunoreactive material deposits, *J Am Coll Cardiol* 31, 645, 1998.

Arbustini E, Morbini P, Pilotto A, et al., Genetics of idiopathic dilated cardiomyopathy, *Herz* 25, 156, 2000.

Arbustini E, Pilotto A, Repetto A, et al., Autosomal dominant dilated cardiomyopathy with atrioventricular block: a lamin A/C defect-related disease, *J Am Coll Cardiol* 39, 981, 2002.

Aretz HT, Myocarditis: the Dallas criteria, *Hum Pathol* 18, 619, 1987.

Aretz HT, Billingham ME, Edwards WD, et al., Myocarditis: a histopathologic definition and classification, *Am J Cardiovasc Pathol* 1, 3, 1987.

Arimany J, Medallo J, Pujol A, Vingut A, Borondo JC, and Valverde JL. Intentional overdose and death with 3,4-methylenedioxyethamphetamine (MDEA; "Eve"): a case report, *Am J Forensic Med Pathol* 19, 148, 1998.

Armour AJ, Myocardial ischaemia and the cardiac nervous system, *Cardiovasc Res* 41, 41, 1999.

Aronow WS and Cassidy J, Effect of marijuana and placebo-marijuana smoking on angina pectoris, *N Engl J Med* 291, 65, 1974.

Aronow WS and Cassidy J, Effect of smoking marijuana and of a high nicotine cigarette on angina pectoris, *Clin Pharmacol Ther* 17, 549, 1975.

Arzola-Castaner D and Johnson C, Cocaine-induced myocardial infarction associated with severe reversible systolic dysfunction and pulmonary edema, *P R Health Sci J* 23, 319, 2004.

Ascione R, Caputo M, and Angelini GD, Off pump coronary artery bypass grafting not a flash in the pan, *Ann Thorac Surg* 75, 306, 2003.

Asensio JA, Soto SN, Forno W, et al., Penetrating cardiac injuries: a complex challenge, *Injury*, 32, 533, 2001.

Asfaw I and Arbulu A, Penetrating wounds of the pericardium and heart, *Surg Clin North Am*, 57, 37, 1977.

Askanase AD, Friedman DM, Copel J, et al., Spectrum and progression of conduction abnormalities in infants born to mothers with anti-SSA/Ro-SSB/La antibodies, *Lupus* 11, 145, 2002.

Attie F, Rosas M, Rijlaasdam M, et al., The adult patient with Ebstein anomaly: outcome in 72 unoperated patients, *Medicine* 79, 27, 2000.

Avner JR, Shaw KN, and Chin AJ, Atypical presentation of Kawasaki disease with giant coronary artery aneurysms, *J Pediatr* 114, 605, 1989.

Axelrod FB, Goldberg JD, Ye XY, et al., Survival in familial dysautonomia: impact of early intervention, *J Pediatr* 141, 518, 2002.

Bachs L and Morland H, Acute cardiovascular fatalities following cannabis use, *Forensic Sci Int* 124, 200, 2001.

Bajanowski T, Rossi L, Biondo B, et al., Prolonged QT interval and sudden infant death: report of two cases, *Forensic Sci Int* 115, 147, 2001.

Bajanowski T, Teige K, and Brinkmann B, Ausgewählte pathologische Befunde beim plötzlichen Kindstod, in *Prävention des SID*, Trowitzsch E, Schlüter B, and Andler W, Eds., Arcon, Berlin, 1993, 165–171.

Baker AM, Craig BR, and Lonergan GJ, Homicidal commotio cordis: the final blow in a battered infant, *Child Abuse & Neglect*, 27, 125, 2003.

Bakker A, Thone F, and Jacob W, Contact sites between the inner and outer mitochondrial membrane in standard versus hibernating myocardium, *Cardiovasc Pathol* 4, 195, 1995.

Balaji S, Wiles HB, Sens MA, and Gillette PC, Immunosuppressive treatment for myocarditis and borderline myocarditis in children with ventricular ectopic rhythm, *Br Heart J* 72, 354, 1994.

Balakrishnan KG, Jaiswal PK, Thrakan JM, et al., Clinical course of patients in Kerela, in *Endomyocarial Fibrosis*, Valiathan MS, Somers K, and Kartha CC, Eds., Oxford University Press, Delhi, 1993, pp. 20–29.

Baldwin JJ and Edwards JE, Clinical conference: rupture of right ventricle complicating closed chest cardiac massage, *Circulation*, 53, 562, 1976.

Balser JR, The cardiac sodium channel: gating function and molecular pharmacology, *J Mol Cell Cardiol* 33, 599, 2001.

Banchi A, Morfologia delle arterie coronarie cordis, *Arch Ital Anat Embryol* 3, 87, 1904.

Baoroudi G, Carbonneau E, Pouliot V, and Chahine M, SCN5A mutation (T1620M) causing Brugada syndrome exhibits different phenotypes when expressed in Xenopus oocytes and mammalian cells, *FEBS Lett* 467, 12–16, 2000.

Barak M, Herschkowitz S, Shapiro I, et al., Familial combined sinus node and atrioventricular conduction dysfunctions, *Int J Cardiol* 15, 231, 1987.

Barber MJ, Mueller TM, Henry DP, et al., Transmural myocardial infarction in the dog produces sympathectomy in noninfarcted myocardium, *Circulation* 67, 787, 1983.

Barbero-Marcial M, Verginelli G, Sirera JC, et al., Surgical treatment of coarctation of the aorta in the first year of life: immediate and late results in 35 patients, *Thorac Cardiovasc Surg* 30, 75, 1982.

Bardales RH, Hailey SL, Su Su Xie, et al., *In situ* apoptosis assay for detection of early acute infarction, *Am J Pathol* 149, 821, 1996.

Barilli F, De Vincentis G, Maugieri E, et al., Recovery of contractility of viable myocardium during inotropic stimulation is not dependent on an increase of myocardial blood flow in the absence of collateral filling, *J Am Coll Cardiol* 33, 697, 1999.

Barnett HJ, Boughner DR, Taylor DW, et al., Further evidence relating mitral-valve prolapse to cerebral ischemic events, *N Engl J Med* 302, 139, 1980.

Barnett JM and Codd G, Sudden, unexpected death of a cannabis bodypacker, due to perforation of the rectum, *J Clin Forensic Med* 9, 82, 2002.

Baroldi G, Acute coronary occlusion as a cause of myocardial infarct and sudden coronary heart death, *Am J Cardiol* 16, 859, 1965.

Baroldi G, High resistance of the human myocardium to shock and red blood cell aggregation (sludge), *Cardiologia* 54, 271, 1969.

Baroldi G, The coronary arteries in cor pulmonale, *Acta Cardiol* 26, 602, 1971.

Baroldi G, Different types of myocardial necrosis in coronary heart disease: a pathophysiological review of their functional significance, *Am Heart J* 89, 742, 1975a.

Baroldi G, Limitation of infarct size: theoretical aspects, in *Cardiology: An International Perspective*, Vol. I, Chazov EI, Smirnov VN, and Organov RG, Eds., Plenum Press, New York, 1984, p. 517.

Baroldi G, Diseases of the extramural coronary arteries, in *Cardiovascular Pathology*, 2nd ed., Silver MD, Ed., Churchill Livingstone, New York, 1991a, p. 487.

Baroldi G, Morphologic forms of myocardial necrosis related to myocardial cell function, in *Cardiovascular Pathology*, 2nd ed., Silver MD, Ed., Churchill Livingstone, New York, 1991b.

Baroldi G, The cardiovascular pathologist: a pathologist awaiting a better definition, *Cardiovasc Pathol* 2, 215, 1993a.

Baroldi G, Myocardial cell death, including ischemic heart disease and its complications, in *Cardiovascular Pathology*, 3rd ed., Silver MD, Gotlieb A, and Schoen FJ, Eds., Churchill Livingstone, New York, 2001a.

Baroldi G, Structural background and clinical imaging: is pathology still needed?, in *Understanding Cardiac Imaging Techniques from Basic Pathology to Image Fusion*, NATO Science Series, Vol. 332, Marzullo P, Ed., 10S Press, Amsterdam, 2001b, p. 14.

Baroldi G, Camerini F, and Goodwin JF, *Advances in Cardiomyopathies*, Springer, Heidelberg, 1990a.

Baroldi G, Corallo S, Moroni M, et al., Focal lymphocytic myocarditis in acquired immunodeficiency syndrome (AIDS): a correlative morphologic and clinical study in 26 consecutive fatal cases, *J Am Coll Cardiol* 12, 463, 1998a.

Baroldi G, Di Pasquale G, Silver MD, et al., Type and extent of myocardial injury related to brain damage and its significance in heart transplantation: a morphometric study, *J Heart Lung Transplant* 16 994, 1997a.

Baroldi G, Falzi G, and Lampertico P, The nuclear patterns of cardiac muscle fiber, *Cardiologia* 51, 109, 1967.

Baroldi G, Falzi G, and Mariani F, Sudden coronary death: a postmortem study in 208 selected cases compared to 97 "control" subjects, *Am Heart J* 98, 20, 1979.

Baroldi G, Falzi G, Mariani F, et al., Morphology, frequency and significance of intramural arterial lesions in sudden coronary death, *J Ital Cardiol* 10, 644, 1980.

Baroldi G and Manion WC, Microcirculatory disturbances and human myocardial infarction, *Am Heart J* 74, 171, 1967.

Baroldi G, Mantero O, and Scomazzoni G, The collaterals of the coronary arteries in normal and pathologic hearts, *Circ Res* 4, 223, 1956.

Baroldi G, Marzilli M, L'Abbate A, et al., Coronary occlusion: cause or consequence of acute myocardial infarction? A case report, *Clin Cardiol* 13, 49, 1990b.

Baroldi G, Milam JD, Wukash DC, Sandiford FM, Romagnoli A, and Cooley DA. Myocardial cell damage in "stone heart," *J Mol Cell Cardiol* 6, 395, 1974a.

Baroldi G, Mittleman RE, Parolini M, et al., Myocardial contraction bands: definition, quantification and significance in forensic pathology, *Int J Legal Med* 115, 142, 2001.

Baroldi G, Oliveira SJM, and Silver MD, Sudden and unexpected death in clinically "silent" Chagas' disease: a hypothesis, *Int J Cardiol* 58, 263, 1997b.

Baroldi G, Parravicini C, and Gaiera G, Sudden cardiac death in a "silent" case of acquired immunodeficiency syndrome (AIDS), *J Ital Cardiol* 23, 353, 1993b.

Baroldi G, Radice F, Schmid C, et al., Morphology of acute myocardial infarction in relation to coronary thrombosis, *Am Heart J* 87, 65, 1974b.

Baroldi G, Rapezzi C, De Maria R, et al., The clinicopathologic spectrum of hypertrophic cardiomyopathy: the experience of the Italian Heart Transplant Program, in *Advances in Cardiomyopathies*, Camerini F, Gavazzi A, and De Maria R, Eds., Springer-Verlag, Heidelberg, 1998b, 60.

Baroldi G and Scomazzoni G, Coronary Circulation in the Normal and Pathologic Heart, American Registry of Pathology, A.F.I.P., Ed., U.S. Government Printing Office, Washington, 1967.

Baroldi G and Silver MD, The healing of myocardial infarcts in man, *J Ital Cardiol* 5, 465, 1975b.

Baroldi G and Silver MD, *Sudden Death in Ischemic Heart Disease: An Alternative View on the Significance of Morphologic Findings*, RG Landes Co, Austin, TX, Springer-Verlag, Heidelberg, 1995.

Baroldi G, Silver MD, De Maria R, et al., Lipomatous metaplasia in left ventricular scar, *Can J Cardiol* 13, 65, 1997c.

Baroldi G, Silver MD, De Maria R, et al., Pathology and pathogenesis of congestive heart failure: a quantitative morphologic study of 144 hearts excised at transplantation, *Pathogenesis* 1, 107, 1998c.

Baroldi G, Silver MD, De Maria R, et al., Frequency and extent of contraction band necrosis in orthotopically transplanted human hearts: a morphometric study, *Int J Cardiol* 88, 267, 2003.

Baroldi G, Silver MD, Lixfield W, et al., Irreversible myocardial damage resembling catecholamine necrosis secondary to acute coronary occlusion in dogs: its prevention by propranolol, *J Mol Cell Cardiol* 9, 687, 1977.

Baroldi G, Silver MD, Mariani F, and Giuliano G, Correlation of morphologic variables in the coronary atherosclerotic plaque with clinical patterns of ischemic heart disease, *Am J Cardiovasc Pathol* 2, 159, 1988.

Baroldi G, Silver MD, Parolini M, Pomara C, Turillazzi E, and Fineschi V. Myofiber break-up: a marker of ventricular fibrillation in sudden cardiac death, *Int J Cardiol* 100, 435, 2005.

Baroldi G and Silver MD, *The Etiopathogenesis of Coronary Heart Disease: A Heretical Theory Based on Morphology,* 2nd ed., Landes Bioscience, Georgetown, Texas, 2004.

Baroldi G and Thiene G, *Biopsia endomiocardica*, Testo Atlante, Piccin, Padova, 1998.

Barrie HJ and Urback PG, The cellular changes in myocardial infarction, *Can Med Assoc J* 77, 106, 1957.

Barry WH, Molecular inotropy: a future approach to the treatment of heart failure? *Circulation* 100, 2303, 1999.

Barth PG, Wanders RJ, Vreken P, et al., X-linked cardioskeletal myopathy and neutropenia (Barth syndrome) (MIM 302060), *J Inherit Metab Dis* 22, 555, 1999.

Basso C, Calabrese F, Corrado D, et al., Postmortem diagnosis in sudden cardiac death victims: macroscopic, microscopic, and molecular findings, *Cardiovasc Res* 50, 290, 2001a.

Basso C, Corrado D, Rossi L, et al., Ventricular preexcitation in children and young adults: atrial myocarditis as a possible trigger of sudden death. *Circulation* 103, 269, 2001b.

Basso C, Corrado D, and Thiene G, Cardiovascular causes of sudden death in young individuals including athletes, *Cardiol Rev* 7, 127, 1999.

Basso C, Fox PR, Meurs KM et al., Arrhythmogenic right ventricular cardiomyopathy causing sudden cardiac death in boxer dogs. A new animal model of human disease, *Circulation* 109, 1180, 2004.

Basso C, Frescura C, Corrado D, et al., Congenital heart disease and sudden death in the young, *Hum Pathol* 26, 1065, 1995.

Basso C and Thiene G. Adipositas cordis, fatty infiltration of the right ventricle, and arrhythmogenic right ventricular cardiomyopathy Just a matter of fat? *Cardiovasc Pathol* 14, 37, 2005.

Basso C, Thiene G, Corrado D, et al., Arrhythmogenic right ventricular cardiomyopathy: dysplasia, dystrophy, or myocarditis? *Circulation* 94, 983, 1996.

Basso C, Thiene G, Corrado D, et al., Hypertrophic cardiomyopathy and sudden death in the young: pathologic evidence of myocardial ischemia, *Hum Pathol* 31, 988, 2000.

Basson CT, Cowley GS, Solomon SD, et al., The clinical and genetic spectrum of Holt-Oram syndrome (heart-hand syndrome), *N Engl J Med* 330, 885, 1994.

Batra AS and Hohn AR, Consultation with the specialist: palpitations, syncope, and sudden cardiac death in children: Who's at risk? *Pediatr Rev* 24, 269, 2003.

Batra AS and Silka MJ, Mechanism of sudden cardiac arrest while swimming in a child with the prolonged QT syndrome, *J Pediatr* 141, 283, 2002.

Batsakis JG, Degenerative lesions of the heart, in *The Pathology of the Heart and Blood Vessels*, 3rd ed., Gould SE, Ed., Charles C Thomas, Springfield, IL, 1968, p. 479.

Baubin M, Rabl W, Pfeiffer KP, et al., Chets injuries after active compression – decompression cardiopulmonary resuscitaion (ACD – CPR) in cadavers, *Resuscitation*, 43, 9, 1999a.

Baubin M, Suman G, Rabl W, et al., Increased frequency of thorax injuries with ACD–CPR, *Resuscitation*, 43, 33, 1999b.

Bauce B, Nava A, Rampazzo A, et al., Familial effort polymorphic ventricular arrhythmias in arrhythmogenic right ventricular cardiomyopathy map to chromosome 1q42-43, *Am J Cardiol* 85, 573, 2000.

Bauce B, Rampazzo A, Basso C, et al., Screening for ryanodine receptor type 2 mutations in families with effort-induced polymorphic ventricular arrhythmias and sudden death: early diagnosis of asymptomatic carriers, *J Am Coll Cardiol* 40, 341, 2002.

Bauman JL and DiDomenico RJ, Cocaine-induced channelopathies: emerging evidence on the multiple mechanisms of sudden death, *J Cardiovasc Pharmacol Ther* 7, 195, 2002.

Bax JJ, Visser FC, and Poldermans D, Time course of functional recovery of stunned and hibernating segments after surgical revascularization, *Circulation* 104, I314, 2001.

Bayer AS, Bolger AF, Taubert KA, et al., Diagnosis and management of infective endocarditis and its complications, *Circulation* 98, 2963, 1998.

Bayés de Luna A and Guindo J, Sudden death in ischemic heart disease, *Rev Port Cardiol* 9, 473, 1990.

Bayés de Luna A, Coumel P, and Leclercq JF, Ambulatory sudden cardiac death: mechanisms of production of fatal arrhythmia on the basis of data from 157 cases, *Am Heart J* 117, 15, 1989.

Bean SL and Saffitz JE, Transmural heterogeneity of norepinephrine uptake in failing human hearts, *J Am Coll Cardiol* 23, 579, 1994.

Bean WB, Bullet wound of the heart with coronary artery ligation, *Am Heart J* 21, 375, 1944.

Becane HM, Bonne G, Varnous S, et al., High incidence of sudden death with conduction system and myocardial disease due to lamins A and C gene mutation, *Pacing Clin Electrophysiol* 23, 1661, 2000.

Becker CG and Murphy GE, Demonstration of contractile proteins in endothelium and cells of the heart valves, endocardium, intima, arteriosclerotic plaques and Aschoff bodies of rheumatic heart disease, *Am J Pathol* 55, 1, 1969.

Beckwith JB, Observations on the pathological anatomy of the sudden infant death syndrome, in *International Conference on Causes of Sudden Death in Infants*, Bergmann AB, Beckwith JB, and Ray CG, Eds., University of Washington Press, Seattle, 1970, 83–139.

Beckwith JB, The mechanism of death in sudden infant death syndrome, in *Sudden Infant Death Syndrome. Medical Aspects and Physiological Management,* Culbertson JL, Krous HK, and Bendell RD, Eds., Edward Arnold, London, 1989, 48–61.

Bedell SE and Fulton EJ, Unexpected findings and complications at autopsy after cardiopulmonary resuscitation (CPR), *Arch Intern Med*, 146, 1725, 1986.

Behrendt H and Boffin H, Myocardial cell lesions caused by anabolic hormone, *Cell Tissue Res* 181, 423, 1977.

Beitzke A, Becker H, Rigler B, et al., Development of aortic aneurysms in familial supravalvar aortic stenosis, *Pediatr Cardiol* 6, 227, 1986.

Beltrami AP, Urbanek K, Kaistura J, et al., Evidence that human cardiac myocytes divide after myocardial infarction, *N Engl J Med* 344, 1750, 2001.

Beltrami CA, Finato N, Rocco M, et al., Structural basis of end stage failure in ischemic cardiomyopathy in humans, *Circulation* 89, 151, 1994.

Bemis CE, Gorlin R, Kemp HG, et al., Progression of coronary artery disease: a clinical arteriographic study, *Circulation* 47, 455, 1973.

Benditt EP, Evidence for a monoclonal origin of human atherosclerotic plaques and some implications, *Circulation* 50, 650, 1974.

Bengel FM, Neberfuhr P, Ziegler SI, et al., Serial assessment of sympathetic reinnervation after orthotopic heart transplantation: a longitudinal study using PET and C-11 hydroxyephedrine, *Circulation* 99, 1866, 1999.

Bennet HS, Luft JH, and Hampton JC, Morphological classification of vertebrate blood capillaries, *Am J Physiol* 196, 381, 1959.

Benshop RJ, Nieuwhenhuis EE, Tromp EA, et al., Effects of β-adrenergic blockade on immunologic and cardiovascular changes induced by mental stress, *Circulation* 89, 762, 1994.

Benzaquen BS, Cohen V, and Eisenberg MJ, Effects of cocaine on the coronary arteries, *Am Heart J* 142, 402, 2001.

Bernstein D, Finkbeiner WE, Soifer S, and Teitel D, Perinatal myocardial infarction: a case report and review of the literature, *Pediatr Cardiol* 6, 313, 1986.

Berry W, *The Unsettling of America: Culture and Agriculture*, Sierra Club Books, San Francisco, 1977.

Bertrand ME, LaBlanche JM, Tilmant P, et al., Frequency of provoked coronary arterial spasm in 1089 consecutive patients undergoing coronary arteriography, *Circulation* 65, 129, 1982.

Bevilacqua M, Norbiato G, Vago T, et al., Alterations in norepinephrine content and beta adrenoceptor regulation in myocardium bordering aneurysm in human heart: their possible role in the genesis of ventricular tachycardia, *Eur J Clin Invest* 16, 163, 1986.

Bezzina C, Veldkamp MW, van Den Berg MP, et al., A single Na(+) channel mutation causing both long-QT and Brugada syndromes, *Circ Res* 85, 1206, 1999.

Bharati S, Pathology of the conduction system, in *Cardiovascular Pathology*, 3rd ed., Silver MD, Gotlieb AI, and Schoen FJ, Eds., Churchill Livingstone, New York, 2001, chap. 20.

Bharati S, Bauernfeind R, Miller LB, et al., Sudden death in three teenagers: conduction system studies, *J Am Coll Cardiol* 1, 879, 1983.

Bharati S, Bicoff JP, Fridman JL, et al., Sudden death caused by benign tumor of the atrioventricular node, *Arch Intern Med* 136, 224, 1976.

Bharati S, Bump FT, Bauernfeind R, et al., Dystrophica myotonia: correlative electrocardiographic, electrophysiologic, and conduction system study, *Chest* 86, 444, 1984.

Bharati S, Dreifus L, Bucheleres G, et al., The conduction system in patients with a prolonged QT interval, *J Am Coll Cardiol* 6, 1110, 1985.

Bharati S, Karp R, and Lev M, The conduction system in truncus arteriosus and its surgical significance: a study of five cases, *J Thorac Cardiovasc Surg* 104, 954, 1992a.

Bharati S, Krongrad E, and Lev M, Study of the conduction system in a population of patients with sudden infant death syndrome, *Pediatr Cardiol* 6, 29–40, 1985.

Bharati S and Lev M, The pathologic changes in the conduction system beyond the age of ninety, *Am Heart J* 124, 486, 1992b.

Bharati S and Lev M, Conduction system in sudden unexpected death a considerable time after repair of atrial septal defect, *Chest* 94, 142, 1988.

Bharati S and Lev M, Cardiac conduction system involvement in sudden death of obese young people, *Am Heart J* 129, 273, 1995.

Bharati S, Molthan ME, Veasy LG, et al., Conduction system in two cases of sudden death two years after the Mustard procedure, *J Thorac Cardiovasc Surg* 77, 101, 1979.

Bharati S, Rosen K, Steinfield L, et al., The anatomic substrate for preexcitation in corrected transposition, *Circulation* 62, 831, 1980.

Bharati S, Surawicz B, Vidaillet HJ Jr, et al., Familial congenital sinus rhythm anomalies: clinical and pathological correlations, *Pacing Clin Electrophysiol* 15, 1720, 1992c.

Bieber CP, Stinson EB, Shumway NE, et al., Cardiac transplantation in man: VII Pathology, *Am J Cardiol* 25, 84, 1970.

Bigi R, Cortigiani L, Palombo P, et al., Prognostic and clinical correlates of angiographically diffuse non-obstructive coronary lesions, *Heart* 89, 1009, 1999.

Billinger M, Fleish M, Eberli FR, et al., Is the development of myocardial tolerance to repeated ischemia in humans due to preconditioning or to collateral recruitment? *J Am Coll Cardiol* 33, 1027, 1999.

Billingham ME, The postsurgical heart: the pathology of cardiac transplantation, *Am J Cardiovasc Pathol* 1, 319, 1988.

Billingham ME, Role of endomyocardial biopsy in diagnosis and treatment of heart disease, in *Cardiovascular Pathology*, 2nd ed., Silver MD, Ed., Churchill Livingstone, New York, 1991, p. 1465.

Billingham ME and Tazelaar HD, The morphological progression of viral myocarditis, *Postgrad Med J* 62, 581, 1986.

Bing RJ, Tillmanns H, Fauvel JM, et al., Effect of prolonged alcohol administration on calcium transport in heart muscle of the dog, *Circ Res* 35, 33, 1974.

Bini RM, Westaby S, Bargeron LM Jr, et al., Investigation and management of primary cardiac tumors in infants and children, *J Am Coll Cardiol* 2, 351, 1983.

Bird LM, Billman GF, Lacro RV, et al., Sudden death in Williams syndrome: report of ten cases, *J Pediatr* 129, 926, 1996.

Bird LM, Krous HF, Eichenfield LF, et al., Female infant with oncocytic cardiomyopathy and microphthalmia with linear skin defects (MLS): a clue to the pathogenesis of oncocytic cardiomyopathy? *Am J Med Genet* 53, 141, 1994.

Birnie D, Tometzki A, Curzio J, et al., Outcomes of transposition of the great arteries in the ear of atrial inflow correction, *Heart* 80, 170, 1998.

Bisognano JD, Weiberger HD, Bohmeyer TS, et al., Myocardial-direct overexpression of the human 1-adrenergic receptor in transgenic mice, *J Mol Coll Cardiol* 32, 817, 2000.

Bisset GS III, Seigel SF, Gaum WE, et al., Chaotic atrial tachycardia in childhood, *Am Heart J* 101, 268, 1981.

Bjerregaard P and Gussak I, Short QT syndrome: mechanisms, diagnosis and treatment, *Nat Clin Pract Cardiovasc Med* 2, 84, 2005.

Black A, Black MM, Gensini G, and Di Giorgi S, Exertion and acute coronary artery injury, *Circulation* 32 (Suppl 2), 3, 1965.

Blackstone EH and Kirklin JW, Death and other time-related events after valve replacement, *Circulation* 72, 753, 1985.

Blaes N and Boissel JF, Growth-stimulating effect of catecholamines on rat aortic smooth muscle cells in culture, *J Cell Physiol* 116, 167, 1983.

Blake HA, Manion WC, Mattingly TW, et al., Coronary artery anomalies, *Circulation* 30, 927, 1964.

Blautass AH, Production of ventricular hypertrophy simulating idiopathic hypertrophic subaortic stenosis (IHSS) by subhypertensive infusion of norepinephrine (NE) in the conscious dog (abstract), *Clin Res* 23, 77, 1975.

Bleyl SB, Mumford BR, Thompson V, et al., Neonatal, lethal noncompaction of the left ventricular myocardium is allelic with Barth syndrome, *Am J Hum Genet* 61, 868, 1997a.

Bleyl SB, Mumford BR, Brown-Harrison MC, et al., Xq28-linked noncompaction of the left ventricular myocardium: prenatal diagnosis and pathologic analysis of affected individuals, *Am J Med Genet* 72, 257, 1997b.

Blumgart HL, Schlesinger MJ, and Davis D, Studies on the relation of the clinical manifestations of angina pectoris, coronary thrombosis and myocardial infarction to the pathologic findings with particular reference to the significance of the collateral circulation, *Am Heart J* 19, 1, 1940.

Bodenheimer MM, Banka VS, Hermann GA, et al., The effect of severity of coronary artery obstructive disease and the coronary collateral circulation on local histopathologic and electrographic observations in man, *Am J Med* 63, 193, 1977.

Bodily K and Fischer RP, Aortic rupture and right ventricular rupture induced by closed chest cardiac massage, *Minnesota Med*, 62, 225, 1979.

Boglioli LR, Taff ML, and Harleman G, Child homicide caused by commotio cordis, *Pediatr Cardiol* 19, 436, 1998.

Bohles H, Singer H, Ruitenbeek W, et al., Foamy myocardial transformation in a child with a disturbed respiratory chain, *Eur J Pediatr* 146, 582, 1987.

Bohm N and Krebs G, Solitary rhabdomyoma of the heart: clinically silent case with sudden, unexpected death in an 11-month-old boy, *Eur J Pediatr* 134, 167, 1980.

Bohn M, La Rosee K, Schwinger RHG, and Erdmann E, Evidence for reduction of norepinephrine uptake sites in the failing human heart, *J Am Coll Cardiol* 25, 146, 1995.

Bonfiglio TA, Botti RE, and Hagstrom JWC, Coronary arteritis, occlusion and myocardial infarction due to lupus erythematosus, *Am Heart J* 83, 153, 1972.

Bonhoeffer P, Bonnet D, Piechaud JF, et al., Coronary artery obstruction after the arterial switch operation for transposition of the great arteries in newborns, *J Am Coll Cardiol* 29, 202, 1997.

Bonin S, Petrera F, Niccolini B, et al., PCR analysis in archival postmortem tissues. *Mol Pathol* 56, 184, 2003.

Bonnet D, Martin D, De Lonlay P, et al., Arrhythmias and conduction defects as presenting symptoms of fatty acid oxidation disorders in children, *Circulation* 100, 2248, 1999.

Bontonyric P, Bussy C, Hayaz D, et al., Local pulse pressure and regresion of arterial wall hypertrophy during long-term antihypertensive treatment, *Circulation* 101, 2601, 2000.

Borges M, Thoné F, Wouters L, et al., Structural correlates of regional myocardial dysfunction in patients with critical coronary stenosis: chronic hibernation? *Cardiovasc Pathol* 2, 237, 1993.

Bork K, Uber die Kranzadersklerose, *Virchows Arch Pathol Anat* 262, 646, 1926.

Born GUR, Honour AJ, and Mitchell JRA, Inhibition by adenosine and by 2-chloroadenosine of the formation and embolization of platelet thrombi, *Nature* 4934, 761, 1964.

Borrie J and Lichter I, Pericardial rupture from blunt chest trauma, *Thorax* 29, 329, 1979.

Bos I, Johannisson R, and Djonlagic H, Morphologic alterations in the long Q–T syndrome. Light and electron microscopic observations in the conduction system and in sympathetic trunks, *Pathol Res Pract* 180, 691, 1985.

Bosi G, Lintermans JP, Pellegrino PA, et al., The natural history of cardiac rhabdomyoma with and without tuberous sclerosis, *Acta Paediatr* 85, 928, 1996.

Boucek RJ and Takeshita RB, Intimal hypertrophy in coronary arteries and considerations of the papillary muscle arteries (man), *Anat Rec* 153, 243, 1965.

Boucek RJ, Takeshita R, and Fojaco R, Relation between microanatomy and functional properties of the coronary arteries (dog), *Anat Rec* 147, 199, 1963.

Bouchardy B and Majno G, A new approach to the histologic diagnosis of early myocardial infarcts, *Cardiology* 56, 327, 1971–1972.

Bowles NE, Ni J, Marcus F, and Towbin J, The detection of cardiotropic viruses in the myocardium of patients with arrhythmogenic right ventricular dysplasia/cardiomyopathy, *J Am Coll Cardiol* 39, 892, 2002.

Boyd W, *A Textbook of Pathology: Structure and Function in Disease*, 7th ed., Lea & Febiger, Philadelphia, 1961.

Brackett JC, Sims HF, Steiner RD, et al., A novel mutation in medium chain acyl-CoA dehydrogenase causes sudden neonatal death, *J Clin Invest* 94, 1477, 1994.

Bradbury S, Thirty years after ligation of the anterior descending branch of the left coronary artery, *Am Heart J* 24, 562, 1942.

Brain MC, Dacie JV, and Hourihane DOB, Microangiopathic haemolytic anaemia: the possible role of vascular lesions in pathogenesis, *Brit J Hematol* 8, 358, 1962.

Brandenburg RO, Cardiomyopathies and their role in sudden death, *J Am Coll Cardiol* 5, 185B, 1985.

Brandt RL, Foley WJ, Fink GH, et al., Mechanism of perforation of the heart with producyion of hydropericardium by a venous catheter and its prevention, *Am J Surg*, 119, 311, 1970.

Brathwaite CEM, Rodriguez A, Turney SZ, et al., Blunt traumatic cardiac rupture: a five-year experience, *Ann Surg*, 212, 701, 1990.

Braunwald E, Lambrew CT, Rockoff D, et al., Idiopathic hypertrophic subaortic stenosis, *Circulation* 30(Suppl IV), 217, 1964.

Braunwald E, Shattuck lecture: Cardiovascular medicine at the turn of the millenium: triumphs concerns and opportunities, *N Engl J Med* 337, 1360, 1997.

Braunwald E, *Textbook of Cardiology*, 3rd ed., 1990.

Braunwald E and Kloner RA, The stunned myocardium: prolonged, postischemic ventricular dysfunction, *Circulation* 66, 1146, 1982.

Brechenmacher C, Coumel P, Fauchier JP, et al., De Subitaneis Mortibus. XXII. Intractable paroxysmal tachycardias which proved fatal in type A Wolff-Parkinson-White Syndrome, *Circulation* 55, 408, 1977.

Brechenmacher C, Coumel P, and James TN, De Subitaneis Mortibus. XVI. Intractable tachycardia in infancy, *Circulation* 53 377, 1976.

Brecklin CS, Gopaniuk-Folga A, Sabah S, Singh A, Arruda JAL, and Dunea G. Prevalence of hypertension in chronic cocaine users, *Am J Hypert* 11, 1279, 1998.

Bridgen W, Uncommon myocardial diseases: the noncoronary cardiomyopathies, *Lancet* 2, 1179, 1957.

Brink PA, Ferreira A, Moolman JC, et al., Gene for progressive familial heart block type I maps to chromosome 19q13, *Circulation* 91, 1633, 1995.

Brinkmann B, Sepulchre MA, and Fechner G, The application of selected histochemical and immunohistochemical markers and procedures to the diagnosis of early myocardial damage, *Int J Legal Med* 106, 135, 1993.

Bristow MR and Gilbert EM, Improvement in cardiac myocyte function by biological effect of medical therapy: a new concept in the treatment of heart failure, *Eur Heart J* 16 (Suppl F), 20, 1995.

Brody WR, Angeli WW, and Kosek JC, Histologic fate of the venous coronary artery bypass in dog, *Am J Pathol* 66, 111, 1972.

Brody SL, Slovis CM, and Wrenn KD, Cocaine related medical problems: consecutive series of 233 patients, *Am J Med* 88, 325, 1990.

Brogan WC, Lange RA, Glamann DB, et al., Recurrent coronary vasoconstriction caused by intranasal cocaine: possible role for metabolites, *Ann Intern Med* 116, 556, 1992.

Bromberg BI, Lindsay BD, Cain ME, and Cox, JL, Impact of clinical history and electrophysiologic characterization of accessory pathways on management strategies to reduce sudden death among children with Wolff-Parkinson-White syndrome, *J Am Coll Cardiol* 27, 690, 1996.

Brown CE and Richter IM, Medial coronary sclerosis in infancy, *Arch Pathol* 31, 449, 1941.

Brown G, Albers JJ, Fisher LD, et al., Regression of coronary artery disease as a result of intensive lipid-lowering therapy in men with high levels of apolipoprotein B, *N Engl J Med* 323, 1289, 1990.

Brucato A, Doria A, Frassi M, et al., Pregnancy outcome in 100 women with autoimmune diseases and anti-Ro/SSA antibodies: a prospective controlled study, *Lupus* 11, 716, 2002.

Brugada J, Brugada R, Antzelevitch C, et al., Long-term follow-up of individuals with the electrocardiographic pattern of right bundle-branch block and ST-segment elevation in precordial leads V1 to V3, *Circulation* 105, 73, 2002.

Brugada J, Brugada R, and Brugada P, Determinants of sudden cardiac death in individuals with the electrocardiographic pattern of Brugada syndrome and no previous cardiac arrest, *Circulation* 108, 3092, 2003.

Brugada J, Brugada R, and Brugada P, Brugada syndrome, *Arch Mal Coer Vaiss* 92, 847–850, 1999.

Brugada P and Brugada J, Right bundle branch block, persistent ST segment elevation and sudden cardiac death: a distinct clinical and electrocardiographic syndrome, *J Am Coll Cardiol* 20, 1391, 1992.

Brugada P, Brugada R, and Brugada J, The Brugada syndrome, *Curr Cardiol Rep* 2, 507, 2000.

Brugada P, Brugada R, Mont L, et al., Natural history of Brugada syndrome: the prognostic value of programmed electrical stimulation of the heart, *J Cardiovasc Electrophysiol* 14, 455, 2003.

Brugada R, Brugada J, Antzelevitch C, et al., Sodium channel blockers identify risk for sudden death in patients with ST-segment elevation and right bundle branch block but structurally normal hearts, *Circulation* 101, 510, 2000.

Brugada R, Hong K, Cordeiro JM, et al., Short QT syndrome, CMAJ doi: 10.1503/cmaj.050596, 2005.

Brugada R, Tapscott T, Czernuszewicz GZ, et al., Identification of a genetic locus for familial atrial fibrillation, *N Engl J Med* 336, 905, 1997.

Brymer JF, Coronary ostial stenosis: a complication of aortic valve replacement, *Circulation* 49, 530, 1974.

Brundage SI, Harruff R, Jurkovich GJ, et al., The epidemiology of thoracic aortic injuries in pedestrians, *J Trauma Inj Infect Crit Care*, 45, 1010, 1998.

Bruschi G, Agati S, Iorio F, et al., Papillary muscle rupture and pericardial injuries after blunt chest trauma, *Eur J Cardio – Thorac Surg*, 20, 200, 2001.

Bucher O, Ueber den Bau Blutgefasse des menschlichen Herzens, *Acta Anat* 3, 162, 1947.

Buchner F, Patologische Anatomie der Herzinsuffizienz, *Verh Deutsch Ges Kreislaufforsch* 16, 26, 1950.

Buffon A, Santini SA, Ramazzotti V, et al., Large sustained lipid peroxidation and reduced antioxidant capacity in the coronary circulation after brief episodes of myocardial ischemia, *J Am Coll Cardiol* 35, 633, 2000.

Buja ML, Poliner LR, Parkey RW, et al., Clinicopathologic study of persistently positive technetium-99 stannous pyrophosphate myocardial scintigrams and myocytolytic degeneration after myocardial infarction, *Circulation* 56, 1016, 1977.

Bulkley BH and Hutchins GM, Accelerated "atherosclerosis": a morphologic study of 97 saphenous vein coronary artery bypass grafts, *Circulation* 55, 163, 1977.

Bulkley BH, Klacsmann PG, Hutchins GM, Angina pectoris, myocardial infarction and sudden cardiac death with normal coronary arteries: a clinicopathologic study of 9 patients with progressive systemic sclerosis, *Am Heart J* 95, 563, 1978.

Burch GE and De Pasquale NP, Arteriosclerosis in high pressure and low pressure coronary arteries, *Am Heart J* 63, 720, 1962.

Burch M, Sharland M, Shinebourne E, Smith G, Patton M, and McKenna W. Cardiologic abnormalities in Noonan syndrome: phenotypic diagnosis and echocardiographic assessment of 118 patients, *J Am Coll Cardiol* 22, 1189, 1993.

Burch M, Siddiqi SA, Celermajer DS, et al., Dilated cardiomyopathy in children: determinants of outcome, *Br Heart J* 72, 246, 1994.

Burgess C, O'Donohoe A, and Gill M, Agony and ecstasy: a review of MDMA effects and toxicity, *Eur Psychiatry* 15, 287, 2000.

Burgmeister G, Infektiöse Myokarditis im Kindesalter, *Dtsch Gesundh Wes* 18, 26, 1963.

Burke AP, Afzal MN, Barnett DS, and Virmani R, Sudden death after a cold drink: case report, *Am J Forensic Med Pathol* 20, 37, 1999a.

Burke AP, Anderson PG, Virmani R, et al., Tumor of the atrioventricular nodal region: a clinical and immunohistochemical study, *Pathol Lab Med* 114, 1057, 1990.

Burke AP, Farb A, Malcolm GT, et al., Coronary risk factors and plaque morphology in men with coronary disease who died suddenly, *N Engl J Med* 336, 1276, 1997a.

Burke AP, Farb A, Malcolm GT, et al., Plaque rupture and sudden death related to exertion in men with coronary artery disease, *JAMA* 281, 921, 1999b.

Burke AP, Farb A, Tang A, et al., Fibromuscular dysplasia of small coronary arteries and fibrosis in the basilar ventricular septum in mitral valve prolapse, *Am Heart J* 134, 282, 1997b.

Burke AP, Farb A., Tashko G, et al., Arrhythmogenic right ventricular cardiomyopathy and fatty replacement of the right ventricular myocardium: are they different diseases? *Circulation* 97, 1571, 1998a.

Burke AP, Farb A, Virmani R, et al., Sports-related and non-sports-related sudden cardiac death in young adults, *Am Heart J* 121 (2 Pt), 568, 1991a.

Burke AP, Kalra P, Li, L, et al., Infectious endocarditis and sudden unexpected death: incidence and morphology of lesions in intravenous addicts and non-drug abusers, *J Heart Valve Dis* 6, 198, 1997c.

Burke AP, Ribe JK, Bajaj AK, et al., Hamartoma of mature cardiac myocytes, *Hum Pathol* 29, 904, 1998b.

Burke AP, Rosado-de-Christenson M, Templeton PA, and Virmani R, Cardiac fibroma: clinicopathologic correlates and surgical treatment, *J Thorac Cardiovasc Surg* 108, 862, 1994.

Burke AP, Subramanian R, Smialek J, and Virmani R, Nonatherosclerotic narrowing of the atrioventricular node artery and sudden death, *J Am Coll Cardiol* 21, 117, 1993.

Burke AP and Virmani R, Cardiac rhabdomyoma: a clinicopathologic study, *Mod Pathol* 4, 70, 1991b.

Burke AP and Virmani R, Intramural coronary artery dysplasia of the ventricular septum and sudden death, *Hum Pathol* 29, 1124, 1998.

Burns CJ and Manion WC, Sudden unexpected death of a two-year-old child from thrombosis of both coronary arteries with aneurysmal dilatation of the vessels, *Med Ann District of Columbia* 38, 381, 1969.

Burns JC, Shike H, Gordon JB, et al., Sequelae of Kawasaki disease in adolescents and young adults, *J Am Coll Cardiol* 28, 253, 1996.

Butterworth JS and Poindexter CA, Papiloima of cusp of the aortic valve: report of a patient with sudden death, *Circulation* 48, 213, 1973.

Buyon JP, Hiebert R, Copel J, et al., Autoimmune-associated congenital heart block: demographics, mortality, morbidity and recurrence rates obtained from a national neonatal lupus registry, *J Am Coll Cardiol* 31, 1658, 1998.

Byard RW, Idiopathic arterial calcification and unexpected infant death, *Pediatr Pathol Lab Med* 16, 985, 1996.

Byard RW, Bourne AJ, and Adams PS, Subarterial ventricular septal defect in an infant with sudden unexpected death: cause or coincidence? *Am J Cardiovasc Pathol* 3, 333, 1990a.

Byard RW, Edmonds JF, Silverman E, and Silver MM, Clinical conference: respiratory distress and fever in a 2-month-old infant, *J Pediatr* 118, 306, 1991a.

Byard RW, Gilbert J, James R, and Lokan RJ, Amphetamine derivative fatalities in South Australia: is "Ecstasy" the culprit? *Am J Forensic Med Pathol* 19, 261, 1998.

Byard RW, Keeley FW, and Smith CR, Type IV Ehlers-Danlos syndrome presenting as sudden infant death, *Am J Clin Pathol* 93, 579, 1990b.

Byard RW and Moore L, Total anomalous pulmonary venous drainage and sudden death in infancy, *Forensic Sci Int* 51, 197, 1991.

Byard RW, Smith NM, and Bourne AJ, Association of right coronary artery hypoplasia with sudden death in an eleven-year-old child, *J Forensic Sci* 36, 1234, 1991b.

Byard RW, Smith NM, and Bourne AJ, Incidental cardiac rhabdomyomas: a significant finding necessitating additional investigation at the time of autopsy, *J Forensic Sci* 36, 1229, 1991c.

Cabin HS, Clubb KS, Vita N, et al., Regional dysfunction by equilibrium radionuclide angiocardiography: a clinicopathologic study evaluating the relation of degree of dysfunction to the presence and extent of myocardial infarction, *J Am Coll Cardiol* 4, 743, 1987.

Cachecho R, Grindlinger GA, and Lee VW, The clinical significance of myocardial contusion, *J Trauma*, 33, 68, 1992.

Caforio A, Mahon N, Tona F, and McKenna W, Circulating cardiac autoantibodies in dilated cardiomyopathy and myocarditis: pathogenetic and clinical significance, *Eur J Heart Fail* 4, 411, 2002.

Cagle PT, Kim HS, and Titus JL, Congenital stenotic arteriopathy with medial dysplasia, *Hum Pathol* 16, 528, 1985.

Calabrese F, Basso C, Carturan E, et al., Arrhythmogenic right ventricular cardiomyopathy/dysplasia: is there a role for viruses? *Cardiovascular Pathol* 5, 1, 2006.

Calabrò P, Willerson JT, and Yeh ETH, Inflammatory cytokines stimulated C-reactive protein production by human coronary artery smooth muscle cells, *Circulation* 108, 1930, 2003.

Calafiore AM, Di Mauro M, Canosa C, et al., Early and late outcome of myocardial revascularization with and without cardiopulmonary bypass in high risk patients (Euroscore >6), *Eur J Cardiothorac Surg* 23, 360, 2003.

Calif RM, Abdelmeguid AE, Kuntz RE, et al., Myonecrosis after revascularization procedures, *J Am Coll Cardiol* 31, 241, 1998.

Calkins H, Arrythmogenic right-ventricular dysplasia/cardiomyopathy, *Curr Opin Cardiol* 21, 55, 2006.

Calkins H, Allman K, Bolling S, et al., Correlation between scintigraphic evidence of regional sympathetic neuronal dysfunction and ventricular refractoriness in human heart, *Circulation* 89, 172, 1993.

Camerini F, Gavazzi A, and De Maria R, *Advances in Cardiomyopathies*, Springer-Verlag, Heidelberg, 1998.

Camici P, Wijn W, Borgers M, et al., Pathophysiologic mechanism of chronic reversible left ventricular dysfunction due to coronary artery disease (hibernating myocardium), *Circulation* 96, 3205, 1997.

Campbell NC, Thomson SR, Muckart DJ, et al., Review of 1189 cases of penetrating cardiac trauma, *Br J Surg*, 84, 1737, 1977.

Candelle J, Valle V, Paya J, et al., Post-traumatic coronary occlusion and early ventricular aneurysm, *Am Heart J* 97, 509, 1979.

Cannon WB, "Voodoo" death, *Am Anthropol* 44, 169, 1942.

Canun S, Peraz N, and Beirana LG, Andersen syndrome autosomal dominant in three generations, *Am J Med Genet* 85, 147, 1999.

Cao JM, Fishbein MC, Han JB, et al., Relationship between regional cardiac hyperinnervation and ventricular arrhythmia, *Circulation* 101, 1960, 2000.

Cao W, Hashibe M, Rao JY, et al., Comparison of methods for DNA extraction from paraffin embedded tissues and buccal cells, *Cancer Detect Prev* 27, 397, 2003.

Carleton RA and Boyd T, Traumatic laceration of the anterior coronary artery treated by ligation without myocardial infarction: report of a case with a review of the literature, *Am Heart J* 56, 136, 1958.

Carley S, Ali B, and Mackway-Jones K, Acute myocardial infarction in cocaine induced chest pain presenting as an emergency, *Emerg Med J* 20, 174, 2003.

Case CL, Gillette PC, and Crawford FA, Cardiac rhabdomyomas causing supraventricular and lethal ventricular arrhythmias in an infant, *Am Heart J* 122, 1484, 1991.

Castro VJ and Nacht R, Cocaine-induced bradyarrhythmia, *Chest* 117, 275, 2000.

Caves PK, Schultz WP, Dong E, Stinson EB, and Shumway NE, New instrument for transvenous cardiac biopsy, *Am J Cardiol* 33, 264, 1974.

Cebelin MS and Hirsch CS, Human stress cardiomyopathy: myocardial lesions in victims of homicidal assaults without internal injuries, *Hum Pathol* 11, 123, 1980.

Celermajer DS, Bull C, and Till JA, Ebstein's anomaly: presentation and outcome from fetus to adult, *J Am Coll Cardiol* 23, 170, 1994.

Celermajer DS, Sholler GF, Howman-Giles R, and Celermajer JM, Myocardial infarction in childhood: clinical analysis of 17 cases and medium term follow up of survivors, *Br Heart J* 65, 332, 1991.

Chalmers RA, Stanley CA, English N, et al., Mitochondrial carnitine-acylcarnitine translocase deficiency presenting as sudden neonatal death, *J Pediatr* 131, 220, 1997.

Chambers JW, Denes P, Dahl W, et al., Familial sudden death syndrome with an abnormal signal-averaged electrocardiogram as a potential marker, *Am Heart J* 130, 318, 1995.

Chan K-L, Early clinical course and long-term outcome of patients with infective endocarditis complicated by perivalvular abscess, *CMAJ* 167, 19, 2002.

Chan KY, Iwahara M, Benson LN, et al., Immunosuppressive therapy in the management of acute myocarditis in children: a clinical trial, *J Am Coll Cardiol* 17, 458, 1991.

Chandler AB, Chapman I, Erhardt LR, et al., Coronary thrombosis in myocardial infarction: report of a workshop on the role of coronary thrombosis in the pathogenesis of acute myocardial infarction, *Am J Cardiol* 34, 823, 1974.

Chandrasekar B, Doucet S, Bilodeau L, et al., Complications of cardiac catheterization in the current era: a single-center experience, *Catheter Cardiovasc Interv*, 52, 289, 2000.

Chang RA and Rossi NF, Intermittent cocaine use associated with recurrent dissection of the thoracic and abdominal aorta, *Chest* 108, 1758, 1995.

Chapman DW, The cumulative risks of prolapsing mitral valve: 40 years of follow-up, *Tex Heart Inst J* 21, 267, 1994.

Chapman I, The cause–effect relationship between recent coronary artery occlusion and acute myocardial infarction, *Am Heart J* 87, 267, 1974.

Chareconthaitawee P, Christian TF, Hirose K, et al., Relation of initial infarct size to extent of left ventricular remodeling in the year after acute myocardial infarction, *J Am Coll Cardiol* 25, 567, 1995.

Charles R, Holt S, and Kirkham N, Myocardial infarction and marijuana, *Clin Toxicol* 14, 433, 1979.

Chauvaud SM, Brancaccio G, and Carpentier AF, Cardiac arrhythmia in patients undergoing surgical repair of Ebstein's anomaly, *Ann Thorac Surg* 71, 1547, 2001.

Cheitlin MD, De Castro CM, and McAllister HA, Sudden death as a complication of anomalous left coronary origin from the anterior sinus of Valsalva: a not-so-minor congenital anomaly, *Circulation* 50, 780, 1974.

Cheitlin MD, McAllister HA, and De Castro CM, Myocardial infarction without atherosclerosis, *JAMA* 231, 951, 1975.

Chen C, Ma L, Linfert DR, et al., Myocardial cell death and apoptosis in hibernating myocardium, *J Am Coll Cardiol* 30, 1407, 1997.

Chen Q, Kirsch GE, Zhang D, et al., Genetic basis and molecular mechanism for idiopathic ventricular fibrillation, *Nature* 392, 293, 1998.

Chen Q, Zhang D, Gingell RL, et al., Homozygous deletion in KVLQT1 associated with Jervell and Lange-Nielsen syndrome, *Circulation* 99, 1344, 1999.

Chenard AA, Becane HM, Tertrain F, et al., Ventricular arrhythmia in Duchenne muscular dystrophy: prevalence, significance and prognosis, *Neuromuscul Disord* 3, 201, 1993.

Cheng TO, Bashour T, Shing BK, et al., Myocardial infarction in the absence of coronary arteriosclerosis, *Am J Cardiol* 30, 680, 1972.

Cheng W, Li B, Kajstura L, et al., Stretch-induced programmed myocyte cell death, *J Clin Invest* 96, 2247, 1995.

Chesler E, King RA, and Edwards JE, The myxomatous mitral valve and sudden death, *Circulation* 67, 632, 1983.

Cheung YF, Cheng VY, Yung TC, et al., Cardiac rhythm and symptomatic arrhythmia in right atrial isomerism, *Am Heart J* 144, 159, 2002.

Chevalier P, Dacosta A, Defaye P, et al., Arrhythmic cardiac arrest due to isolated coronary artery spasm: long term outcome of seven resuscitated patients, *J Am Coll Cardiol* 31, 57, 1998.

Chiang C-E and Roden DM, The long QT syndromes: genetic basis and clinical implications, *J Am Coll Cardiol* 36, 1, 2000.

Chiasson DA, Ipp M, and Silver MM, Clinical conference: acute heart failure in an 8-year-old diabetic girl, *J Pediatr* 116, 472, 1990.

Chilian WM, Mass HJ, Williams SM, et al., Microvascular occlusions promote coronary collateral growth, *Am J Physiol* 258 (*Heart Circ Physiol* 27), H1103, 1990.

Chimenti C, Pieroni M, Maseri A, et al., Histologic findings in patients with clinical and instrumental diagnosis of sporadic arrhythmogenic right ventricular dysplasia, *J Am Coll Cardiol* 43, 2306, 2004.

Chopra P and Bhatia ML, Chronic rheumatic heart disease in India: a reappraisal of pathologic changes, *J Heart Valve Dis* 1, 92, 1992.

Chow LTC, Chow, DH, Lee JCK, and Lie JT, Tuberculous aortitis with coronary ostial and left ventricular outflow obstruction: unusual cause of sudden unexpected death, *Cardiovasc Pathol* 5, 133, 1996.

Chug SS, Senashova O, Watts A, et al., Postmortem molecular screening in unexplained sudden death, *J Am Coll Cardiol*, 43, 1625, 2004.

Chugh SS, Jui J, Gunson K, et al., Current burden of sudden cardiac death: multiple source surveillance versus retrospective death certificate-based review in a large U.S. community, *J Am Coll Cardiol* 44, 1268, 2004.

Ciszewski A, Bilinska ZT, Lubiszewska B, et al., Dilated cardiomyopathy in children: clinical course and prognosis, *Pediatr Cardiol* 15, 121, 1994.

Ciuffo AA, Ouyang P, Becker LC, et al., Reduction of sympathetic inotropic response after ischemia in dog: contributor to stunned myocardium, *J Clin Invest* 75, 1504, 1985.

Clancy CE and Rudy Y, Na+ channel mutation that causes both Brugada and long-QT syndrome phenotypes: a simulation study of mechanism, *Circulation* 105, 1208, 2002.

Clancy CE, Tateyama M, and Kass RS, Insights into the molecular mechanisms of bradycardia-triggered arrhythmias in long QT-3 syndrome, *J Clin Invest* 110, 1251, 2002.

Claudon DG, Claudon DB, and Edwards JE, Primary dissecting aneurysm of coronary artery: a cause of acute myocardial ischemia, *Circulation* 45, 259, 1972.

Clausell N, Butany J, Gladstone P, et al., Myocardial vacuolization, a marker of ischemic injury, in surveillance cardiac biopsies post transplant: correlations with morphologic vascular disease and endothelial dysfunction, *Cardiovasc Pathol* 5, 29, 1996.

Clusin WT, Bristow MR, Karagueuzian HS, et al., Do calcium-dependent ionic currents mediate ischemic ventricular fibrillation? *Am J Cardiol* 49, 606, 1982.

Coady MA, Davies RR, Roberts M, et al., Familial patterns of thoracic aortic aneurysms, *Arch Surg* 134, 361, 1999.

Cobb LA, Baum RS, and Schaffer WA, Resuscitation from out-of-hospital ventricular fibrillation: 4 years follow-up, *Circulation* 51–52 (Suppl 3), 223, 1975.

Cobb DK, High KP, Sawyer RG, et al., A controlled trial of scheduled replacement of central venous and pulmonary-artery catheters, *N Engl J Med*,327, 1062, 1992.

Cobb LA, Werner JA, and Trobaugh GB, Sudden cardiac death: I: A decade's experience with out-of-hospital resuscitation, *Modern Concepts Cardiovasc Dis* 6, 31, 1980.

Codd MB, Sugrue DD, Gersh BJ, and Melton LJ III, Epidemiology of idiopathic dilated and hypertrophic cardiomyopathy. A population-based study in Olmsted County, Minnesota, 1975–1984, *Circulation* 80(3), 564, 1989.

Cohen M, Fuster V, Steele PM, et al., Coarctation of the aorta: long-term follow-up and prediction of outcome after surgical correction, *Circulation* 80, 840, 1989.

Cohle SD, Balraj E, and Bell M, Sudden death due to ventricular septal defect, *Pediatr Dev Pathol* 2, 327, 1999.

Cohle SD and Lie JT, Dissection of the aorta and coronary arteries associated with acute cocaine intoxication, *Arch Pathol Lab Med* 116, 1239, 1992.

Cohle SD and Sampson BA, The negative autopsy: sudden cardiac death or other?, *Cardiovasc Pathol* 10, 219, 2001.

Cohle SD, Suarez-Mier MP, and Aguilera B, Sudden death resulting from lesions of the cardiac conduction system, *Am J Forensic Med Pathol* 23, 83, 2002.

Cohle SD, Titus JL, Espinola A, and Jachimczyk JA, Sudden unexpected death due to coronary giant cell arteritis, *Arch Pathol Lab Med* 106, 171, 1982.

Cohn PF, *Silent Myocardial Ischemia and Infarction*, 2nd ed., Marcel Dekker, New York, 1989.

Cohn JN, Ferrari R, Sharpe N, et al., Cardiac remodelling: concepts and clinical implications: a consensus paper from an international forum on cardiac remodelling, *J Am Coll Cardiol* 35, 569, 2000.

Cohn LH, Kosek J, and Angell WW, Pulmonary arteriosclerosis produced by hyperoxemic normotensive perfusion, *Circulation* 42 (Suppl 3), 114, 1970.

Cohn PF, Maddox DE, Holman BL, and See JR, Effect of coronary collateral vessels on regional myocardial blood flow in patients with coronary artery disease: relation of collateral circulation to vasodilator reserve and left ventricular function, *Am J Cardiol* 46, 359, 1980.

Cohnheim J and von Schulthess-Rechberg A, Ueber die Folgen der Kranzarterienverschliessung fur das Herz, *Virchow Arch Pathol Anat* 85, 503, 1881.

Coleman DL, Ross TF, and Naughton JL, Myocardial ischemia and infarction related to recreational cocaine use, *West J Med* 136, 444, 1982.

Collier PE and Goodman GB, Cardiac tamponade caused by central venous catheter perforation of the heart: a preventable complication, *J Am Coll Surg*, 181, 459, 1995.

Collier PE, Ryan JJ, and Diamond DL, Cardiac tamponade from central venous catheters: report of a case and review of the English literature, *Angiology*, 35, 595, 1984.

Collins JS, Higginson JD, Boyle DM, et al., Myocardial infarction during marijuana smoking in a young female, *Eur Heart J* 6, 637, 1985.

Colucci WS, Apoptosis in the heart, *N Engl J Med* 335, 1224, 1996.

Communal C, Singh K, Pimentel DR, and Colucci WS, Norepinephrine stimulates apoptosis in adult rat ventricular myocytes by activation of the β-adrenergic pathway, *Circulation* 98, 1329, 1998.

Conces DJ and Holden RW, Aberrant location and complications in initial palcement of subclavian vein catheters, *Arch Surg*, 119, 293, 1984.

Connelly MS, Liu PP, Williams WG, et al., Congenitally corrected transposition of the great arteries in the adult: functional status and complications, *J Am Coll Cardiol* 27, 1238, 1996.

Connor RCR, Focal myocytolysis and fuchsinophilic degeneration of the myocardium of patients dying with various brain lesions, *Ann NY Acad Sci* 156, 261, 1969.

Constantinides P, The role of endothelial injury in arterial thrombosis and atherogenesis, in *Thrombosis and Coronary Heart Disease*, Halonen PI and Louhita A, Eds., Karger, Basel Advances in Cardiology 4, 67, 1970.

Conte MR, Bonfiglio G, Orzan F, et al., Hypertrophic obstructive cardiomyopathy in a patient with Turner syndrome [article in Italian], *Cardiologia* 40, 947, 1995.

Cooke RA and Chambers JB, QT interval in anorexia nervosa, *Br Heart J* 72, 69, 1994.

Cooley DA, Bloodwell RD, Hallman GL, et al., Human cardiac transplantation, *Circulation* 39, I3, 1969.

Corbalan R, Verrier R, and Lown B, Psychological stress and ventricular arrhythmias during myocardial infarction in the conscious dog, *Am J Cardiol* 34, 692, 1974.

Corcos AP, Tzivoni D, and Medina A, Long QT syndrome and complete situs inversus: preliminary report of a family, *Cardiology* 76, 228, 1989.

Corday E, Heng MK, Meerbaum S, et al., Derangements of myocardial metabolism preceding onset of ventricular fibrillation after coronary occlusion, *Am J Cardiol* 39, 880, 1977.

Corrado D, Basso C, Buja G, et al., Right bundle branch block, right precordial St-segment elevation, and sudden death in young people, *Circulation* 103, 710, 2001a.

Corrado D, Basso C, Nava A, et al., Sudden death in young people with apparently isolated mitral valve prolapse, *J Ital Cardiol* 27, 1097, 1997a.

Corrado D, Basso C, and Thiene G, Arrhythmogenic right ventricular cardiomyopathy: diagnosis, prognosis, and treatment, *Heart* 83, 588, 2000.

Corrado D, Basso C, Thiene G, et al., Spectrum of clinicopathologic manifestations of arrhythmogenic right ventricular cardiomyopathy/dyspalsia: a multicenter study, *J Am Coll Cardiol* 30, 1512, 1997b.

Corrado D, Basso C, and Thiene G, Pathologic findings in victims of sport-related sudden cardiac death, in *La dimensione medico-legale della medicina dello sport* (Sports Medicine: A Forensic Approach), Turillazzi E, Ed., Edizione Colosseum, Rome, 1998, p. 9.

Corrado D, Basso C, and Thiene G, Sudden cardiac death in young people with apparently normal heart, *Cardiovasc Res* 50, 399, 2001b.

Corrado D, Fontaine G, Marcus FI, et al., Arrhythmogenic right ventricular dysplasia/cardiomyopathy: need for an international registry, *Circulation* 101, 101, 2000.

Corrado G, Lissoni A, Beretta S, et al., Prognostic value of electrocardiograms, ventricular late potentials, ventricular arrhythmias, and left ventricular systolic dysfunction in patients with Duchenne muscular dystrophy, *Am J Cardiol* 89, 838, 2002.

Corrado D, Nava A, Buja G, et al., Familial cardiomyopathy underlies syndrome of right bundle branch block, ST segment elevation and sudden death, *J Am Coll Cardiol* 27, 443, 1996.

Corrado D, Thiene G, Cocco P, and Frescura C, Non-atherosclerotic coronary artery disease and sudden death in the young, *Br Heart J* 68, 601, 1992.

Corrado D, Thiene G, and Pennelli N, Sudden death as the first manifestation of coronary artery disease in young people (less than or equal to 35 years), *Eur Heart J* 9 (Suppl N), 139, 1988.

Corti R, Osende JI, and Fayad ZA, *In vivo* non-invasive detection and age definition of arterial thrombus by MRI, *J Am Coll Cardiol* 39, 1366, 2002.

Côté A, Russo P, and Michaud J, Sudden unexpected deaths in infancy: what are the causes? *J Pediatr* 135, 437, 1999.

Cowley MJ, Disciascio G, Rehr RB, et al., Angiographic observations and clinical relevance of coronary thrombus in unstable angina pectoris, *Am J Cardiol* 63, 108E, 1989.

Cowley MJ, Hastillo A, Vetrovec GW, and Hess ML, Effects of intracoronary streptokinase in acute myocardial infarction, *Am Heart J* 102, 1149, 1981.

Coy KM, Park JC, Fishbein MC, et al., *In vitro* validation of three-dimensional intravascular ultrasound for the evaluation of arterial injury after balloon angioplasty, *J Am Coll Cardiol* 20, 692, 1992.

Crainicianu A, Anatomische Studien uber die Koronararterien und experimentelle Untersuchungen uber ihre Durchgangigkeit, *Virch Arch Pathol Anat* 238, 1, 1922.

Cribier A, Korsatz L, Koning R, et al., Improved myocardial ischemic response and enhanced collateral circulation with long repetitive coronary occlusion during angioplasty: a prospective study, *J Am Coll Cardiol* 20, 578, 1992.

Cruickshank JM, Pennert K, Sornan AE, et al., Low mortality from all causes including myocardial infarction, in well-controlled hypertensives treated with a beta-blocker plus other antihypertensives, *J Hypertension* 5, 489, 1987.

Cuadros CL, Hutchinson JE, and Mogtader AH, Laceration of a mitral papillary muscle and the aortic root as a result of blunt trauma to the chest, *J Thorac Cardiovasc Surg*, 88, 134, 1984.

Curfman GD, Fatal impact — concussion of the heart, *New Engl J Med*, 338, 25, 1841, 1998.

Curran ME, Splawski I, Timothy KW, Vincent GM, Green ED, and Keatling MT, A molecular basis for cardiac arrhythmia: HERG mutations cause long QT syndrome, *Cell* 80, 795–803, 1995.

Czegledy-Nagy EN, Cutz E, and Becker LE, Sudden death in infants under one year of age, *Pediatr Pathol* 13, 671, 1993.

D'Adamo P, Fassone L, Gedeon A, et al., The X-linked gene G4.5 is responsible for different infantile dilated cardiomyopathies, *Am J Hum Genet* 61, 862, 1997.

D'Amore PA and Thompson RW, Mechanisms of angiogenesis, *Ann Rev Physiol* 49, 453, 1987.

Dabizzi RP, Caprioli G, Aiazzi L, et al., Distribution and anomalies of coronary arteries in tetralogy of Fallot, *Circulation* 61, 95, 1980.

Dae MW, Herre JM, O'Connel WJ, et al., Scintigraphic assessment of sympathetic innervation after transmural versus nontransmural myocardial infarction, *J Am Coll Cardiol* 17, 1416, 1991.

Dae MW, Lee RJ, Ursell PC, et al., Heterogeneous sympathetic innervation in Germans shepherd dogs with inherited ventricular arrhythmia and sudden cardiac death, *Circulation* 63, 1337, 1997.

Dalal D, Nasir K, Bomma C, et al., Arrhythmogenic right ventricular dysplasia: a United States experience, *Circulation* 112, 3823, 2005.

Daliento L, Somerville J, Presbitero P, et al., Eisenmenger syndrome: factors relating to deterioration and death, *Eur Heart J* 19, 1845, 1998.

D'Amati G, Leone O, di Gioia CR, et al., Arrhythmogenic right ventricular cardiomyopathy: clinicopathologic correlation based on a revised definition of pathologic patterns, *Hum Pathol* 32, 1078, 2001.

Damotte D, Rambaud C, Cheron G, Nassif X, Burgard M, Lavaud J, Canioni D, Brousse N, and Rudler M, Frequency of meningitis in cot deaths. *Third SIDS International Conference*, Stavanger, program and abstracts, 1994, 156.

Danieli GA and Rampazzo A, Genetics of arrhythmogenic right ventricular cardiomyopathy, *Curr Opin Cardiol* 17, 218, 2002.

Dantzig JM, Delemarre BJ, Bot H, and Visser A, Left ventricular thrombus in acute myocardial infarction, *Eur Heart J* 17, 1640, 1996.

Daoud AS, Pankin D, Tulgan H, et al., Aneurysm of the coronary artery: report of ten cases and review of the literature, *Am J Cardiol* 11, 228, 1963.

Darke S and Zador D, Fatal heroin overose: a review, *Addiction* 91, 1765, 1996.

Darok M, Schmid CB, Gatterning R, et al., Sudden death from myocardial contusion following an isolated blunt force trauma to the chest, *Int J Legal Med* 115, 85, 2001.

Datta BN, Khattri HN, Bidwas PS, et al., Infective endocarditis at autopsy in northern India, *Jpn Heart J* 23, 329, 1982.

Dauchin N, Vaur L, Genes N, et al., Treatment of acute myocardial infarction by primary coronary angioplasty or intravenous thrombolysis in the "real world": one-year results from a nationwide French survey, *Circulation* 99, 2639, 1999.

David D and Blumberg RM, Subintimal aortic dissection with occlusion after blunt abdominal trauma, *Arch Surg* 100, 302, 1970.

David TE, Armstrong S, Ivanov J, et al., Results of aortic valve-sparing operations, *J Thorac Cardiovasc Surg* 122, 39, 2001.

David TE and Feindel CM, An aortic valve sparing operation for patients with aortic incompetence and aneurysm of the ascending aorta, *J Thorac Cardiovasc Surg* 103(4), 617, 1992.

Davies MJ, The investigation of sudden cardiac death, *Histopathology* 34, 93, 1999.

Davies MJ, A macro and micro view of coronary vascular insult in ischemic heart disease, *Circulation* 82 (Suppl II), 32, 1990.

Davies MJ and McKenna WJ, Hypertrophic cardiomyopathy — pathology and pathogenesis, *Histopathology* 26, 493, 1995.

Davies MJ and Thomas A, Thrombosis and acute coronary artery lesions in sudden cardiac ischemic death, *N Engl J Med* 310, 1137, 1984.

Davies MJ, Thomas AC, Knapman PA, et al., Intramyocardial platelet aggregation in patients with unstable angina suffering sudden ischemic cardiac death, *Circulation* 73, 418, 1986.

Davies MJ, Woolf N, Robertson WB, Pathology of acute myocardial infarction with particular reference to occlusive coronary thrombi, *Br Heart J* 38, 659, 1976.

Davis AM, Gow RM, McCrindle BW, and Hamilton RM, Clinical spectrum, therapeutic management, and follow-up of ventricular tachycardia in infants and young children, *Am Heart J* 131, 186, 1996a.

Davis AM, Silver MM, and Freedom RM, Benign congenital cardiac tumour associated with "malignant" cardiac malformation, *Cardiol Young* 6, 84, 1996b.

Davis GG and Swalwell CI, Acute aortic dissections and ruptured berry aneurysms associated with methamphetamine abuse, *J Forensic Sci* 39, 1481, 1994.

Day JD, Rayburn BK, Gaudin PB, et al., Cardiac allograft vasculopathy: the central pathogenetic role of ischemia-induced endothelial injury, *J Heart Lung Transplant* 14, S142, 1995.

Delaney K and Hoffman RS, Pulmonary infarction associated with crack cocaine use in a previously healthy 23 year old woman, *Am J Med* 91, 92, 1991.

de la Torre R, Farre M, Roset PN, et al., Human pharmacology of MDMA: pharmacokinetics, metabolism, and disposition, *Ther Drug Monit* 26, 137, 2004.

de Leeuw N, Melchers WJ, Balk AH, et al., Study on microbial persistence in end-stage idiopathic dilated cardiomyopathy, *Clin Infect Dis* 29, 522, 1999.

Deloche A, Jebara VA, Relland JYM, et al., Valve repair with Carpentier techniques: the second decade, *J Thorac Cardiovasc Surg* 99, 990, 1990.

De Marco T, Dae M, Yuen-Green MSF, et al., Iodine-123 metaiodobenzylguanidine scintigraphic assessment of the transplanted human heart: evidence for late innervation, *J Am Coll Cardiol* 25, 927, 1995.

De Maria R, Parodi O, Baroldi G, et al., Morphological bases for thallium-201 uptake in cardiac imaging and correlates with myocardial blood flow distribution, *Eur Heart J* 17, 951, 1996.

De Piccoli B, Giada F, Benettin A, et al., Anabolic steroid use in body builders: an echocardiographic study of left ventricle morphology and function, *Int J Sports Med* 12, 408, 1991.

De Ponti F, Poluzzi E, Cavalli A, et al., Safety of non-antiarrhythmic drugs that prolong the QT interval or induce torsade de pointes: an overview, *Drug Saf* 25, 263, 2002.

de Rooij PD and Haarman HJTM, Herniation of the stomach into the pericardial sac combined with cardiac luxation caused by blunt trauma: a case report, *J Trauma*, 34, 453, 1993.

de Virgilio C, Nelson RJ, Milliken J, et al., Ascending aortic dissection in weight lifters with cystic medial degeneration, *Ann Thorac Surg* 49, 638, 1990.

De Wood MA, Spores J, Notske R, et al., Prevalence of total coronary occlusion during the early hours of transmural myocardial infarction, *N Engl J Med* 303, 897, 1980.

De Wood MA, Stifter WF, Simpson CS, et al., Coronary arteriographic findings soon after non-Q-wave myocardial infarction, *N Engl J Med* 315, 417, 1986.

Deal BJ, Keane JF, Gillette PC, et al., Wolff-Parkinson-White syndrome and supraventricular tachycardia during infancy: management and follow-up, *J Am Coll Cardiol* 5, 130, 1985.

Deanfield JE, Ho SY, Anderson RH, McKenna WJ, et al., Late sudden death after repair of tetralogy of Fallot: a clinicopathologic study, *Circulation* 67, 626, 1983.

Debich DE, Williams KE, and Anderson RH, Congenital atresia of the orifice of the left coronary artery and its main stem, *Int J Cardiol* 22, 398, 1989.

Dec GW Jr, Palacios IF, Fallon JT, and Aretz HT, Active myocarditis in the spectrum of acute dilated cardiomyopathies: clinical features, histologic correlates, and clinical outcome, *N Engl J Med* 312, 885, 1985.

Del Monte F, Harding SE, Schmidt U, et al., Restoration of contractile function in isolated cardiomyocytes from failing human hearts by gene transfer of serca 2a, *Circulation* 100, 2308, 1999.

Demetriades D, Penetrating injuries to the thoracic great vessels, *J Card Surg*, 12, 173, 1997.

Denfield SW, Rosenthal G, Gajarski RJ, et al., Restrictive cardiomyopathies in childhood: etiologies and natural history, *Tex Heart Inst J* 24, 38, 1997.

Denton JS and Kalelkar MB. Homicidal commotio cordis in two children, *J Forensic Sci*, 45, 734, 2000.

Deschenes I, Baroudi G, Berthet M, et al., Electrophysiological characterization of SCN5A mutations causing long QT (E1784K) and Brugada (R1512W and R1432G) syndromes, *Cardiovasc Res* 46, 55, 2000.

Detrano RC, Wong ND, Doherty JM, et al., Coronary calcium does not accurately predict near-term future coronary events in high-risk adults, *Circulation* 99, 2633, 1999.

Dettmeyer R, Schlamann M, and Madea B, Immunohistochemical techniques improve the diagnosis of myocarditis in cases of suspected sudden infant death syndrome (SIDS), *Forensic Sci Int* 105, 83, 1999.

Dettmeyer R, Schlamann M, and Madea B, Myokarditis und plötzlicher Kindstod: Konventionelle histologische und immunhistochemische Unter-uchungen, *Rechtsmedizin* 8(Supp I), A60, 1998.

Deutsch E, Berger M, Kussmaul WG, et al., Adaptation to ischemia during percutaneous transluminal coronary angioplasty: clinical, hemodynamic and metabolic features, *Circulation* 82, 2044, 1990.

Devlin G, Lazzam L, and Schwartz L, Moratlity related to diagnostic cardiac catheterization. The importance of left main coronary disease and catheter induced trauma, *Int J Card Imag*, 13, 379, 1997.

Dhurandhar RW, Watt DL, Silver MD, et al., Printzmetal's variant form of angina with arteriographic evidence of coronary arterial spasm, *Am J Cardiol* 30, 902, 1972.

Di Bello MG, Masini E, Ioannides C, et al., Histamine release from rat mast cells induced by the metabolic activation of drugs of abuse into free radicals, *Inflamm Res* 47, 122, 1998.

Di Carli MF, Bianco-Batles M, Lauda ME, et al., Effects of neuropathy on coronary blood flow in patients with diabetes mellitus, *Circulation* 100, 813, 1999.

Dick WF, Baskett PJ, Grande C, et al., Recommendations for uniform reporting of data following major trauma — the Utstein style (as of July 17, 1999). An international trauma anaesthesia and critical care society (ITACCS), *Acta Anaesthesiol Belg*, 51, 18, 2000.

Dickens P, Poon CS, and Wat MS, Sudden death associated with solitary intracavitary right atrial metastatic tumor deposit, *Forensic Sci Int* 57, 169, 1992.

Dickerman R, Schaller F, Prather I, and McConathy WJ, Sudden cardiac death in a 20-year-old bodybuilder using anabolic steroids, *Cardiology* 86, 172, 1995.

Dickerman RD, McConathy WJ, and Zachariah NY, Testosterone, sex hormone–binding globulin, lipoproteins, and vascular disease risk, *J Cardiovasc Risk* 4, 363, 1997.

Dietl CA, Cazzaniga ME, Dubner SJ, et al., Life-threatening arrhythmias and RV dysfunction after surgical repair of tetralogy of Fallot: comparison between transventricular and transatrial approaches, *Circulation* 90, 117, 1994.

Dietrich JB, Mangeol A, Revel MO, Burgun C, Aunis D, and Zwiller J, Acute or repeated cocaine administration generates reactive oxygen species and induces antioxidant enzyme activity in dopaminergic rat brain structures, *Neuropharmacol* 48, 965, 2005.

Dietz WA, Tobis JM, and Isner JM, Failure of angiography to accurately depict the extent of coronary artery narrowing in three fatal cases of percutaneous transluminal coronary angioplasty, *J Am Coll Cardiol* 19, 1261, 1992.

Dion R, Complete arterial revascularization with the internal thoracic arteries, in *Operative Techniques in Cardiac and Thoracic Surgery: A Comparative Atlas*, Vol. 1, Book 2, Cox JL and Sundt TM, Eds., 1996, p. 84.

Di Paolo M, Luchini D, Bloise R, and Priori SG, Postmortem molecular analysis in victims of sudden unexplained death, *Am J Forensic Med Pathol* 25, 182, 2004.

Di Paolo N, Fineschi V, Di Paolo M, Wetly CW, Garosi G, Del Vecchio MT, and Bianciardi G, Kidney vascular damage and cocaine, *Clin Nephrol* 47, 298, 1997.

Dipchand AI, Tein I, Robinson B, and Benson LN, Maternally inherited hypertrophic cardiomyopathy: a manifestation of mitochondrial DNA mutations — clinical course in two families, *Pediatr Cardiol* 22, 14, 2001.

Dispersyn GD, Ausma J, Thone F, et al., Cardiomyocyte remodeling during myocardial hibernation and atrial fibrillation: prelude to apoptosis, *Cardiovasc Res* 43, 947, 1999.

Dock W, The predilection of atherosclerosis for the coronary arteries, *JAMA* 131, 875, 1946.

Dollar AL and Roberts WC, Morphologic comparison of patients with mitral valve prolapse who died suddenly with patients who died from severe valvular dysfunction or other conditions, *J Am Coll Cardiol* 17, 921, 1991.

Dominguez FE, Tate LG, and Robinson MJ, Familial fibromuscular dysplasia presenting as sudden death, *Am J Cardiovasc Pathol* 2, 269, 1988.

Donald DE, Myocardial performance after excision of the extrinsic nerves in the dog, *Circ Res* 34, 317, 1974.

Dorros G, Cowley MI, Janke L, et al., In hospital mortality rate in the National Heart, Lung and Blood Institutes percutaneous transluminal coronary angioplasty registry, *Am J Cardiol* 53, 7C, 1984.

Dowling G, McDonough E, and Bost R, Eve and Ecstasy: a report of five deaths associated with the use of MDEA and MDMA, *JAMA* 257, 1615, 1987.

Dowling GP and Buja ML, Sudden death due to left coronary artery occlusion in infective endocarditis, *Arch Pathol Lab Med* 112, 932, 1988.

Doyon S, The many faces of ecstasy, *Curr Opin Pediatr* 13, 170, 2001.

Dressler FA, Malekzadeh S, and Roberts WC, Quantitative analysis of amounts of coronary artery disease in cocaine addicts, *Am J Cardiol* 65, 303, 1990.

Driscoll DJ and Edwards WD, Sudden unexpected death in children and adolescents, *J Am Coll Cardiol* 5 (6 Suppl), 118B, 1985.

Drory Y, Turetz Y, Hiss Y, et al., Sudden unexpected death in persons less than 40 years of age, *Am J Cardiol* 68, 1388, 1991.

Drut RM and Drut R, Focal giant-cell cardiomyopathy, *Pediatr Pathol* 7, 467, 1987.

Du X, Cox HS, Dart AM, and Esler MD, Sympathetic activation triggers ventricular arrhythmias in rat heart with chronic infarction and failure, *Cardiovasc Res* 43, 919, 1999.

Dudorkinova D and Bouska I, Histochemistry of the atrioventricular conducting system during postnatal development, *Ped Pathol* 13, 191–201, 1993.

Duflou J and Mark A, Aortic dissection after ingestion of "ecstasy" (MDMA), *Am J Forensic Med Pathol* 21, 261, 2000.

Dunning DW, Kahn JK, Hawkins ET, et al., Iatrogenic coronary artery dissections extending into and involving the aortic root, *Catheter Cardiovasc Interv*, 51, 387, 2000.

Durak D, Cardiac rupture following blunt trauma, *J Forensic Sci*, 46, 171, 2000.

Duren DR and Becker AE, Focal myocytolysis mimicking the electrocardiographic pattern of transmural anteroseptal myocardial infarction, *Chest* 4, 506, 1976.

Duren DR, Becker AE, and Dunning AJ, Long-term follow-up of idiopathic mitral valve prolapse in 300 patients: a prospective study, *J Am Coll Cardiol* 11, 42, 1988.

Eagle KA, Isselbacher EM, and DeSanctis RW, International registry for Aortic Dissection (IRAD) Investigators: cocaine-related aortic dissection in perspectives, *Circulation* 105, 1529, 2002.

Eddy DD and Farber EM, Pseudoxanthoma elasticum: internal manifestations — a report of cases and a statistical review of the literature, *Arch Dermatol* 86, 729, 1962.

Edston E and van Hage-Hamsten M, Anaphylactoid shock — a common cause of death in heroin addicts? *Allergy* 52, 950, 1997.

Edwards AD, Vickers MA, and Morgan CJ, Infective endocarditis affecting the eustachian valve, *Br Heart J* 56, 561, 1986.

Edwards JE, Congenital malformations of the heart and great vessels, in *Pathology of the Heart and Blood Vessels*, 3rd ed., Gould SE, Ed., Charles C. Thomas, Springfield, IL, 1968, p. 262.

Edwards JE, Floppy mitral valve syndrome, in *Contemporary Issues in Cardiovascular Pathology*, Waller BF, Ed., Philadelphia, FA Davis, 1988, p. 249.

Edwards JE, Burnsides C, Swarm RL, and Lansing AL, Arteriosclerosis in the intramural and extramural portions of coronary arteries in the human heart, *Circulation* 13, 235, 1956.

Egashira K, Pipers F, and Morgan JP, Effects of cocaine on epicardial coronary artery reactivity in miniature swine after endothelial injury and high cholesterol feeding: *in vivo* and *in vitro* analysis, *J Clin Invest* 88, 1307, 1991.

Ehrich W, De la Chapelle C, and Cohn AE, Anatomical ontogeny (B)Man: a study of the coronary arteries, *Am J Anat* 49, 241, 1931.

Eisenberg S, Blood viscosity and fibrinogen concentration following cerebral infarction, *Circulation* 33–34 (Suppl 2), 10, 1966.

Eisenberg SJ, Scheinman MM, Dullet NK, et al., Sudden cardiac death and polymorphous ventricular tachycardia in patients with normal QT intervals and normal systolic cardiac function, *Am J Cardiol* 75, 687, 1995.

Ekelund LG, Morberg A, Olsson AG, and Oro L, Recent myocardial infarction and the conduction system: a clinicopathological correlation, *Br Heart J* 34, 774, 1972.

El-Maraghi N and Genton E, The relevance of platelet and fibrin thromboembolism of the coronary microcirculation, with special reference to sudden cardiac death, *Circulation* 62, 936, 1980.

Eliot RS, *From Stress to Strength*, Bantam Books, New York, 1994.

Eliot RS, Baroldi G, and Leone A, Necropsy studies in myocardial infarction with minimal or no coronary luminal reduction due to atherosclerosis, *Circulation* 49, 1127, 1974.

Eliot RS and Bratt G, The paradox of myocardial ischemia and necrosis in young women with normal coronary arteriograms: relation to abnormal hemoglobin–oxygen dissociation, *Am J Cardiol* 23, 633, 1969.

Eliot RS and Buell JC, Role of emotions and stress in the genesis of sudden death, *J Am Coll Cardiol* 5, 95B, 1985.

Ellaway CJ, Sholler G, Leonard H, et al., Prolonged QT interval in Rett syndrome, *Arch Dis Child* 80, 470, 1999.

Elliott PM, Gimeno Blanes JR, Mahon NG, et al., Relation between severity of left-ventricular hypertrophy and prognosis in patients with hypertrophic cardiomyopathy, *Lancet* 357, 420, 2001.

Elliott PM, Poloniecki J, Dickie S, et al., Sudden death in hypertrophic cardiomyopathy: identification of high risk patients, *J Am Coll Cardiol* 36, 2212, 2000.

Elsasser A, Schlepper M, Klovekorn WP, et al., Hibernating myocardium: an incomplete adaptation to ischemia, *Circulation* 96, 2920, 1997.

Elshershari H, Okutan V, and Celiker A, Isolated noncompaction of ventricular myocardium, *Cardiol Young* 11, 472, 2001.

Elsner M and Zeiher AM, Perforation and rupture of coroanry arteries, *Hertz*, 23, 311, 1998.

Emanuel R, Ng RA, Marcomichelakis J, et al., Formes frustes of Marfan's syndrome presenting with severe aortic regurgitation: clinicogenetic study of 18 families, *Br Heart J* 39, 190, 1977.

Emberson JW and Muir AR, Changes in ultrastructure of rat myocardium induced by hypokalaemia, *J Exp Physiol* 54, 36, 1969.

Emdin M, Marin Neto A, Carpeggiani C, et al., Heart rate variability and cardiac denervation in Chagas' disease, *J Ambul Monit* 5, 251, 1992.

Emery AE, X-linked muscular dystrophy with early contractures and cardiomyopathy (Emery-Dreifuss type), *Clin Genet* 32, 360, 1987.

Emmrich K, Herbst M, Trenckmann H, et al., Severe late complications after operative correction of aortic coarctation by interposition of prosthesis, *J Cardiovasc Surg* (Torino) 23, 205, 1982.

Engel G, Sudden and rapid death during psychological stress: folklore or folk wisdom, *Ann Int Med* 74, 771, 1971.

Enjuto M, Francino A, Navarro-Lopez F, et al., Malignant hypertrophic cardiomyopathy caused by Arg723Gly mutation in beta-myosin heavy chain gene, *J Mol Cell Cardiol* 32, 2307, 2000.

Entman ML and Ballantyne CM, Inflammation in acute coronary syndromes, *Circulation* 88, 800, 1993.

Entman ML, Hackel DB, Martin AM, et al., Prevention of myocardial lesions during hemorrhagic shock in dogs by pronethalol, *Arch Pathol* 83, 392, 1967.

Erhardt L, Biochemical markers in acute myocardial infarction: the beginning of a new era, *Eur Heart J* 17, 1781, 1996.

Ericson K, Saldeen TGP, Lindquist O, et al., Relationship of *Chlamydia pneumoniae* infection to severity of human coronary atherosclerosis, *Circulation* 101, 2568, 2000.

Eronen M, Siren MK, Ekblad H, et al., Short- and long-term outcome of children with congenital complete heart block diagnosed in utero or as a newborn, *Pediatrics* 106, 86, 2000.

Erwin MB, Hoyle JR, Smith CH, and Deliargyris EN, Cocaine and accelerated atherosclerosis: insights from intravascular ultrasound, *Int J Cardiol* 93, 301, 2004.

Escobedo LG and Zack MM, Comparisons of sudden and nonsudden coronary deaths in the United States, *Circulation* 93, 2033, 1996.

Eskander KE, Brass NS, and Gelfand ET, Cocaine abuse and coronary artery dissection, *Ann Thorac Surg* 71, 340, 2001.

Espinola-Klein C, Rupprecht HJ, Blankenberg S, et al., Impact of infectious burden on extent and long-term prognosis of atherosclerosis, *Circulation* 105, 15, 2002.

Estanol BV, Loyo MV, Mateos HJ, et al., Cardiac arrhythmias in experimental subarachnoid hemorrhage, *Stroke* 8, 440, 1977.

Esterly JA, Glacow S, and Ferguson DJ, Morphogenesis of intimal obliterative hyperplasia of small arteries in experimental pulmonary hypertension: an ultrastructural study of the role of smooth-muscle cells, *Am J Pathol* 52, 325, 1968.

Etheridge SP and Judd VE, Supraventricular tachycardia in infancy: evaluation, management, and follow-up, *Arch Pediatr Adolesc Med* 153, 267, 1999.

Fabian TC, Richardson JD, Croce MA, et al., Prospective study of blunt aortic injury: multicenter trial of the American Association for the Surgery of Trauma, *J Trauma Inj Infect Crit Care*, 42, 374, 1997.

Factor SM, Smooth muscle contraction bands in the media of coronary arteries: a postmortem marker of antemortem coronary spasm, *J Am Coll Cardiol* 6, 1326, 1985.

Factor SM and Bache RJ, Pathophysiology of myocardial ischemia, in *The Heart*, Hurst JW, Ed., McGraw-Hill, New York, 1994, p. 1119.

Factor SM, Butany J, Sole MJ, et al., Pathologic fibrosis and matrix connective tissue in the subaortic myocardium of patients with hypertrophic cardiomyopathy, *J Am Coll Cardiol* 17, 1343, 1991.

Factor SM, Minase T, Cho S, et al., Microvascular spasm in the cardiomyopathic Syrian hamster: a preventable cause of focal myocardial necrosis, *Circulation* 66, 342, 1982.

Faenchick G and Adelman S, Myocardial infarction associated with anabolic steroid use in a previously healthy 37-year-old weight lifter, *Am Heart J* 124, 507, 1992.

Faerman I, Faccio E, Mile J, et al., Autonomic neuropathy and painless myocardial infarction in diabetic patients, *Diabetes* 26, 1147, 1977.

Fairweather D, Kaya Z, Shellam GR, et al., From infection to autoimmunity, *J Autoimmun* 16, 175, 2001.

Falk E, Unstable angina with fatal outcome: dynamic coronary thrombosis leading to infarction and/or sudden death. Autopsy evidence of recurrent mural thrombosis with peripheral embolization culminating in total vascular occlusion, *Circulation* 71, 699, 1985.

Falk RH, Rubinow A, and Cohen AS, Cardiac arrhythmias in systemic amyloidosis: correlation with echocardiographic abnormalities, *J Am Coll Cardiol* 3, 107, 1984.

Falsaperla R, Romeo G, Sciacca P, et al., Cardiologic study of 10 patients with Duchenne muscular dystrophy(DMD): personal experience [article in Italian], *Pediatr Med Chir* 23, 57, 2001.

Famularo G, Polchi S, Di Bona G, et al., Acute aortic dissection after cocaine and sildenafil abuse, *J Emerg Med* 21, 78, 2001.

Fang W, Huang CC, Chu NS, et al., Childhood-onset autosomal-dominant limb-girdle muscular dystrophy with cardiac conduction block, *Muscle Nerve* 20, 286, 1997.

Fangman RJ and Hellwig CA, Histology of coronary arteries in newborn infants, *Am J Pathol* 23, 901, 1947.

Faraci RM and Westcott JL, Dissecting hematoma of the aorta secondary to blunt chest trauma, *Radiology*, 123, 569, 1977.

Farb A, Tang AL, Burke AL, et al., Sudden coronary death: frequency of active coronary lesions, inactive coronary lesions and myocardial infarction, *Circulation* 92, 1701, 1995.

Farrer-Brown G, Normal and diseased vascular pattern of myocardium of human heart: I. Normal pattern in the left ventricular free wall, *Br Heart J* 30, 527, 1968.

Farrer-Brown G and Rowles PM, Vascular supply of interventricular septum of human heart, *Br Heart J* 31, 727, 1969.

Faruqui AMA, Maloy WC, Felner JM, et al., Symptomatic myocardial bridging of coronary artery, *Am J Cardiol* 41, 1305, 1978.

Fatkin D, MacRae C, Sasaki T, et al., Missense mutations in the rod domain of the lamin A/C gene as causes of dilated cardiomyopathy and conduction-system disease, *N Engl J Med* 341, 1715, 1999.

Favaloro RG, *J Thorac Cardiovasc Surg* 58, 178, 1969.

Feldman AM, Combes A, Wagner D, et al., The role of tumor necrosis factor in the pathophysiology of heart failure, *J Am Coll Cardiol* 35, 537, 2000a.

Feldman LJ, Himbert D, Juliard JM, et al., Reperfusion syndrome: relationship of coronary blood flow reserve to left ventricular function and infarct size, *J Am Coll Cardiol* 35, 1162, 2000b.

Feldman S, Glagov S, Wissler RW, and Hughes RH, Postmortem delineation of infarcted myocardium: coronary perfusion with Nitro Blue Tetrazolium, *Arch Pathol Lab Med* 100, 55, 1976.

Felker GM, Shaw LK, and O'Connor CM, A standardized definition of ischemic cardiomyopathy for use in clinical research, *J Am Coll Cardiol* 39, 210, 2002.

Felmeden D, Singh SP, and Lip GY, Anomalous coronary arteries of aortic origin, *Int J Clin Pract* 54, 390, 2000.

Ferjani M, Droc G, Dreux S, et al., Circulating cardiac troponin T in myocardial contusion, *Chest*, 111, 427, 1997.

Fermanis GG, Ekangaki AK, Salmon AP, et al., Twelve year experience with the modified Blalock-Taussig shunt in neonates, *Eur J Cardiothorac Surg* 6, 586, 1992.

Ferrans V and Rodriguez RJ, Ultrastructure of normal heart, in *Cardiovascular Pathology*, 2nd ed., Silver MD, Ed., Churchill Livingstone, New York, 1991, p. 78.

Ferrans VJ, Morrow AG, and Roberts WC, Myocardial ultrastructure in idiopathic hypertrophic subaortic stenosis: a study of operatively excised left ventricular outflow tract muscle in 14 patients, *Circulation* 45, 769, 1972.

Ferrari R, Alfieri O, Curello S, et al., Occurrence of oxidative stress during reperfusion of the human heart, *Circulation* 8, 201, 1990.

Ferris JAJ, Hypoxic changes in conducting tissue of the heart in sudden death in infancy syndrome, *BMJ* 2, 23–25, 1973.

Ferris JAJ, The heart in sudden infant death, *J Forens Sci Soc* 12, 591–596, 1972.

Feyter de PJ, Ozaki Y, Baptista J, et al., Ischemia-related lesion characteristics in patients with stable or unstable angina: a study with intracoronary angioscopy and ultrasound, *Circulation* 92, 1408, 1995.

Fieguth HG, Wahlers T, Trappe HJ, and Borst HG, Arrhythmogenic mortality in heart-transplant candidates, *Transplant Int* 9, S219, 1996.

Filiano JJ and Kinney HC, Arcuate nucleus hypoplasia in the sudden infant death syndrome, *J Neuropathol Exp Neurol* 51, 394–403, 1992.

Fineschi V, Agricola E, Baroldi G, et al., Myocardial morphology of acute carbon monoxide toxicity: a human and experimental morphometric study, *Int J Legal Med* 113, 262, 2000.

Fineschi V and Baroldi G, Eds., *Patologia cardiaca e morte improvvisa*, CEDAM, Padova, 2004.

Fineschi V, Baroldi G, Centini F, et al., Cardiac oxidative stress to intraperitoneal cocaine exposure and morphologic markers indicative of myocardial injury: an experimental study in rats, *Int J Legal Med* 114, 323, 2001a.

Fineschi V, Baroldi G, Monciotti F, et al., Anabolic steroid abuse and cardiac sudden death: a pathologic report of two cases, *Arch Pathol Lab Med* 2, 253, 2001b.

Fineschi V, Cecchi R, Centini F, Paglicci Reattelli L, and Turillazzi E, Immunohistochemical quantification of pulmonary mast cell tryptase and post-mortem blood dosages of tryptase and eosinophil cationic protein in 48 heroin-related deaths, *Forensic Sci Int* 120, 189, 2001c.

Fineschi V, Centini F, Mazzeo E, and Turillazzi E, Adam (MDMA) and Eve (MDEA) misuse: an immunohistochemical study on three fatal cases, *Forensic Sci Int* 104, 65, 1999a.

Fineschi V, Centini F, Monciotti F, and Turillazzi E, The cocaine "body stuffer" syndrome: a fatal case, *Forensic Sci Int* 126, 7, 2002.

Fineschi V and Masti A, Fatal poisoning by MDMA (ecstasy) and MDEA: a case report, *Int J Legal Med* 108, 272, 1996.

Fineschi V, Paglicci Reattelli L, and Baroldi G, Coronary artery aneurysms in a young adult: a case of sudden death. A late sequelae of Kawasaki disease? *Int J Legal Med* 112, 120, 1999b.

Fineschi V, Riezzo I, Centini F, et al., Sudden cardiac death during anabolic steroid abuse: morphologic and toxicologic findings in two fatal cases of bodybuilders, *Int J Legal Med* 15, 1, 2005.

Fineschi V, Silver MD, Karch SB, et al., Myocardial disarray: an architectural disorganization possibly linked with adrenergic stress, *Int J Cardiol* 99, 277, 2005.

Fineschi V, Wetly CV, Di Paolo M, and Baroldi G, Myocardial necrosis and cocaine: a quantitative morphologic study in 26 cocaine-associated deaths, *Int J Leg Med* 3, 164, 1997.

Finsterer J, Bittner RE, and Grimm M, Cardiac involvement in Becker's muscular dystrophy, necessitating heart transplantation, 6 years before apparent skeletal muscle involvement, *Neuromuscul Disord* 9, 598, 1999.

Fish FA, Gillette PC, and Benson DW Jr, Proarrhythmia, cardiac arrest and death in young patients receiving encainide and flecainide: the Pediatric Electrophysiology Group, *J Am Coll Cardiol* 18, 356, 1991.

Fishbein MC and Siegel RJ, How big are coronary atherosclerotic plaques that rupture? *Circulation* 94, 2662, 1996.

Fishbein MC, Siegel RJ, Thompson CE, Hopkins LC, et al., Sudden death of a carrier of X-linked Emery-Dreifuss muscular dystrophy, *Ann Intern Med* 119, 900, 1993.

Fisher BA, Ghuran A, Vadamalai V, et al., Cardiovascular complications induced by cannabis smoking: a case report and review of the literature, *Emerg Med J* 22, 679, 2005.

Fisher JD, Krikler D, and Hallidie-Smith KA, Familial polymorphic ventricular arrhythmias: a quarter century of successful medical treatment based on serial exercise-pharmocologic testing, *J Am Coll Cardiol* 34, 2015, 1999.

Fitzgerald K, The "reduce the risks" campaign, SIDS International, the global strategy task force and the european society for study and prevention of infant death, in *Sudden Infant Death Syndrome. Problems, Progress & Possibilities,* Byard RW and Krous HF, Eds., Edward Arnold, London, 2001, 310–318.

Fitzpatrick AP, Shapiro LM, Rickards AF, et al., Familial restrictive cardiomyopathy with atrioventricular block and skeletal myopathy, *Br Heart J* 63, 114, 1990.

Flaherty JT, Pierce JE, Ferrans VJ, et al., Endothelial nuclear pattern in the canine arterial tree with particular reference to hemodynamic events, *Circ Res* 30, 23, 1972.

Flameng W, Van Belle H, Vanhaecke J, et al., Relation between coronary artery stenosis and myocardial purine metabolism, histology and regional function in humans, *J Am Coll Cardiol* 9, 1235, 1987.

Fleckenstein A, Janke J, Doring HJ, et al., *Myocardial Fiber Necrosis due to Intracellular Ca^{++} Overload: A New Principle in Cardiac Pathophysiology. Recent Advances in Studies on Cardiac Structure and Metabolism,* Vol. 4, Dhallas NS, Ed., University Park Press, Baltimore, MD, 1975, p. 563.

Fleisch M, Billinger M, Eberly FR, et al., Physiologically assessed coronary collateral flow and intracoronary growth factor concentration in patients with 1 to 3 vessel coronary disease, *Circulation* 100, 943, 1999.

Flesh M, Maack C, Cremers B, et al., Effect of β-blockers on free radical-induced cardiac contractile dysfunction, *Circulation* 100, 346, 1999.

Flores ED, Lange RA, Cigarroa RG, et al., Effect of cocaine on coronary artery dimensions in atherosclerotic coronary artery disease: enhanced vasoconstriction at sites of significant stenoses, *J Am Coll Cardiol* 16, 74, 1990.

Flynn MS, Kern MJ, Donohue TJ, et al., Alterations of coronary collateral blood flow velocity during intraaortic balloon pumping, *Am J Cardiol* 71, 1451, 1993.

Fodstad H, Swan H, Auberson M, et al., Loss-of-function mutations of the K(+) channel gene KCNJ2 constitute a rare cause of long QT syndrome, *J Mol Cell Cardiol* 37, 593, 2004.

Folts JD, Gallagher K, and Rowe GG, Blood flow reduction in stenosed canine coronary arteries: vasospasm or platelet aggregation? *Circulation* 65, 248, 1982.

Fontaine G, Fontaliran F, Hebert JL, et al., Arrhythmogenic right ventricular dysplasia, *Annu Rev Med* 50, 17, 1999.

Fontaliran F, Fontaine G, Fillette F, et al., Nosologic frontiers of arrhythmogenic dysplasia: quantitative variations of normal adipose tissue of the right heart ventricle, *Arch Mal Coeur Vaiss* 84, 33, 1991.

Forauer AR, Narasimham LD, Gemmete JJ, et al., Pericardial tamponade complicating central venous interventions, *J Vasc Interv Radiol*, 14, 255, 2003.

Ford SE, Congenital cystic tumors of the atrio-ventricular node: successful demonstration by an abbreviated dissection of the conduction system, *Cardiovasc Pathol* 8, 233, 1999.

Forfang K, Rostad H, Sorland S, and Levorstad K, Late sudden death after surgical correction of coarctation of the aorta: importance of aneurysm of the ascending aorta, *Acta Med Scand* 206, 375, 1979.

Fornes P, Ratel S, and Lecompte D, Pathology of arrhythmogenic right ventricular cardiomyopathy/dysplasia: an autopsy study of 20 forensic cases, *J Forensic Sci* 43, 777, 1998.

Forrest AR, Galloway JH, Marsh ID, Strachan GA, and Clark JC, A fatal overdose with 3,4-methylenedioxyamphetamine derivatives, *Forensic Sci Int* 64, 57, 1994.

Forrester JS, Price MJ, and Makkar RR, Stem cell repair of infarcted myocardium: an overview for clinicians, *Circulation* 108, 1139, 2003.

Fortuin NJ, Pitt B, and Kaihara S, The distribution of regional myocardial blood flow in the dog, *Circulation* 38 (Suppl VI), 1968.

Fortunato G, Berruti R, Brancadoro V, et al., Identification of a novel mutation in the ryanodine receptor gene (RYR1) in a malignant hyperthermia Italian family, *Eur J Hum Genet* 8, 149, 2000.

Fosse E and Lindberg H, Left ventricular rupture following external chest compression, *Acta Anaesthesiol Scand*, 40, 502, 1996.

Fowles RE, *Cardiac Biopsy*, Future Publishing, Mount Kisco, NY, 1992.

Fox PR, Maron BJ, Basso C, et al., Spontaneously occurring arrhythmogenic right ventricular cardiomyopathy in the domestic cat: A new animal model similar to the human disease, *Circulation* 102, 1863, 2000.

Fozzard HA, Electromechanical dissociation and its possible role in sudden cardiac death, *J Am Coll Cardiol* 5 (Suppl 6), 31B, 1985.

Franciosi RA and Singh A, Oncocytic cardiomyopathy syndrome, *Hum Pathol* 19, 1361, 1988.

Frank O, Zur dynamik des Herzwurder, *Z Biol* 32, 370, 1895.

Frazee RC, Mucha P, Farnell MB, et al., Objective evaluation of blunt cardiac trauma, *J Trauma* 26, 510, 1986.

Frazer M and Mirchandani H, Commotio cordis, revisited, *Am J Forensic Med Pathol*, 5, 249, 1984.

Freant LJ and Hopkins RA, Aschoff bodies in an operatively excised mitral valve, *Cardiovasc Pathol* 6, 231, 1997.

Freasure-Smith N, Lesperance F, Gravel G, et al., Social support, depression and mortality during the first year after myocardial infarction, *Circulation* 101, 1919, 2000.

Freifeld AG, Schuster EH, and Bulkley BH, Nontransmural versus transmural myocardial infarction: a morphological study, *Am J Med* 75, 423, 1983.

Frescura C, Basso C, Thiene G, et al., Anomalous origin of coronary arteries and risk of sudden death: a study based on an autopsy population of congenital heart disease, *Hum Pathol* 29, 689, 1998.

Fried K, Beer S, Vure E, Algom M, and Shapira Y, Autosomal recessive sudden unexpected death in children probably caused by a cardiomyopathy associated with myopathy, *J Med Genet* 16, 341–346, 1979.

Friedl K, Effects of anabolic steroids on physical health, in *Anaboloic Steroids in Sport and Exercise*, Yesalis CE, Ed., Human Kinetics, Champaign, IL, 1993.

Friedrich SP, Berman AD, Baim DS, et al., Myocardial perforation in the cardiac catheterization laboratory: incidence, presentation, diagnosis, and management, *Cathet Cardiovasc Diagn*, 32, 99, 1994.

Frink RJ, Trowbridge JO, and Roney PA, Jr, Nonobstructive coronary thrombosis in sudden cardiac death, *Am J Cardiol* 42, 48, 1978.

Frishman WH, Del Vecchio A, Sanal S, et al., Cardiovascular manifestations of substance abuse part 1: cocaine, *Heart Dis* 5, 187, 2003.

Frishman WH, Del Vecchio A, Sanal S, et al., Cardiovascular manifestations of substance abuse: part 2: alcohol, amphetamines, heroin, cannabis, and caffeine, *Heart Dis* 5, 253, 2003.

Frohn-Mulder IM, Meilof JF, Szatmari A, et al., Clinical significance of maternal anti-Ro/SS-A antibodies in children with isolated heart block, *J Am Coll Cardiol* 23, 1677, 1994.

Frustaci A, Chimenti C, Bellocci F, et al., Histological substrate of atrial biopsies in patients with lone atrial fibrillation, *Circulation* 96, 1180, 1997.

Fry DL, Certain chemorheologic considerations regarding the blood vascular interface with particular reference to coronary artery disease, *Circulation* 40 (Suppl 4), 38, 1969.

Fu C, Jasani B, Vujanic GM, Leadbeatter S, Berry PJ, Knight BH, The immunocytochemical demonstration of a relative lack of nerve fibres in the atrioventricular node and bundle of His in the sudden infant death syndrome (SIDS), *Forensic Sci Int* 66, 175–185, 1994.

Fujino N, Shimizu M, Ino H, et al., A novel mutation Lys273Glu in the cardiac troponin T gene shows high degree of penetrance and transition from hypertrophic to dilated cardiomyopathy, *Am J Cardiol* 89, 29, 2002.

Fulda G, Brathwaite CEM, and Rodriquez A, Blunt traumatic rupture of the heart and pericardium: a ten year experience (1979 – 1989), *J Trauma* 31, 167, 1991.

Fulda GJ, Giberson F, Hailstone D, et al., An evaluation of serum troponin T and signal-averaged electrocardiography in predicting electrocardiographic abnormalities after blunt chest trauma, *J Trauma*, 43, 304, 1997.

Fujiwara H, Onodera T, Tanaka M, et al., A clinicopathologic study of patients with hemorrhagic myocardial infarction treated with selective thrombolysis with urokinase, *Circulation* 73, 749, 1986.

Fung AY and Rabkin SW, Beneficial effects of streptokinase on left ventricular function after myocardial reoxygenation and reperfusion following global ischemia in the isolated rabbit heart, *J Cardiovasc Pharmacol* 6, 429, 1984.

Fuster V, Steele PM, Edwards WD, et al., Primary pulmonary hypertension: natural history and the importance of thrombosis, *Circulation* 70, 580, 1984.

Fyfe B, Loth E, Winters GI, et al., Heart transplantation: associated perioperative ischemic myocardial injury: morphologic features and clinical significance, *Circulation* 93, 1133, 1996.

Gallo P, Baroldi G, Thiene G, et al., When and why do heart transplant recipients die? A 7 year experience of 1069 cardiac transplants, *Virchows Arch A Pathol Anat* 422, 453, 1993.

Gallo P and d'Amati G, Cardiomyopathies, in *Cardiovascular Pathology*, Silver MD, Gotlieb AI, and Schoen FJ, Eds., Churchill Livingstone, Philadelphia, 2001, p. 308.

Gallo P, d'Amati G, and Pelliccia F, Pathologic evidence of extensive left ventricular involvement in arrhythmogenic right ventricular cardiomyopathy, *Hum Pathol* 23, 948, 1992.

Galvin JE, Hemric ME, Kosanke SD, et al., Induction of myocarditis and valvulitis in Lewis rates by different epitopes of cardiac myosin and its implications in rheumatic carditis, 2002.

Galvin JE, Hemric ME, Ward K, and Cunningham MW, Cytotoxix mAb from rheumatic carditis recognizes heart valve and laminin, *J Clin Invest* 106, 217, 2000.

Gammie JS, Shah AS, Hattler BG, et al., Traumatic aortic rupture: diagnosis and management, *Ann Thorac Surg* 66, 1295, 1998.Gammie JS, Katz WE, Swanson ER, and Peitzman AB, Acute aortic dissection after blunt chest trauma, *J Trauma Inj Infect Crit Care* 40, 1, 126, 1996.

Gamouras GA, Monir G, Plunkitt K, et al., Cocaine abuse: repolarization abnormalities and ventricular arrhythmias, *Am J Med Sci* 320, 9, 2000.

Ganote C and Armstrong S, Ischemia and the myocyte cytoskeleton: review and speculations, *Cardiovasc Res* 27, 1387, 1993.

Ganz W, Buchbinder N, Marcus H, et al., Intracoronary thrombolysis in acute myocardial infarction: experimental background and clinical experience, *Am Heart J* 102, 1145, 1981.

Garan AR, Maron BJ, Wang PJ, et al., Role of streptomycin-sensitive stretch-activated channel in chest wall impact induced sudden death (commotio cordis), *J Cardiovasc Electrophysiol*, 16, 433, 2005.

Garfia A, Rodriguez M, Chavarria H, and Garrido M, Sudden cardiac death during exercise due to an isolated multiple anomaly of the left coronary artery in a 12-year-old girl: clinicopathologic findings, *J Forensic Sci* 42, 330, 1997.

Garg A, Finneran W, and Feld GK, Familial sudden cardiac death associated with a terminal QRS abnormality on surface 12-lead electrocardiogram in the index case, *J Cardiovasc Electrophysiol* 9, 642, 1998.

Garson A Jr, Bink-Boelkens M, Hesslein PS, et al., Atrial flutter in the young: a collaborative study of 380 cases, *J Am Coll Cardiol* 6, 871, 1985.

Garson A Jr, Dick M II, Fournier A, et al., The long QT syndrome in children: an international study of 287 patients, *Circulation* 87, 1866, 1993.

Garson A Jr and McNamara DG, Sudden death in a pediatric cardiology population, 1958 to 1983: relation to prior arrhythmias, *J Am Coll Cardiol* 5 (6 Suppl), 134B, 1985.

Garson A, Porter CBJ, Gillette PC, et al., Induction of ventricular tachycardia during electrophysiological study after repair of tetralogy of Fallot, *J Am Coll Cardiol* 1, 1493, 1983.

Gatalica Z, Gibas Z, and Martinez-Hernandez A, Dissecting aortic aneurysm as a complication of generalized fibromuscular dysplasia, *Hum Pathol* 23, 586, 1992.

Gates JD, Clair DG, and Hectman DH, Thoracic aortic dissection with renal artery involvment following blunt thoracic trauma: case report, *J Trauma*, 36, 430, 1994.

Gates RN, Laks H, Drinkwater DC Jr., et al., The Fontan procedure in adults, *Ann Thorac Surg* 63, 1085, 1997.

Gatzoulis MA, Balaji S, Webber SA, et al., Risk factors for arrhythmia and sudden cardiac death late after repair of tetralogy of Fallot: a multicentre study, *Lancet* 356, 975, 2000.

Gavazzi A, De Maria R, Porcu M, et al., Cardiomiopatia dilatativa: una nuova storia naturale? L'esperienza dello Studio Policentrico Italiano Cardiomiopatie (SPIC), *J Ital Cardiol* 25, 1109, 1995.

Gavazzi A, De Maria R, Renosto G, et al., The spectrum of left ventricular size in dilated cardiomyopathy: clinical correlates and prognostic implications, *Am Heart J* 125, 410, 1993.

Gedeon AK, Wilson MJ, Colley AC, et al., X linked fatal infantile cardiomyopathy maps to Xq28 and is possibly allelic to Barth syndrome, *J Med Genet* 32, 383, 1995.

Geiringer E, The mural coronary, *Am Heart J* 41, 359, 1951.

Geisterfer-Lowrance AA, Kass S, Tanigawa G, et al., A molecular basis for familial hypertrophic cardiomyopathy: a beta-cardiac myosin heavy chain gene missense mutation, *Cell* 62, 999, 1990.

Gelatt M, Hamilton RM, McCrindle BW, et al., Arrhythmia and mortality after the Mustard procedure: a 30-year single-centre experience, *J Am Coll Cardiol* 29, 194, 1997.

Gembruch U, Hansmann M, Redel DA, et al., Fetal complete heart block: antenatal diagnosis, significance and management, *Eur J Obstet Gynecol Reprod Biol* 31, 9, 1989.

Geng YJ and Libby P, Evidence for apoptosis in advanced human atheroma: colocalization with interleukin-1-converting enzyme, *Am J Pathol* 147, 251, 1995.

Gensini GG and Kelly AE, Incidence and progression of coronary artery disease: an angiographic correlation in 1263 patients, *Arch Intern Med* 129, 814, 1972.

Gentles TL, Calder AL, Clarkson PM, and Neutze JM, Predictors of long-term survival with Ebstein's anomaly of the tricuspid valve, *Am J Cardiol* 69, 377, 1992.

Gerber BL, Wijns W, Vanoverschelde JJ, et al., Myocardial perfusion and oxygen consumption in reperfused noninfarcted dysfunctional myocardium after unstable angina, *J Am Coll Cardiol* 34, 1939, 1999.

Gerdes MA, Kellerman SE, Moore AJ, et al., Structural remodeling of cardiac myocytes in patients with ischemic cardiomyopathy, *Circulation* 86, 426, 1992.

George CH, Higgs GV, and Lai FA, Ryanodine receptor mutations associated with stress-induced ventricular tachycardia mediate increased calcium release in stimulated cardiomyocytes, *Circ Res* 93, 531, 2003.

Ghidoni JJ, Liotta D, and Thomas H, Massive subendocardial damage accompanying prolonged ventricular fibrillation, *Am J Pathol* 56, 15, 1969.

Ghuran A and Nolan J, Recreational drug misuse: issues for the cardiologist, *Heart* 83, 627, 2000.

Gibbons LW, Cooper KH, Meyer BM, et al., The acute cardiac risk of strenuous exercise, *JAMA* 244, 1799, 1980.

Gibson MC, Ryan KA, Murphy SA, et al., Impaired coronary blood flow in nonculprit arteries in the setting of acute myocardial infarction, *J Am Coll Cardiol* 34, 974, 1999.

Gicquel C, Cabrol S, Schneid H, et al., Molecular diagnosis of Turner's syndrome, *J Med Genet* 29, 547, 1992.

Gill JR, Hayes JA, deSouza IS, Marker E, and Stajic M, Ecstasy (MDMA) deaths in New York City: a case series and review of the literature, *J Forensic Sci* 47, 121, 2002.

Gillan JE, Costigan DC, Keeley FW, Rose T, Cutz E, and Rose V. Spontaneous dissecting aneurysm of the ductus arteriosus in an infant with Marfan syndrome, *J Pediatr* 105, 952, 1984.

Gillette PC and Garson A Jr, Sudden cardiac death in the pediatric population, *Circulation* 85 (1 Suppl), I64, 1992.

Gilligan DM, Chan WL, Joshi J, et al., A double-blind, placebo-controlled crossover trial of nadolol and verapamil in mild and moderately symptomatic hypertrophic cardiomyopathy, *J Am Coll Cardiol* 21, 1672, 1993.

Gilon D, Buonanno FS, Joffe MM, et al., Lack of evidence of an association between mitral-valve prolapse and stroke in young patients, *N Engl J Med* 341, 8–13, 1999.

Gladman G, Silverman ED, Yuk-Law, et al., Fetal echocardiographic screening of pregnancies of mothers with anti-Ro and/or anti-La antibodies, *Am J Perinatol* 19, 73, 2002.

Glagov S, Weisemberg E, Zarnes CK, Stankunavicius R, and Kolettis GJ, Compensatory enlargement of human atherosclerotic coronary arteries, *N Engl J Med* 316, 1371, 1987.

Glancy DL, Morrow AG, Simon AL, and Roberts WC, Juxtaductal aortic coarctation: analysis of 84 patients studied hemodynamically, angiographically, and morphologically after age 1 year, *Am J Cardiol* 51, 537, 1983.

Glatter KA, Chiamvimonvat N, Viitasalo M, et al., Risk stratification in Brugada syndrome, *Lancet* 366, 530, 2005.

Glatter KA, Wang Q, Keating M, et al., Effectiveness of sotalol treatment in symptomatic Brugada syndrome, *Am J Cardiol* 93, 1320, 2004.

Gledhill-Hoyt J, Lee H, Strote J, and Wechsler H, Increased use of marijuana and other illicit drugs at US colleges in the 1990s: results of three national surveys, *Addiction* 95, 1655, 2000.

Glew RH, Varghese FP, Krovetz JL, Dorst JP, and Rowe RD, Sudden death in congenital aortic stenosis: a review of eight cases with an evaluation of premonitory clinical features, *Am Heart J* 78, 615, 1969.

Glickstein JS, Axelrod FB, and Friedman D, Electrocardiographic repolarization abnormalities in familial dysautonomia: an indicator of cardiac autonomic dysfunction, *Clin Auton Res* 9, 109, 1999.

Goldberg SM, Pizzarello RA, Goldman MA, and Padmanabhan VT, Aortic dilatation resulting in chronic aortic regurgitation and complicated by aortic dissection in a patient with Turner's syndrome, *Clin Cardiol* 7, 233, 1984.

Goldfrank L, Flomenbaum N, Lewin N, Weisman R, and Howland M, *Goldfrank's Toxicologic Emergencies*, 6th ed., Appleton-Lange, Norwalk, CT, 1998.

Goldman BS, Williams WG, Hill T, et al., Permanent cardiac pacing after open heart surgery: congenital heart disease, *Pacing Clin Electrophysiol* 8, 732, 1985.

Goldstein RE, Borer JS, and Epstein SE, Augmentation of contractility following ischemia in the isolated supported heart, *Am J Cardiol* 29, 265, 1972.

Goldstein S, Landis JR, Leighton R, et al., Characteristics of the resuscitated out-of-hospital cardiac arrest victim with coronary heart disease, *Circulation* 64, 977, 1981.

Gollob MH, Green MS, Tang AS, et al., Identification of a gene responsible for familial Wolff-Parkinson-White syndrome, *N Engl J Med* 344, 1823, 2001.

Goodwin JF, The frontiers of cardiomyopathy, *Brit Heart J* 48, 1, 1982.

Goodwin JF and Krikler DM, Arrhythmia as a cause of sudden death in hypertophic cardiomyopathy, *Lancet* 2, 937, 1976.

Goodwin JF and Krikler DM, Sudden death in cardiomyopathy, in *Sudden Coronary Death*, Manninen V and Halonen PI, Eds., Karger, Basel, *Adv Cardiol* 25, 98, 1978.

Goraya TY, Jacobsen SJ, Kottke TE, et al., Coronary heart disease death and sudden cardiac death, *Am J Epidemiol* 157, 763, 2003.

Gordon PA, Rosenthal E, Khamashta MA, et al., Absence of conduction defects in the electrocardiograms [correction of echocardiograms] of mothers with children with congenital complete heart block, *J Rheumatol* 28, 366, 2001.

Gore I and Kline IK, Myocarditis, in Gould SE, Ed., *Pathology of the Heart and Blood Vessels*, 3rd ed., Charles C Thomas, Springfield, IL, 1968, p. 731.

Gore SM, Fatal uncertainty: death-rate from use of ecstasy or heroin, *Lancet* 354, 1265, 1999.

Gorgels AP, Al Fadley F, Zaman L, et al., The long QT syndrome with impaired atrioventricular conduction: a malignant variant in infants, *J Cardiovasc Electrophysiol* 9, 1225, 1998.

Gorlin R, Fuster V, and Ambrose JA, Anatomic-physiologic links between acute coronary syndromes, *Circulation* 74, 6, 1986.

Gotlieb A, Masse S, Allard J, and Huang SN, Concentric hemorrhagic necrosis of the myocardium: a morphological and clinical study, *Hum Pathol* 8, 27, 1977a.

Gotlieb AI, Chan M, Palmer WH, and Huang SN, Ventricular preexcitation syndrome: accessory left atrioventricular connection and rhabdomyomatous myocardial fibers, *Arch Pathol Lab Med* 101, 486, 1977b.

Gotoh K, Minamino T, Katoh O, et al., The role of intracoronary thrombus in unstable angina: angiographic assessment and thrombolytic therapy during ongoing anginal attacks, *Circulation* 77, 526, 1988.

Gott VL, Laschinger JC, Cameron DE, et al., The Marfan syndrome and the cardiovascular surgeon, *Eur J Cardiothorac Surg* 10, 149, 1996.

Gottlieb R, Burleson KO, Kloner RA, et al., Reperfusion injury induces apoptosis in rabbit cardiomyocytes, *J Clin Invest* 94, 1621, 1994.

Gotway MB, Marder SR, Hanks DK, et al., Thoracic complications of illicit drug use: an organ system approach, *Radiographics 22, S119,* 2002.

Gotzsche CO, Krag-Olsen B, Nielsen J, et al., Prevalence of cardiovascular malformations and association with karyotypes in Turner's syndrome, *Arch Dis Child* 71, 433, 1994.

Goudevenos JA, Katsouras CS, Graekas G, et al., Ventricular pre-excitation in the general population: a study on the mode of presentation and clinical course, *Heart* 83, 29, 2000.

Gouley BA, Bellet S, and McMillan TM, Tuberculosis of the myocardium: report of six cases with observations on involvement of the coronary arteries, *Arch Intern Med* 51, 244, 1933.

Goverde P, Van Schil P, d'Archambeau O, et al., Traumatic type B aortic dissection, *Acta Chir Belg*, 96, 233, 1996.

Gowda RM, Punukollu G, Khan IA, et al., Ibutilide-induced long QT syndrome and torsade de pointes, *Am J Ther* 9, 527, 2002.

Gowing LR, Henry-Edwards SM, Irvine RJ, and Ali RL, The health effects of ecstasy: a literature review, *Drug Alcohol Rev* 21, 53, 2002.

Graber HL, Unverferth DV, Baker PB, et al., Evolution of a hereditary cardiac conduction and muscle disorder: a study involving a family with six generations affected, *Circulation* 74, 21, 1986.

Grace F, Sculthorpe N, Baker J, and Davies B, Blood pressure and rate pressure product response in males using high-dose anabolic androgenic steroids (AAS), *J Sci Med Sport* 6, 307, 2003.

Grant AO, Carboni MP, Neplioueva V, et al., Long QT syndrome, Brugada syndrome, and conduction system disease are linked to a single sodium channel mutation, *J Clin Invest* 110, 1201, 2002.

Grant RT, Development of the cardiac coronary vessels in the rabbit, *Heart* 13, 261, 1926.

Grant RT and Regnier M, The comparative anatomy of the cardiac coronary vessels, *Heart* 13, 285, 1926.

Gravanis MB, Robinson K, Santoian EC, et al., The reparative phenomena at the site of balloon angioplasty in human and experimental models, *Cardiovasc Pathol* 2, 263, 1993.

Gregg DE, *Coronary Circulation in Health and Disease*, Lea Febiger, Philadelphia, 1950.

Gregg DE, The microcirculation of the heart in reduced flow states, in *Microcirculation as Related to Shock,* Shepro D and Fulton GP, Eds., Academic Press, NY, 1968.

Gregg DE, The natural history of collateral development, *Circ Res* 35, 335, 1974.

Gregg DE and Patterson RE, Functional importance of the coronary collaterals, *New Engl J Med* 303, 1404, 1980.

Griffith LSC, Achuff SC, Conti R, et al., Changes in intrinsic coronary circulation and segmental ventricular motion after saphenous-vein coronary bypass graft surgery, *New Engl J Med* 288, 590, 1973.

Grogan M, Redfield MM, Bailey KR, et al., Long-term outcome of patients with biopsy-proved myocarditis: comparison with idiopathic dilated cardiomyopathy, *J Am Coll Cardiol* 26, 80, 1995.

Gross L, *The Blood Supply to the Heart in Its Anatomical and Clinical Aspects*, Paul B. Hoeber Inc., New York, 1921.

Gross L, Kugel MA, and Epstein EZ, Lesions of the coronary arteries and their branches in rheumatic fever, *Am J Pathol* 11, 253, 1935.

Grotenhermen F, Pharmacokinetics and pharmacodynamics of cannabinoids, *Clin Pharmacokinet* 42, 327, 2003.

Grumbach IM, Heim A, Pring-Akerblom P, et al., Adenoviruses and enteroviruses as pathogens in myocarditis and dilated cardiomyopathy, *Acta Cardiol* 54, 83, 1999.

Grunig E, Tasman JA, Kucherer H, et al., Frequency and phenotypes of familial dilated cardiomyopathy, *J Am Coll Cardiol* 31, 186, 1998.

Guan DW, Ohshima T, Jia JT, et al., Morphological findings of 'cardiac concussion' due to experimental blunt impact to the precordial region, *Forensic Sc Int*, 100, 211, 1999.

Guenthard J, Wyler F, Fowler B, and Baumgartner R, Cardiomyopathy in respiratory chain disorders, *Arch Dis Child* 72, 223, 1995.

Guerra S, Leri A, Wang X, et al., Myocyte death in the failing human heart is gender dependent, *Circ Res* 85, 856, 1999.

Guideri F, Acampa M, Hayek G, et al., Reduced heart rate variability in patients affected with Rett syndrome: a possible explanation for sudden death, *Neuropediatrics* 30, 146, 1999.

Guilherme L, Dulphy N, Douay C, et al., Molecular evidence for antigen-driven immune responses in cardiac lesions of rheumatic heart disease patients, *Int Immunol* 12, 1063, 2000.

Guimaraes A, Natural history and current status in Brazil, in *Endomyocarial Fibrosis*, Valiathan MS, Somers K, and Kartha CC, Eds., Oxford University Press, Delhi, 1993, 37.

Gunja-Smith Z, Morales AR, Romanelli R, and Woessner JF, Remodeling of human myocardial collagen in idiopathic dilated cardiomyopathy: role of metalloproteinase and pyridimoline cross-links, *Am J Pathol* 148, 1639, 1996.

Guntheroth W and Spiers P, Long QT syndrome and sudden infant death syndrome, *Am J Cardiol* 96, 1034, 2005.

Guntheroth W, Komarniski C, Atkinson W, and Fligner CL, Criterion for fetal primary spongiform cardiomyopathy: restrictive pathophysiology, *Obstet Gynecol* 99, 882, 2002.

Guo D, Hasham S, Kuang SQ, et al., Familial thoracic aortic aneurysms and dissections: genetic heterogeneity with a major locus mapping to 5q13-14, *Circulation* 103, 2461, 2001.

Gupta BD, Jani CB, and Shah PH, Fatal "bhang" poisoning, *Med Sci Law* 41, 349, 2001.

Gussak I, Brugada P, Brugada J, et al., Idiopathic short QT interval: a new clinical syndrome? *Cardiology* 94, 99, 2000.

Gutgesell HP, Mullins CE, Gillette PC, et al., Transient hypertrophic subaortic stenosis in infants of diabetic mothers, *J Pediatr* 89, 120, 1976.

Gutierrez-Roelens I, Sluysmans T, Gewillig M, et al., Progressive AV-block and anomalous venous return among cardiac anomalies associated with two novel missense mutations in the CSX/NKX2-5 gene, *Hum Mutat* 20, 75, 2002.

Hackel DB, Pathology of primary congenital complete heart block, *Mod Pathol* 1, 114, 1988.

Hackel DB and Reimer KA, *Sudden Death Cardiac and Other Causes*, Carolina Academic Press, Durham, NC, 1993.

Hackett D, Davies G, Chierchia S, and Maseri A, Intermittent coronary occlusion in acute myocardial infarction: value of combined thrombolytic and vasodilator therapy, *N Engl J Med* 317, 1055, 1987.

Hackett D, Verwilghen J, Davies G, et al., Coronary stenoses before and after acute myocardial infarction, *Am J Cardiol* 63, 1517, 1989.

Haerem JW, Platelet aggregates in intramyocardial vessels of patients dying suddenly and unexpectedly of coronary artery disease, *Atherosclerosis* 15, 199, 1972.

Haerem JW, Myocardial lesions in sudden unexpected coronary death, *Am Heart J* 90, 562, 1975.

Haft JI and Al-Zarka AM, Comparison of the natural history of irregular and smooth coronary lesions: insights into the pathogenesis, progression and prognosis of coronary atherosclerosis, *Am Heart J* 126, 551, 1993.

Hahn IH and Hoffman RS, Cocaine use and acute myocardial infarction, *Emerg Med Clin North Am* 19, 493, 2001.

Haitas B, Baker SG, Meyer TE, et al., Natural history and cardiac manifestations of homozygous familial hypercholesterolaemia, *Q J Med* 76, 731, 1990.

Hale SL, Alker KJ, Rezkalla S, et al., Adverse effects of cocaine on cardiovascular dynamics, myocardial blood flow, and coronary artery diameter in an experimental model, *Am Heart J* 118, 927, 1989.

Halperin IC, Penny JL, and Kennedy RJ, Single coronary artery: antemortem diagnosis in a patient with congestive heart failure, *Am J Cardiol* 19, 424, 1967.

Halpert I, Goldberg D, Levine AB, et al., Reinnervation of the transplanted human heart as evidenced from heart rate variability studies, *Am J Cardiol* 77, 180, 1996.

Hals J, Ek J, and Sandnes K, Cardiac myxoma as the cause of death in an infant, *Acta Pediatr Scand* 79, 999, 1990.

Hamilton SJ, Sunter JP, and Cooper PN, Commotio cordis — a report of three cases, *Int J Legal Med*, 119, 88, 2005.

Hammer A, Ein Fall von trombotischen Verschlusse einer der Kranzarterien des Herzens, *Wiener Medizinische Wochenschrift* 5, 83, 1878.

Hamperl H, Zur fragmentatio miocardii, *Beitr Pathol Anat* 82, 597, 1929.

Hand DKM, Haudenshild CC, Hong MK, et al., Evidence for apoptosis in human atherogenesis and in rat vascular injury model, *Am J Pathol* 147, 267, 1995.

Harada M, Shimizu A, Murata M, et al., Relation between microvolt-level T-wave alternans and cardiac sympathetic nervous system abnormality using iodine-123 metaiodobenzylguanidine imaging in patients with idiopathic dilated cardiomyopathy, *Am J Cardiol* 92, 998, 2003.

Hardaway RM, McKay DG, and Hollowell OW, Vascular spasm and disseminated intravascular coagulation: influence of the phenomena one on the other, *Arch Surg* 83, 183, 1961.

Harris LS and Adelson I, Fatal coronary embolism from a myxomatous polyp of the aortic valve: an unusual cause of sudden death, *Am J Clin Pathol* 43, 61, 1965.

Harrison CV, Giant-cell or temporal arteritis: a review, *Am J Clin Pathol* 1, 197, 1948.

Harrison DA, Connelly M, Harris L, et al., Sudden cardiac death in the adult with congenital heart disease, *Can J Cardiol* 12, 1161, 1996.

Harrison EG and McCormack LJ, Pathologic classification of renal arterial disease in renovascular hypertension, *Mayo Clin Proc* 46, 161, 1971.

Hartgens F and Kuipers H, Effects of androgenic-anabolic steroids in athletes, *Sports Med* 34, 513, 2004.

Hartmann F, Ziegler S, Nguyen N, and Schwaiger M, Imaging of myocardial autonomic innervation in patients with congestive heart failure: methods and clinical implications, *Heart Failure Rev* 1, 15, 1996.

Hassler O, The origin of the cell constituting arterial intima thickening: an experimental autoradiographic study with the use of H^3-thymidine, *Lab Invest* 2, 286, 1970.

Hasumi M, Sekiguchi M, Yu ZX, et al., Analysis of histopathologic findings in cases with dilated cardiomyopathy with special reference to formulating diagnostic criteria on the possibility of postmyocarditic change, *Jpn Circ J* 50, 1280, 1986.

Hausmann R, Hammer S, and Betz P, Performance enhancing drugs and sudden death: a case report and review of the literature, *Int J Legal Med* 111, 261, 1998.

Hawkey CM, Production of cardiac muscle abnormalities in offspring of rats receiving triiodothyroacetic acid (triac) and the effect of beta-adrenergic blockade, *Cardiovasc Res* 15, 196, 1981.

Hayes CJ and Gersony WM, Arrhythmias after the Mustard operation for transposition of the great arteries: a long-term study, *J Am Coll Cardiol* 7, 133, 1986.

Hayes SN, Moyer TP, Morley D, et al., Intravenous cocaine causes epicardial coronary vasoconstriction in the intact dog, *Am Heart J* 121, 1639, 1991.

Heath D, Cardiac fibroma, *Br Heart J* 31, 656, 1969.

Hebe J, Ebstein's anomaly in adults. Arrhythmias: diagnosis and therapeutic approach, *Thorac Cardiovasc Surg* 48, 214, 2000.

Hecht GM, Klues HG, Roberts WC, and Maron BJ, Coexistence of sudden cardiac death and end-stage heart failure in familial hypertrophic cardiomyopathy, *J Am Coll Cardiol* 22, 489, 1993.

Hecht H, Electrophysiologic principles of cardiac fibers, in *Pathology of the Heart and Blood Vessels*, Gould SE, Ed., Charles C Thomas, Springfield, IL, 1968, p. 168.

Heifetz SA, Robinowitz M, Mueller KH, and Virmani R, Total anomalous origin of the coronary arteries from the pulmonary artery, *Pediatr Cardiol* 7, 11, 1986.

Hein S and Schaper J, The cytoskeleton of cardiomyocytes is altered in the failing human heart, *Heart Failure* 12, 128, 1996.

Hein S, Sheffold T, and Schaper J, Ischemia induces early changes to cytoskeletal and contractile proteins in diseased human myocardium, *J Thorac Cardiovasc Surg* 110, 89, 1995.

Heistad D, Unstable coronary-artery plaques, *N Engl J Med* 349, 2285, 2003.

Helfant RH, Kemp HG, and Gorlin R, The interrelation between extent of coronary artery disease, presence of collaterals, ventriculographic abnormalities and hemodynamics, *Am J Cardiol* 25, 102, 1970.

Helft G, Worthley S, Fuster V, et al., Progression and regression of atherosclerotic lesions monitoring with serial non-invasive magnetic resonance imaging, *Circulation* 105, 993, 2002.

Hellerstein HK and Santiago-Stevenson D, Atrophy of the heart: a correlative study of 85 proved cases, *Circulation* 1, 93, 1959.

Hellstrom HR, The injury-spasm (ischemia-induced hemostatic vasoconstrictive) and vascular autoregolatory hypothesis of ischemic disease: resistance vessel-spasm hypothesis of ischemic disease, *Am J Cardiol* 49, 802, 1982.

Helton E, Darragh R, Francis P, et al., Metabolic aspects of myocardial disease and a role for L-carnitine in the treatment of childhood CM, *Pediatrics* 105, 1260, 2000.

Helwig FC and Wilhelmy EW, Sudden and unexpected death from acute interstitial myocarditis: a report of three cases, *Am Intern Med* 13, 107, 1939.

Henry JA, Jeffrey KJ, and Dawling S, Toxicity and deaths from 3,4-methylenedioxymethamphet-amine ("ectasy"), *Lancet* 340, 384, 1992.

Henry JA, Oldfield WLG, and Kon OM, Comparing cannabis with tobacco, *BMJ* 326, 942, 2003.

Hermans WRM, Rensing BJ, Foley DP, et al., Therapeutic dissection after successful coronary balloon angioplasty: no influence on restenosis or on clinical outcome in 693 patients, *J Am Coll Cardiol* 20, 767, 1992.

Herrick JB, Clinical features of sudden obstruction of the coronary arteries, *JAMA* 59, 87, 1912.

Herrick JB, Thrombosis of the coronary arteries, *JAMA* 72, 93, 1919.

Hesayen AA, Azevedo ER, Newton GE, and Parker JD, The effects of dobutamine on cardiac sympathetic activity in patients with congestive heart failure, *J Am Coll Cardiol* 39, 1269, 2002.

Heusch A, Kuhl U, Rammos S, et al., Complete AV-block in two children with immunohistochemical proven myocarditis, *Eur J Pediatr* 155, 633, 1996.

Heuser RR, Outpatient coronary angiography: indications, safety, and complications, *Hertz*, 23, 21, 1998.

Hickman M, Carnwath Z, Madden P, et al., Drug-related mortality and fatal overdose risk: pilot cohort study of heroin users recruited from specialist drug treatment sites in London, *J Urban Health* 80, 274, 2003.

Higgins CB and De Roos A, *Cardiovascular MRI & MRA*, Lippincott, Williams & Wilkins, Phila-delphia, 2002.

Higuchi R, Simple and rapid preparation of samples for PCR, in *PCR Technology: Principles and Applications for DNA Amplification*, Ehrlich HA, Ed., Stockton Press, New York, 1989, 31–38.

Hijmering ML, Stroes ESG, Olijhoek J, et al., Sympathetic activation markedly reduces endothelium-dependent, flow-mediated vasodilation, *J Am Coll Cardiol* 39, 683, 2002.

Hiley CR and Ford WR, Cannabinoid pharmacology in the cardiovascular system: potential pro-tective mechanisms through lipid signalling, *Biol Rev Camb Philos Soc* 79, 187, 2004.

Hillard CJ, Endocannabinoids and vascular function, *J Pharmacol Exp Ther*, 294, 27, 2000.

Hina K, Kusachi S, Iwasaki K, et al., Progression of left ventricular enlargement in patients with hypertrophic cardiomyopathy: incidence and prognostic value, *Clin Cardiol* 16, 403, 1993.

Hirschl MM, Gwechenberger M, Binder T, et al., Assessment of myocardial injury by serum tumour necrosis factor alpha measurements in acute myocardial infarction, *Eur Heart J* 17, 1852, 1996.

Ho SY and Anderson RH, Conduction tissue and SIDS. *Ann NY Acad Sci* 533, 176–190, 1988.

Ho SY, Esscher E, Anderson RH, et al., Anatomy of congenital complete heart block and relation to maternal Anti-Ro antibodies, *Am J Cardiol* 58, 291, 1986.

Ho SY, Rossi MB, Mehta AV, et al., Heart block and atrioventricular septal defect, *Thorac Cardiovasc Surg* 33, 362, 1985.

Hodgson JM, Reddy KG, Suneja R, et al., Intracoronary ultrasound imaging: correlation of plaque morphology with angiography, clinical syndrome and procedural results in patients under-going coronary angioplasty, *J Am Coll Cardiol* 21, 35, 1993.

Hofbek M, Ulmer H, Beinder E, et al., Prenatal findings in patients with prolonged QT interval in the neonatal period, *Heart* 77, 198, 1997.

Hokanson JS and Moller JH, Significance of early transient complete heart block as a predictor of sudden death late after operative correction of tetralogy of Fallot, *Am J Cardiol* 87, 1271, 2001.

Hollister DW and Hollister WG, The "long-thumb" brachydactyly syndrome, *Am J Med Genet* 8, 5, 1981.

Holpster DJ, Milroy CM, Burns J, and Roberts NB, Necropsy study of the association between cardiac death, cardiac isoenzymes and contraction band necrosis, *J Clin Pathol* 49, 403, 1996.

Holsinger DR, Ormundson PJ, and Edwards JE, The heart in periarteritis nodosa, *Circulation* 25, 610, 1962.

Honda Y, Yokota Y, and Yokoyama M, Familial aggregation of dilated cardiomyopathy — evaluation of clinical characteristics and prognosis, *Jpn Circ J* 59, 589, 1995.

Honig CR, Kirk ES, and Myers WW, Transmural distributions of blood flow, oxygen tension and metabolism in myocardium: mechanism and adaptations, in *The Coronary Circulation and Energetics of the Myocardium*, Marchetti G and Taccardi B, Eds., Karger, Basel, 1967, p. 31.

Hood WB, Experimental myocardial infarction, III: recovery of left ventricular function in the healing phase — contribution of increased fiber shortening in noninfarcted myocardium, *Am Heart J* 79, 531, 1970.

Hori M, Gotoh K, Kitakaze M, et al., Role of oxygen-derived free radicals in myocardial edema and ischemia in coronary microvascular embolization, *Circulation* 84, 828, 1991.

Hort W, Mikroskopische Beobachtung an menschlichen Infarktherzen, *Virchow Arch Pathol Anat* 345, 61, 1968.

Hosoda T, Komuro I, Shiojima I, et al., Familial atrial septal defect and atrioventricular conduction disturbance associated with a point mutation in the cardiac homeobox gene CSX/NKX2-5 in a Japanese patient, *Jpn Circ J* 63, 425, 1999.

Hossack KF, Neutze JM, Lowe JB, and Barratt-Boyes BG, Congenital valvar aortic stenosis: natural history and assessment for operation, *Br Heart J* 43, 561, 1980.

Howell C and Bergin JD, A case report of pacemaker lead perforation causing late pericardial effusione and subacute cardiac tamponade, *J Cardiovasc Nurs*, 20, 271, 2005.

Hsue PY, Salinas CL, Bolger AF, et al., Acute aortic dissection related to crack cocaine, *Circulation* 105, 1592, 2002.

Hufnagel G, Pankuweit S, Richter A, et al., The European Study of Epidemiology and Treatment of Cardiac Inflammatory Diseases (ESETCID): first epidemiological results, *Herz* 25, 279, 2000.

Hughes SE, The pathology of hypertrophic cardiomyopathy, *Histopathology* 44, 412, 2004.

Huhta JC, Maloney JD, Ritter DG, et al., Complete atrioventricular block in patients with atrioventricular discordance, *Circulation* 67, 1374, 1983.

Huie M, An acute myocardial infarction occurring in an anabolic steroid user, *Med Sci Sports Exerc* 26, 408, 1994.

Hulot JS, Jouven X, Empana JP, et al., Natural history and risk stratification of arrhythmogenic right ventricular dysplasia/cardiomyopathy, *Circulation* 110, 1879, 2004.

Hunt CE, Relationship between infant sleep position and SIDS, in *Sudden Infant Death Syndrome: New Trends in the Nineties,* Rognum TO, Ed., Scandinavian University Press, Oslo, 1995, 106–108.

Hunt D and Gore I, Myocardial lesion following experimental intracranial hemorrhage: prevention with propranolol, *Am Heart J* 83, 232, 1972.

Hurst JW, Notes on teaching how some words prevent teaching, *JAMA* 74, 858, 1967.

Hutchins GM, Moore GW, and Skoog DK, The association of floppy mitral valve with disjunction of the mitral annulus fibrosus, *N Engl J Med* 314, 535, 1986.

Huttner I, More RH, and Rona G, Fine structural evidence of specific mechanism for increased endothelial permeability in experimental hypertension, *Am J Pathol* 61, 395, 1970.

Hwa J, Richards JG, Huang H, et al., The natural history of aortic dilatation in Marfan syndrome, *Med J Aust* 158, 558, 1993.

Hwang TH, Lee WH, Kimura A, et al., Early expression of a malignant phenotype of familial hypertrophic cardiomyopathy associated with a Gly716Arg myosin heavy chain mutation in a Korean family, *Am J Cardiol* 82, 1509, 1998.

Hynes MS, Veinot JP, and Chan KL, Occurrence of a second primary papillary fibroelastoma, *Can J Cardiol* 18, 753, 2002.

Iafolla AK, Thompson RJ, and Roe CR, Medium-chain acyl-coenzyme A dehydrogenase deficiency: clinical course in 120 affected children, *J Pediatr* 124, 409, 1994.

Ichida F, Hamamichi Y, Miyawaki T, et al., Clinical features of isolated noncompaction of the ventricular myocardium: long-term clinical course, hemodynamic properties, and genetic background, *J Am Coll Cardiol* 34, 233, 1999.

Iida K, Yutani C, Imakita M, and Ishibashi-Ueda H, Comparison of percentage area of myocardial fibrosis and disarray in patients with classical form and dilated phase of hypertrophic cardiomyopathy, *J Cardiol* 32, 173, 1998.

Imanura M, Yokomara S, and Kikuchi K, Coronary fibromuscular dysplasia presenting as sudden infant death, *Arch Pathol Lab Med* 121, 159, 1997.

Ino T, Sherwood WG, Benson LN, et al., Cardiac manifestations in disorders of fat and carnitine metabolism in infancy, *J Am Coll Cardiol* 11, 1301, 1988.

Inoue H and Zipes DP, Results of sympathetic denervation in the canine heart: supersensitivity that may be arrhythmogenic, *Circulation* 75, 877, 1987.

Irniger W, Histologische Alterbestimmung von Thrombosen und Embolien, *Virch Arch Pathol Anat* 336, 220, 1963.

Ishizawa A, Oho S, Dodo H, et al., Cardiovascular abnormalities in Noonan syndrome: the clinical findings and treatments, *Acta Pediatr Jpn* 38, 84, 1996.

Isner JM, Kearney M, Bortman S, and Passeri J, Apoptosis in human atherosclerosis and restenosis, *Circulation* 91, 2703, 1995.

Isobe M, Oka T, Takenaka H, et al., Familial sick sinus syndrome with atrioventricular conduction disturbance, *Jpn Circ J* 62, 788, 1998.

Jacoby RM and Nesto RW, Acute myocardial infarction in the diabetic patient: pathophysiology, clinical course and prognosis, *J Am Coll Cardiol* 20, 736, 1992.

Jaffe BD, Broderick TM, and Leier CV, Cocaine-induced coronary-artery dissection, *N Engl J Med* 330, 510, 1994.

Jahangiri M, Shinebourne EA, Ross DB, et al., Long-term results of relief of subaortic stenosis in univentricular atrioventricular connection with discordant ventriculoarterial connections, *Ann Thorac Surg* 71, 907, 2001.

James CL, Keeling JW, Smith NM, and Byard RW, Total anomalous pulmonary venous drainage associated with fatal outcome in infancy and early childhood: an autopsy study of 52 cases, *Pediatr Pathol* 14, 665, 1994.

James TN, *Anatomy of the Coronary Arteries*, PB Hoeber, New York, 1961.

James TN, Pathology of small coronary arteries, *Am J Cardiol* 20, 679, 1967.

James TN, Sudden death in babies: new observations in the heart, *Am J Cardiol* 22, 479–506, 1968.

James TN, De Subitaneis Mortibus. VIII. Coronary arteries and conduction system in scleroderma heart disease, *Circulation* 50, 844, 1974.

James TN, De Subitaneis Mortibus. XIX. On the cause of sudden death in pheochromocytoma, with special reference to the pulmonary arteries, the cardiac conduction system, and the aggregation of platelets, *Circulation* 54, 348, 1976.

James TN, De Subitaneis Mortibus. XXIII. Rheumatoid arthritis and ankylosing spondylitis, *Circulation* 55, 669, 1977a.

James TN, De Subitaneis Mortibus. XXIV. Ruptured interventricular septum and heart block, *Circulation* 55, 934, 1977b.

James TN, De Subitaneis Mortibus. XXV. Sarcoid heart disease, *Circulation* 56, 320, 1977c.

James TN, De Subitaneis Mortibus. XXVIII. Apoplexy of the heart, *Circulation* 57, 385, 1978.

James TN, Morphologic substrates of sudden death: summary, *J Am Coll Cardiol* 5, 81B, 1985.

James TN, Degenerative lesions of a coronary chemoreceptor and nearby neural elements in the heart of victims of sudden death, *J Am Coll Cardiol* 8, 12A, 1986.

James TN, Armstrong RS, Silverman J, and Marshall TK, De Subitaneis Mortibus. VI. Two young soldiers, *Circulation* 49, 1239, 1974a.

James TN, Beeson CW II, Sherman EB, and Mowry RW, De Subitaneis Mortibus. XIII. Multifocal Purkinje cell tumors of the heart, *Circulation* 52, 333, 1975a.

James TN, Carson DJ, and Marshall TK, De subitaneis mortibus. I. Fibroma compressing His bundle, *Circulation* 48, 428, 1973a.

James TN, Carson NAJ, and Froggatt P, De Subitaneis Mortibus. IV. Coronary vessels and conduction system in homocystinuria, *Circulation* 49, 367, 1974b.

James TN, Frame B, and Coates EO, De Subitaneis Mortibus. III. Pickwickian syndrome, *Circulation* 48, 1311, 1973b.

James TN, Froggatt P, Atkinson WJ Jr, et al., De subitaneis mortibus. XXX. Observations on the pathophysiology of the long QT syndromes with special reference to the neuropathology of the heart, *Circulation* 57, 1221, 1978a.

James TN, Froggatt P, and Marshall TK, De Subitaneis Mortibus. II. Coronary embolism in the fetus, *Circulation* 48, 890, 1973c.

James TN and Galakhov I, De Subitaneis Mortibus. XXVI. Fatal electrical instability of the heart associated with benign congenital polycystic tumor of the atrioventricular node, *Circulation* 56, 667, 1977.

James TN, Hackel DB, and Marshall TK, De Subitaneis Mortibus. V. Occluded A-V node artery, *Circulation* 49, 772, 1974c.

James TN and Haubrich WS, De Subitaneis Mortibus. XIV. Bacterial arteritis in Whipple's disease, *Circulation* 52, 722, 1975b.

James TN and Jackson DA, De Subitaneis Mortibus. XXVII. Histological abnormalities in the sinus node, atrioventricular node and His bundle associated with coarctation of the aorta, *Circulation* 56, 1094, 1977.

James TN, Marilley RJ Ir, and Marriott HJ, De Subitaneis Mortibus. XI. Young girl with palpitations, *Circulation* 51, 743, 1975c.

James TN and Marshall TK, De Subitaneis Mortibus. XII. Asymmetrical hypertrophy of the heart, *Circulation* 51, 1149, 1975.

James TN and Marshall TK, De Subitaneis Mortibus. XVII. Multifocal stenoses due to fibromuscular dysplasia of the sinus node artery, *Circulation* 53, 736, 1976a.

James TN and Marshall TK, De Subitaneis Mortibus. XVIII. Persistent fetal dispersion of the atrioventricular node and His bundle within the central fibrous body, *Circulation* 53, 1026, 1976b.

James TN, Marshall ML, and Craig MW, De Subitaneis Mortibus. VII. Disseminated intravascular coagulation and paroxysmal atrial tachycardia, *Circulation* 50, 395, 1974.

James TN, Marshall TK, and Edwards JE, De Subitaneis Mortibus. XX. Cardiac electrical instability in the presence of a left superior vena cava, *Circulation* 54, 689, 1976a.

James TN, McKone RC, and Hudspeth AS, De Subitaneis Mortibus. X. Familial congenital heart block, *Circulation* 51, 379, 1975d.

James TN, Pearce WN, and Givhan EG, Sudden death while driving: role of sinus perinodal degeneration and cardiac neural degeneration and ganglionitis, *Am J Cardiol* 45, 1095, 1980.

James TN and Puech P, De Subitaneis Mortibus. IX. Type A Wolff-Parkinson-White Syndrome, *Circulation* 50, 1264, 1974.

James TN and Riddick L, Sudden death due to isolated acute infarction of the His bundle, *J Am Coll Cardiol* 15, 1183, 1990.

James TN, Robertson BT, Waldo AL, and Branch CE, De Subitaneis Mortibus. XV. Hereditary stenosis of the His bundle in pug dogs, *Circulation* 52, 1152, 1975e.

James TN, Schlant RC, and Marshall TK, De Subitaneis Mortibus. XXIX. Randomly distributed focal myocardial lesions causing destruction in the His bundle or a narrow-origin left bundle branch, *Circulation* 57, 816, 1978b.

James TN, Spencer MS, and Kloepfer JC, De Subitaneis Mortibus. XXI. Adult onset syncope, with comments on the nature of congenital heart block and the morphogenesis of the human atrioventricular septal junction, *Circulation* 54, 1001, 1976b.

James TN, St Martin T, Willis PW, et al., Apoptosis as a possible cause of gradual development of complete heart block and fatal arrhythmias associated with absence of the AV node, sinus node and internodal pathways, *Circulation* 93, 1424, 1996.

Janousek J and Paul T, Safety of oral propafenone in the treatment of arrhythmias in infants and children (European retrospective multicenter study). Working Group on Pediatric Arrhythmias and Electrophysiology of the Association of European Pediatric Cardiologists, *Am J Cardiol* 81, 1121, 1998.

Janson JT, Harris DG, Pretorious J, et al., Pericardial rupture and cardiac herniation after blunt chest trauma, *Ann Thorac Surg*, 75, 581, 2003.

Jennings RB, Sommers HM, Herdson PB, and Kaltenbach JP, Early phase of myocardial ischemic injury and infarction, *Ann NY Acad Sci* 156, 61, 1969.

Jentzen JM, Cocaine-induced myocarditis, *Am Heart J* 117, 1398, 1989.

Jerome KR, Vallant C, and Jaggi R, The TUNEL assay in the diagnosis of graft-versus-host disease: caveats for interpretation, *Pathology* 32, 136, 2000.

Jervell A and Nielsen FL, Congenital deaf-mutism, functional heart disease withprolongation of the QT interval, and sudden death, *Am Heart J* 54, 59–68, 1957.

Jetter WW and White PD, Rupture of the heart in patients in mental institutions, *Ann Intern Med* 21, 783, 1944.

Ji S, Cesario D, Valderrabano M, et al., The molecular basis of cardiac arrhythmias in patients with cardiomyopathy, *Curr Heart Fail Rep* 1, 98, 2004.

Jiang C, Atkinson D, Towbin JA, Splawski I, Lehmann MH, Li H, Timothy KW, Taggart RT, Schwartz PJ, Vincent GM, Moss AJ, and Keating MT, Two long QT syndrome loci map to chromosomes 3 and 7 with evidence for further heterogeneity, *Nature Genet* 8, 141–147, 1994.

Jing HL and Hu BJ, Sudden death caused by stricture of the sinus node artery, *Am J Forensic Med Pathol* 18, 360, 1997.

Johnson DW, Flemma RJ, and Lepley D, Direct reconstruction of flow to small distal coronary arteries, *Am J Cardiol* 25, 105, 1970.

Johnson JR and Di Palma JR, Intramyocardial pressure and its relation to aortic blood pressure, *Am J Physiol* 125, 234, 1939.

Johnsson H, Ivert T, Brodin LA, and Jonasson R, Late sudden deaths after repair of tetralogy of Fallot: electrographic findings associated with survival, *Scand J Thorac Cardiovasc Surg* 29, 131, 1995.

Joho S, Asanoi H, Remah HA, et al., Time-varying spectral analysis of heart rate and left ventricular pressure variability during balloon coronary occlusion in humans, *J Am Coll Cardiol* 34, 1924, 1999.

Jokl E and Greenstein J, Fatal coronary sclerosis in a boy of ten years, *Lancet* 247, 659, 1944.

Jones AM and Weston JT, The examination of the sudden infant death syndrome infant: investigative and autopsy protocols, *J Forensic Sci* 21, 833–841, 1976.

Jones CE, Devous MD, Thomas JX, et al., The effect of chronic cardiac denervation on infarct size following acute coronary occlusion, *Am Heart J* 95, 738, 1978.

Jones FL, Transmural myocardial necrosis after nonpenetrating cardiac trauma, *Am J Cardiol* 26, 419, 1970.

Jones JA, Salmon JE, Djuretic T, Nichols G, George RC, and Gill ON. An outbreak of serious illness and death among injecting drug users in England during 2000, *J Med Microbiol* 51, 978, 2002.

Jones JW, Hewitt RL, and Drapanas T, Cardiac contusion: A capricious syndrome, *Ann Surg*, 181, 567, 1975.

Jones RT, Cardiovascular system effects of marijuana, *J Clin Pharmacol* 42(11 Suppl), 58S, 2002.

Jorgensen L, Haerem JW, Chandler BA, et al., The pathology of acute coronary death, *Acta Anesthesiol Scand* 29, 193, 1968.

Jorgensen L, Hovig T, Rowsell HC, and Mustard JF, Adenosine-diphosphate-induced platelet aggregation and vascular injury in swine and rabbits, *Am J Pathol* 61, 161, 1970.

Jorgensen L, Rowsell HC, Hovig T, et al., Adenosine-diphosphate-induced platelet aggregation and myocardial infarction in swine, *Lab Invest* 17, 616, 1967.

Joseph G, Chandy ST, Krishnaswami S, et al., Mechanisms of cardiac perforation leading to tamponade in ballon mitral valvuloplasty, *Cathet Cardiovasc Diagn*, 42, 138, 1997.

Josephson ME, *Clinical Cardiac Electrophysiology*, Lippincott, Williams & Wilkins, Philadelphia, 2002.

Josselson A, Bagnall JW, and Virmani R, Acute rheumatic carditis causing sudden death, *Am J Forensic Med Pathol* 5, 151, 1984.

Jost S, Deckers JW, Nikutta P, et al., Progression of coronary artery disease is dependent on anatomic location and diameter, *J Am Coll Cardiol* 2, 1339, 1993.

Jugdutt BI, Ventricular remodelling after infarction and the extracellular collagen matrix, *Circulation* 108, 1395, 2003.

Julkunen H, Kaaja R, Siren MK, et al., Immune-mediated congenital heart block (CHB): identifying and counseling patients at risk for having children with CHB, *Semin Arthritis Rheum* 28, 97, 1998.

Junttila MJ, Raatikainen MJ, Karjalainen J, et al., Prevalence and prognosis of subjects with Bruga-datype ECG pattern in a young and middle-aged Finnish population, *Eur Heart J* 25, 847, 2004.

Jureidini SB, Marino CJ, Singh GK, et al., Main coronary artery and coronary ostial stenosis in children: detection by transthoracic color flow and pulsed Doppler echocardiography, *J Am Soc Echocardiogr* 13, 255, 2000.

Kahn AH, Pericarditis of myocardial infarction: review of the literature with case presentation, *Am Heart J* 90, 788, 1975.

Kahn JP, Perumal AS, Gully RJ, et al., Correlation of type A behaviour with adrenergic receptor density: implications for coronary artery disease pathogenesis, *Lancet* 24, 937, 1987.

Kakishita M, Kurita T, Matsuo K, et al., Mode of onset of ventricular fibrillation in patients with Brugada syndrome detected by implantable cardioverter defibrillator therapy. *J Am Coll Cardiol* 36, 1646, 2000.

Kakou Guikahue M, Sidi D, Kachaner J, et al., Anomalous left coronary artery arising from the pulmonary artery in infancy: is early operation better? *Br Heart J* 60, 522, 1988.

Kalant H, Adverse effects of cannabis on health: an update of the literature since 1996, *Progress in Neuro-Psychopharmacology & Biological Psychiatry* 28, 849, 2004.

Kalant H, The pharmacology and toxicology of "ecstasy" (MDMA) and related drugs, *CMAJ* 165, 917, 2001.

Kamisago M, Sharma SD, DePalma SR, et al., Mutations in sarcomere protein genes as a cause of dilated cardiomyopathy, *N Engl J Med* 343, 1688, 2000.

Kanda M, Shimizu W, Matsuo K, et al., Electrophysiologic characteristics and implications of induced ventricular fibrillation in symptomatic patients with Brugada syndrome, *J Am Coll Cardiol* 39, 1799, 2002.

Kannel WB and Thom TJ, Incidence prevalence and mortality of cardiovascular disease, in *The Heart*, Hurst JW, Ed., McGraw-Hill, New York, 1994, p. 200.

Kanoh M, Takemura G, Misao J, et al., Significance of myocytes with positive DNA *in situ* nick end-labeling (Tunel) in hearts with dilated cardiomyopathy: not apoptosis but DNA repair, *Circulation* 99, 2757, 1999.

Kaplan BM, Miller AJ, Bharati S, et al., Complete AV block following mediastinal radiation therapy: electrocardiographic and pathologic correlation and review of the world literature, *J Interv Cardiol Electrophysiol* 1, 175, 1997.

Kaplan J, The influence of sympathetic activation and psychosocial stress on coronary artery atherosclerosis, in *Hypertension — The Tip of an Iceberg*, Therapeutics Today Series, Vol. 7, Adis Press, Manchester, 1988, p. 9.

Kaplan JA, Karofsky PS, and Volturo GA, Commotio cordis in two amateur ice hockey players despite the use of commercial chest protectors: case reports, *J Trauma* 34, 151, 1993.

Kaplan JR, Manuck SB, Adams MR, et al., The effects of beta-adrenergic blocking agents on atherosclerosis and its complications, *Eur Heart J* 8, 928, 1987.

Kapur S, Kuehl KS, Midgely FM, and Chandra RS, Focal giant cell cardiomyopathy with Beckwith-Wiedemann syndrome, *Pediatr Pathol* 3, 261, 1985.

Karch SB, Resuscitation-induced myocardial necrosis: catecholamines and defibrillation, *Am J Forensic Med Pathol* 8, 3, 1987.

Karch SB, *The Pathology of Drug Abuse*, 2nd ed., CRC Press, Boca Raton, FL, 1996.

Karch SB, Cocaine cardiovascular toxicity, *South Med J* 98, 794, 2005.

Karch SB and Billingham ME, Myocardial contraction bands revisited, *Hum Pathol* 17, 9, 1986.

Karch SB and Billingham ME, Morphologic effects of defibrillation: a preliminary report, *Crit Care Med* 12, 920–921, 1984.

Karch SB, Green GS, and Young S, Myocardial hypertrophy and coronary artery disease in male cocaine users, *J Forensic Sci* 40, 591, 1995.

Karin E, Greenberg R, Avital S, et al., The management of stab wounds to the heart with laceration of the left anterior descending coronary artery, *Eur J Emerg Med*, 8, 321, 2001.

Karibe A, Tobacman LS, Strand J, et al., Hypertrophic cardiomyopathy caused by a novel alpha-tropomyosin mutation (V95A) is associated with mild cardiac phenotype, abnormal calcium binding to troponin, abnormal myosin cycling, and poor prognosis, *Circulation* 103, 65, 2001.

Karmy-Jones R, van Wijngaarden MH, Talwar MK, et al., Penetrating cardiac injuries, *Injury*, 28, 57, 1997.

Karsner HT and Bayless F, Coronary arteries in rheumatic fever, *Am Heart J* 9, 557, 1934.

Kasanuki H, Ohnishi S, Ohtuka M, et al., Idiopathic ventricular fibrillation with vagal activity in patients without obvious heart disease, *Circulation* 95, 2277, 1997.

Kaski JC, Crea F, Meran D, et al., Local coronary supersensitivity to diverse vasoconstrictive stimuli in patients with variant angina, *Circulation* 74, 1255, 1986.

Kasper EK, Agema WR, Hutchins GM, et al., The causes of dilated cardiomyopathy: a clinicopathologic review of 673 consecutive patients, *J Am Coll Cardiol* 23, 586, 1994.

Katsanos KH, Pappas CJ, Patsouras D, et al., Alarming atrioventricular block and mitral valve prolapse in the Kearns-Sayre syndrome, *Int J Cardiol* 83, 179, 2002.

Katz AM, Cellular mechanisms in congestive heart failure, *Am J Cardiol* 62, 3A, 1988.

Katz AM, The cardiomyopathy of overload: an unnatural growth response, *Eur Heart J* 16 (Suppl O), 110, 1995.

Kaufman ES, Priori SG, Napolitano C, et al., Electrocardiographic prediction of abnormal genotype in congenital long QT syndrome: experience in 101 related family members, *J Cardiovasc Electrophysiol* 12, 455, 2001.

Kaunitz PE, Origin of left coronary artery from pulmonary artery: review of the literature and report of two cases, *Am Heart J* 33, 182, 1947.

Kawai C and Abelmann WH, *Pathogenesis of Myocarditis and Cardiomyopathies*, University of Tokyo Press, 1987.

Kaye DM, Lambert GW, Lefkovits J, et al., Neurochemical evidence of cardiac sympathetic activation and increased central nervous system norepinephrine turnover in severe congestive heart failure, *J Am Coll Cardiol* 23, 570, 1994.

Keane JF, Driscoll DJ, Gersony WM, et al., Second natural history study of congenital heart defects: results of treatment of patients with aortic valvar stenosis, *Circulation* 87 (Suppl), I16, 1993.

Kearney DL, Titus JL, Hawkins EP, et al., Pathologic features of myocardial hamartomas causing childhood tachyarrhythmias, *Circulation* 75, 705, 1987.

Keating MT, Genetic approaches to cardiovascular disease: supravalvular aortic stenosis, Williams syndrome and long-QT syndrome, *Circulation* 92, 142, 1995.

Keating MT, The long QT syndrome: a review of recent molecular genetic and physiologic discoveries, *Medicine* 75, 1, 1996.

Keating MT, Atkinson D, Dunn C, Timothy KW, Vincent GM, and Leppert M, Linkage of a cardiac arrhythmia, the long QT syndrome, and the Harvey ras 1-gene, *Science* 252, 704–706, 1991.

Keeley EC, Velez CA, O'Neill W, and Safian RD, Long-term clinical outcome and predictors of major adverse cardiac events after percutaneous interventions on saphenous vein grafts, *J Am Coll Cardiol* 38, 659, 2001.

Keeling JW and Knowles SA, Sudden death in childhood and adolescence, *J Pathol* 159, 221, 1989.

Keenan RL and Boyan CP, Cardiac arrest due to anesthesia: a study of incidence and causes, *JAMA* 253, 2373, 1985.

Kegel SM, Dorsey TJ, Rowen M, and Taylor WF, Cardiac death in mucocutaneous lymph node syndrome, *Am J Cardiol* 40, 282, 1977.

Keller BB, Mehta AV, Shamszadeh M, et al., Oncocytic cardiomyopathy of infancy with Wolff-Parkinson-White syndrome and ectopic foci causing tachydysrhythmias in children, *Am Heart J* 114, 782, 1987.

Keller DI, Carrier L, and Schwartz K, Genetics of familial cardiomyopathies and arrhythmias, *Swiss Med Wkly* 132, 401, 2002.

Kelley ST, Malekan R, Gorman JH, et al., Restraining infarct expansion preserves left ventricular geometry and function after acute anterolateral infarction, *Circulation* 99, 135, 1999.

Kelly DH, Pathak A, and Meny R, Sudden severe bradycardia in infancy, *Pediatr Pulm* 10, 199–204, 1991.

Kelm M, Perings SM, Jax T, et al., Incidence and clinical outcome of iatrogenic femoral arteriovenous fistulas, *J Am Coll Cardiol,* 40, 291, 2002.

Kempen PM and Allgod R, Right ventricular rupture during closed-chest cardiopulmonary resuscitation after pneumonectomy with pericardiotomy: a case report, *Crit Care Med*, 27, 1378, 1999.

Kendeel SR and Ferris JAJ, Fibrosis of the conducting tissue in infancy, *J Path* 117, 123–130, 1974.

Kennedy HL and Das SK, Postmyocardial infarction (Dressler's) syndrome: report of a case with immunologic and viral studies, *Am Heart J* 91, 233, 1976.

Kennedy MC and Lawrence C, Anabolic steroid abuse and cardiac death, *Med J Aust* 158, 346, 1993.

Kent SP, Diffusion of myoglobin in the diagnosis of early myocardial ischemia, *Lab Invest* 46, 270, 1982.

Keren A, Goldberg S, Gottlieb S, et al., Natural history of left ventricular thrombi: their appearance and resolution in the hospitalization period of acute myocardial infarction, *J Am Coll Cardiol* 15, 790, 1990.

Kern WH, Dermer GB, and Lindesmith GG, The intimal proliferation in aortic-coronary saphenous vein grafts: light and electron microscopic studies, *Am Heart J* 84, 771, 1972.

Khouri EM, Gregg DE, and Lowesohn HS, Flow in the major branches of the left coronary artery during experimental coronary insufficiency in the unanesthetized dog, *Circ Res* 23, 99, 1968.

Khouri EM, Gregg DE, and McGranahan GM, Regression and reappearance of coronary collaterals, *Am J Physiol* 220, 655, 1971.

Khoury Z, Keren A, Benhorin J, and Stern S, Aborted sudden death in a young patient with isolated granulomatous myocarditis, *Eur Heart J* 15, 397, 1994.

Kim J and Busuttil A, Traumatic rupture of the aorta, *J Clinic Forensic Med*, 3, 123, 1996.

Kimmelstiel P, Gilmour MT, and Hodges HH, Degeneration of elastic fibers in granulomatous giant cell arteritis (temporal arteritis), *AMA Arch Pathol* 54, 157, 1952.

Kini A, Marmur JD, Kini S, et al., Creatine kinase-MB elevation after coronary intervention correlates with diffuse atherosclerosis and low-to-medium level elevation has a benign clinical course, *J Am Coll Cardiol* 34, 663, 1999.

Kirschner RH, Eckner FA, and Baron RC, The cardiac pathology of sudden, unexplained nocturnal death in Southeast Asian refugees, *JAMA* 256, 2700, 1986.

Kisch B, The capillaries of the human myocardium: an electron microscopic study, *Rev Can Biol* 22, 317, 1963.

Kishi S, Yoshida A, Yamauchi T, et al., Torsade de pointes associated with hypokalemia after anthracycline treatment in a patient with acute lymphocytic leukemia, *Int J Hematol* 71, 172, 2000.

Kitada M, Nakagawa T, and Yamaguchi Y, A survey of sudden death among school children in Osaka Prefecture, *Jpn Circ J* 54, 401, 1990.

Kleinert S, Weintraub RG, Wilkinson JL, and Chow CW, Myocarditis in children with dilated cardiomyopathy: incidence and outcome after dual therapy immunosuppression, *J Heart Lung Transplant* 16, 1248, 1997.

Klima T, Spjut HJ, Coelho A, et al., The morphology of ascending aortic aneurysms, *Hum Pathol* 14, 810, 1983.

Klintschar M, Darok M, and Radner H, Massive injury to the heart after attempted active compression–decompression cardiopulmonary resuscitation, *Int J Legal Med*, 111, 93, 1998.

Kloner RA, Allen J, Zheng Y, et al., Myocardial stunning following exercise treadmill testing in man, *J Am Coll Cardiol* 15, 203A, 1990.

Kloner RA, Bolli R, Marban E, et al., Medical and cellular implications of stunning, hibernating and preconditioning, *Circulation* 97, 1848, 1998.

Kloner RA, Ganote CE, and Jennings RB, The "no-reflow" phenomenon after temporary coronary occlusion in the dog, *J Clin Invest* 54, 1496, 1974.

Kloner RA, Hale S, Alker K, and Rezkalla S, The effects of acute and chronic cocaine use on heart, *Circulation* 85, 407, 1992.

Kloner RA, Przyklenk K, and Patel B, Altered myocardial states: the stunned and hibernating myocardium, *Am J Med* 86 (Suppl 1A), 14, 1989.

Kloner RA and Rezkalla SH, Cocaine and the heart, *N Engl J Med* 348, 487, 2003.

Knisely MH, The settling of sludge during life: first observations, evidences and significance — a contribution to the biophysics of disease, *Acta Anat* 44, 7, 1961.

Kobayashi Y, Yazawa T, Baba T, et al., Clinical, electrophysiological, and histopathological observations in supraventricular tachycardia, *Pacing Clin Electrophysiol* 11, 1154, 1988.

Kochar M and Hosko MJ, Electrocardiographic effects of marijuana, *JAMA* 225, 25, 1973.

Kochi K, Takebayashi S, Hikori T, et al., Significance of adventitial inflammation of the coronary artery in patients with unstable angina: results at autopsy, *Circulation* 71, 709, 1985.

Koesters SC, Rogers PD, and Rajasingham CR, MDMA ("ecstasy") and other "club drugs": the new epidemic, *Pediatr Clin North Am* 49, 415, 2002.

Kolodgie FD, Gold HK, Burke AP, et al., Intraplaque hemorrhage and progression of coronary atheroma, *N Engl J Med* 394, 2316, 2003.

Kolodgie FD, Virmani R, Cornhill JF, et al., Increase in atherosclerosis and adventitial mast cells in cocaine abusers: an alternative mechanism of cocaine-associated coronary vasospasm and thrombosis, *J Am Coll Cardiol* 17, 1553, 1991.

Konishi Y, Tatsuta N, Kumada K, et al., Dissecting aneurysm during pregnancy and the puerperium, *Jpn Circ J* 44, 726, 1980.

Konomi N, Lebwohl E, and Zhang D, Comparison of DNA and RNA extraction methods for mummified tissues, *Mol Cell Probes* 16, 445, 2002.

Kopf GS, Hellenbrand W, Kleinman C, et al., Repair of aortic coarctation in the first three months of life: immediate and long-term results, *Ann Thorac Surg* 41, 425, 1986.

Koponen MA and Siegel R, Hamartomatous malformation of the left ventricle associated with sudden death, *J Forensic Sci* 40, 495, 1995.

Koponen MA and Siegel RJ, Histiocytoid cardiomyopathy and sudden death, *Hum Pathol* 27, 420, 1996.

Korb G and Totovic V, Electron microscopical studies on experimental ischemic lesions of the heart, *Ann NW Acad Sci* 105, 135, 1969.

Kosek JC, Chartrand C, Hurley EJ, et al., Arteries in canine cardiac homografts: ultrastructure during acute rejection, *Lab Invest* 21, 328, 1969.

Kosior DA, Filipiak KJ, Stolarz P, et al., Paroxysmal atrial fibrillation in a young female patient following marijuana intoxication — a case report of possible association, *Med Sci Monit* 6, 386, 2000.

Kosior DA, Filipiak KJ, Stolarz P, et al., Paroxysmal atrial fibrillation following marijuana intoxication: a two-case report of possible association, *Int J Cardiol* 78, 183, 2001.

Kouwenhoven W, Jude JR, and Knickerbocher CG, Closed-chest cardiac massage, *JAMA*, 173, 1064, 1960.

Kozakewich HPW, McManus BM, and Vawter GF, The sinus node in sudden infant death syndrome, *Circulation* 65, 1242–1246, 1982.

Kral BG, Becker LC, Blumenthal RS, et al., Exaggerated reactivity to mental stress is associated with exercise-induced myocardial ischemia in an asymptomatic high-risk population, *Circulation* 96, 42, 1997.

Kramer CM, Nicol PD, Rogers WJ, et al., Reduced sympathetic innervation underlies adjacent noninfarcted region dysfunction during left ventricular remodelling, *J Am Coll Cardiol* 30, 1079, 1997.

Krehl B (1891), quoted in Baroldi G and Silver MD, *Sudden Death in Ischemic Heart Disease: An Alternative View on the Significance of Morphologic Findings*, RG Landes Co., Austin, TX, Springer-Verlag, Heidelberg, 1995.

Krischer JP, Fine EG, Davis JH, et al., Complications of cardiac resuscitation, *Chest*, 92, 287, 1987.

Kubler W, The sympathetic system in evolution and in ischemic heart disease: a controversy? *Eur Heart J* 13, 1301, 1992.

Kubler W and Strasser RH, Signal transduction in myocardial ischemia, *Eur Heart J* 15, 437, 1994.

Kugelmass AD, Shannon RP, Yeo EL, and Ware JA, Intravenous cocaine induces platelet activation in the conscious dog, *Circulation* 91, 1336, 1995.

Kugelmeier J, Egloff L, Real F, et al., Congenital aortic stenosis: early and late results of aortic valvulotomy, *Thorac Cardiovasc Surg* 30, 91, 1982.

Kuhl U, Lauer B, Souvatzoglu M, et al., Antimyosin scintigraphy and immunohistologic analysis of endomyocardial biopsy in patients with clinically suspected myocarditis — evidence of myocardial cell damage and inflammation in the absence of histologic signs of myocarditis, *Am Coll Cardiol* 32, 1371, 1998.

Kuhn FE, Gillis RA, Virmani R, et al., Cocaine produces coronary artery vasoconstriction independent of an intact endothelium, *Chest* 102, 581, 1992.

Kuhn FE, Johnson MN, Gillis RA, et al., Effects of cocaine on the coronary circulation and systemic hemodynamics in dogs, *J Am Coll Cardiol* 16, 1481, 1990.

Kuisma M, Suominen P, and Korpela R, Paediatric out-of-hospital cardiac arrests — epidemiology and outcome, *Resuscitation* 30, 141, 1995.

Kulan K, Komsuoglu B, Tuncer C, and Kulan C, Significance of QT dispersion on ventricular arrhythmias in mitral valve prolapse, *Int J Cardiol* 54, 251, 1996.

Kunos G, Járai Z, Batkai S, et al., Endocannabinoids as cardiovascular modulators, *Chem Phys Lipids* 108, 159, 2000.

Kurita T, Shimizu W, Inagaki M, et al., The electrophysiologic mechanism of ST-segment elevation in Brugada syndrome, *J Am Coll Cardiol* 40, 330, 2002.

Kurosawa H, Wagenaar SS, and Becker AE, Sudden death in a youth: a case of quadricuspid aortic valve with isolation of origin of left coronary artery, *Br Heart J* 46, 211, 1981.

Kyndt F, Probst V, Potet F, et al., Novel SCN5A mutation leading either to isolated cardiac conduction defect or Brugada syndrome in a large French family, *Circulation* 104, 3081, 2001.

L'Abbate A and Parodi O, *La cardiologia nucleare, dalla fisiopatologia all'interpretazione*, Ciba-Geigy Edizioni, Milan, 1992.

La Canna G, Alfieri O, Giubbini R, et al., Echocardiography during infusion of dobutamine for identification of reversible dysfunction in patients with chronic coronary artery disease, *J Am Coll Cardiol* 23, 617, 1994.

Laham RJ, Chronos NA, PIke M, et al., Intracoronary basic fibroblast growth factor (FGF-2) in patients with severe ischemic heart disease: results of a phase I open-label dose escalations study, *J Am Coll Cardiol* 36, 2132, 2000.

Lahat H, Eldar M, Levy-Nissenbaum E, et al., Autosomal recessive catecholamine- or exercise-induced polymorphic ventricular tachycardia: clinical features and assignment of the disease gene to chromosome 1p13-21, *Circulation* 103, 2822, 2001.

Lai TI, Hwang JJ, Fang CC, and Chen WJ, Methylene 3,4-dioxymethamphetamine-induced acute myocardial infarction, *Ann Emerg Med* 42, 759, 2003.

Laine P, Naukkarinen A, Heikkila L, et al., Adventitial mast cells connect with sensory nerve fibers in atherosclerotic coronary arteries, *Circulation* 101, 1665, 2000.

Laitinen PJ, Brown KM, Piippo K, et al., Mutations of the cardiac ryanodine receptor (RyR2) gene in familial polymorphic ventricular tachycardia, *Circulation* 103, 485, 2001.

Lakkis NM, Nagueh SF, Dunn JK, et al., Nonsurgical septal reduction therapy for hypertrophic obstructive cardiomyopathy: one-year follow-up, *J Am Coll Cardiol* 36, 852, 2000.

Laks H, Kaiser GC, Mudd JG, et al., Revascularization of the right coronary artery, *Am J Cardiol* 43, 1109, 1979.

Laks MM, Myocardial hypertrophy produced by chronic infusion of subhypertensive doses of norepinephrine in the dog, *Chest* 64, 75, 1973.

Lambert EC, Menon VA, Wagner HR, and Vlad P, Sudden unexpected death from cardiovascular disease in children, *Am J Cardiol* 34, 89, 1974.

Lamberti C, Kruse R, Ruelfs C, et al., Microsatellite instability — a useful diagnostic tool to select patients at high risk for hereditary nonpolyposis colorectal cancer: a study in different groups of patients with colorectal cancer, *Gut* 44, 839, 1999.

Lameris TW, Zeeuw de S, Albert G, et al., Time course and mechanism of myocardial catecholamine release during transient ischemia *in vivo*, *Circulation* 101, 2645, 2000.

Lampros TD and Cobanoglu A, Discrete subaortic stenosis: an acquired heart disease, *Eur J Cardiothorac Surg* 14, 296, 1998.

Lancisi GM, *Opera Omnia, Tomus I De Subitaneis Mortibus*, Libri duo, Tertia editione, Palladis Roma 1745. (Translated by White PD and Boursey AV, St. John's University Press, New York, 1971.)

Land RN, Hamilton AY, and Fuchs PC, Sudden death in a young athlete due to an anomalous commissural origin of the left coronary artery, and focal intimal proliferation of aortic valve leaflet at the adjacent commissure, *Arch Pathol Lab Med* 118, 931, 1994.

Landry MJ, MDMA: a review of epidemiologic data, *J Psychoactive Drugs* 34, 163, 2002.

Laner MS, Blackstone EH, Young JB, and Topol EJ, Cause of death in clinical research: time for a reassessment? *J Am Coll Cardiol* 34, 618, 1999.

Lange RA, Cigarroa RG, Yancy CW Jr, et al., Cocaine-induced coronary-artery vasoconstriction, *N Engl J Med* 321, 1557, 1989.

Lange RA and Hillis D, Cardiovascular complications of cocaine use, *N Engl J Med* 345, 351, 2001.

Langille LB, Hemodynamic factors and vascular disease, in *Cardiovascular Pathology*, 2nd ed., Silver MD, Ed., Churchill Livingstone, New York, 1991, p. 131.

Lanza GA, Pedrotti P, Pasceri V, et al., Autonomic changes associated with spontaneous coronary spasm in patients with variant angina, *J Am Coll Cardiol* 28, 1249, 1996.

Larson EW and Edwards WD, Risk factors for aortic dissection: a necropsy study of 161 cases, *Am J Cardiol* 53, 849, 1984.

Larsson E, Wesslen L, Lindquist O, et al., Sudden unexpected cardiac deaths among young Swedish orienteers — morphological changes in hearts and other organs, *APMIS* 107, 325, 1999.

Lashus AG, Case CL, and Gillette PC, Catheter ablation treatment of supraventricular tachycardia-induced cardiomyopathy, *Arch Pediatr Adolesc Med* 151, 264, 1997.

Lasky II, Traumatic nonpenetrating cardiac disorders, *Legal Med Ann*, 161, 192, 1972.

Laurain AR and Inoshita T, Sudden death from pericardial tamponade: unusual complication of nonbacterial thrombotic endocarditis, *Arch Pathol Lab Med* 109, 171, 1985.

Lautsch EV, Functional morphology of heart valves, in *Functional Morphology of the Heart*, Bajusz E and Jasmin G, Eds., Karger, Basel, *Meth Achievm Exp Pathol* 5, 214, 1971.

Lazarides MK, Arvanitis DP, Liatas AC, et al., Iatrogenic and noniatrogenic arterial trauma: a comparative study, *Eur J Surg*, 157, 17, 1991.

Lazarides MK, Tsoupanos SS, Georgopoulos SE, Chronopoulos AV, Arvanitis DP, Doundoulakis NJ, and Dayantas JN, Incidence and patterns od iatrogenic arterial injuries. A decade's experience, *J Cardiovasc Surg* 39, 3, 281, 1998.

Leachman RD, Cokkinos DV, Zamalloa O, and Del Rio C, Intercoronary artery steal, *Cardiovasc Res Cent Bull* 10, 71, 1972.

Leadbeatter S, Nawman HM, and Jasani B, Further evaluation of immunocytochemical staining in the diagnosis of early ischaemic/hypoxic damage, *Forensic Sci Int* 45, 135, 1990.

Leary T, Coronary spasm as a possible factor in producing sudden death, *Am Heart J* 10, 338, 1935.

Lee LA, Pickrell MB, and Reichlin M, Development of complete heart block in an adult patient with Sjogren's syndrome and anti-Ro/SS-A autoantibodies, *Arthritis Rheum* 39, 1427, 1996.

Leenen FHH, Cardiovascular consequence of sympathetic hyperactivity, *Can J Cardiol* 40 (Suppl A), 2A, 1999.

Leenhardt A, Glaser E, Burguera M, et al., Short-coupled variant of torsade de pointes: a new electrocardiographic entity in the spectrum of idiopathic ventricular tachyarrhythmias, *Circulation* 89, 206, 1994.

Leenhardt A, Lucet V, Denjoy I, et al., Catecholaminergic polymorphic ventricular tachycardia in children: a 7-year follow-up of 21 patients, *Circulation* 91, 1512, 1995.

Lehnart SE, Wehrens XH, Kushnir A, et al., Cardiac ryanodine receptor function and regulation in heart disease, *Ann N Y Acad Sci* 1015, 144, 2004.

Leicht JW, McElduff P, Dolson A, and Heller R, Outcome with calcium channel antagonist after myocardial infarction: a community-based study, *J Am Coll Cardiol* 31, 11, 1998.

Lemma M, Mangini A, Gelpi G, and Antona C, Are composite Y-grafts able to fully respond to the left coronary system flow demand early after coronary bypass graft? *Ann Thorac Surg* 76(4), 1339, 2003.

Lenders JW, De Macker PN, Vos JA, et al., Deleterious effects of anabolic steroids on serum lipoproteins, blood pressure and liver function in amateur body builders, *Int J Sports Med* 9, 19, 1988.

Leor J, Patterson M, Quinones MJ, et al., Transplantation of fetal myocardial tissue into the infarcted myocardium of rat: a potential method for repair of infarcted myocardium? *Circulation* 94, II332, 1996.

Leor J, Poole KW, and Kloner RA, Sudden cardiac death triggered by an earthquake, *N Engl J Med* 334, 413, 1996.

Lepper W, Sieswerda GT, Franke A, et al., Repeated assessment of coronary flow velocity pattern in patients with first acute myocardial infarction, *J Am Coll Cardiol* 39, 1283, 2002.

Lesch M and Kehoe RF, Predictability of sudden cardiac death: a partially fulfilled promise, *N Engl J Med* 310, 255, 1984.

Lev M and Bharati S, A method of study of the pathology of the conduction system for the electrocardiographic and His bundle electrogram correlations, *Anat Rec* 201, 43, 1981.

Lev M and Watne AL, Method for routine histopathologic study of human sinoatrial node, *Arch Pathol* 57, 168, 1954.

Lev M, Widran J, and Erickson EE, A method for histopathological study of A-V node bundle and branches, *Arch Pathol* 52, 73, 1951.

Levi G, Scalvini S, Volterrani M, et al., Coxsackie virus heart disease: 15 years after, *Eur Heart J* 9, 1303, 1988.

Levin DC and Fallon JT, Significance of the angiographic morphology of localized coronary stenoses: histopathologic correlations, *Circulation* 66, 316, 1982.

Levine HD and Welles RE, Blood sludging observed at the bed side: flexibility and limitations of examination by ophthalmoscopy, *Am J Cardiol* 13, 48, 1964.

Lewis DR, Bullbulia RA, Murphy P, et al., Vascular surgical intervention for complications of cardiovascular radiology: 13 years' experience in a single center, *Ann R Coll Surg Engl*, 81, 23, 1999.

Lewis W and Silver MD, Adverse effects of drugs on the cardiovascular system, in *Cardiovascular Pathology*, 3rd ed., Silver MD, Gotlieb AI, and Schoen FJ, Eds., Churchill Livingstone, Philadelphia, 2001, p. 541.

Leyton RA and Sonnenblick EH, The ultrastructure of the failing heart, *Am J Med Sci* 258, 304, 1969.

Li D, Ahmad F, Gardner MJ, et al., The locus of a novel gene responsible for arrhythmogenic rightventricular dysplasia characterized by early onset and high penetrance maps to chromosome 10p12–p14, *Am J Hum Genet* 66, 148, 2000.

Li D, Czernuszewicz GZ, Gonzalez O, et al., Novel cardiac troponin T mutation as a cause of familial dilated cardiomyopathy, *Circulation* 104, 2188, 2001.

Li RK, Jia ZQ, Weisel RD, et al., Smooth muscle cell transplantation into myocardial scar tissue improves heart function, *J Mol Cell Cardiol* 31, 513, 1999.

Li ST, Tack CH, Fananapazir L, and Goldstein DS, Myocardial perfusion and sympathetic innervation in patients with hypertrophic cardiomyopathy, *J Am Coll Cardiol* 35, 1867, 2000a.

Li W, Su J, Sehgal S, Altura BT, and Altura BM, Cocaine-induced relaxation of isolated rat aortic rings and mechanisms of action: possible relation to cocaine-induced aortic dissection and hypotension, *Eur J Pharmacol* 496, 151, 2004.

Li Y, Bourlet T, Andreoletti L, et al., Enteroviral capsid protein VP1 is present in myocardial tissues from some patients with myocarditis or dilated cardiomyopathy, *Circulation* 101, 231, 2000b.

Libby P, Molecular bases of the acute coronary syndromes, *Circulation* 91, 2844, 1995.

Liberthson RR, Sudden death from cardiac causes in children and young adults, *N Engl J Med* 334, 1039, 1996.

Lie JT, Histopathology of the conduction system in sudden death from coronary heart disease, *Circulation* 51, 446, 1975.

Lie JT, Holley KE, Kampa WR, and Titus JL, New histochemical method for morphologic diagnosis of early stages of myocardial ischemia, *Mayo Clin Proc* 46, 316, 1971.

Lie JT, Lawrie GM, Morris GC, et al., Hemorrhagic myocardial infarction associated with aortocoronary bypass revascularization, *Am Heart J* 96, 295, 1978.

Lie JT, Lufschanowski R, and Erickson EE, Heterotopic epithelial replacement (so-called "mesothelioma") of the atrioventricular node, congenital heart block, and sudden death, *Am J Forensic Med Pathol* 1, 131, 1980.

Lie JT, Rosenberg HS, and Erickson EE, Histopathology of the conduction system in the sudden infant death syndrome, *Circulation* 53, 3–8, 1976.

Liedtke AJ and DeMuth WE, Nonpenetrating cardiac injuries: A collective review, *Am Heart J,* 86, 687, 1973.

Liggett SB, Tepe NM, Lorenz JN, et al., Early and delayed consequences of β2-adrenergic receptor overexpression in mouse hearts: critical role for expression level, *Circulation* 101, 1707, 2000.

Likar IN, Robinson RW, and Gouvelis A, Microthrombi and intimal thickening in bovine coronary arteries, *Arch Pathol* 87, 146, 1969.

Lin AE, Lippe BM, Geffner ME, et al., Aortic dilation, dissection, and rupture in patients with Turner syndrome, *J Pediatr* 109, 820, 1986.

Lin AE, Lippe B, and Rosenfeld RG, Further delineation of aortic dilation, dissection, and rupture in patients with Turner syndrome, *Pediatrics* 102, e12, 1998.

Lindsay AC, Foale RA, Warren O, et al., Cannabis as a precipitant of cardiovascular emergencies, *Int J Cardiol* 104, 230, 2005.

Lindsay S, The cardiovascular system in gargoylism, *Br Heart J* 12, 17, 1950.

Lindsey D and Navin TR, Transient elevation of serum activity of MB isoenzyme of creatine phosphokinase in drivers involved in automobile accidents, *Chest,* 74, 15, 1978.

Lindstaedt M, Germing A, von Dryander S, et al., Acute and long-term clinical significance of myocardial contusion following blunt thoracic trauma: results of a prospective study, *J Trauma Inj Infect Crit Care,* 52, 479, 2002.

Link MS, Maron BJ, VanderBrink BA, et al., Impact directly over the cardiac silhouette is necessary to produce ventricular fibrillation in an experimental model of commotio cordis, *JACC,* 37, 649, 2001.

Link MS, Wang PJ, Pandian NG, et al., An experimental model of sudden death due to low-energy chest-wall impact (commotio cordis), *New Engl J Med,* 338, 1805, 1998.

Linzbach AJ, Mikrometrische und histologische Analyse hypertropher menschlicher Herzen, *Virchows Arch Pathol Anat* 314, 534, 1947.

Linzbach AJ, Heart failure from the point of view of quantitative anatomy, *Am J Cardiol* 5, 370, 1960.

Lipsett J, Byard RW, Carpenter BF, et al., Anomalous coronary arteries arising from the aorta associated with sudden death in infancy and early childhood: an autopsy series, *Arch Pathol Lab Med* 115, 770, 1991.

Lipsett J, Cohle SD, Berry PJ, et al., Anomalous coronary arteries: a multicenter pediatric autopsy study, *Pediatr Pathol* 14, 287, 1994.

Lipsitt LP, Sturner WQ, Oh W, Barrett J, and Truex RC, Wolff-Parkinson-White and sudden-infant-death syndromes, *New Engl J Med* 300, 1111, 1979.

Lisse JR, Davis CP, and Thurmond-Anderle M, Cocaine abuse and deep venous thrombosis, *Ann Intern Med* 110, 571, 1989.

Litsey SE, Noonan JA, O'Connor WN, et al., Maternal connective tissue disease and congenital heart block, *New Engl J Med* 312, 98, 1985.

Little WC, Constantinescu M, Applegate RJ, et al., Can coronary angiography predict the site of a subsequent myocardial infarction in patients with mild-to-moderate coronary artery disease? *Circulation* 78, 1157, 1988.

Liuzzo G, Buffon A, Biasucci LM, et al., Enhanced inflammatory response to coronary angioplasty in patients with severe unstable angina, *Circulation* 98, 2370, 1998.

Lobban CDR, The human dive reflex as a primary cause of SIDS. A review of the literature, *Med J Australia* 155, 561–563, 1991.

Lobo FV, Silver MD, Butany J, et al., Left ventricular involvement in right ventricular dysplasia/cardiomyopathy, *Can J Cardiol* 15, 1239, 1999.

Lobo FV, Heggtveit HA, Butany J, et al., Right ventricular dysplasia: morphological findings in 13 cases, *Can J Cardiol* 8, 261, 1992.

Lobo FVO and Heggtviet HA, Cardiovascular trauma, in *Cardiovascular Pathology*, 3rd ed., Silver MD, Gotlieb AI, and Schoen FJ, Eds., Churchill Livingstone, Philadelphia, 2001, p. 562.

Locati EH, Zareba W, Moss AJ, et al., Age- and sex-related differences in clinical manifestations in patients with congenital long-QT syndrome: findings from the International LQTS Registry, *Circulation* 97, 2237, 1998.

Lodge-Patch I, The aging of cardiac infarct and its influence on cardiac rupture, *Br Heart J* 13, 37, 1951.

Lora-Tamayo C, Tena T, and Rodriguez A, Amphetamine derivative related deaths, *Forensic Sci Int* 85, 149, 1997.

Louie EK and Maron BJ, Hypertrophic cardiomyopathy with extreme increase in left ventricular wall thickness, *J Am Coll Cardiol* 8, 57, 1986.

Lowe JE, Cummings RG, Adams DH, et al., Evidence that ischemic cell death begins in the subendocardium independent of variations in collateral flow or wall tension, *Circulation* 68, 190, 1983.

Lown B, Sudden cardiac death: the major challenge confronting contemporary cardiology, *Am J Cardiol* 43, 313, 1979.

Lown B, Verrier RL, and Rabinowitz SH, Neural and psychologic mechanisms and the problem of sudden cardiac death, *Am J Cardiol* 39, 890, 1977.

Lu CW, Wu MH, and Chu SH, Paroxysmal supraventricular tachycardia in identical twins with the same left lateral accessory pathways and innocent dual atrioventricular pathways, *Pacing Clin Electrophysiol* 23, 1564, 2000.

Lubiszewska B, Rozanski J, Szufladowicz M, et al., Mechanical valve replacement in congenital heart disease in children, *J Heart Valve Dis* 8, 74, 1999.

Ludwig G, Capillary pattern of the myocardium, in Functional Morphology of the Heart, Bajusz E and Jasmin G, Eds., *Meth Achiev Exp Pathol* 5, 238, 1971.

Luisi AJ, Fallavollita JA, Suzuki G, et al., Spatial inhomogeneity of sympathetic nerve function in hibernating myocardium, *Circulation* 106, 779, 2002.

Luke JL, Farb A, Virmani R, and Sample RHB, Sudden cardiac death during exercise in a weight lifter using anabolic-androgenic steroids: pathological and toxicological findings, *J Forensic Sci* 35, 1441, 1990.

Lum M, Stevenson WG, Stevenson LW, et al., Diverse mechanisms of unexpected cardiac arrest in advanced heart failure, *Circulation* 80, 1675, 1989.

Lunetta P, Levo A, Laitinen PJ, et al., Molecular screening of selected long QT syndrome (LQTS) mutations in 165 consecutive bodies found in water, *Int J Legal Med* 117, 115, 2003.

Lurie KG, Lindo C, and Chin J, CPR: the P stands for plumber's helper, *J Am Med Assoc*, 264, 1661, 1990.

Lynch JJ, Paskewitz DA, Gimbel KS, and Thomas SA, Psychological aspects of cardiac arrhythmia, *Am Heart J* 93, 645, 1977.

Machado MV, Tynan MJ, Curry PV, et al., Fetal complete heart block, *Br Heart J* 60, 512, 1988.

Machii M, Inaba H, Nakae H, et al., Cardiac rupture by penetration of fractured sternum: a rare complication of cardiopulmonary resuscitation, *Resuscitation*, 43, 151, 2000.

MacIsaac AI, Thomas JD, and Topol EJ, Toward the quiescent coronary plaque, *J Am Coll Cardiol* 22, 1228, 1993.

MacRae CA, Ghaisas N, Kass S, et al., Familial hypertrophic cardiomyopathy with Wolff-Parkinson-White syndrome maps to a locus on chromosome 7q3, *J Clin Invest* 96, 1216, 1995.

Madea B and Grellner W, Long-term cardiovascular effects of anabolic steroids, *Lancet* 352, 33, 1998.

Madiba TE, Thomson SR, and Mdlalose N, Penetrating chest injuries in the firearm era, *Injury*, 32, 13, 2001.

Mahnke PF, Epidemiologie und spezielle Pathologie des plötzlichen Todes im Kindesalter, *Dtsch Gesundh Wes* 21, 2188–2191, 1966.

Mahon NG, Zal B, Arno G, et al., Absence of viral nucleic acids in early and late dilated cardiomyopathy, *Heart* 86, 687, 2001.

Mahowald JM, Blieden LC, Coe JI, et al., Ectopic origin of a coronary artery from the aorta; sudden death in 3 of 23 patients, *Chest* 89, 668, 1986.

Maisch B, Ristic AD, Hufnagel G, and Pankuweit S, Pathophysiology of viral myocarditis: the role of humoral immune response, *Cardiovasc Pathol* 11, 112, 2002.

Maisel AS, Beneficial effects of metaprolol treatment in congestive heart failure: reversal of sympathetic-induced alterations of immunologic function, *Circulation* 90, 1774, 1994.

Majno G, Ames A, Chaing J, et al., No-reflow after cerebral ischemia, *Lancet* 2, 569, 1967.

Majno G and Joris I, Apoptosis, oncosis and necrosis: an overview of cell death, *Am J Pathol* 146, 3, 1995.

Mak T and Weiglicki WB, Protection by beta-blocking agents against free radical-mediated sarcolemmal lipid peroxidation, *Circ Res* 63, 262, 1988.

Maki T, Niimura I, Nishikawa T, and Sekiguchi M, An atypical case of cardiomyopathy in a child: hypertrophic or restrictive cardiomyopathy? *Heart Vessels Suppl* 5, 84, 1990.

Malhotra V, Ferrans VJ, and Virmani R, Infantile histiocytoid cardiomyopathy: three cases and literature review, *Am Heart J* 128, 1009, 1994.

Mallat Z, Tedgui A, Fontaliran F, et al., Evidence of apoptosis in arrhythmogenic right ventricular dysplasia, *N Eng J Med* 335, 1190, 1996.

Malliani A, Schwartz PJ, and Zanchetti A, A sympathetic reflex elicited by experimental coronary occlusion, *Am J Physiol* 217, 703, 1979.

Mallory GK, White PD, and Salcedo-Salgar J, The speed of healing of myocardial infarction: a study of the pathologic-anatomy of seventy-two cases, *Am Heart J* 18, 647, 1939.

Mammarella A, Paradiso M, Antonini G, et al., Natural history of cardiac involvement in myotonic dystrophy (Steinert's disease): a 13-year follow-up study, *Adv Ther* 17, 238, 2000.

Mandal AK and Sanusi M, Penetrating chest wounds: 24 years experience, *World J Surg*, 25, 1145, 2001.

Mandorla S and Martino C, Familial atrial septal defect with atrioventricular conduction defects, *J Ital Cardiol* 28, 294, 1998.

Manning WJ and Pennell DJ, *Cardiovascular Magnetic Resonance*, Churchill Livingstone, London, 2002.

Manojkumar R, Sharma A, and Grover A, Secondary lymphoma of the heart presenting as recurrent syncope, *Indian Heart J* 53, 221, 2001.

Mantini E, Tanaka S, and Lillehei CW, Analysis of the earliest effects of mammary artery implantation on the ischemic ventricle, *Ann Thorac Surg* 5, 393, 1968.

Manuel-Rimbau E, Lozano P, Gomez A, et al., Iatrogenic vascular lesions after cardiac catheterization, *Rev Esp Cardiol*, 51, 750, 1998.

Marcus F, Towbin JA, Zareba W, et al., Arrhythmogenic right ventricular dysplasia/cardiomyopathy (ARVD/C). A multidisciplinary study: design and protocol, *Circulation* 107, 2975, 2003.

Marcus FI, Electrocardiographic features of inherited diseases that predispose to the development of cardiac arrhythmias, long QT syndrome, arrhythmogenic right ventricular cardiomyopathy/dysplasia, and Brugada syndrome, *J Electrocardiol* 33 (Suppl 1), 1, 2000.

Marcus FI and Fontaine G, Arrhythmogenic right ventricular dysplasia — cardiomyopathy: a review, *Pacing Clin Electrophysiol* 18, 1298, 1995.

Marcus FI, Fontaine GH, Guiraudon G, et al., Right ventricular dysplasia: a report of 24 adult cases, *Circulation* 65, 384, 1982.

Maresi E, Becchina G, Ottoveggio G, et al., Arrhythmic sudden cardiac death in a 3-year-old child with intimal fibroplasia of coronary arteries, aorta, and its branches, *Cardiovasc Pathol* 10, 43, 2001a.

Maresi E, Passantino R, Midulla R, et al., Sudden infant death caused by a ruptured coronary aneurysm during acute phase of atypical Kawasaki disease, *Hum Pathol* 32, 1407, 2001b.

Marian AJ, Pathogenesis of diverse clinical and pathological phenotypes in hypertrophic cardiomyopathy, *Lancet* 355, 58, 2000.

Marino B, Gaglardi MG, Digilio MC, et al., Noonan syndrome: structural abnormalities of the mitral valve causing subaortic obstruction, *Eur J Pediatr* 154, 949, 1995.

Marino TA and Kane BM, Cardiac atrioventricular junctional tissues in hearts from infants who died suddenly, *JACC* 5, 1178–1184, 1985.

Mark DB, Naylor DC, Heatki MA, et al., Use of medical resorce and quality of life after acute myocardial infarction in Canada and United States, *N Engl J Med* 331, 1130, 1994.

Marks AR, Priori S, Memmi M, et al., Involvement of the cardiac ryanodine receptor/calcium release channel in catecholaminergic polymorphic ventricular tachycardia, *J Cell Physiol* 190, 1, 2002.

Marks ML, Whisler SL, Clericuzio C, et al., A new form of long QT syndrome associated with syndactyly, *J Am Coll Cardiol* 25, 59, 1995.

Maron BJ, Right ventricular cardiomyopathy: another cause of sudden death in the young, *N Engl J Med* 318, 178, 1988.

Maron BJ, Ventricular arrhythmias, sudden death, and prevention in patients with hypertrophic cardiomyopathy, *Curr Cardiol Rep* 2, 522, 2000.

Maron BJ, Hypertrophic cardiomyopathy: a systematic review, *JAMA* 287, 1308, 2002.

Maron BJ, Bonow RO, Cannon RO, et al., Hypertrophic cardiomyopathy: interrelations of clinical manifestations, pathophysiology, and therapy, *N Engl J Med* 316, 780, 1987.

Maron BJ, Clark CE, Goldstein RE, and Epstein SE, Potential role of QT interval prolongation in sudden infant death syndrome, *Circulation* 54, 423–430, 1976.

Maron BJ and Epstein SE, Hypertrophic cardiomyopathy, *Am J Cardiol* 45, 141, 1980.

Maron BJ and Epstein SE, Hypertrophic cardiomyopathy: a discussion of nomenclature, *Am J Cardiol* 43, 1242, 1979.

Maron BJ and Fananapazir L, Sudden cardiac death in hypertrophic cardiomyopathy, *Circulation* 85 (1 Suppl), I57, 1992.

Maron BJ and Fisher RS, Sudden infant death syndrome (SIDS): cardiac pathologic observations in infants with SIDS, *Am Heart J* 93, 762–766, 1977.

Maron BJ, Gardin JM, Flack JM, et al., Prevalence of hypertrophic cardiomyopathy in a general population of young adults, *Circulation* 92, 785, 1995.

Maron BJ, Gohman TE, Kyle SB, et al., Clinical profile and spectrum of commotio cordis, *JAMA* 287, 1142, 2002.

Maron BJ, Link MS, Wang PJ, et al., Clinical profile of commotio cordis: an unappreciated cause of sudden death in the young during sports and other activities, *J Cardiovasc Electrophysiol* 10, 114, 1999.

Maron BJ, Olivotto I, Spirito P, et al., Epidemiology of hypertrophic cardiomyopathy-related death: revisited in a large non-referral-based patient population, *Circulation* 102, 858, 2000.

Maron BJ, Poliac LC, Kaplan JA, et al., Blunt impact to the chest leading to sudden death from cardiac arrest during sports activities, *N Engl J Med* 333, 337, 1995.

Maron BJ and Roberts WC, Quantitative analysis of cardiac muscle cell disorganization in the ventricular septum of patients with hypertrophic cardiomyopathy, *Circulation* 59, 689, 1979.

Maron BJ, Roberts WC, Edwards JE, et al., Sudden death in patients with hypertrophic cardiomyopathy: characterization of 26 patients without functional limitation, *Am J Cardiol* 41, 803, 1978.

Maron BJ, Roberts WC, and Epstein SE, Sudden death in hypertrophic cardiomyopathy: a profile of 78 patients, *Circulation* 65, 1388, 1982.

Maron BJ, Roberts WC, McAllister HA, et al., Sudden death in young athletes, *Circulation* 62, 218, 1980.

Maron BJ, Shen WK, Link MS, et al., Efficacy of implantable cardioverter-defibrillators for the prevention of sudden death in patients with hypertrophic cardiomyopathy, *N Engl J Med* 342, 365, 2000.

Maron BJ, Wentzel DC, Zenovich AG, et al., Death in a young athelete due to commotio cordis despite prompt external defibrillation, *Heart Rhythm*, 2, 991, 2005.

Maron BJ, Wolfon JK, and Roberts WC, Relation between extent of cardiac muscle cell disorganization and left ventricular wall thickness in hypertrophic cardiomyopathy, *Am J Cardiol* 70, 785, 1992.

Marron K, Wharton J, Sheppard MN, et al., Distribution, morphology and neurochemistry of endocardial and epicardial nerve terminal arborizations in the human heart, *Circulation* 92, 2343, 1995.

Marrott PK, Newcombe KD, Becroft DM, and Friedlander DH, Idiopathic infantile arterial calcification with survival to adult life, *Pediatr Cardiol* 5, 119, 1984.

Marsalese DL, Moodie DS, Lytle BW, et al., Cystic medial necrosis of the aorta in patients without Marfan's syndrome: surgical outcome and long-term follow-up, *J Am Coll Cardiol* 16, 68, 1990.

Marti MC, Bouchardy B, and Cox JN, Aorto-coronary bypass with autogenous saphenous vein grafts: histopathological aspects, *Virchow Arch Pathol Anat* 152, 255, 1971.

Martin AM and Hackel DB, The myocardium of the dog in hemorrhagic shock: an histochemical study, *Lab Invest* 12, 77, 1963.

Martin AM and Hackel DB, An electron microscopic study of the progression of myocardial lesions in the dog after hemorrhagic shock, *Lab Invest* 15, 243, 1966.

Martin SJ, Apoptosis: suicide, execution or murder? *Trends Cell Biol* 3, 141, 1993.

Martini B, Basso C, and Thiene G, Sudden death in mitral valve prolapse with Holter monitoring-documented ventricular fibrillation: evidence of coexisting arrhythmogenic right ventricular cardiomyopathy, *Int J Cardiol* 49, 274, 1995.

Marzilli M, Sambuceti G, Fedele S, and L'Abbate A, Coronary microcirculatory vasoconstriction during ischemia in patients with unstable angina, *J Am Coll Cardiol* 35, 327, 2000.

Marzullo P, *Guida all'imaging cardiaco: dall'anatomia alla fusione di immagini. I quaderni del CNR*, Edizioni Primula, Pisa, 2003.

Maseri A, L'Abbate A, Baroldi G, et al., Coronary vasospasm as a possible cause of myocardial infarction: a conclusion derived from the study of "preinfarction" angina, *N Engl J Med* 299, 1271, 1978a.

Maseri A, Severi S, De Nes M, et al., "Variant" angina: one aspect of a continuous spectrum of vasospastic myocardial ischemia, *Am J Cardiol* 42, 1019, 1978b.

Mason JK, The aircraft accidents as an example of a major disaster, in *The Pathology of Violent Injury,* Mason JK, Ed., Edward Arnold, London, 1978, 56–74.

Mason JW and O'Connel JB, Clinical merit of endomyocardial biopsy, *Circulation* 79, 971, 1989.

Massin M, Leroy P, Misson JP, and Lepage P, Catecholaminergic polymorphic ventricular tachycardia in a child: an often unrecognized diagnosis, *Arch Pediatr* 10, 524, 2003.

Mathey DG, Kuck KH, Tilsner V, et al., Nonsurgical coronary artery recanalization in acute transmural myocardial infarction, *Circulation* 63, 489, 1981.

Mathur A, Sims HF, Gopalakrishnan D, et al., Molecular heterogeneity in very-long-chain acyl-CoA dehydrogenase deficiency causing pediatric cardiomyopathy and sudden death, *Circulation* 99, 1337, 1999.

Matitiau A, Perez-Atayde A, Sanders SP, et al., Infantile dilated cardiomyopathy: relation of outcome to left ventricular mechanics, hemodynamics, and histology at the time of presentation, *Circulation* 90, 1310, 1994.

Matsubara O, Kuwata T, Nemoto T, et al., Coronary artery lesions in Takayasu arteritis: pathological considerations, *Heart Vessels Suppl* 7, 26, 1992.

Matsuda H, Seo Y, and Takahama K, A medico-legal approach to the myocardial changes caused by transthoracic direct current countershock, *Jap J Legal Med* 51, 11–17, 1997.

Matsunari I, Schricke U, Bengel FM, et al., Extent of cardiac sympathetic neuronal damage is determined by the area of ischemia in patients with acute coronary syndromes, *Circulation* 101, 2579, 2000.

Matsuo K, Kurita T, Inagaki M, et al., The circadian pattern of the development of ventricular fibrillation in patients with Brugada syndrome, *Eur Heart J* 20, 465, 1999.

Mattila S, Silvola H, and Ketonen P, Traumatic rupture of the pericardium with luxation of the heart, *J Thorac Cardiovasc Surg*, 70, 495, 1975.

Mattioli AV, Rossi R, Annicchiarico E, et al., Causes of death in patients with unipolar single chamber ventricular pacing: prevalence and circumstances in dependence on arrhythmias leading to pacemaker implantation, *Pacing Clin Electrophysiol* 18, 11, 1995.

Matturi L, Ottaviani G, Ramos SG, et al., Sudden infant death syndrome (SIDS): a study of cardiac conduction system, *Cardiovasc Pathol* 9, 137, 2000.

Mavroudis C, Backer CL, Muster AJ, et al., Expanding indications for pediatric coronary artery bypass, *J Thorac Cardiovasc Surg* 111, 181, 1996.

May AK, Patterson M., Rue LW, et al., Combined blunt and pericardiale rupture: review of the literature and report of a new diagnostic algorithm, *Am Surg*, 65, 568, 1999.

Mazzanti L and Cacciari E, Congenital heart disease in patients with Turner's syndrome: Italian Study Group for Turner Syndrome (ISGTS), *J Pediatr* 133, 688, 1998.

Mazzanti L, Prandstraller D, Tassinari D, et al., Heart disease in Turner's syndrome, *Helv Paediatr Acta* 43, 25, 1988.

Mazzone A, De Servi S, Ricevuti G, et al., Increased expression of neutrophil and monocyte adhesion molecules in unstable coronary artery disease, *Circulation* 88, 358, 1993.

McAllister HA and Fenoglio JJ, Tumors of the cardiovascular system, in *Atlas of Tumor Pathology*, Hartmann WH and Cowan WH, Eds., Armed Forces Institute of Pathology, Washington, DC, 1978, 3.

McCain FH, Kline EM, and Gilson JS, A clinical study of 281 autopsy reports on patients with myocardial infarction, *Am Heart J* 39, 263, 1950.

McCance AJ, Thompson PA, and Forfar JC, Increased cardiac sympathetic nervous activity in patients with unstable coronary heart disease, *Eur Heart J* 14, 751, 1993.

McCarthy RE III, Boehmer JP, Hruban RH, et al., Long-term outcome of fulminant myocarditis as compared with acute (nonfulminant) myocarditis, *N Engl J Med* 342, 690, 2000.

McClellan JT and Jokle E, Congenital anomalies of coronary arteries as cause of sudden death associated with physical exertion, *Am J Clin Pathol* 50, 229, 1968.

McDonald ML, Orszulak TA, Bannon MP, et al., Mitral valve injury after blunt chest trauma, *Ann Thorac Surg*, 61, 1024, 1996.

McDonald PC, Wilson JE, McNeill S, et al., The challenge of defining normality for human mitral and aortic valves: geometrical and compositional analysis, *Cardiovasc Pathol* 11, 193, 2002.

McElhinney DB, Carpentieri DF, Bridges ND, et al., Sarcoma of the mitral valve causing coronary arterial occlusion in children, *Cardiol Young* 11, 539, 2001.

McKay DG, Diseases of hypersensitivity: disseminated intravascular coagulation, *Arch Inter Med* 116, 83, 1965.

McKenna WJ, England D, Doi YL, et al., Arrhythmia in hypertrophic cardiomyopathy, *Br Heart J* 46, 168, 1981.

McKenna WJ, Stewart JT, Nihoyannopoulos P, et al., Hypertrophic cardiomyopathy without hypertrophy: 2 families with myocardial disarray in the absence of increased myocardial mass, *Br Heart J* 63, 287, 1990.

McKenna WJ, Thiene G, Nava A, et al., Diagnosis of arrhythmogenic right ventricular dysplasia/cardiomyopathy, *Br Heart J* 71, 215, 1994.

McLean L, Sharma S, and Maycher B, Mycotic pulmonary arterial aneurysms in an intravenous drug user, *Can Respir J* 5, 307, 1998.

McLean RF, Devitt JH, McLellan BA, et al., Significance of myocardial contusion following blunt chest trauma, *J Trauma*, 33, 240, 1992.

McLeod AL, McKenna CJ, and Northridge DB, Myocardial infarction following the combined recreational use of Viagra and cannabis, *Clin Cardiol* 25133, 2002.

McNutt R, Ferenchick G, Kirlin P, and Hamlin N, Acute myocardial infarction in a 22-year-old world class weight lifter using anabolic steroids, *Am J Cardiol* 62, 164, 1988.

Mechanik N, Das Venensystem der Herzwande, *Z Anat Entwicklungsgesch* 103, 813, 1934.

Meerson F, *Adaptive Medicine: Protective Cross-Effect of Adaptation*, Russian Academy of Medical Sciences, Moscow, 1993.

Meerson FZ, The myocardium in hyperfunction, hypertophy and heart failure, *Circ Res* 25 (Suppl 2), 1, 1969.

Meerson FZ, Kagan VE, Kozlov YP, et al., The role of lipid peroxidation in pathogenesis of ischemic damage and the antioxidant protection of the heart, *Basic Res Cardiol* 77, 465, 1982.

Mehmet CO, Roberts CS, and Lemole GM, Role of lymphostasis in accelerated atherosclerosis in transplanted heart, *Am J Cardiol* 60, 430, 1987.

Mehta AV, Rhabdomyoma and ventricular preexcitation syndrome, *Am J Dis Child* 147, 669, 1993.

Meier R, van Griensven M, Pape HC, et al., Effects of cardiac contusion in isolated perfused rat hearts, *Shock*, 19, 123, 2003.

Melberg A, Oldfors A, Blomstrom-Lundqvist C, et al., Autosomal dominant myofibrillar myopathy with arrhythmogenic right ventricular cardiomyopathy linked to chromosome 10q, *Ann Neurol* 46, 684, 1999.

Melchert RB and Welder AA, Cardiovascular effects of androgenic-anabolic steroids, *Med Sci Sports Exerc* 27, 1252, 1995.

Melillo G, Ruggieri MP, Magni G, et al., Malignant cardiac involvement in a family with myotonic dystrophy, *J Ital Cardiol* 26, 853, 1996.

Merino JL, Carmona JR, Fernandez-Lozano I, et al., Mechanisms of sustained ventricular tachycardia in myotonic dystrophy: implications for catheter ablation, *Circulation* 98, 541, 1998.

Merlevede K, Vermander D, Theys P, et al., Cardiac involvement and CTG expansion in myotonic dystrophy, *J Neurol* 249, 693, 2002.

Mevorach D, Raz E, Shalev O, et al., Complete heart block and seizures in an adult with systemic lupus erythematosus: a possible pathophysiologic role for anti-SS-A/Ro and anti-SS-B/La autoantibodies, *Arthritis Rheum* 36, 259, 1993.

Michaëlsson M, Jonzon A, and Riesenfeld T, Isolated congenital complete atrioventricular block in adult life: a prospective study, *Circulation* 92, 442, 1995.

Michalodimitrakis EN and Tsatsakis AM, Vehicular accidents and cardiac concussion, *Am J Forensic Med Pathol* 18, 282, 1997.

Michalodimitrakis EN, Tsiftsis DDA, Tsatsakis AM, et al., Sudden cardiac death and right ventricular dysplasia, *Am J Forensic Med Pathol* 22, 19, 2001.

Michaud K, Romain N, Brandt-Casadevall C, et al., Sudden death related to small coronary artery disease, *Am J Forensic Med Pathol* 22, 225, 2001.

Michele DE, Albayya FP, and Metzger JM, Direct convergent hypersensitivity of calcium-activated force generation produced by hypertrophic cardiomyopathy mutant α-tropomyosins in adult cardiac monocytes, *Nature Med* 5, 1413, 1999.

Milewicz DM, Chen H, Park ES, et al., Reduced penetrance and variable expressivity of familial thoracic aortic aneurysms/dissections, *Am J Cardiol* 82, 474, 1998.

Miller AJ, *Lymphatics of the Heart*, Raven Press, New York, 1982.

Miller P and Plant M, Heavy cannabis use among UK teenagers: an exploration, *Drug Alcohol Depend* 65, 235, 2002.

Milroy CM, Clark JC, and Forrest AR, Pathology of deaths associated with "ecstasy" and "eve" misuse, *J Clin Pathol* 49, 149, 1996.

Milstein S, Buetikofer J, Lesser J, et al., Cardiac asystole: a manifestation of neurally mediated hypotension-bradycardia, *J Am Coll Cardiol* 14, 1626, 1989.

Mimasaka S, Yajima Y, Hashiyada M, et al., A case of aortic dissection caused by blunt chest trauma, *Forensic Sci Int*, 132, 5, 2003.

Minkowski WL, The coronary arteries of infants, *Am J Med Sci* 214, 623, 1947.

Minor RL Jr, Scott BD, Brown DD, and Winniford MD, Cocaine-induced myocardial infarction in patients with normal coronary arteries, *Ann Intern Med* 115, 797, 1991.

Mints GS, Painter JA, Pichard AD, et al., Atherosclerosis in angiographically "normal" coronary artery reference segments: an intravascular ultrasound study with clinical correlation, *J Am Coll Cardiol* 25, 1479, 1995.

Mitrani RD, Klein SL, Miles WH, et al., Regional sympathetic denervation in patients with ventricular tachycardia in absence of coronary artery disease, *J Am Coll Cardiol* 22, 1344, 1993.

Mittal V, McAleese P, Young S, et al., Penetrating cardiac injuries, *Am Surg*, 65, 444, 1999.

Mittleman MA, Lewis RA, Maclure M, et al., Triggering myocardial infarction by marijuana, *Circulation* 103, 2805, 2001.

Mittleman MA, Maclure M, Tofler GH, Sherwood JB, Goldberg RJ, and Muller JE, For the Determinants of Myocardial Infarction Onset Study Investigators, Triggering of acute myocardial infarction by heavy exertion: protection against triggering by regular exertion, *N Engl J Med* 329, 1677, 1993.

Mittleman MA, Mintzer D, Maclure M, Tofler GH, Sherwood JB, and Muller JE. Triggering of myocardial infarction by cocaine, *Circulation* 99, 2737, 1999.

Mizuno K, Satomura K, Miyamoto A, et al., Angioscopic evaluation of coronary artery thrombi in acute coronary syndrome, *N Engl J Med* 326, 287, 1992.

Moak JP, Barron KS, Hougen TJ, et al., Congenital heart block: development of late-onset cardiomyopathy, a previously underappreciated sequela, *J Am Coll Cardiol* 37, 238, 2001.

Moberg A, Anastomoses between extracardiac vessels and coronary arteries, *Acta Med Scand Suppl* 485, 1, 1968.

Mogayzel C, Quan L, Graves JR, et al., Out-of-hospital ventricular fibrillation in children and adolescents: causes and outcomes, *Ann Emerg Med* 25, 484, 1995.

Mohammed W and Murphy A, Cardiac fibroma presenting as sudden death in a six-month-old infant, *West Indian Med J* 46, 28, 1997.

Moir TW, Subendocardial distribution of coronary blood flow and the effect of antianginal drugs, *Circ Res* 30, 624, 1972.

Molander N, Sudden natural death in later childhood and adolescence, *Arch Dis Child* 57, 572, 1982.

Molina JE, Coronary stenosis following aortic valve replacement, *Am Thorac Surg* 35, 473, 1993.

Moliterno DJ, Willard JE, Lange RA, et al., Coronary-artery vasoconstriction induced by cocaine, cigarette smoking, or both, *N Engl J Med* 330, 454, 1994.

Monnier N, Romero NB, Lerale J, et al., Familial and sporadic forms of central core disease are associated with mutations in the C-terminal domain of the skeletal muscle ryanodine receptor, *Hum Mol Genet* 10, 2581, 2001.

Moolman JC, Corfield VA, Posen B, et al., Sudden death due to troponin T mutations, *J Am Coll Cardiol* 29, 549, 1997.

Moon HD and Rinehart JF, Histogenesis of coronary arteriosclerosis, *Circulation* 6, 481, 1952.

Moore JB, Moore EE, and Harken AH, Emergency department thoracotomy, in *Trauma*, Moore EE, Mattox KL, and Feliciano DV, Eds., Appleton and Lange, East Norwalk, CT, 1991.

Moore JW, Kirby WC, Rogers WM, and Poth MA, Partial anomalous pulmonary venous drainage associated with 45,X Turner's syndrome, *Pediatrics* 86, 273, 1990.

Moore L and Byard RW, Sudden and unexpected death in infancy associated with a single coronary artery, *Pediatr Pathol* 12, 231, 1992.

Moothart RW, Pryor R, Hawley RL, et al., The heritable syndrome of prolonged QT interval, syncope, and sudden death, *Chest* 70, 263, 1976.

Morales AR, Romanelli R, and Boucek RJ, The mural left anterior descending coronary artery, strenuous exercise and sudden death, *Circulation* 62, 230, 1980.

Morales AR, Romanelli R, Tate LG, et al., Intramural left anterior descending coronary artery: significance of the depth of the muscular tunnel, *Hum Pathol* 24, 693, 1993.

Morentin B, Aguilera B, Garamendi PM, and Suarez-Mier MP, Sudden unexpected non-violent death between 1 and 19 years in north Spain, *Arch Dis Child* 82, 456, 2000.

Morgan AD, *The Pathogenesis of Coronary Occlusion*, Blackwell Scientific, Oxford, 1956.

Morimoto S, Hiramitsu S, Yamada K, et al., Clinical and pathologic features of chronic myocarditis: four autopsy cases presenting as dilated cardiomyopathy in life, *Am J Cardiovasc Pathol* 4, 181, 1992.

Moritz AR and Atkins JP, Cardiac contusion: an experimental and pathological study, *Arch Pathol* 25, 445, 1938.

Moritz AR and Zamcheck N, Sudden and unexpected death of young soldiers: diseases responsible for such death during World War II, *Arch Pathol* 42, 459, 1946.

Moritz F, Monteil C, Isabelle M, et al., Role of reactive oxygen species in cocaine-induced cardiac dysfunction, *Cardiovasc Res* 59, 834, 2003.

Morris JN and Gardner MS, Epidemiology of ischemic heart disease, *Am J Med* 46, 674, 1969.

Morrow DA, Rifai N, Antman EM, et al., C-reactive protein is a potent predictor of mortality independently of and in combination with troponin T in acute coronary syndromes: a TIMI 11A substudy, *J Am Coll Cardiol* 31, 1460, 1998.

Moschcowitz E, Hyaline thrombosis of the terminal arterioles and capillaries: a hitherto undescribed disease, *Proc NY Pathol Soc* 24, 21, 1924.

Moschos CB, Heider B, Khan Y, et al., Relation of platelets to catecholamine induced myocardial injury, *Cardiov Res* 12, 243, 1978.

Moss AJ, The QT interval and torsade de pointes, *Drug Saf* 21 (Suppl 1), 5, 1999.

Moss AJ, Robinson JL, Gessman L, et al., Comparison of clinical and genetic variables of cardiac events associated with loud noise versus swimming among subjects with the long QT syndrome, *Am J Cardiol* 84, 876, 1999.

Moss AJ, Schwartz PJ, Crampton RS, et al., The long QT syndrome: prospective longitudinal study of 328 families, *Circulation* 84, 1136, 1991.

Moss AJ, Schwartz PJ, Crampton RS, et al., The long QT syndrome: a prospective international study, *Circulation* 71, 17, 1985.

Mouchet A, *Les arteres coronaires du coeur chez l'homme*, Norbert Maloine, Paris, 1933.

Moura C, Hillion Y, Daikha-Dahmane F, et al., Isolated non-compaction of the myocardium diagnosed in the fetus: two sporadic and two familial cases, *Cardiol Young* 12, 278, 2002.

Moussouttas M, Cannabis use and cerebrovascular disease, *Neurologist* 10, 47, 2004.

Müller G, *Der plötzliche Kindstod,* Thieme, Stuttgart, 1963.

Muller G, Ulmer HE, Hagel KJ, and Wolf D, Cardiac dysrhythmias in children with idiopathic dilated or hypertrophic cardiomyopathy, *Pediatr Cardiol* 16, 56, 1995.

Munger TM, Packer DL, Hammill SC, et al., A population study of the natural history of Wolff-Parkinson-White syndrome in Olmstead County, Minnesota, 1953–1989, *Circulation* 87, 866, 1993.

Muntoni F, Catani G, Mateddu A, et al., Familial cardiomyopathy, mental retardation and myopathy associated with desmin-type intermediate filaments, *Neuromusc Disorders* 4, 233, 1994.

Murray CA and Edwards JE, Spontaneous laceration of ascending aorta, *Circulation* 47, 848, 1973.

Murry CE, Jennings RB, and Reimer KA, Preconditioning with ischemia: a delay of lethal cell injury in ischemic myocardium, *Circulation* 74, 1124, 1986.

Mustard JF and Packham MA, Platelet function and myocardial infarction, *Circulation* 40 (Suppl 4), 20, 1969.

Myasnikov AL, Chazov EI, Hoshevnikova TL, et al., Some new data on the occurrence of coronary thrombosis in conjuction with atherosclerosis, *J Atheroscl Res* 1, 401, 1961.

Myerberg RJ and Castellanos A, Cardiac arrest and sudden death, in *Heart Disease*, 4th ed., Braunwald E, Ed., Philadelphia, W.B. Saunders, 1992, 756.

Mygind T, Ostergaard L, Birkelund S, et al., Evaluation of five DNA extraction methods for purification of DNA from atherosclerotic tissue and estimation of prevalence of *Chlamydia pneumoniae* in tissue from a Danish population undergoing vascular repair, *BMC Microbiol* 3, 19, 2003.

Nabel EG, Genomic medicine: cardiovascular disease, *N Engl J Med* 349, 60, 2003.

Nachlas MM and Shnitka TK, Macroscopic identification of early myocardial infarcts by alterations in dehydrogenase activity, *Am J Pathol* 42, 379, 1963.

Nademanee K, Veerakul G, Nimmannit S, et al., Arrhythmogenic marker for the sudden unexpected death syndrome in Thai men, *Circulation* 96, 2595, 1997.

Naeye RL, Whalen P, Ryser M, and Fisher R, Cardiac and other abnormalities in the suden infant death syndrome, *Am J Pathol* 82, 1–8, 1976.

Nagueh SF, Mikati J, Weilbaecher D, et al., Relation of contractile reserve of hibernating myocardium to myocardial structure in humans, *Circulation* 100, 490, 1999.

Naheed ZJ, Strasburger JF, Deal BJ, et al., Fetal tachycardia: mechanisms and predictors of hydrops fetalis, *J Am Coll Cardiol* 27, 1736, 1996.

Naidoo DP, Naicker S, Vythylingum S, et al., Isolated tricuspid valve infective endocarditis: a report of 6 cases, *S Afr Med J* 78, 34, 1990.

Najm HK, Williams WG, Coles JG, et al., Pulmonary atresia with intact ventricular septum: results of the Fontan procedure, *Ann Thorac Surg* 63, 669, 1997.

Nakagawa M, Sato A, Okagawa H, et al., Detection and evaluation of asymptomatic myocarditis in schoolchildren: report of four cases, *Chest* 116, 340, 1999.

Nakata T, Miyamoto K, Doi A, et al., Cardiac death prediction and impaired cardiac sympathetic innervation assessed by MIBG in patients with failing and nonfailing hearts, *J Nucl Cardiol* 5, 579, 1998.

Narkiewicz K, De Borne van PJH, Hausberg M, et al., Cigarette smoking increases sympathetic outflow in humans, *Circulation* 98, 528, 1998.

Narula J, Heider N, Virmani N, et al., Apoptosis in myocytes in end-stage heart failure, *N Eng J Med* 335, 1182, 1996.

Nashed G, French B, Gallagher D, et al., Right ventricular perforation with cardiac tamponade associated with use of a temporary pacing wire and abciximab during complex coronary angioplasty, *Catheter Cardiovasc Interv*, 48, 388, 1999.

Nashef SA, Roques F, Michel P, et al., European system for cardiac operative risk evaluation (Euro-SCORE), *Eur J Cardiothorac Surg* 16, 9, 1999.

Nasser FN, Walls JT, Edwards WD, et al., Lidocaine-induced reduction in size of experimental myocardial infarction, *Am J Cardiol* 46, 967, 1980.

Natali J and Benhamou C, Iatrogenic vascular injuries. A review of 125 cases (excluding angiographic injuries), *J Cardiovasc Surg*, 20, 169, 1979.

Natelson BH, Suarez RV, Terrence CF, and Turizo R, Patients with epilepsy who die suddenly have cardiac disease, *Arch Neurol* 55, 857, 1998.

Nava A, Bauce B, Basso C, et al., Clinical profile and long-term follow-up of 37 families with arrhythmogenic right ventricular cardiomyopathy, *J Am Coll Cardiol* 36, 2226, 2000.

Nava A, Rossi L, and Thiene G, Eds., *Arrhythmogenic Right Ventricular Cardiomyopathy/Dysplasia*, Elsevier, Amsterdam, 1997.

Nei M, Ho RT, and Sperling MR, EKG abnormalities during partial seizures in refractory epilepsy, *Epilepsia* 41, 542, 2000.

Neri E, Toscano T, Massetti M, Capannini G, Frati G, and Sassi C. Cocaine-induced intramural hematoma of the ascending aorta, *Tex Heart Inst J* 28, 218, 2001.

Neri Serneri GG, Abbate R, Gori AM, et al., Transient intermittent lymphocyte activation is responsible for the instability of angina, *Circulation* 86, 790, 1992.

Neri Serneri GG, Boddi M, Arata L, et al., Silent ischemia in unstable angina is related to an altered cardiac norepinephrine handling, *Circulation* 87, 1928, 1993.

Nesto R and Kowalchuk G, The ischemic cascade: temporal sequence of hemodynamic, electrocardiographic and symptomatic expression of ischemia, *Am J Cardiol* 59, 23C, 1987.

Neufeld NH, Lester RG, Adams P, et al., Congenital communications of a coronary artery with a cardiac chamber or the pulmonary trunk (coronary artery fistula), *Circulation* 24, 171, 1961.

Neuspiel DR and Kuller LH, Sudden and unexpected natural death in childhood and adolescence, *JAMA* 254, 1321, 1985.

Newbury-Ecob RA, Leanage R, Raeburn R, and Young ID, Holt-Oram syndrome: a clinical genetic study, *J Med Genet* 33, 300, 1996.

Neyroud N, Tesson F, Denjoy I, et al., A novel mutation in the potassium channel gene KVLQT1 cause Jervell and Lange-Nielson cardioauditory syndrome, *Nat Genet* 15, 186, 1997.

Nguyen HH, Wolfe JT III, Holmes DR Jr, et al., Pathology of the cardiac conduction system in myotonic dystrophy: a study of 12 cases, *J Am Coll Cardiol* 11, 662, 1988.

Nicod P, Bloor C, Godfrey M, et al., Familial aortic dissecting aneurysm, *J Am Coll Cardiol* 13, 811, 1989.

Nield LE, Silverman ED, Smallhorn JF, et al., Endocardial fibroelastosis associated with maternal anti-Ro and anti-La antibodies in the absence of atrioventricular block, *J Am Coll Cardiol* 40, 796, 2002.

Nieminem MS, Ramo MP, Vittasalo M, et al., Serious cardiovascular side effects of large doses of anabolic steroids in weight lifters, *Eur Heart J* 17, 1576, 1996.

Nishikawa T, Ishiyama S, Nagata M, et al., Programmed cell death in the myocardium of arrhythmogenic right ventricular cardiomyopathy in children and adults, *Cardiovasc Pathol* 84, 185, 1999.

Nishikawa T, Tanaka Y, Sasaki Y, et al., A case of pediatric cardiomyopathy with severely restrictive physiology, *Heart Vessels* 7, 206, 1992.

Nishimura RA and Holmes DR, Hypertrophic obstructive cardiomyopathy, *N Engl J Med* 350, 1320, 2004.

Niwa K, Perloff JK, Kaplan S, Child JS, and Miner PD, Eisenmenger syndrome in adults: ventricular septal defect, truncus arteriosus, univentricular heart, *J Am Coll Cardiol* 34, 223, 1999.

Noakes TD, Opie LH, and Rose AG, Autopsy-proved coronary atherosclerosis in marathon runners, *N Engl J Med* 301, 86, 1979.

Noble J, Bourassa MG, Peticlerc R, et al., Myocardial bridging and milking effect of the left anterior descending coronary artery: normal variant or obstruction? *Am J Cardiol* 37, 993, 1976.

Noffsinger AE, Blisard KS, and Balko MG, Cardiac laceration and pericardial tamponade due to cardiopulmonary resuscitation after myocardial infarction, *J Forensic Sci*, 36, 1760, 1991.

Noh CI, Song JY, Kim HS, et al., Ventricular tachycardia and exercise related syncope in children with structurally normal hearts: emphasis on repolarisation abnormality, *Br Heart J* 73, 544, 1995.

Nora JJ, Torres FG, Sinha AK, and McNamara DG, Characteristic cardiovascular anomalies of XO Turner syndrome, XX and XY phenotype and XO-XX Turner mosaic, *Am J Cardiol* 25, 639, 1970.

Noris KC, Thornhill-Joynes M, Robinson C, Alperson BL, Witana SC, and Ward HJ. Cocaine use, hypertension, and end-stage renal disease, *Am J Kidney Dis* 38, 523, 2001.

Norman MG, Taylor GP, and Clarke LA, Sudden, unexpected, natural deaths in childhood, *Pediatr Pathol* 10, 769, 1990.

Obata H, Mitsuoka T, Kikuchi Y, et al., Twenty-seven-year follow-up of arrhythmogenic right ventricular dysplasia, *Pacing Clin Electrophysiol* 24, 510, 2001.

O'Connell JB, Fowles RE, and Robinson JA, Clinical and pathologic findings of myocarditis in two families with dilated cardiomyopathy, *Am Heart J* 107, 127, 1984.

O'Connor WN, Davis JB Jr, Geissler R, et al., Supravalvular aortic stenosis: clinical and pathologic observations in six patients, *Arch Pathol Lab Med* 109, 179, 1985.

O'Laughlin MP, Slack MC, Grifka RG, et al., Implantation and intermediate-term follow-up of stents in congenital heart disease, *Circulation* 88, 605, 1993.

O'Reilly RJ and Spellberg RD, Rapid resolution of coronary arterial emboli: myocardial infarction and subsequent normal coronary arteriograms, *Ann Intern Med* 81, 348, 1974.

Oechslin EN, Harrison DA, Connelly MS, et al., Mode of death in adults with congenital heart disease, *Am J Cardiol* 86, 1111, 2000.

Ogden JA, Congenital anomalies of the coronary arteries, *Am J Cardiol* 25, 474, 1970.

Oh JK, Holmes DR Jr, Hayes DL, et al., Cardiac arrhythmias in patients with surgical repair of Ebstein's anomaly, *J Am Coll Cardiol* 6, 1351, 1985.

Ohkawa S, Hackel DB, and Ideker RE, Correlation of the width of the QRS complex with the pathologic anatomy of the cardiac conduction system in patients with chronic complete atrioventricular block, *Circulation* 63, 938, 1981.

Ohman ME, Armstrong PW, Christenson RH, et al., Cardiac troponin T levels for risk stratification in acute myocardial ischemia, *N Engl J Med* 335, 1333, 1996.

Ohshima T, Lin Z, and Sato Y, Unexpected sudden death of a 12-year-old male with congenital single coronary artery, *Forensic Sci Int* 82, 177, 1996.

Oka M and Angrist A, Histoenzymatic studies of vessels in hypertensive rats, *Lab Invest* 16, 25, 1967.

Okada R and Kawai S, Histopathology of the conduction system in sudden cardiac death, *Jpn Circ J* 47, 573, 1983.

Okajima Y, Tanabe Y, Takayanagi M, et al., A follow up study of myocardial involvement in patients with mitochondrial encephalomyopathy, lactic acidosis, and stroke-like episodes (MELAS), *Heart* 80, 292, 1998.

Okishige K, Sasano T, Yano K, et al., Serious arrhythmias in patients with apical hypertrophic cardiomyopathy, *Intern Med* 40, 396, 2001.

Oliva PB and Breckinridge JC, Arteriographic evidence of coronary arterial spasm in acute myocardial infarction, *Circulation* 56, 366, 1977.

Oliva PB, Hammill SC, and Edwards WD, Cardiac rupture, a clinically predictable complication of acute myocardial infarction: report of 70 cases with clinicopathologic correlations, *J Am Coll Cardiol* 22, 720, 1993.

Oliva PB, Potts DE, and Plus RG, Coronary arterial spasm in Prinzmetal angina, *N Engl J Med* 288, 745, 1973.

Oliver JM, Gonzalez A, Gallego P, et al., Discrete subaortic stenosis in adults: increased prevalence and slow rate of progression of the obstruction and aortic regurgitation, *J Am Coll Cardiol* 38, 835, 2001.

Oliver MF, Oral contraceptives and myocardial infarction, *Br Med J* 2, 210, 1970.

Olsen EG, The pathogenesis of dilated cardiomyopathy, *Postgrad Med J* 68 (Suppl 1), S7–10, 1992.

Olson TM, Doan TP, Kishimoto NY, et al., Inherited and de novo mutations in the cardiac actin gene cause hypertrophic cardiomyopathy, *J Mol Cell Cardiol* 32, 1687, 2000.

Om A, Warner M, Sabri N, et al., Frequency of coronary artery disease and left ventricular dysfunction in cocaine users, *Am J Cardiol* 69, 1549, 1992.

Ometto R, Thiene G, Corrado D, et al., Enhanced A-V nodal conduction (Lown-Ganong-Levine syndrome) by congenitally hypoplastic A-V node, *Eur Heart J* 13, 1579, 1992.

Ommen SR and Nishimura RA, Hypertrophic cardiomyopathy, *Curr Probl Cardiol* 29, 233, 2004.

Onat A, Onat T, and Domanic N, Discrete subaortic stenosis as part of a short stature syndrome, *Hum Genet* 65, 331, 1984.

Ono M, Yagyu K, Furuse A, et al., A case of Standford type A acute aortic dissection caused by blunt chest trauma, *J Trauma Inj Infect Crit Care*, 44, 543, 1998.

Oparil S, Morphologic substrates of sudden death: pathogenesis of ventricular hypertrophy, *J Am Coll Cardiol* 5, 57B, 1985.

Opeskin K, Thomas A, and Berkovic SF, Does cardiac conduction pathology contribute to sudden unexpected death in epilepsy? *Epilepsy Res* 40, 17, 2000.

Opie L, Sudden death and sport, *Lancet* 1, 263, 1975.

Opie LH, The mechanism of myocyte death in ischaemia, *Eur Heart J* 14 (Suppl G), 31, 1993.

Orbison JL, Morphology of thrombotic thrombocytopenic purpura with demonstration of aneurysm, *Am J Pathol* 28, 129, 1952.

Oren A, Bar-Schlomo B, and Stern S, Acute coronary occlusion following blunt injury to the chest in the absence of coronary atherosclerosis, *Am Heart J* 92, 501, 1976.

Orita M, Iwahana H, Kanazawa H, Hayashi K, and Sekiya T, Detection of polymorphisms of human DNA by gel electrophoresis as single-strand conformation polymorphisms, *Proc Natl Acad Sci USA* 86, 2766–2770, 1989.

Orlic D, Kajstura JM, Chimenti S, et al., Bone marrow cells regenerate infarcted myocardium, *Nature* 410, 701, 2001.

Ottaviani G, Rossi L, Ramos SG, and Matturri L, Pathology of the heart and conduction system in a case of sudden death due to a cardiac fibroma in a 6-month-old child, *Cardiovasc Pathol* 8, 109, 1999.

Ozcan C, Jahangir A, Friedman PA, et al., Sudden death after radiofrequency ablation of the atrioventricular node in patients with atrial fibrillation, *J Am Coll Cardiol* 40, 105, 2002.

Pacher P, Ba´tkai S, and Kunos G, Blood pressure regulation by endocannabinoids and their receptors, *Neuropharmacology* 48, 1130, 2005.

Packer M, Sudden unexpected death in patients with congestive heart failure: a second frontier, *Circulation* 72, 681, 1985.

Paessens R and Borchard F, Morphology of cardiac nerves in experimental infarction of rat hearts. I. Fluorescence microscopical findings, *Virchows Arch A Pathol Anat Histol* 386, 265, 1980.

Page DL, Caulfield JB, Kastor JA, et al., Myocardial changes associated with cardiogenic shock, *N Engl J Med* 285, 133, 1971.

Pagenstecher H, Weiterer Beitrag zur Herzchirurgie: Die Unterbindung der verletzen Arteria Coronaria, *Dtsch Med Wochenschr* 4, 56, 1901.

Pahl E, Zales VR, Fricker FJ, and Addonizio LJ, Posttransplant coronary artery disease in children: a multicenter national survey, *Circulation* 90, II56, 1994.

Palmiere C, Burkhardt S, Staub C, et al., Thoracic aortic dissection associated with cocaine abuse, *Forensic Sci Int* 141, 137, 2004.

Park KY, Dalakas MC, Semino-Mora C, et al., Sporadic cardiac and skeletal myopathy caused by a de novo desmin mutation, *Clin Genet* 57, 423, 2000.

Parker KM and Embry JH, Sudden death due to tricuspid valve myxoma with massive pulmonary embolism in a 15-month old male, *J Forensic Sci* 42, 524, 1997.

Parks WJ, Ngo TD, Plauth WH Jr, et al., Incidence of aneurysm formation after Dacron patch aortoplasty repair for coarctation of the aorta: long-term results and assessment utilizing magnetic resonance angiography with three-dimensional surface rendering, *J Am Coll Cardiol* 26, 266, 1995.

Parmeley LF, Mattingly TW, and Manion WC, Penetrating wounds of the heart and aorta, *Circulation* 17, 953, 1958.

Parmeley LF and Symbas PN, Traumatic heart diseases, in *The Heart,* Vol. 2, Hurst J.W., Ed., McGraw-Hill, New York, 1978, 1683.

Parmeley WW, Factors causing arrhythmias in chronic congestive heart failure, *Am Heart J* 114, 1267, 1987.

Parodi O, De Maria R, Oltrona L, et al., Myocardial blood flow distribution in patients with ischemic heart disease or dilated cardiomyopathy undergoing heart transplantation, *Circulation* 88, 509, 1993.

Parravicini C, Baroldi G, Gaiera G, and Lazzarini A, Phenotype of intramyocardial leukocytic infiltrates in acquired immunodeficiency syndrome (AIDS): a postmortem immunohistochemical study in 34 consecutive cases, *Mod Pathol* 4, 559, 1991.

Parums D and Mitchinson MJ, Demonstration of immunoglobulin in the neighbourhood of advanced atherosclerotic plaques, *Atherosclerosis* 38, 211, 1981.

Pasic M, Ewert R, Engel M, et al., Aortic rupture and concomitant transection of the left bronchus after blunt chest trauma, *Chest*, 117, 1508, 2000.

Passarino G, Ciccone G, Siragusa R, Tappero P, and Mollo F, Histopathological findings in 851 autopsies of drug addicts, with toxicologic and virologic correlations, *Am J Forensic Med Pathol* 26, 106, 2005.

Pasternac A, Tubau JF, Puddu PE, et al., Increased plasma catecholamine levels in patients with symptomatic mitral valve prolapse, *Am J Med* 73, 783, 1982.

Pate JW and Richardson RL, Penetrating wounds of the cardiac valves, *JAMA*, 207, 309, 1969.

Patel VS, Lim M, Massin EK, et al., Sudden cardiac death in cardiac transplant recipients, *Circulation* 94 (9 Suppl), 273, 1996.

Patten BM, *Human Embryology*, McGraw-Hill, New York, 1968.

Paul T, Bertram H, Bokenkamp R, et al., Supraventricular tachycardia in infants, children and adolescents: diagnosis, and pharmacological and interventional therapy, *Paediatr Drugs* 2, 171, 2000.

Pawel BR, de Chadarevian J-P, Wolk JH, et al., Sudden death in childhood due to right ventricular dysplasia: report of two cases, *Pediatr Pathol* 14, 987, 1994.

Payne RM, Johnson MC, Grant JW, and Stauss AW, Toward a molecular understanding of congenital heart disease, *Circulation* 91, 494, 1995.

Pearce PZ, Commotio cordis: sudden death in a young hockey player, *Curr Sports Med Rep*, 4, 157, 2005.

Pearl W and Choi YS, Marijuana as a cause of myocardial infarction, *Int J Cardiol* 34, 353, 1992.

Pearson AC, Schiff M, Mrosek D, et al., Left ventricular function in weightlifters, *Am J Cardiol* 58, 1254, 1986.

Pelliccia A, Maron BJ, De Luca R, et al., Remodeling of left ventricular hypertrophy in elite athletes long-term deconditioning, *Circulation* 105, 944, 2002.

Penninger JM and Bachmaier K, Review of microbial infections and the immune response to cardiac antigens, *J Infect Dis* 181 (Suppl 3), S498, 2000.

Pepin M, Schwarze U, Superti-Furga A, and Byers PH, Clinical and genetic features of Ehlers-Danlos syndrome type IV, the vascular type, *N Engl J Med* 342, 673, 2000.

Perera R, Kraebber A, and Schwartz MJ, Prolonged QT interval and cocaine use, *J Electrocardiol* 30, 337, 1997.

Perez Riera AR, Ferreira C, Dubner SJ, et al., Brief review of the recently described short QT syndrome and other channellopathies, *Ann Noninvasive Electrocardiol*, 10, 371, 2005.

Perron AD and Gibbs M, Thoracic aortic dissection secondary to crack cocaine ingestion, *Am J Emerg Med* 15, 507, 1997.

Perry YY, Triedman JK, Gauvreau K, et al., Sudden death in patients after transcatheter device implantation for congenital heart disease, *Am J Cardiol* 85, 992, 2000.

Peter J, Kirchner A, Kuhlisch E, et al., The relevance of the detection of troponins to the forensic diagnosis of cardiac contusion, *Forensic Sci Int*, 2005, Oct 31, Epub ahead of print.

Peters C, Schulz T, and Michna H, Biomedical Side Effects of Doping, Project of the European Union, Verlag Sport und Buch Straub, Köln, 2002.

Peters RW, Mitchell LB, Brooks MM, et al., Circadian pattern of arrhythmic death in patients receiving encainide, flecainide or moricizine in the cardiac arrhythmia suppression trial (CAST), *J Am Coll Cardiol* 23, 283, 1994.

Petsas AA, Anastassiades LC, Constantinou EC, and Antonopoulos AG, Familial discrete subaortic stenosis, *Clin Cardiol* 21, 63, 1998.

Picano E, *Ecocardiografia da stress*, Springer-Verlag, Berlin, 1992.

Pierard LA, De Landsheere CM, Berthe C, et al., Identification of viable myocardium by echocardiography during dobutamine infusion in patients with myocardial infarction after thrombolytic therapy: comparison with positron emission tomography, *J Am Coll Cardiol* 15, 1021, 1990.

Piippo K, Laitinen P, Swan H, et al., Homozygosity for a HERG potassium channel mutation causes a severe form of long QT syndrome: identification of an apparent founder mutation in the Finns, *J Am Coll Cardiol* 35, 1919, 2000.

Pinamonti B, Miani D, Sinagra G, et al., Familial right ventricular dysplasia with biventricular involvement and inflammatory infiltration, *Heart* 76, 66, 1996.

Pinar Bermudez E, Garcia-Alberola A, Martinez Sanchez J, et al., Spontaneous sustained monomorphic ventricular tachycardia after administration of ajmaline in a patient with Brugada syndrome, *Pacing Clin Electrophysiol* 23, 407, 2000.

Piper C, Bilger J, Henrichs EM, et al., Is myocardial Na^+/Ca^{++} exchanger transcription a marker for different stages of myocardial dysfunction? Quantitative polymerase chain reaction of the messenger RNA in endomyocardial biopsies of patients with heart failure, *J Am Coll Cardiol* 36, 233, 2000.

Pitts WR, Vongpatanasin W, Cigarroa JE, Hillis LD, and Lange RA, Effects of the intracoronary infusion of cocaine on left ventricular systolic and diastolic function in humans, *Circulation* 97, 1270, 1998.

Planche C, Bruniaux J, Lacour-Gayet F, et al., Switch operation for transposition of the great arteries in neonates: a study of 120 patients, *J Thorac Cardiovasc Surg* 96, 354, 1988.

Pohlgeers A and Villafane J, Ventricular fibrillation in two infants treated with amiodarone hydrochloride, *Pediatr Cardiol* 16, 82, 1995.

Polacek P, Relation of myocardial bridges and loops on the coronary arteries to coronary occlusion, *Am Heart J* 61, 44, 1961.

Polak PE, Zijlstra F, and Roelandt JR, Indications for pacemaker implantation in the Kearns-Sayre syndrome, *Eur Heart J* 10, 281, 1989.

Poletti PA, Platon A, Shanmuganathan K, et al., Aymptomatic traumatic pericardial rupture with partial right atrial herniation: case report, *J Trauma*, 58, 1068, 2005.

Pollak S and Stellwag-Carion C, Delayed cardiac rupture due to blunt chest trauma, *Am J Forensic Med Pathol*, 12, 153, 1991.

Pollanen MS, Chiasson DA, Cairns J, et al., Sudden unexplained death in Asian immigrants: recognition of a syndrome in metropolitan Toronto, *CMAJ* 155, 537, 1996.

Pool-Wilson PA, Relation of pathophysiologic mechanisms to outcome in heart failure, *J Am Coll Cardiol* 22 (Suppl A), A22, 1993.

Potkin BN, Bartorelli AL, Gessert BS, et al., Coronary artery imaging with intravascular high-frequency ultrasound, *Circulation* 81, 1575, 1990.

Potter DJ and Hopkins GM, Reversible paraplegia and acute renal failure due to occlusive disease of the abdominal aorta, *Ann Intern Med*, 69, 777, 1968.

Poupa O and Carlsten A, Experimental cardiomyopathies in poikilotherms, *Recent Adv Stud Cardiac Struct Metab*, 2, 321, 1973.

Poupa O and Ostdal B, Experimental cardiomegalies and "Cardiomegalies" in free-living animals, *Ann NY Acad Sci* 156, 445, 1969.

Poupa O, Rakusan K, and Ostadal B, *The Effects of Physical Activity upon the Heart of Vertebrates: Medicine and Sport, Vol. 4: Physical Activity and Aging*, Karger, Basel, 1970.

Powner DJ, Holcombe PA, and Mello LA, Cardiopulmonary resuscitation-related injuries, *Crit Care Med*, 12, 54, 1984.

Prandstraller D, Mazzanti L, Picchio FM, et al., Turner's syndrome: cardiologic profile according to the different chromosomal patterns and long-term clinical follow-up of 136 nonpreselected patients, *Pediatr Cardiol* 20, 108, 1999.

Prasquier R, Gibert C, Witchits S, et al., Acute mitral valve obstruction during infective endocarditis, *Br Med* 1 (6104), 9, 1978.

Price WH and Wilson J, Dissection of the aorta in Turner's Syndrome, *J Med Genet* 20, 61, 1983.

Prinzmetal M, Simkin B, Bergman HC, and Kruger HE, Studies on the coronary circulation. II. The collateral circulation of the normal human heart by coronary perfusion with radioactive erythrocytes and glass spheres, *Am Heart J* 33, 420, 1947.

Priori SG and Napolitano C, Genetics of cardiac arrhythmias and sudden cardiac death, *Ann N Y Acad Sci* 1015, 96, 2004.

Priori SG, Napolitano C, Giordano U, et al., Brugada syndrome and sudden death in children, *Lancet* 355, 808, 2000a.

Priori SG, Napolitano C, Gasparini M, et al., Natural history of Brugada syndrome: insights for risk stratification and management, *Circulation* 105, 1342, 2002.

Priori SG, Napolitano C, Gasparini M, et al., Clinical and genetic heterogeneity of right bundle branch block and ST-segment elevation syndrome, *Circulation* 102, 2509, 2000.

Priori SG, Napolitano C, Memmi M, et al., Clinical and molecular characterization of patients with catecholaminergic polymorphic ventricular tachycardia, *Circulation* 106, 69, 2002.

Priori SG, Napolitano C, and Schwartz PJ, Low penetrance in the long-QT syndrome: clinical impact, *Circulation* 99, 529, 1999.

Priori SG, Napolitano C, Schwartz PJ, et al., The elusive link between LQT3 and Brugada syndrome: the role of flecainide challenge, *Circulation* 102, 945, 2000b.

Priori SG, Napolitano C, Tiso N, et al., Mutations in the cardiac ryanodine receptor gene (hRyR2) underlie catecholaminergic polymorphic ventricular tachycardia, *Circulation* 103, 196, 2001.

Protonotarios N, Tsatsopoulou A, Anastasakis A, et al., Genotype-phenotype assessment in autosomal recessive arrhythmogenic right ventricular cardiomyopathy (Naxos disease) caused by a deletion in plakoglobin, *J Am Coll Cardiol* 38, 1477, 2001.

Provenza DV and Scherlis S, Demonstration of muscle sphincters as a capillary component in the human heart, *Circulation* 20, 35, 1959.

Przyklenk K, Bauer B, Ovize M, et al., Regional ischemic "preconditioning" protects remote virgin myocardium from subsequent sustained coronary occlusion, *Circulation* 87, 893, 1993.

Przyklenk K and Kloner RA, Superoxide dismutase plus catalase improve contractile function in the canine model of the stunned myocardium, *Circ Res* 58, 148, 1986.

Pucci AM, Forbes RD, and Billingham ME, Pathologic features in long-term cardiac allografts, *J Heart Transplant* 9, 339, 1990.

Puley G, Siu S, Connelly M, et al., Arrhythmia and survival of patients >18 years of age after the Mustard procedure for complete transposition of the great arteries, *Am J Cardiol* 83, 1080, 1999.

Qasim A, Townend J, and Davies MK, Ecstasy induced acute myiocardial infarction, *Heart* 85, e10, 2001.

Quaini F, Urbanek K, Beltrami AP, et al., Chimerism of the transplanted heart, *N Engl J Med* 346, 15, 2002.

Quinn A, Kosanke S, Fischetti VA, et al., Induction of autoimmune valvular heart disease by recombinant streptococcal M protein, *Infect Immun* 69, 4072, 2001.

Qin JX, Shiota T, Lever HM, et al., Outcome of patients with hypertrophic obstructive cardiomyopathy after percutaneous transluminal septal myocardial ablation and septal myectomy surgery, *J Am Coll Cardiol* 38, 1994, 2001.

Raab W, Preventive myocardiology, fundamentals and targets, in *Bannerstone Division of American Lectures in Living Chemistry*, Kugelmass NI, Ed., Charles C Thomas, Springfield, IL, 1970, 130.

Rabl W, Baubin M, Broinger G, et al., Serious complications from active compression–decompression cardiopulmonary resuscitation, *Int J Legal Med*, 109, 84, 1996.

Rabl W, Baubin M, Haid C, Pfeiffer KP, and Scheithauer R, Review of active compression-decompression cardiopulmonary resuscitation (ACD-CPR). Analysis of iatrogenic complications and their biomechanical explanation, *Forensic Sci Int* 89, 175, 1997.

Rafflenbeul W, Smith LR, Rogers WL, et al., Quantitative coronary arteriography: coronary anatomy of patients with unstable angina pectoris reexamined 1 year after optimal medical therapy, *Am J Cardiol* 43, 699, 1979.

Rahimtoola SH, The hibernating myocardium, *Am Heart J* 117, 211, 1989.

Rajanayagam SMA, Shou M, Thirumurti V, et al., Intracoronary basic fibroblast growth factor enhances myocardial collateral perfusion in dogs, *J Am Coll Cardiol* 35, 519, 2000.

Rakocevic-Stojanovic V, Grujic M, Seferovic P, et al., Myotonic dystrophy and cardiac disorders, *Panminerva Med* 42, 257, 2000.

Rakusan K, Quantitative morphology of capillaries of the heart: number of capillaries in animal and human hearts under normal and pathological conditions, in Functional Morphology of the Heart, Bajusz E and Rona G, Eds., *Meth Achiev Exp Pathol Karger*, 5, 272, 1971.

Ralevic V, Kendall DA, Randall MD, et al., Cannabinoid modulation of sensory neurotransmission via cannabinoid and vanilloid receptors: roles in regulation of cardiovascular function, *Life Sci* 71, 2577, 2002.

Rambaud C, Campbell P, and Guilleminault C, Interpretation of autopsy findings and definition of SIDS. *Third SIDS International Conference*, Stavanger, program and abstracts, 1994, 82.

Rampazzo A, Nava A, Danieli GA, et al., The gene for arrhythmogenic right ventricular cardiomyopathy maps to chromosome 14q23–q24, *Hum Mol Genet* 3, 959, 1994.

Randall MD, Harris D, Kendall DA, et al., Cardiovascular effects of cannabinoids, *Pharmacol Ther*, 95, 191, 2002.

Randall MD, Kendall DA, and O'Sullivan S, The complexities of the cardiovascular actions of cannabinoids, *Br J Pharmacol* 142, 20, 2004.

Ranganathan N and Burch GE, Gross morphology and arterial supply of the papillary muscles of the left ventricle of man, *Am Heart J* 77, 506, 1969.

Ranganathan N, Lam JHC, Wigle ED, and Silver MD, Morphology of human mitral valve II: the valve leaflets, *Circulation* 41, 459, 1970.

Rashid J, Eisenberg MJ, and Topol EJ, Cocaine-induced aortic dissection, *Am Heart J* 132, 1301, 1996.

Rashid MA, Wilkström T, and Ortenwall P, Cardiac injuries: a ten-year experience, *Eur J Surg*, 166, 18, 2000.

Rasten-Almquist P, Eksborg S, and Rajs J, Myocarditis and sudden infant death syndrome, *APMIS* 110, 469, 2002.

Razzouk AJ, Freedom RM, Cohen AJ, et al., The recognition, identification of morphologic substrate, and treatment of subaortic stenosis after a Fontan operation: an analysis of twelve patients, *J Thorac Cardiovasc Surg* 104, 938, 1992.

Reagan TJ, Wu CF, Weisse AB, et al., Acute myocardial infarction in toxic cadiomyopathy without coronary obstruction, *Circulation* 51, 453, 1975.

Reduto LA, Freund GC, Gaeta JM, et al., Coronary artery reperfusion in acute myocardial infarction: beneficial effects of intracoronary streptokinase on left ventricular salvage and performance, *Am Heart J* 102, 1168, 1981.

Reichenbach D and Benditt EP, Myofibrillar degeneration: a common form of cardiac muscle injury, *Ann NY Acad Sci* 156, 164, 1969.

Reichenbach DD and Benditt EP, Catecholamine and cardiomiopathy: the pathogenesis and potential importance of myofibrillar degeneration, *Hum Pathol* 1, 125, 1970.

Reimer KA and Jennings RB, The changing anatomic reference base of evolving myocardial infarction: underestimation of myocardial collateral blood flow and overestimation of experimental anatomic infarct size due to edema, hemorrhage and acute inflammation, *Circulation* 60, 866, 1979.

Reimer KA, Lowe JE, Rasmussen MM, and Jennings AB, The wavefront phenomenon of ischemic cell death: myocardial infarct size vs duration of coronary occlusion in dog, *Circulation* 56, 786, 1977.

Reimer KA, Rasmussen MM, and Jennings RB, On the nature of protection by propranolol against myocardial necrosis after temporary coronary occlusion in dogs, *Am J Cardiol* 37, 520, 1976.

Reinders JG, Heijmen BJ, Olofsen-van Acht MJ, et al., Ischemic heart disease after mantlefield irradiation for Hodgkin's disease in long-term follow-up, *Radiother Oncol* 51, 35, 1999.

Reinecke H, Zhang M, Bartosek T, and Murry CE, Survival, integration, and differentiation of cardiomyocyte grafts: a study in normal and injured rat hearts, *Circulation* 100, 193, 1999.

Reiner L, Gross examination of the heart, in *Pathology of the Heart and Blood Vessels*, Gould SE, Ed., Charles C Thomas, Springfield, IL, 1968, p. 1136.

Reissman P, Rivkind A, Jurim O, et al., Case report: the management of penetrating cardiac trauma with major coronary artery injury — is cardiopulmonary bypass essential?, *J Trauma*, 33, 773, 1992.

Rella JG and Murano T, Ecstasy and acute myocardial infarction, *Ann Emerg Med* 44, 550, 2004.

Remme CA, Wever EF, Wilde AA, et al., Diagnosis and long-term follow-up of the Brugada syndrome in patients with idiopathic ventricular fibrillation, *Eur Heart J* 22, 400, 2001.

Renkin EM, Blood flow and transcapillarity exchange in skeletal and cardiac muscle, in *Coronary Circulation and Energetics of the Myocardium*, Marchetti G and Taccardi B, Eds., Karger, Basel, 1967, p. 18.

Rentrop KP, Feit F, Sherman W, and Thornton JC, Serial angiographic assessment of coronary artery obstruction and collateral flow in acute myocardial infarction: report from the second Mount Sinai-New York University reperfusion trial, *Circulation* 80, 1166, 1989.

Rentrop KP, Thornton JC, Feit F, and Van Buskirk M, Determinants and protective potential of coronary arterial collaterals as assessed by an angioplasty model, *Am J Cardiol* 61, 677, 1988.

Rentrop PK, Thrombi in acute coronary syndromes: revisited and revised, *Circulation* 101, 1619, 2000.

Report of World Health Organization Expert Committee, WHO Technical Report Series 697: Cardiomyopathies, Geneva, 1984.

Report of World Health Organization/International Society and Federation of Cardiology Task Force on the definition and classification of cardiomyopathies, *Circulation* 93, 841, 1996.

Repossini A, Arena V, Alamanni F, Di Matteo S, Antona C, and Biglioli P. Straddling endoventricular pericardial patch in type I myocardial rupture prevention, *Ann Thorac Surg* 56, 163, 1993.

Repossini A, Moriggia S, Cianci V, et al., The last operation is safe and effective: MIDCABG clinical and angiographic evaluation, *Ann Thorac Surg* 70, 74, 2000.

Rey JM and Tennant CC, Cannabis and mental health, *BMJ* 325, 1183, 2002.

Rhee PM, Foy H, Kaufmann C, et al., Penetrating cardiac injuries: a population-based study, *J Trauma*, 45, 366, 1998.

Ricci MA, Trevisani GT, and Pilcher DB, Vascular complications of cardiac catheterization, *Am J Surg*, 167, 375, 1994.

Rice WG and Wittstruck KP, Acute hypertension and delayed traumatic rupture of the aorta, *JAMA*, 147, 915, 1951.

Rich NM and Spencer FC, *Vascular Trauma*, WB Saunders, Philadelphia, 1978.

Richens D, Field M, Neale M, et al., The mechanism of injury in blunt traumatic rupture of the aorta, *Eur J Cardio –Thorac Surg* 21, 288, 2003.

Riddle MA, Geller B, and Ryan N, Another sudden death in a child treated with desipramine, *J Am Acad Child Adolesc Psychiatry* 32, 792, 1993.

Ridker PM, On evolutionary biology, inflammation, infection, and the causes of atherosclerosis, *Circulation* 105, 2, 2002.

Rigle DA, Dexter RD, and McGee MB, Cardiac rhabdomyoma presenting as sudden infant death syndrome, *J Forensic Sci* 34, 694, 1989.

Rigo P, Becker LC, Griffith LSC, et al., Influence of coronary collateral vessels on the results of thallium-201 myocardial stress imaging, *Am J Cardiol* 44, 452, 1979.

Rinaldo P, Stanley CA, Hsu BY, et al., Sudden neonatal death in carnitine transporter deficiency, *J Pediatr* 131, 304, 1997.

Ristic AD and Maisch B, Cardiac rhythm and conduction disturbances: what is the role of auto-immune mechanisms? *Herz* 25, 181, 2000.

Riße M and Weiler G, Quantitative myokardiale Mastzellbefunde bei Säuglingen und Kleinkindern zur Ermittlung altersabhängiger Normwerte und als Grundlage differentialdiagnostischer Überlegungen, *Rechtsmed* 7, 49-72, 1997.

Rivens SM, Kearney DL, Smith EO'B, et al., Sudden death and cardiovascular collapse in children with restrictive cardiomyopathy, *Circulation* 102, 876, 2000.

Roberts R, Genomics and cardiac arrhythmias, *J Am Coll Cardiol* 47, 1, 2006.

Roberts S, Kosanke S, Terrence Dunn S, et al., Pathogenic mechanisms in rheumatic carditis: focus on valvular endothelium, *J Infect Dis* 183, 597, 2001.

Roberts WC, Congenital cardiovascular abnormalities usually "silent" until adulthood: morphological features of the floppy mitral valve, valvular aortic stenosis, hypertrophic cardiomyopathy, sinus of Valsalva aneurysm and the Marfan syndrome, *Cardiovasc Clin* 10/1, 407, 1979.

Roberts WC, Qualitative and quantitative comparison of amounts of narrowing by atherosclerotic plaques in the major epicardial coronary arteries at necropsy in sudden coronary death, transmural acute myocardial infarction, transmural healed myocardial infarction and unstable angina pectoris, *Am J Cardiol* 64, 324, 1989.

Roberts WC, Sudden cardiac death: a diversity of causes with focus on atherosclerotic coronary artery disease, in *Sudden Cardiac Death*, Josephson ME, Ed., Blackwell, Boston, 1993, 1.

Roberts WC, Curry RC, Isner JM, et al., Sudden death in Prinzmetal's angina with coronary spasm documented by angiography: analysis of three necropsy patients, *Am J Cardiol* 50, 203, 1982a.

Roberts WC and Jones AA, Quantitation of coronary arterial narrowing at necropsy in sudden coronary death: analysis of 31 patients and comparison with 25 control subjects, *Am J Cardiol* 44, 39, 1979.

Roberts WC, Siegel RJ, and Zipes DP, Origin of the right coronary artery from the left sinus of Valsalva and its functional consequences: analysis of 10 necropsy patients, *Am J Cardiol* 49, 863, 1982b.

Rodbard S, Vascular modifications induced by flow, *Am Heart J* 51, 926, 1956.

Rodbard S, Evidence that vascular conductance is regulated at the capillary, *Hypertension* 13, 160, 1965.

Rodbard S, Physical factors in arterial sclerosis and stenosis, *Angiology* 22, 267, 1971.

Roden DM, Pharmacogenetics and drug-induced arrhythmias, *Cardiovasc Res*, 50, 24, 2001.

Roelandt JRTC, Di Mario C, Pandian NG, et al., Three-dimensional reconstruction of intracoronary ultrasound images: rationale, approaches, problems and directions, *Circulation* 90, 1044, 1994.

Rogers FB, Osler TM, and Shackford SR, Aortic dissection after trauma, *J Trauma Inj Infect Crit Care*, 41, 906, 1996.

Roget's International Thesaurus, 5th ed., Robert L. Chapman Collins, 1992.

Romano C, Gemme G, and Pongiglione R, Aritmie cardiace rare delleta pedriatica: accessi per fibrillazione ventricolare parossistica, *Clin Ped* 45, 656–683, 1963.

Rona G, Chappel CI, Balazs T, et al., An infarct-like myocardial lesion and other toxic manifestations produced by isoproterenol in rat, *Arch Pathol* 67, 443, 1959.

Rona G and Kahn DS, Experimental studies on the healing of cardiac necrosis, *Ann NY Acad Sci* 156, 177, 1969.

Ronneberger DL, Hausmann R, and Betz P, Sudden death associated with myxomatous transformation of the mitral valve in an 8-year-old boy, *Intl J Legal Med* 111, 199, 1998.

Rose AG and Uys CJ, Pathology of cardiac transplantation, in *Cardiovascular Pathology*, 2nd ed., Silver MD, Ed., Churchill Livingstone, New York, 1991, p. 1649.

Rosenberg HG, Systemic arterial disease with myocardial infarction: report, on two infants, *Circulation* 47, 270, 1973.

Rosenblatt A and Selzer A, The nature and clinical features of myocardial infarction with normal coronary arteriogram, *Circulation* 55, 578, 1977.

Rosenschein U, Ellis SG, Handenschild CC, et al., Comparison of histopathologic coronary artery lesions obtained from directional atherectomy in stable angina versus acute coronary syndromes, *Am J Cardiol* 73, 508, 1994.

Ross MJ, Heart block, sudden death, and atrioventricular node mesothelioma, *Am J Dis Child* 131, 1209, 1977.

Ross R, The arterial wall and atherosclerosis, *Ann Rev Med* 30, 1, 1979.

Ross R, The pathogenesis of atherosclerosis: a perspective for the 1990s, *Nature* 362, 801, 1993.

Rossi L, Bulbo-spinal pathology in neurocardiac sudden death of adults:a pragmatic approach to a neglected problem, *Int J Legal Med* 112, 83–90, 1999.

Rossi L, Structural and non-structural disease underlying high-risk cardiac arrhythmias relevant to sports medicine, *J Sports Med Phys Fitness* 35, 79–80, 1995.

Rossi L, Cardioneuropathy and extracardiac neural disease, *J Am Coll Cardiol* 5 (Suppl 6), 66B, 1985.

Rossi L and Matturri L T-lymphocytic leptomeningitis of the ventral medullary surface and nucleus arcuatus hypoplasia in SIDS: first report of a case, *Ann Esp Ped* (Supp) 95, 55, 1997.

Rossi L and Matturri L, Anatomohistological features of the heart's conduction system and innervation in SIDS, in *Sudden Infant Death Syndrome: New Trends in the Nineties,* Rognum TO, Ed., Scandinavian University Press, Oslo, 1995, 207–212.

Rossi MA, Microvascular changes as a cause of chronic cardiomyopathy in Chagas' disease, *Am Heart J* 120, 233, 1990.

Rowley AH, Gonsalez-Cruzzi F, Gidding SS, et al., Incomplete Kawasaki disease with coronary artery involvement, *J Pediatr* 110, 409, 1987.

Rozanski A, Blumenthal JA, and Kaplan J, Impact of psychological factors on the pathogenesis of cardiovascular diseases and implication for therapy, *Circulation* 99, 2196, 1999.

Rubin E and Farber JL, *Pathology,* JB Lippincott, Philadelphia, 1988.

Ruffer MA, Arterial lesions found in Egyptian mummies, *J Pathol Bact* 15, 453, 1911.

Ruiz de la Fuente S and Prieto F, Heart-hand syndrome. III. A new syndrome in three generations, *Hum Genet* 55, 43, 1980.

Rundqvist B, Elam M, Bergmann-Sverridottir Y, et al., Increased cardiac adrenergic drive precedes generalized sympathetic activation in human heart failure, *Circulation* 95, 169, 1997.

Rushmer RF, *Cardiovascular Dynamics,* 2nd ed., W.B. Saunders, Philadelphia, 1963.

Saavedra WF, Tunin RS, Paolocci N, et al., Reverse remodelling and enhanced adrenergic reserve from passive external support in experimental dilated heart failure, *J Am Coll Cardiol* 39, 2069, 2002.

Sabbah HN, Goldberg AD, Schoels W, et al., Spontaneous and inducible ventricular arrhythmias in a canine model of chronic heart failure: relation to haemodynamics and sympathoadrenergic activation, *Eur Heart J* 13, 1562, 1992.

Saber RS, Edwards WD, Bailey KR, et al., Coronary embolization after balloon angioplasty or thrombolytic therapy: an autopsy study of 32 cases, *J Am Coll Cardiol* 22, 1283, 1993.

Sabia PJ, Powers ER, and Jayaweera AR, Functional significance of collateral blood flow in patients with recent acute myocardial infarction: a study using myocardial contrast echocardiography, *Circulation* 85, 2080, 1992.

Sabiston DC and Gregg DE, Effect of cardiac contraction on coronary blood flow, *Circulation* 15, 14, 1957.

Sadeh D, Shannon DC, Abboud S, et al., Altered cardiac repolarization in some victims of sudden infant death syndrome, *N Engl J Med* 317, 1501, 1987.

Sader MA, Griffiths KA, McCredie RJ, Handelsman DJ, and Celermajer DS, Androgenic anabolic steroids and arterial structure and function in male bodybuilders, *J Am Coll Cardiol* 37, 224, 2001.

Sakata K, Miura F, Sugino H, et al., Assessment of regional sympathetic nerve activity in vasospastic angina: analysis of iodine-123-labeled metaiodobenzylguanidine scintigraphy, *Am Heart J* 133, 484, 1997a.

Sakata K, Shirotani M, Yoshida H, and Kurata C, Iodine-123 metaiodobenzylguanidine cardiac imaging to identify and localize vasospastic angina without significant coronary artery narrowing, *Am J Coll Cardiol* 30, 370, 1997b.

Sakata Y, Kodama K, Kitanaza M, et al., Different mechanism of ischemic adaptation to repeated coronary occlusions in patients with and without recruitable collateral circulation, *J Am Coll Cardiol* 30, 1679, 1997c.

Sakka SG, Huettemann E, Giebe W, et al., Late cardiac arrhythmias after blunt chest trauma, *Intensive Care Med*, 26, 792, 2000.

Salomon V, Niemela M, Miettinen H, et al., Relationship of socioeconomic status to the incidence and prehospital, 28 days and 1-year mortality rates of acute events in the Finmonica myocardial register study, *Circulation* 101, 1913, 2000.

Sambuceti G, Giorgetti A, Corsiglia L, et al., Perfusion-contraction mismatch during inotropic stimulation in hibernating myocardium, *J Nucl Med* 39, 396, 1998.

Sambuceti G, Marzilli M, Maraccini P, et al., Coronary vasoconstriction during myocardial ischemia induced by rises in metabolic demand in patients with coronary artery disease, *Circulation* 95, 2652, 1997.

Sanatani S, Saul JP, Walsh EP, et al., Spontaneously terminating apparent ventricular fibrillation during transesophageal electrophysiological testing in infants with Wolff-Parkinson-White syndrome, *Pacing Clin Electrophysiol* 24, 1816, 2001.

Sanbar SS, Cardiac and vascular trauma, *Trauma*, 1989, 31, 51, 1989.

Sand NPR, Rehling M, Bagger JP, et al., Functional significance of recruitable collaterals during temporary occlusion evaluated by ^{99m}Tc-sestamibi single-photon emission computerized tomography, *J Am Coll Cardiol* 35, 624, 2000.

Sangiorgi G, Rumberger JA, Severson A, et al., Arterial calcification and not lumen stenosis is highly correlated with atherosclerotic plaque burden in human: a histologic study of 723 coronary artery segments using nondecalcifying method, *J Am Coll Cardiol* 31, 126, 1998.

Sansone V, Griggs RC, Meola G, et al., Andersen's syndrome: a distinct periodic paralysis, *Ann Neurol* 42, 305, 1997.

Santorelli FM, Schlessel JS, Slonim AE, and DiMauro S, Novel mutation in the mitochondrial DNA tRNA glycine gene associated with sudden unexpected death, *Pediatr Neurol* 15, 145, 1996.

Sanyal SK and Johnson WW, Cardiac conduction abnormalities in children with Duchenne's progressive muscular dystrophy: electrocardiographic features and morphologic correlates, *Circulation* 66, 853, 1982.

Saram M, Ueber die azellulare Entstehung von Narben bei Durchblutunzgsstorungen in Herzmuskel, *Beitr Pathol Anat Allg Pathol* 118, 275, 1957.

Saririan M and Eisenberg MJ, Myocardial laser revascularization for the treatment of end-stage coronary artery disease, *J Am Coll Cardiol* 41, 173, 2003.

Sarkozy A and Brugada P, Sudden cardiac death: what is inside our genes? *Can J Cardiol* 21, 1099, 2005.

Sarkozy A and Brugada P, Sudden cardiac death and inherited arrhythmia syndromes, *J Cardiovasc Electrophysiol* 16(Suppl 1), S8, 2005.

Sasano H, Virmani R, Patterson RH, Robinowitz M, and Guccion JG, Eosinophilic products lead to myocardial damage, *Hum Pathol* 20, 850, 1989.

Sasse L, Wagner R, and Murray FE, Transmural myocardial infarction during pregnancy, *Am J Cardiol* 35, 448, 1975.

Sato Y, Sugie R, Tsuchiya B, et al., Comparison of the DNA extraction methods for polymerase chain reaction amplification from formalin-fixed and paraffin-embedded tissues, *Diagn Mol Pathol* 10, 265, 2001.

Satran A, Bart BA, Henry CR, et al., Increased prevalence of coronary artery aneurysms among cocaine users, *Circulation* 111, 2424, 2005.

Saumarez RC, Camm AJ, and Panagos A, Ventricular fibrillation in hypertrophic cardiomyopathy is associated with increased fractionation of paced right ventricular electrograms, *Circulation* 86, 467, 1992.

Schaeffers M, Lerch H, Wichter T, et al., Cardiac sympathetic innervation in patients with idiopathic ventricular outflow tract tachycardia, *J Am Coll Cardiol* 32, 181, 1998.

Scharfman WB, Wallach JB, and Angrist A, Myocardial infarction due to syphilitic coronary ostial stenosis, *Am Heart J* 40, 603, 1950.

Schechtmann VL, Harper RM, Kluge KA, Wilson A, Hoffmann HJ, and Southall DP, Cardiac and respiratory patterns in normal infants and victims of the sudden infant death syndrome. *Sleep* 11, 413–424, 1988.

Schechtmann VL, Harper RM, Kluge KA, Wilson A, Hoffmann HJ, and Southall DP, Heart rate variation in normal infants and victims of the sudden infant death syndrome, *Early Hum Dev* 19, 167–181, 1989.

Scheinman MM, Is the Brugada syndrome a distinct clinical entity? *J Cardiovasc Electrophysiol* 8, 332, 1997.

Schiebler GL, Loring AE, Brogdon BG, et al., Cardiovascular manifestation of Hurler's syndrome, *Circulation* 26, 782, 1962.

Schifano F, A bitter pill: overview of ecstasy (MDMA, MDA) related fatalities, *Psychopharmacology* 173, 242, 2004.

Schifano F, Potential human neurotoxicity of MDMA ("Ecstacy"): subjective self-reports, evidence from an Italian drug addiction center and clinical case studies, *Neuropsychobiology* 42, 25, 2000.

Schifano F, Oyefeso A, Corkery J, et al., Death rates from ecstasy (MDMA, MDA) and polydrug use in England and Wales 1996–2002, *Hum Psychopharmacol* 18, 519, 2003a.

Schifano F, Oyefeso A, Webb L, et al., Review of deaths related to taking ecstasy, England and Wales, 1997–2000, *BMJ* 326, 80, 2003b.

Schionning JD, Frederiksen P, and Kristensen IB, Arrhythmogenic right ventricular dysplasia as a cause of sudden death, *Am J Forensic Med Pathol* 18, 345, 1997.

Schlesinger MJ, An injection plus dissection study of coronary artery occlusions and anastomoses, *Am Heart J* 15, 528, 1938.

Schlesinger MJ, Relation of anatomic pattern to pathological conditions of the coronary arteries, *Arch Pathol* 30, 403, 1940.

Schlesinger MJ and Reiner L, Focal myocytolysis of the heart, *Am J Pathol* 31, 443, 1955.

Schlesinger MJ, Zoll PM, and Wessler S, The conus artery, a third coronary artery, *Am Heart J* 38, 823, 1949.

Schoen FJ, Pathology of heart valve substitution with mechanical and tissue valves, in *Cardiovascular Pathology*, 3rd ed., Silver MD, Gotlieb AI, and Schoen FJ, Eds., Churchill Livingstone, Philadelphia, 2001, p. 629.

Schoen FJ and Edwards WD, Valvular heart disease: general principles and stenosis, in *Cardiovascular Pathology*, 3rd ed., Silver MD, Gotlieb AI, and Schoen FJ, Eds., Churchill Livingstone, Philadelphia, 2001, p. 402.

Schoenmackers J, Vergleichende quantitative Untersuchungen uber den Faserbestand des Herzens bei Herz-und Herzklappenfehlern sowie Hochdruck, *Virchows Archiv* 331, 3, 1958.

Schofer J, Spielman R, Schuchert A, et al., Iodine-123 meta-iodobenzylguanidine scintigraphy: a noninvasive method to demonstrate myocardial adrenergic nervous system disintegrity in patients with idiopathic dilated cardiomyopathy, *J Am Coll Cardiol* 12, 1252, 1988.

Schomig A, Ndrepepa G, Mehilli J, et al., Therapy-dependent influence of time-to-treatment interval on myocardial salvage in patients with acute myocardial infarction treated with coronary artery stenting or thrombolysis, *Circulation* 108, 1084, 2003.

Schott JJ, Carpentier F, Peltier S, Foley P, Drouin E, Bouhour JB, Donelly P, Vergnaud G, Bachner L, Moisan JP, Le Marec H, and Pascal O, Mapping of a gene for long QT syndrome to chromosome 4q25-q27, *Am J Hum Genet* 57, 1114–1122, 1995.

Schroder ES, Sirna SJ, Kieso RA, et al., Sensitization of reperfused myocardium to subsequent coronary flow reductions: an extension of the concept of myocardial stunning, *Circulation* 78, 717, 1988.

Schulman SP, Thiemann DR, Ouyang P, et al., Effects of acute hormone therapy on recurrent ischemia in postmenopausal women with unstable angina, *J Am Coll Cardiol* 39, 231, 2002.

Schwartz CJ and Mitchell JRA, Cellular infiltration of the human arterial adventitia associated with atheromatous plaque, *Circulation* 26, 73, 1962.

Schwartz ER, Schoendube FA, Kostin S, et al., Prolonged myocardial hibernation exacerbates cardiomyocyte degeneration and impairs recovery of function after revascularization, *J Am Coll Cardiol* 31, 118, 1998a.

Schwartz ER, Speakman MT, Patterson M, et al., Evaluation of the effects of intramyocardial injection of DNA expressing vascular endothelial growth factor (VEGF) in a myocardial infarction model in the rat: angiogenesis and angioma formation, *J Am Coll Cardiol* 35, 1323, 2000a.

Schwartz PJ, The cardiac theory and sudden infant death syndrome, in Sudden Infant Death Syndrome. Medical Aspects and Physiological Management, Culbertson JL, Krous HK, and Bendell RD, Eds., Edward Arnold, London, 1989, 121–138.

Schwartz PJ, The quest for the mechanisms of the sudden infant death syndrome: doubts and progress, *Circulation* 75, 677–683, 1987.

Schwartz PJ, Cardiac sympathetic innervation and the sudden infant death syndrome. A possible pathogenic link, *Am J Med* 60, 167–172, 1976.

Schwartz PJ and Segantini A, Cardiac innervation, neonatal electrocardiography, and SIDS. A key for a novel preventive strategy? *Ann NY Sci* 533, 210–220, 1988.

Schwartz PJ, La Rovere MT, and Vanoli E, Autonomic nervous system and sudden cardiac death: experimental basis and clinical observations for post-myocardial infarction risk stratification, *Circulation* 85 (Suppl 1), 1, 1992.

Schwartz PJ, Montemerlo M, Facchini M, et al., The QT interval throughout the first 6 months of life: a prospective study, *Circulation* 66, 496, 1982.

Schwartz PJ, Priori SG, Dumaine R, et al., A molecular link between the sudden infant death syndrome and the long-QT syndrome, *New Engl J Med* 343, 262, 2000b.

Schwartz PJ, Priori SG, Spazzolini C, et al., Genotype-phenotype correlation in the long-QT syndrome: gene-specific triggers for life-threatening arrhythmias, *Circulation* 103, 89, 2001.

Schwartz PJ, Stamba-Badiale M, Segantini A, et al., Prolongation of the QT interval and the sudden infant death syndrome, *N Engl J Med* 338, 1709, 1998b.

Schwartz SM, Campbell GR, and Campbell JA, Replication of smooth muscle cells in vascular disease, *Circ Res* 58, 487, 1986.

Scomazzoni G, Baroldi G, and Mantero O, Studio anatomo-clinico su di un singolare caso di aneurisma dissecante del miocardio, *Ospedale Maggiore* 45, 1, 1957.

Scorsin M, Hagege AA, Marotte F, et al., Does transplantation of cardiomyocytes improve function of infarcted myocardium? *Circulation* 96 (Suppl II), II188, 1997.

Scott WL, Complications associated with central venous catheters. A survey, *Chest*, 94, 1221, 1988.

Seguchi M, Hino Y, Aiba S, et al., Ostial stenosis of the left coronary artery as a sole clinical manifestation of Takayasu's arteritis: a possible cause of unexpected sudden death, *Heart Vessels* 5, 188, 1990.

Seidman C and Sampson B, Genetic causes of diseases affecting the heart and great vessels *Cardio-vascular Pathology*, 3rd ed., Silver MD, Gotlieb AI, and Schoen FJ, Eds., Churchill Livingstone, Philadelphia, 2001, p. 763.

Seidman JG and Seidman C, The genetic basis for cardiomyopathy: from mutation identification to mechanistic paradigms, *Cell* 104, 557, 2001.

Seifert G, zur Pathologie der Virusmyokarditis (insbeso. durch Coxsachie-Viren) im Säuglings- u. Kindesalter, *Zbl allg path Anat* 102, 274, 1961.

Seiler C, Jenni R, Vassali G, et al., Left ventricular chamber dilatation in hypertrophic cardiomyopathy: related variables and prognosis in patients with medical and surgical therapy, *Br Heart J*, 74, 508, 1995.

Shan K, Bick RJ, Poindexter BJ, et al., Relation to tissue Doppler derived myocardial velocities to myocardial structure and beta-adrenergic receptor density in humans, *J Am Coll Cardiol* 36, 891, 2000.

Shan K, Bick RJ, and Poindexter BJ, Altered adrenergic receptor density in myocardial hibernation in humans: a possible mechanism of depressed myocardial function, *Circulation* 102, 2599, 2000.

Shanes JG, Ghali J, Billinghan ME, et al., Interobserver variability in the pathologic interpretation of endomyocardial biopsy results, *Circulation* 75, 401, 1987.

Shannon RP, Manders T, and Shen YT, Role of blood doping in the coronary vasoconstrictor response of cocaine, *Circulation* 92, 97, 1995.

Sharff JA, Renal infarction associated with intravenous cocaine use, *Ann Emerg Med* 13, 1145, 1984.

Sharma B, Asinger R, Francis G, et al., Demonstration of exercise-induced painless myocardial ischemia in survivors of out-of-hospital ventricular fibrillation, *Am J Cardiol* 59, 740, 1987.

Shatz A, Hiss J, and Arensburg B, Myocarditis misdiagnosed as sudden infant death syndrome (SIDS), *Med Sci Law* 37, 16, 1997.

Shepherd JT and Vanhoutte PM, Mechanism responsible for coronary vasospasm, *J Am Coll Cardiol* 8, 1, 1986.

Shields LBE, Hunsaker DM, and Hunsaker JC, Iatrogenic catherer-related cardiac tamponade: a case report of fatal hydropericardium following subcutaneous implantation of a chemotherapeutic injection port, *J Forensic Sci*, 48, 414, 2003.

Shintani S, Murohara T, Ikeda H, et al., Mobilization of endothelial progenitor cells in patients with acute myocardial infarction, *Circulation* 103, 2776, 2001.

Shirani J, Berezowski K, and Roberts WC, Quantitative measurement of normal and excessive (coradiposum) subepicardial adipose tissue, its clinical significance, and its effect on electrocardiographic QRS voltage, *Am J Cardiol* 76, 414, 1995.

Shirani J, Pick R, Roberts WC, and Maron BJ, Morphology and significance of the left ventricular collagen network in young patients with hypertrophic cardiomyopathy and sudden cardiac death, *J Am Coll Cardiol* 35, 6, 2000.

Shirey EK, Hawk WA, Murkerj D, and Effler DB, Percutaneous myocardial biopsy of the left ventricle: experience in 198 patients, *Circulation* 46, 112, 1972.

Shvalev VN, Virkhert AM, Stropus RA, et al., Changes in neural and humoral mechanisms of the heart in sudden death due to myocardial abnormalities, *J Am Coll Cardiol* 8, 55A, 1986.

Sicari R, Picano E, and Cortigiani L, Prognostic value in myocardial viabilità recognized by low-dose dobutamine echocardiography in chronic ischemic left ventricular dysfunction, *Am J Cardiol* 92, 1263, 2003.

Sidney S, Cardiovascular consequences of marijuana use, *J Clin Pharmacol* 42(11 Suppl), 64S, 2002.

Siegel RJ and Dunton SF, Systemic occlusive arteriopathy with sudden death in a 10-year-old boy, *Hum Pathol* 22, 197, 1991.

Silka MJ, Hardy BG, Menashe VD, and Morris CD, A population-based prospective evaluation of risk of sudden death after operation for common congenital heart defects, *J Am Coll Cardiol* 32, 245, 1998.

Silka MJ, Kron J, Dunnigan A, et al., Sudden cardiac death and the use of implantable cardioverter-defibrillators in pediatric patients: the Pediatric Electrophysiology Society, *Circulation* 87, 800, 1993.

Silver MD, Medial hemorrhage and dissection in a coronary artery: an unusual cause of coronary occlusion, *CMAJ* 32, 99, 1968.

Silver MD, The healed and sealed aortic intimal tear, *Cardiovasc Pathol* 6, 315, 1997.

Silver MD, Baroldi G, and Mariani F, The relationship between acute occlusive coronary thrombi and myocardial infarction studied in 100 consecutive patients, *Circulation* 61, 219, 1980.

Silver MD, Butany J, and Chiasson DA, The pathology of myocardial infarction and its mechanical complications, in *Mechanical Complications of Myocardial Infarction*, David TE, Ed., RG Landes Company, Austin, TX, 1993, p. 4.

Silver MD, Wigle ED, Trimble AS, et al., Iatrogenic coronary ostial stenosis, *Arch Pathol* 88, 73, 1969.

Silver MM, Sudden cardiac death in infants and children, in *Sports Medicine: A Forensic Approach*, Fineschi V, Ed., Edizioni Colosseum, Rome, 1998, p. 195.

Silver MM, Benson LN, and Wigle ED, Clinicopathological conference: cardiomegaly in a young infant, *Cardiovasc Pathol* 5, 271, 1996a.

Silver MM, Burns JE, Sethi RK, and Rowe RD, Oncocytic cardiomyopathy in an infant with onco-cytosis in exocrine and endocrine glands, *Hum Pathol* 11, 598, 1980.

Silver MM and Freedom RM, Gross examination and structure of the heart, in *Cardiovascular Pathology*, 2nd ed., Silver MD, Ed., Churchill Livingstone, New York, 1991, p. 1.

Silver MM, Perrin D, and Freedom RM, Tissue iron storage patterns in fetal hydrops associated with congestive heart failure, *Pediatr Pathol Lab Med*, 16, 563, 1996b.

Silver MM and Silver MD, Pathology of cardiomyopathies in childhood, *Prog Pediatr Cardiol* 1, 8, 1992.

Silvestri G, Bertini E, Servidei S, et al., Maternally inherited cardiomyopathy: a new phenotype associated with the A to G AT nt.3243 of mitochondrial DNA (MELAS mutation), *Muscle Nerve* 20, 221, 1997.

Sims HF, Brackett JC, Powell CK, et al., The molecular basis of pediatric long chain 3-hydroxyacyl-CoA dehydrogenase deficiency associated with maternal acute fatty liver of pregnancy, *Proc Natl Acad Sci USA* 92, 841, 1995.

Sinclair W and Nitsch E, Polyarteritis nodosa of the coronary arteries: report of a case with rupture of an aneurysm and intrapericardial hemorrhage, *Am Heart J* 38, 898, 1949.

Skinner JE, Regulation of cardiac vulnerability by the cerebral defense system, *J Am Coll Cardiol* 5, 88B, 1985.

Skinner JR, Manzoor A, Hayes AM, et al., A regional study of presentation and outcome of hyper-trophic cardiomyopathy in infants, *Heart* 77, 229, 1997.

Sladden RA, Coronary arteriosclerosis and calcification in infancy, *J Clin Pathol* 5, 175, 1952.

Slade AK, Saumarez RC, and McKenna WJ, The arrhythmogenic substrate — diagnostic and ther-apeutic implications: hypertrophic cardiomyopathy, *Eur Heart J* 14 (Suppl E), 84, 1993.

Slavin RE, Saeki K, Bhagavan B, and Maas AE, Segmental arterial mediolysis: a precursor of fibro-muscular dysplasia, *Mod Pathol* 8, 287, 1995.

Smedira NG, Zikri M, Thomas JD, et al., Blunt traumatic rupture of a mitral papillary muscle head, *Ann Thorac Surg*, 61, 1526, 1996.

Smith HW III, Liberman HA, Brody SL, Battey LL, Donohue BC, and Morris DC, Acute myocardial infarction temporally related to cocaine use: clinical, angiographic, and pathophysiologic observations, *Ann Intern Med* 107, 13, 1987a.

Smith M, Kichuk MR, and Ratcliff NB, Clinical and pathologic study of two siblings with arrhythmogenic right ventricular cardiomyopathy, *Cardiovasc Pathol* 8, 273, 1999.

Smith NM, Bourne AJ, Clapton WK, and Byard RW, The spectrum of presentation at autopsy of myocarditis in infancy and childhood, *Pathology* 24, 129, 1992.

Smith SH, Kirling JK, Geer JC, et al., Arteritis in cardiac rejection after transplantation, *Am J Cardiol* 59, 1171, 1987b.

Smits JP, Eckardt L, Probst V, et al., Genotype-phenotype relationship in Brugada syndrome: electrocardiographic features differentiate *SCN5A*-related patients from non-*SCN5A*-related patients, *J Am Coll Cardiol* 40, 350, 2002.

Sokolove PE, Willis-Shore J, and Panacek EA, Exsanguination duet to right ventricular rupture during closed-chest cardiopulmonary resuscitation, *J Emerg Med*, 23, 161, 2002.

Sommers HM and Jennings RB, Ventricular fibrillation and myocardial necrosis after transient ischemia: effect of treatment with oxygen, procainamide, reserpine and propranolol, *Arch Int Med* 129, 780, 1972.

Song Y, Laaksonen H, Saukko P, et al., Histopathological findings of cardiac conduction system of 150 Finns, *Forensic Sci Intl* 119, 310, 2001.

Song Y, Zhu J, Laaksonen H, et al., A modified method for examining the cardiac conduction system, *Forensic Sci Intl* 86, 135, 1997.

Sorland SJ, Rostad H, Forfang K, and Abyholm G, Coarctation of the aorta: a follow-up study after surgical treatment in infancy and childhood, *Acta Paediatr Scand* 69, 113, 1980.

Southall DP, Arrowsmith WA, Oakley JR, McEnery G, Anderson RH, and Shinebourne EA, Prolonged QT interval and cardiacarrhythmias in two neonates: sudden infant death syndrome in one case, *Arch Dis Child* 62, 721–726, 1979.

Southall DP, Arrowsmith WA, Stebbens V, et al., QT interval measurements before sudden infant death syndrome, *Arch Dis Child* 61, 327, 1986.

Southall DP, Richards JM, de Swiet M, Arrowsmith WA, Cree JE, Fleming PJ, Franklin AJ, Orme RLE, Radford MJ, Wilson AJ, Shannon DC, Alexander JR, Brown NJ, and Shinebourne EA, Identification of infants destined to die unexpectedly during infancy: evaluation of predictive importance of prolonged apnoe and disorders of cardiac rhythm or conduction. First report of a multicentred prospective study into the sudden infant death syndrome, *BMJ* 286, 1092–1096, 1983a.

Southall DP, Richards JM, Shinebourne EA, Franks CI, Wilson AJ, and Alexander JR, Prospective population-based studies into heart rate and breathing patterns in newborn infants: prediction of infants at risk of SIDS? in Sudden Infant Death Syndrome, Tildon JT, Roeder LM, and Steinschneider A, Eds., Academic Press, New York, 1983b, 621–651.

Southall DP, Talbert DG, Alexander JR, Stevens AV, and Wilson AJ, Recordings of cardiorespiratory activity in relation to the problem of SIDS, in Sudden Infant Death Syndrome. Risk Factors and Basic Mechanisms, Harper RM and Hoffmann HJ, Eds., PMA Publishing, New York, 1988, 447–458.

Spain DM, Bradess VA, Iral P, et al., Intercoronary anastomotic channels and sudden unexpected death from advanced coronary atherosclerosis, *Circulation* 27, 12, 1963.

Spalteholz W, *Die Arterien der Herzwand. Anatomische Untersuchungen and Meschen und Tieren*, S Hirzel, Leipzig, 1924.

Spalteholz W and Hockrein M, Untersuchungen am Koronarsystem: die anatomische und funktionelle Beschaffenheit der Koronarterienwand, *Arch Exp Path u Pharmakol* 163, 333, 1931.

Spicer RL, Rocchini AP, Crowley DC, et al., Chronic verapamil therapy in pediatric and young adult patients with hypertrophic cardiomyopathy, *Am J Cardiol* 53, 1614, 1984.

Spijkerman IJ, van Ameijden EJC, Mientjes GHC, Coutinho RA, and van den Hoek A, Human immunodeficiency virus infection and other risk factors for skin abscesses and endocarditis among injection drug users, *J Clin Epidemiol* 49, 1149, 1996.

Spinnler MT, Lombardi F, Moretti C, et al., Evidence of functional alterations in sympathetic activity after myocardial infarction, *Eur Heart J* 14, 1334, 1993.

Spirito P, Bellone P, Harris KM, et al., Magnitude of left ventricular hypertrophy and risk of sudden death in hypertrophic cardiomyopathy, *N Engl J Med* 342, 1778, 2000.

Spirito P, Seidman CE, McKenna WJ, et al., The management of hypertrophic cardiomyopathy, *N Engl J Med* 336, 775, 1997.

Spiro D, Lattes RG, and Wiener J, The cellular pathology of experimental hypertension. I. Hyperplastic arteriosclerosis, *Am J Pathol* 47, 19, 1965.

Spiro D, Spotnitz H, and Sonnenblick EH, The relation of cardiac fine structure to function, in *Pathology of the Heart and Blood Vessels*, Gould SE, Ed., Charles C Thomas, Springfield, IL, 1968, p. 131.

Splawski I, Shen J, Timothy KW, et al., Spectrum of mutations in long-QT syndrome genes: KVLQT1, HERG, SCN5A, KCNE1, and KCNE2, *Circulation* 102, 1178, 2000.

Splawski I, Timothy KW, Vincent GM, et al., Molecular basis of the long-QT syndrome associated with deafness, *N Engl J Med* 336, 1562, 1997.

Spring DA and Thomsen JH, Severe atherosclerosis in the single coronary artery: report of a previously undescribed pattern, *Am J Cardiol* 31, 662, 1973.

Staemmler M, Hertz, in *Lehrbuch der speziellen pathologische Anatomie*, Kaufmann E, Ed., Walter deGruyter, Berlin, 1961, p. 3.

Stahl J, Santos LD, and Byard RW, Coronary artery thromboembolism and unexpected death in childhood and adolescence, *J Forensic Sci* 40, 599, 1995.

Stahl RD, Liu JC, and Walsh JF, Blunt cardiac trauma: atrioventricular valve disruption and ventricular septal defect, *Ann Thorac Surg*, 64, 1466, 1997.

Stanley CA, Carnitine disorders, *Adv Pediatr* 42, 209, 1995.

Stanton MS, Tuli MM, Radtke NL, et al., Regional sympathetic denervation after myocardial infarction in humans detected noninvasively using I-123-metaiodobenzyeguanidine, *J Am Coll Cardiol* 14, 1519, 1989.

Steinberger J, Lucas RV, Edwards JE, and Titus JL, Causes of sudden unexpected cardiac death in the first two decades of life, *Am J Cardiol* 77, 992, 1996.

Steiner I, Nonbacterial thrombotic versus infective endocarditis: a necropsy study of 320 cases, *Cardiovasc Pathol* 4, 207, 1995.

Stenberg RG, Winniford MD, Hillis LD, et al., Simultaneous acute thrombosis of two major coronary arteries following intravenous cocaine use, *Arch Pathol Lab Med* 113, 521, 1989.

Stéphan E, de Meeus A, Bouvagnet P, et al., Hereditary bundle branch defect: right bundle branch blocks of different causes have different morphologic characteristics, *Am Heart J* 133, 249, 1997.

Stetler-Stevenson WG, Dynamics of matrix turnover during pathologic remodeling of the extracellular matrix, *Am J Pathol* 148, 1345, 1996.

Stevensen MJ, Raffael DM, Allman KC, et al., Cardiac sympathetic dysinnervation in diabetes: implication for enhanced cardiovascular risk, *Circulation* 98, 961, 1998.

Stewart JR, Paton BC, Blount SG Jr, and Swan H, Congenital aortic stenosis: ten to 22 years after valvulotomy, *Arch Surg* 113, 1248, 1978.

Still WJS, The pathogenesis of the intimal thickenings produced by hypertension in large arteries in the rat, *Lab Invest* 19, 84, 1968.

Stramba-Badiale M, Priori SG, Napolitano C, et al., Gene-specific differences in the circadian variation of ventricular repolarization in the long QT syndrome: a key to sudden death during sleep? *Ital Heart J* 1, 323, 2000.

Strauss AW and Johnson MC, The genetic basis of pediatric cardiovascular disease, *Semin Perinatol* 20, 564, 1996.

Strauss AW, Powell CK, Hale DE, et al., Molecular basis of human mitochondrial very-long-chain acyl-CoA dehydrogenase deficiency causing cardiomyopathy and sudden death in childhood, *Proc Natl Acad Sci USA* 92, 10496, 1995.

Strom EH, Skjorten F, and Stokke ES, Polycystic tumor of the atrioventricular nodal region in a man with Emery-Dreifuss muscular dystrophy, *Pathol Res Pract* 189, 960, 1993.

Stryker WA, Arterial calcification in infancy with special reference to the coronary arteries, *Am J Pathol* 22, 1007, 1946.

Sturtz CL, Abt AB, Leuenberger UA, and Damiano R, Hamartoma of mature cardiac myocytes: a case report, *Mod Pathol* 11, 496, 1998.

Su J, Li J, Li W, Altura B, and Altura B, Cocaine induces apoptosis in primary cultured rat aortic vascular smooth muscle cells: possible relationship to aortic dissection, atherosclerosis, and hypertension, *Int J Toxicol* 23, 233, 2004.

Suarez RV and Riemersma R, "Ecstasy" and sudden cardiac death, *Am J Forensic Med Pathol* 9, 339, 1988.

Suarez-Mier MP, Fernandez-Simon L, Gawallo C, et al., Pathologic changes of the cardiac conduction tissue in sudden cardiac death, *Am J Forensic Med Pathol* 16, 193, 1995.

Suarez-Mier MP and Gamallo C, Atrioventricular node fetal dispersion and His bundle fragmentation of the cardiac conduction system in sudden cardiac death, *J Am Coll Cardiol* 32, 1885, 1998.

Suarez-Mier MP, Sanchez-de-Leon S, and Cohle SD, An unusual site for the AV node tumor: report of two cases, *Cardiovasc Pathol* 8, 325, 1999.

Suarez-Mier PM and Aguilera B, Histopathology of the conduction system in sudden infant death, *Forensic Sci Int* 93, 143–154, 1998.

Sugai M, Kono R, and Kunita Y, A morphologic study on human conduction system of heart considering influences of some disorders of individuals, *Acta Pathol Jpn* 31, 13, 1981.

Sugiura M, Ohkawa S, Hiraoka K, et al., A clinicopathological study on the sick sinus syndrome, *Jpn Heart J* 17, 731, 1976.

Sugiura M, Ohkawa S, Watanabe C, et al., A clinicopathologic study of the accessory bypass tracts in six cases of Wolff-Parkinson-White syndrome, *Jpn Heart J* 30, 313, 1989.

Sullivan ML, Martinez C, and Gallagher J, Atrial fibrillation and anabolic steroids, *J Emerg Med* 17, 851, 1999.

Sullivan ML, Martinez CM, Gennis P, and Gallagher EJ, The cardiac toxicity of anabolic steroids, *Prog Cardiovasc Dis* 41, 1, 1998.

Suma H, Gastroepiploic artery graft: coronary artery bypass graft in patients with diseased ascending aorta using an aortic no-touch technique, in *Operative Techniques in Cardiac and Thoracic Surgery: A Comparative Atlas*, Vol. 1, No. 2, Cox JL and Sundt TM, Eds., 1996, p. 185.

Sun CC, Jacot J, and Brenner JI, Sudden death in supravalvular aortic stenosis: fusion of a coronary leaflet to the sinus ridge, dysplasia and stenosis of aortic and pulmonic valves, *Pediatr Pathol* 12, 751, 1992.

Sun SC, Burch GE, and De Pasquale NP, Histochemical and electron microscopy study of heart muscle after beta-adrenergic blockade, *Am Heart J* 74, 340, 1967.

Surawicz B, Ventricular fibrillation, *J Am Coll Cardiol* 5 (Suppl 6), 42B, 1985.

Suzuki H, Torigoe K, Numata O, et al., Infant case with a malignant form of Brugada syndrome, *J Cardiovasc Electrophysiol* 11, 1277, 2000.

Swalwell CI, Benign intracardiac teratoma: a case of sudden death, *Arch Pathol Lab Med* 117, 739, 1993.

Swalwell CI, Reddy SK, and Rao VJ, Sudden death due to unsuspected coronary vasculitis, *Am J Forensic Pathol* 12, 306, 1991.

Swan H, Piippo K, Viitasalo M, et al., Arrhythmic disorder mapped to chromosome 1q42-q43 causes malignant polymorphic ventricular tachycardia in structurally normal hearts, *J Am Coll Cardiol* 34, 2035, 1999.

Sweeney MG, Bundey S, Brockington M, et al., Mitochondrial myopathy associated with sudden death in young adults and a novel mutation in the mitochondrial DNA leucine transfer RNA(UUR) gene, *Q J Med* 86, 709, 1993.

Sybert VP, Cardiovascular malformations and complications in Turner syndrome, *Pediatrics* 101, E11, 1998.

Sybrandy KC, Cramer MJM, and Burgersdijk C, Diagnosing cardiac contusion: old wisdom and new insights, *Heart,* 89, 485, 2003.

Symbas PN, *Cardiothoracic Trauma,* WB Saunders, Philadelphia, 1989.

Szakacs JE and Cannon A, L-Norepinephrine myocarditis, *Am J Clin Pathol* 30(5), 425, 1958.

Szakacs JE, Dimmette RM, and Gowart EC, Pathologic implication of the catecholamines, epinephrine and norepinephrine, *U.S. Armed Forces Med J* 10, 908, 1959.

Tada H, Aihara N, Ohe T, et al., Arrhythmogenic right ventricular cardiomyopathy underlies syndrome of right bundle branch block, ST-segment elevation, and sudden death, *Am J Cardiol* 81, 5, 1998.

Takach TJ, Reul GJ, Ott DA, and Cooley DA, Primary cardiac tumors in infants and children: immediate and long-term operative results, *Ann Thorac Surg* 62, 559, 1996.

Takano H, Nakamua T, Satou T, et al., Regional myocardial sympathetic denervation in patients with coronary spasm, *Am J Cardiol* 75, 324, 1997.

Taki J, Yasuda T, Gold HK, et al., Characteristic of transient left ventricular dysfunction detected by ambulatory left ventricular function monitoring device in patients with coronary artery disease (abstract), *Circulation* 76 (Suppl 4), 366, 1987.

Taylor AJ, Rogan KM, and Virmani R, Sudden cardiac death associated with isolated congenital coronary artery anomalies, *J Am Coll Cardiol* 20, 640, 1992a.

Taylor AL, Murphree S, Buja ML, et al., Segmental systolic response to brief ischemia and reperfusion in the hypertrophied canine left ventricle, *J Am Coll Cardiol* 20, 994, 1992b.

Taylor D, Parish D, Thompson L, and Cavaliere M, Cocaine induced prolongation of the QT interval, *Emerg Med J* 21, 252, 2004.

Taylor DA, Hruban R, Rodriguez R, and Goldschmidt-Clermont PJ, Cardiac chimerism as a mechanism for self-repair, *Circulation* 106, 2, 2002.

Tazelaar HD and Billingham ME, Myocardial lymphocytes: fact, fancy, or myocarditis? *Am J Cardiovasc Pathol* 1, 47, 1987.

Tazelaar HD, Locke TJ, and McGregor CG, Pathology of surgically excised primary cardiac tumors, *Mayo Clin Proc* 67, 957, 1992.

Teare D, Asymmetrical hypertrophy of the heart in young adults, *Br Heart J* 20, 1, 1958.

Tenzer ML, The spectrum of myocardial contusion: a review, *J Trauma*, 25, 620, 1985.

Tesson F, Richard P, Charron P, et al., Genotype-phenotype analysis in four families with mutations in the beta-myosin heavy chain gene responsible for familial hypertrophic cardiomyopathy, *Hum Mutat* 12, 385, 1998.

Therrien J, Siu SC, Harris L, et al., Impact of pulmonary valve replacement on arrhythmia propensity late after repair of tetralogy of Fallot, *Circulation* 103, 2489, 2001.

Thiblin I, Eksborg S, Petersson A, Fugelstad A, and Rajs J, Fatal intoxication as a consequence of intranasal administration (snorting) or pulmonary inhalation (smoking) of heroin, *Forensic Sci Int* 139, 241, 2004.

Thiblin I, Lindquist O, and Rajs J, Cause and manner of death among users of anabolic androgenic steroids, *J Forensic Sci* 45, 16, 2000.

Thiene G and Basso C, Arrhythmogenic right ventricular cardiomyopathy: an update, *Cardiovasc Pathol* 10, 109, 2001.

Thiene G, Basso C, Calabrese F, et al., Pathology and pathogenesis of arrhythmogenic right ventricular cardiomyopathy, *Herz* 25, 210, 2000.

Thiene G, Basso C, and Corrado D, Cardiovascular causes of sudden death, in *Cardiovascular Pathology*, 3rd ed., Silver MD, Gotlieb AI, and Schoen FJ, Eds., Churchill Livingstone, Philadelphia, 2001, p. 326.

Thiene G, Gambino A, Corrado D, et al., The pathological spectrum underlying sudden death in athletes, *New Trends in Arrhythmias* 3, 323, 1985.

Thiene G and Ho SY, Aortic root pathology and sudden death in youth: review of anatomical varieties, *Appl Pathol* 4, 237, 1986.

Thiene G, Nava A, Corrado D, et al., Right ventricular cardiomyopathy and sudden death in young people, *N Engl J Med* 318, 129, 1988.

Thiene G, Pennelli N, and Rossi L, Cardiac conduction system abnormalities as a possible cause of sudden death in young athletes, *Hum Pathol* 14, 704, 1983.

Thomas JL, Dickstein RA, Parker FB Jr, et al., Prognostic significance of the development of left bundle conduction defects following aortic valve replacement, *J Thorac Cardiovasc Surg* 84, 382, 1982.

Thomsen H and Held H, Immunohistochemical detection of C5b-9 in myocardium: an aid in distinguishing infarction-induced ischemic heart muscle necrosis from other form of lethal myocardium injury, *Forensic Sci Int* 71, 87, 1995.

Thomsen H and Saternus K-S Myokardnekrosen bei plötzlichem und unerwarteten Säuglingstod (SIDS)? — Eine Untersuchung mit polyclonalen Antikörpern gegen C5b-9$_{(m)}$— Komlement-Komplex, *Rechtsmed* 5, 6–9, 1994.

Thomson JG, Production of severe atheroma in a transplanted human heart, *Lancet* 2, 1088, 1969.

Thorgeirsson G and Liebman J, Mesothelioma of the AV node, *Pediatr Cardiol* 4, 219, 1983.

Thornback P and Fowler RS, Sudden unexpected death in children with congenital heart disease, *Can Med Assoc J*, 113, 745, 1975.

Thourani VH, Feliciano DV, Cooper WA, et al., Penetrating cardiac trauma at an urban trauma center: a 22-year perspective, *Am Surg*, 65, 811, 1999.

Thygesen K and Alpert JS, Myocardial infarction redefined: a consensus document of the joint European Society of Cardiology/American College of Cardiology committee for the redefinition of myocardial infarction, *J Am Coll Cardiol* 36, 959, 2000.

Tillmanns H, Ikeda S, and Hansen H, Microcirculation in the ventricle of the dog and turtle, *Circ Res* 34, 561, 1974.

Timmermans C, Smeets JL, Rodriguez LM, et al., Aborted sudden death in Wolff-Parkinson-White syndrome, *Am J Cardiol* 76, 492, 1995.

Tobis JM, Mallery J, Mahon D, et al., Intravascular ultrasound assessment of lumen size and wall morphology in normal subjects and patients with coronary artery disease, *Circulation* 84, 1087, 1991.

Todd GL, Baroldi G, Pieper GM, Clayton F, and Eliot RS, Experimental catecholamine-induced myocardial necrosis. I. Morphology, quantification and regional distribution of acute contraction band lesions, *J Mol Cell Cardiol* 17, 317, 1985a.

Todd GL, Baroldi G, Pieper GM, Clayton F, and Eliot RS, Experimental catecholamine-induced myocardial necrosis. II. Temporal development of isoproternol-induced contraction band lesions correlated with ECG, hemodynamic and biochemical changes, *J Mol Cell Cardiol* 17, 647, 1985b.

Togna G, Tempesta E, Togna AR, Dolci N, Cebo B, and Caprino L. Platelet responsiveness and biosynthesis of thromboxane and prostacyclin in response to *in vitro* cocaine treatment, *Haemostasis* 15, 100, 1985.

Toma C, Pittenger MF, Cahill KS, et al., Human mesenchymal stem cells differentiate to a cardiomyocyte phenotype in the adult murine heart, *Circulation* 105, 93, 2002.

Tomoike H, Ross Jr J, Franklin D, et al., Improvement by propranolol of regional myocardial dysfunction and abnormal coronary flow pattern in conscious dogs with coronary narrowing, *Am J Cardiol* 41, 689, 1978.

Toole JC and Silverman ME, Pericarditis of acute myocardial infarction, *Chest* 67, 647, 1975.

Topaz O and Edwards JE, Pathologic features of sudden death in children, adolescents, and young adults, *Chest* 4, 476, 1985.

Topol EJ, A guide to therapeutic decision-making in patients with non-ST segment elevation acute coronary syndromes, *J Am Coll Cardiol* 41 (Suppl S), 123S, 2003.

Topol EJ and Yadav JS, Recognition of the importance of embolization in atherosclerotic vascular disease, *Circulation* 101, 570, 2000.

Toro-Salazar OH, Steinberger J, Thomas W, et al., Long-term follow-up of patients after coarctation of the aorta repair, *Am J Cardiol* 89, 541, 2002.

Torres V, Tepper D, Flowers D, et al., QT prolongation and the antiarrhythmic efficacy of amiodarone, *J Am Coll Cardiol* 7, 42, 1986.

Tow A, Cor biloculare with truncus arteriosus and endocarditis, *Am J Dis Child* 42, 1413, 1931.

Towbin JA, Molecular genetic basis of sudden cardiac death, *Cardiovasc Pathol* 10, 283, 2000.

Towbin JA and Bowles NE, Molecular genetics of left ventricular dysfunction, *Curr Mol Med* 1, 81, 2001.

Toyama M, Amano A, and Kameda T, Familial aortic dissection: a report of rare family cluster, *Br Heart J* 61, 204, 1989.

Tranebjaerg L, Bathen J, Tyson J, et al., Jervall and Lange-Nielson syndrome: a Norwegian perspective, *Am J Med Genet* 89, 137, 1999.

Tremouroux J, Brasseur L, Meersseman F, and Lavenne F, Infarctus myocardique auriculaire et ventriculaire dans un cas de coronaire unique, *Acta Cardiol* 14, 524, 1959.

Trinchero R, Demarie D, Orzan F, et al., Fixed subaortic stenosis: natural history of patients with mild obstruction and follow-up of operated patients, *J Ital Cardiol* 18, 738, 1988.

Trotter SE and Olsen EG, Marfan's disease and Erdheim's cystic medionecrosis: a study of their pathology, *Eur Heart J* 12, 83, 1991.

Truex RC and Angulo AW, Comparative study on the arterial and venous system of the ventricular myocardium with special reference to the coronary sinus, *Anat Rec* 113, 467, 1952.

Tsang VT, Pawade A, Karl TR, and Mee RB, Surgical management of Marfan syndrome in children, *J Cardiovasc Surg* 9, 50, 1994.

Tsao JW, Marder SR, Goldstone J, and Bloom AI, Presentation, diagnosis, and management of arterial mycotic pseudoaneurysms in injection drug users, *Ann Vasc Surg* 16, 652, 2002.

Tse HF, Shek TW, Tai YT, et al., Case report: lysosomal glycogen storage disease with normal acid maltase: an unusual form of hypertrophic cardiomyopathy with rapidly progressive heart failure, *Am J Med Sci* 312, 182, 1996.

Tsung SH, Huang TY, and Chang HH, Sudden death in young athletes, *Arch Pathol Lab Med* 106, 168, 1982.

Tuncel M, Wang Z, Arbique D, Fadel PJ, Victor RJ, and Vongpatanasin W. Mechanism of the blood-pressure-raising effect on cocaine in humans, *Circulation* 105, 1054, 2002.

Tunstall Ph. Kuulasmaa K, Amouyel P, Arveiler D, Rajakangas AM, and Pajak A, Myocardial infarction and coronary deaths in the World Health Organization MONICA Project: registration procedures, event rates, and case-fatality rates in 38 populations from 21 countries in four continents, *Circulation* 90, 583, 1994.

Tunwell RE, Wickenden C, Bertrand BM, et al., The human cardiac muscle ryanodine receptor-calcium release channel: identification, primary structure and topological analysis, *Biochem J* 318, 477, 1996.

Turillazzi E, Di Donato S, and Fineschi V, Selective penetrating injury of the right coronary artery: a fatal case, *Cardiovasc Pathol* 14, 42, 2005.

Turley K, Bove EL, Amato JJ, et al., Neonatal aortic stenosis, *J Thorac Cardiovasc Surg* 99, 679, 1990.

Turner DD and Sommers SC, Accidental passage of a polyethylene catheter from cubital vein to right atrium, *N Engl J Med* 251, 744, 1954.

Tuzcu EM, Moodie DS, Ghazi F, et al., Ebstein's anomaly: natural and unnatural history, *Cleve Clin J Med* 56, 614, 1989.

Tyni T and Pihko H, Long-chain 3-hydroxyacyl-CoA dehydrogenase deficiency, *Acta Paediatr* 88, 237, 1999.

Udelson JE, Coleman PS, Metherall J, et al., Predicting recovery of severe regional ventricular dysfunction: comparison of resting scintigraphy with [201]TL and ^{99m}Tc-sestamibi, *Circulation* 89, 2552, 1994.

Ulgen MS, Biyik I, Karadede A, et al., Relation between QT dispersion and ventricular arrhythmias in uncomplicated isolated mitral valve prolapse, *Jpn Circ J* 63, 929, 1999.

Unger EF, Sheffield CD, and Epstein SE, Creation of anastomoses between an extracardiac artery and the coronary circulation, *Circulation* 82, 1449, 1990.

Ungerer M, Hartmann F, Karoglan, et al., Regional *in vivo* and *in vitro* characterization of autonomic innervation in cardiomyopathic human heart, *Circulation* 97, 174, 1998.

Unnikrishnan D, Dutcher JP, Varshneya N, et al., Torsades de pointes in 3 patients with leukemia treated with arsenic trioxide, *Blood* 97, 1514, 2001.

Urbach J, Glaser J, Balkin J, et al., Familial membranous subaortic stenosis, *Cardiology* 72, 214, 1985.

Urhausen A, Albers T, and Kindermann W, Are the cardiac effects of anabolic steroid abuse in strength athletes reversible? *Heart* 90, 496, 2004.

Vaideeswar P, Shankar V, Deshpande JR, et al., Pathology of the diffuse variant of supravalvular aortic stenosis, *Cardiovasc Pathol* 9, 33, 2001.

Vakeva AP, Azah A, Rollins SA, et al., Myocardial infarction and apoptosis after myocardial ischemia and reperfusion: role of the terminal complements and inhibition by anti-C5 therapy, *Circulation* 97, 2259, 1998.

Valdes-Dapena M, A pathologist's perspective on the sudden infant death syndrome — 1991, *Pathol Ann* 27, 133–164, 1992.

Valdes-Dapena M, Gillane MM, and Catherman R, The question of right ventricular hypertrophy in sudden infant death syndrome, *Arch Pathol Lab Med* 104, 184–186, 1980.

Valdes-Dapena M, Greene M, Basavanand N, Catherman R, and Truex RC, The myocardial conduction system in sudden death in infancy, *N Eng J Med* 289, 1179–1180, 1973.

Valdes-Dapena M, Sudden infant death syndrome — a review of the medical literature 1974–1979, *Pediatrics* 66, 597–614, 1980.

Valdes-Dapena M, The morphology of the sudden infant death syndrome: an overview, in *Sudden Infant Death Syndrome*, Tildon JT, Roeder LM, and Steinschneider A, Eds., Academic Press, New York, 1983, 169–182.

Valente M, Basso C, Thiene G, et al., Fibroelastic papilloma: a not-so-benign cardiac tumor, *Cardiovasc Pathol* 1, 161, 1992.

Valente M, Calabrese F, Thiene G, et al., *In vivo* evidence of apoptosis in arrhythmogenic right ventricular cardiomyopathy, *Am J Pathol* 152, 479, 1998.

Val-Mejias J, Lee WK, Weisse AB, et al., Left ventricular performance during and after sickle cell crisis, *Am Heart J* 97, 585, 1974.

Van Belle E, Lablanche JM, Banters C, et al., Coronary angioscopic findings in the infarct-related vessels within 1 month of acute myocardial infarction: natural history and effect of thrombolysis, *Circulation* 97, 26, 1998.

Van Camp SP, Bloor CM, Mueller FO, Cantu RC, and Olsen HG, Nontraumatic sports deaths in high school and college athletes, *Med Sci Sports Exerc* 27, 641, 1995.

van der Bel-Kahn J, Duren DR, and Becker AE, Isolated mitral valve prolapse: chordal architecture as an anatomic basis in older patients, *J Am Coll Cardiol* 5, 1335, 1985.

Van der Hauwaert LG, Fryns JP, Dumoulin M, and Logghe N, Cardiovascular malformations in Turner's and Noonan's syndrome, *Br Heart J* 40, 500, 1978.

van der Kooi AJ, Ledderhof TM, de Voogt WG, et al., A newly recognized autosomal dominant limb girdle muscular dystrophy with cardiac involvement, *Ann Neurol* 39, 636, 1996.

van der Wal AC, Becker AE, Koch KT, et al., Clinically stable angina pectoris is not necessarily associated with histologically stable atherosclerotic plaques, *Heart* 76, 312, 1996.

van der Wal AC, Das PK, van de Berg DB, et al., Atherosclerotic lesions in humans: *in situ* immunophenotypic analysis suggesting an immune mediated response, *Lab Invest* 61, 166, 1989.

Van Hare GF, Lesh MD, Ross BA, et al., Mapping and radiofrequency ablation of intraatrial reentrant tachycardia after the Senning or Mustard procedure for transposition of the great arteries, *Am J Cardiol* 77, 985, 1996.

van Son JA, Edwards WD, and Danielson GK, Pathology of coronary arteries, myocardium, and great arteries in supravalvular aortic stenosis: report of five cases with implications for surgical treatment, *J Thorac Cardiovasc Surg* 108, 21, 1994.

Vanderwee MA, Humphrey SM, Gavin JB, et al., Changes in the contractile state: fine structure and metabolism of cardiac muscle cells during the development of rigor mortis, *Virchows Arch* (Cell Pathol) 35, 159, 1981.

Vanoverschelde J, Wijns W, Depré C, et al., Mechanisms of chronic regional postischemic dysfunction in humans: new insights from the study of noninfarcted collateral-dependent myocardium, *Circulation* 87, 1513, 1993.

Varnava AM, Elliott PM, Baboonian C, et al., Hypertrophic cardiomyopathy: histopathological features of sudden death in cardiac troponin T disease, *Circulation* 104, 1380, 2001.

Vatta M, Dumaine R, Varghese G, et al., Genetic and biophysical basis of sudden unexplained nocturnal death syndrome (SUNDS), a disease allelic to Brugada syndrome, *Hum Mol Genet* 11, 337, 2002.

Vaughan CJ, Gotto AM, and Basson GT, The evolving role of statins in the management of atherosclerosis, *J Am Coll Cardiol* 35, 1, 1999.

Vaughan CJ, Veugelers M, and Basson CT, Tumors and the heart: molecular genetic advances, *Curr Opin Cardiol* 16, 195, 2001.

Veinot JP, Johnston B, Acharya V, and Healey J, The spectrum of intramyocardial small vessel disease associated with sudden death, *J Forensic Sci* 47, 384, 2002.

Velasquez EM, Anand RC, Newman WP III, Richard SS, and Glancy DL, Cardiovascular complications associated with cocaine use, *J La State Med Soc* 156, 302, 2004.

Veldkamp MW, Viswanathan PC, Bezzina C, Bartscheer A, Wilde AA, and Balser JR, Two distinct congenital arrhythmias evoked by a multidysfunctional Na $^{(+)}$ channel, *Circ Res* 86, E91–97, 2000.

Velican C and Velican D, *Natural History of Coronary Atherosclerosis*, CRC Press, Boca Raton, FL, 1989.

Verloes A, Massin M, Lombet J, et al., Nosology of lysosomal glycogen storage diseases without *in vitro* acid maltase deficiency: delineation of a neonatal form, *Am J Med Genet* 72, 135, 1997.

Vesterby A, Bjerregaard P, Gregersen M, et al., Sudden death in mitral valve prolapse: associated accessory atrioventricular pathways, *Forensic Sci Int* 19, 125, 1982.

Viano DC, Andrzejak DV, and King AI, Fatal chest injury by baseball impact in children, *Clin J Sport Med*, 2, 161, 1992.

Vidaillet HJ Jr, Pressley JC, Henke E, et al., Familial occurrence of accessory atrioventricular pathways (preexcitation syndrome), *N Engl J Med* 317, 65, 1987.

Vienot JP, Ghadially FN, and Walley VM, Light microscopy and ultrastructure of the blood vessels and heart, in *Cardiovascular Pathology*, 3rd ed., Silver MD, Gotlieb AI, and Schoen FJ, Eds., Churchill Livingstone, Philadelphia, 2001, p. 30.

Vignola PA, Aonuma K, Swaye PS, et al., Lymphocytic myocarditis presenting as unexplained ventricular arrhythmias: diagnosis with endomyocardial biopsy and response to immunosuppression, *J Am Coll Cardiol* 4, 812, 1984.

Villain E, Levy M, Kachaner J, and Garson A Jr, Prolonged QT interval in neonates: benign, transient, or prolonged risk of sudden death, *Am Heart J* 124, 194, 1992.

Villain E, Vetter VL, Garcia JM, et al., Evolving concepts in the management of congenital junctional ectopic tachycardia: a multicenter study, *Circulation* 81, 1544, 1990.

Villota JN, Rubio LF, Fores JS, et al., Cocaine-induced coronary thrombosis and acute myocardial infarction, *Int J Cardiol* 96, 481, 2004.

Vincent GM, The molecular genetics of the long QT syndrome: genes causing fainting and sudden death, *Annu Rev Med* 49, 263, 1998.

Vincent GM and Zhang L, The role of genotyping in diagnosing cardiac channelopathies: progress to date, *Mol Diagn* 9, 105, 2005.

Vincent MG, Anderson JL, and Marshall HW, Coronary spasm producing coronary thrombosis and myocardial infarction, *N Engl J Med* 309, 220, 1983.

Vink A, Schoneveld AH, and Richard W, Plaque burden, arterial remodelling and plaque vulnerability: determined by systemic factors? *J Am Coll Cardiol* 38, 718, 2001.

Violette EJ, Hardin NJ, and McQuillen EN, Sudden unexpected death due to asymptomatic cardiac rhabdomyoma, *J Forensic Sci* 26, 599, 1981.

Virchow R. *Phlogose und Thrombose in Gefasse-system: Gesammelte Abhandlungen zur wissenschaftlichen Medicine*, Frankfurt, 1856.

Virmani N, Atkinson JB, and Forman MB, Aortocoronary bypass grafts and extracardiac conduits, in *Cardiovascular Pathology*, 2nd ed., Silver MD, Ed., Churchill Livingstone, New York, 1991, p. 1607.

Virmani R, Atkinson JB, Byrd BF III, et al., Abnormal chordal insertion: a couse of mital valve prolapse, *Am Heart J* 113, 851, 1987.

Virmani R and Burke AP, Nonatherosclerotic diseases of the aorta and miscellaneous diseases of the main pulmonary arteries and large veins, in *Cardiovascular Pathology*, 3rd ed., Gotlieb AI, Schoen FJ, and Silver MD, Eds., Churchill Livingstone, 2001.

Virmani R, Burke AP, and Farb A, Sudden cardiac death, *Cardiovasc Pathol* 10, 211, 2001.

Virmani R, Farb A, and Burke A, Contraction band necrosis: use for an old man, *Lancet* 347, 1710, 1996.

Virmani R, Forman MB, Rabinowitz M, and McAllister HA, Coronary artery dissections, *Cardiol Clin* 2, 633, 1984.

Virmani R, Robinowitz M, and McAllister HA Jr, Nontraumatic death in joggers: a series of 30 patients at autopsy, *Am J Med* 72, 874, 1982.

Virmani R, Robinowitz M, Smialek JE, and Smyth DF, Cardiovascular effects of cocaine: an autopsy study of 40 patients, *Am Heart J* 115, 1068, 1988.

Viskin S, Alla SR, Barron HV, et al., Mode of onset of torsades de pointes in congenital long QT syndrome, *J Am Coll Cardiol* 28, 1262, 1996.

Viskin S and Belhassen B, Polymorphic ventricular tachyarrhythmias in the absence of organic heart disease: classification, differential diagnosis, and implications for therapy, *Prog Cardiovasc Dis* 41, 17, 1998.

Viskin S, Fish R, Eldar M, et al., Prevalence of the Brugada sign in idiopathic ventricular fibrillation and healthy controls, *Heart* 84, 31, 2000.

Vitali-Mazza L, Anversa P, Tedeschi F, et al., Ultrastructural basis of acute left ventricular failure from severe acute aortic stenosis in rabbits, *J Mol Cell Cardiol* 4, 661, 1972.

Vlay SC, Blumenthal DS, Shoback D, et al., Delayed acute myocardial infarction after blunt chest trauma in a young woman, *Am Heart J* 100, 907, 1980.

Vlay SC, Vlay LC, and Coyle PK, Combined cardiomyopathy and skeletal myopathy: a variant with atrial fibrillation and ventricular tachycardia, *Pacing Clin Electrophysiol* 24, 1389, 2001.

Vlodaver Z and Edwards JE, Pathology of coronary atherosclerosis, *Progr Cardiovasc Dis* 14, 256, 1971.

Vlodaver Z and Edwards JE, Rupture of ventricular septum or papillary muscle complicating myocardial infarction, *Circulation* 55, 815, 1977.

Vlodaver Z, Kahn HA, and Neufeld HN, The coronary arteries in early life in three different ethnic groups, *Circulation* 39, 541, 1969.

Vlodaver Z and Neufeld HN, The musculo-elastic layer in the coronary arteries: a histological and hemodynamic concept, *Vasc Dis* 4, 136, 1967.

Vlodaver Z and Neufeld HN, The coronary arteries in coarctation of the aorta, *Circulation* 37, 449, 1968.

Vlodaver Z, Neufeld HN, and Edwards JE, Pathology of angina pectoris, *Circulation* 46, 1048, 1972.

Vockley J, The changing face of disorders of fatty acid oxidation, *Mayo Clin Proc* 69, 249, 1994.

Vockley J and Whiteman DA, Defects of mitochondrial beta-oxidation: a growing group of disorders, *Neuromuscul Disord* 12, 235, 2002.

Vogt BA, Birk PE, Panzarino V, et al., Aortic dissection in young patients with chronic hypertension, *Am J Kidney Dis* 33, 374, 1999.

Voigt J and Agdal N, Lipomatous infiltration of the heart, *Arch Pathol Lab Med* 106, 497, 1982.

von Bernuth G, Bernsau U, Gutheil H, et al., Tachyarrhythmic syncopes in children with structurally normal hearts with and without QT-prolongation in the electrocardiogram, *Eur J Pediatr* 138, 206, 1982.

von Oppell UO, Bautz P, and De Groot M, Penetrating thoracic injuries: what we have learnt, *Thorac Cardiovasc Surg*, 48, 55, 2000.

Von Sohsten R, Kopistansky C, Cohen M, et al., Cardiac tamponade in the new device era: evaluation of 69 consecutive percutaneous coronary interventions, *Am Heart J*, 140, 279, 2000.

von zur Muhlen F, Klass C, Kreuzer H, et al., Cardiac involvement in proximal myotonic myopathy, *Heart* 79, 619, 1998.

Vongpatanasin W, Brickner ME, Hillis LD, et al., The Eisenmenger syndrome in adults, *Ann Intern Med* 128, 745, 1998.

Vongpatanasin W, Mansour Y, Chavoshan B, et al., Cocaine stimulates the human cardiovascular system via a central mechanism of action, *Circulation* 100, 497, 1999.

Vongpatanasin W, Taylor JA, and Victor GR, Effects of cocaine on heart rate variability in healthy subjects, *Am J Cardiol*, 93, 385, 2004.

Wade WG, The pathogenesis of infarction of the right ventricle, *Br Heart J* 21, 545, 1959.

Waller BF, Pathology of new cardiovascular interventional procedures, in *Cardiovascular Pathology*, 2nd ed., Silver MD, Ed., Churchill Livingstone, New York, 1991, 1683.

Waller BF, Carter JB, William HJ Jr, et al., Bicuspid aortic valve: comparison of congenital and acquired types, *Circulation* 48, 1140, 1973.

Waller BF, Gorfinkel J, Rogers FJ, et al., Early and late morphologic changes in major epicardial coronary arteries after percutaneous transluminal coronary angioplasty, *Am J Cardiol* 53, 426, 1984.

Waller BF and Roberts WC, Sudden death while running in conditioned runners aged 40 years or over, *Am J Cardiol* 45, 1292, 1980.

Waller BF, Rothbaum DA, Pinkerton CA, et al., Status of myocardium and infarcted-related artery in 19 necropsy patients with acute recanalization using pharmacologic (streptokinase, r-tissue plasminogen activator) mechanical (percutaneous transluminal coronary angioplasty) or combined types of reperfusion therapy, *J Am Coll Cardiol* 9, 785, 1987.

Walley VM, Antecol DH, Kyrollos AG, et al., Congenitally bicuspid aortic valves: study of a variant with fenestrated raphe, *Can J Cardiol* 10, 535, 1994.

Walley VM, Virmani R, and Silver MD, Pulmonary arterial dissection and ruptures: to be considered in patients with pulmonary arterial hypertension presenting with cardiogenic shock or sudden death, *Pathology* 22, 1, 1990.

Walsh CK and Krongrad E, Terminal cardiac electrical activity in pediatric patients, *Am J Cardiol* 51, 557, 1983.

Wang J, Geng YJ, and Guo B, Near-infrared spectroscopic characterization of human advanced atherosclerotic plaque, *J Am Coll Cardiol* 39, 1305, 2002a.

Wang JF, Zhang J, Min JY, et al., Cocaine enhances myocarditis induced by encephalomyocarditis virus in murine model, *Am J Physiol Heart Circ Physiol* 282, 956, 2002b.

Wang Q, Curran ME, Splawski I, Burn TC, Millholland JM, van Raay TJ, Shen J, Thimothy KW, Vincent GM, de Jager T, Schwartz PJ, Towbin JA, Moss AJ, Atkinson DL, Landes GM, Connos TD, and Keating MT, Positional cloning of a novel potassium channel gene: KVLQT1mutations cause arrhythmias, *Nat Genet* 12, 17–23, 1996.

Wang Q, Li Z, Shen J, et al., Genomic organization of the human *SCN5A* gene encoding the cardiac sodium channel, *Genomics* 34, 9, 1996.

Wang Q, Shen J, Li Z, Thimothy KW, Vincent GM, Priori SG, Schwartz PJ, and Keating MT, Cardiac sodium channel mutations in patients with long QT syndrome, an inherited cardiac arrhythmia, *Hum Mol Genet* 4, 1603–1607, 1995a.

Wang Q, Shen J, Splawski I, Atkinson D, Li H, Robinson JL, Moss AJ, Towbin JA, and Keating MT, SCN5A mutations associated with an inherited cardiac arrhythmia, long QT syndrome, Cell 80, 805–811, 1995b.

Wang YS, Scheinman MM, Chien WW, et al., Patients with supraventricular tachycardia presenting with aborted sudden death: incidence, mechanism and long-term follow-up, *J Am Coll Cardiol* 18, 1711, 1991.

Wanibuchi H, Veda M, Dingermans KP, et al., The response to percutaneous transluminal coronary angioplasty: an ultrastructural study of smooth muscle cells and endothelial cells, *Cardiovasc Pathol* 1, 295, 1992.

Ward OC, A new familial cardiac syndrome in children, *J Irish Med Assoc* 54, 103–109, 1964.

Ware JA and Heistad DD, Platelet–endothelium interactions, *N Engl J Med* 328, 628, 1993.

Wartman WB and Souders JC, Localization of myocardial infarcts with respect to the muscle bundles of the heart, *Arch Pathol* 50, 329, 1950.

Watkins H, Sudden death in hypertrophic cardiomyopathy, *N Engl J Med* 342, 422, 2000.

Watkins H, McKenna WJ, Thierfelder L, et al., Mutations in the genes for cardiac troponin T and alpha-tropomyosin in hypertrophic cardiomyopathy, *N Engl J Med* 332, 1058, 1995.

Watkins H, Rosenzweig A, Hwang DS, et al., Characteristics and prognostic implications of myosin missense mutations in familial hypertrophic cardiomyopathy, *N Engl J Med* 326, 1108, 1992.

Wearn JT, The extent of the capillary bed of the heart, *J Exp Med* 47, 273, 1928.

Wearn JT, Mettier SR, Klumpp TG, and Zschiesche LJ, The nature of the vascular communications between the coronary arteries and the chambers of the heart, *Am Heart J* 9, 143, 1933.

Webb L, Oyefeso A, Schifano F, Cheta S, Pollard M, and Ghodse AH. Cause and manner of death in drug-related fatality: an analysis of drug-related deaths recorded by coroners in England and Wales in 2000, *Drug and Alcohol Dep* 72, 67, 2003.

Weber KT, Cardiac interstitium in health and disease: the fibrillar collagen network, *J Am Coll Cardiol* 13, 1637, 1989.

Webster MW, Chesebro JH, Smith HC, et al., Myocardial infarction and coronary artery occlusion: a prospective 5-year angiographic study, *J Am Coll Cardiol* 15, 218A, 1990.

Wehrens XH, Vos MA, Doevendans PA, et al., Novel insights in the congenital long QT syndrome, *Ann Intern Med* 137, 981, 2002.

Weinmann W and Bohnert L, Lethal monointoxication by overdosage of MDEA, *Forensic Sci Int* 91, 91, 1998.

Weinstein SI and Steinschneider A, QTc and R-R intervals in victims of SIDS, *Am J Dis Child* 139, 987, 1985.

Weiss DL, Subswicz B, and Rubenstein I, Myocardial lesions of calcium deficiency causing irreversible myocardial failure, *Am J Pathol* 48, 653, 1966.

Weisse AB, Lehan PH, Ettinger PO, et al., The fate of experimentally induced coronary thrombosis, *Am J Cardiol* 23, 229, 1969.

Welch R and Chue P, Antipsychotic agents and QT changes, *J Psychiatry Neurosci* 25, 154, 2000.

Wenger NK and Bauer S, Coronary embolism: review of the literature and presentation of fifteen cases, *Am J Med* 25, 549, 1958.

Werf de F, Cardiac troponins in acute coronary syndromes, *N Engl J Med* 335, 1388, 1996.

Werner B, Wroblewska-Kaluzewska M, Pleskot M, et al., Anomalies of the coronary arteries in children, *Med Sci Monit* 7, 1285, 2001.

Wessler S, Ming SC, Gurewich V, et al., A critical evaluation of thromboangitis obliterans: the case against Buerger's disease, *N Engl J Med* 262, 1149, 1960.

Westaby S, Aortic dissection in Marfan's syndrome, *Ann Thorac Surg* 67, 1861, 1999.

White M, Wiechmann RJ, Roden RL, et al., Cardiac -adrenergic neuroeffector systems in acute myocardial dysfunction related to brain injury: evidence for catecholamine-mediated myocardial damage, *Circulation* 92, 2183, 1995.

Wichter T, Hindrcks G, Lerch H, et al., Regional myocardial sympathetic dysinnervation in arrhythmogenic right ventricular cardiomyopathy: an analysis using I123 meta-iodobenzylguanidine scintigraphy, *Circulation* 89, 667, 1994.

Wichter T, Schafers M, Rhodes CG, et al., Abnormalities of cardiac sympathetic innervation in arrhythmogenic right ventricular cardiomyopathy: quantitative assessment of presynaptic norepinephrine reuptake and postsynaptic -adrenergic receptor density with positron emission tomography, *Circulation* 101, 1552, 2000.

Wight TN, Curwen KD, Litrenta MM, et al., Effect of endothelium on glycosaminoglycan accumulation in injured rabbit aorta, *Am J Pathol* 113, 156, 1983.

Wikstrand J and Kendall M, The role of beta receptor blockade in preventing sudden death, *Eur Heart J* 13 (Suppl D), 111, 1992.

Wikstrand J, Warnold I, Olsson G, et al., Primary prevention with metoprolol in patients with hypertension: mortality results from the MAPHY study, *JAMA* 259, 1976, 1988.

Wilde AAM, Antzelevitch C, Borggrefe M, et al., The Study Group on the Molecular Basis of Arrhythmias of the European Society of Cardiology. Proposed diagnostic criteria for the Brugada syndrome: consensus report, *Circulation* 106, 2514, 2002.

Wilens TE, Biederman J, and Spencer TJ, Case study: adverse effects of smoking marijuana while receiving tricyclic antidepressants, *J Am Acad Child Adolesc Psych* 36, 45, 1997.

William SR, Apoptosis and heart failure, *N Engl J Med* 341, 759, 1999.

Williams A, Vawter G, and Reid L, Increased muscularity of the pulmonary circulation in victims of the sudden infant death syndrome, *Pediatrics* 63, 18–23, 1979.

Williams RB and Emery JL, Endocardial fibrosis in apparently normal infant hearts, *Histopathology* 2, 283–288, 1978.

Willinger M, James LS, and Catz C, Defining the sudden infant death syndrome (SIDS): deliberations of an expert panel convened by the National Institute of Child Health and Human Development, *Pediatr Pathol* 11, 677–684, 1991.

Wilson GJ, The pathology of cardiac pacing, in *Cardiovascular Pathology*, 2nd ed., Silver, MD, Ed., Churchill Livingston, New York, 1991, 1429.

Wilson JB, *An Introduction to Scientific Research*, McGraw-Hill, New York, 1952.

Wilson KW and Hutchins GM, Aortic dissecting aneurysms, *Arch Pathol Lab Med*, 106, 175, 1982.

Wilson LD, Rapid progression of coronary artery disease in the setting of chronic cocaine abuse, *J Emerg Med* 16, 631, 1998.

Wilson RF, Laxson DD, and Christensen BV, Regional differences in sympathetic reinnervation after human orthotopic cardiac transplantation, *Circulation* 88, 165, 1993.

Wilson SK and Hutchins GM, Aortic dissecting aneurysms: causative factors in 204 subjects, *Arch Pathol Lab Med* 106, 175, 1982.

Wimmer PJ, Howes DS, Rumoro DP, and Carbone M, Fatal vascular catastrophe in Ehlers-Danlos syndrome: a case report and review, *J Emerg Med* 14, 25, 1996.

Windorfer A and Sitzmann FC, Acute virus myocarditis in infants and children, *Dtsch Med Wschr* 96, 1177, 1971.

Winter SC and Buist NR, Cardiomyopathy in childhood, mitochondrial dysfunction, and the role of L-carnitine, *Am Heart J* 139, S63, 2000.

Winters G and Schoen FJ, Transplanted human heart, in *Cardiovascular Pathology*, 3rd ed., Silver MD, Gotlieb AI, and Schoen FJ, Eds., Churchill Livingstone, Philadelphia, 2001, p. 756.

Winters GL and Mc Manus BM, Myocarditis, in *Cardiovascular Pathology*, 3rd ed., Silver MD, Gotlieb AI, and Schoen FJ, Eds., Churchill Livingstone, Philadelphia, 2001, p. 256.

Wissler RW, The arterial medial cell, smooth muscle of multifunctional mesenchyme? *Circulation* 36, 1, 1967.

Wissler RW, Current status of regression studies, in *Atherosclerosis Review*, Vol. 3, Paoletti R and Gotto AM Jr, Eds., Raven Press, New York, 1978, p. 213.

Witzleben CL, Idiopathic infantile arterial calcification: a misnomer? *Am J Cardiol* 26, 305, 1970.

Witzler DJ and Kaye MP, Myocardial ultrastructural changes induced by administration of nerve growth factors, *Surgery Forum* 27, 295, 1976.

Wolfe RR, Driscoll DJ, Gersony WM, et al., Arrhythmias in patients with valvar aortic stenosis, valvar pulmonary stenosis, and ventricular septal defect: results of 24-hour ECG monitoring, *Circulation* 87 (2 Suppl), I89, 1993.

Wolinsky H, Effects of estrogen and progestogen treatment on the response of the aorta of male rats to hypertension: morphological studies, *Circul Res* 30, 341, 1972.

Wolkoff K, Ueber die Atherosklerose der Coronarterien des Herzens, *Beitr Pathol Anat* 82, 555, 1929.

Wood KA, Drew BJ, and Scheinman MM, Frequency of disabling symptoms in supraventricular tachycardia, *Am J Cardiol* 79, 145, 1997.

World Health Organization (WHO), Cardiomyopathies, Report of a WHO expert committee, WHO Technical Report Series No. 697, Geneva, 1984.

World Health Organization (WHO), Update of the TF report — *Circulation* 93, 341, 1996, *Heartbeat* No. 2, April 1996.

Wren C, O'Sullivan JJ, and Wright C, Sudden death in children and adolescents, *Heart* 83, 410, 2000.

Wu MH, Hsieh FC, Wang JK, et al., A variant of long QT syndrome manifested as fetal tachycardia and associated with ventricular septal defect, *Heart* 82, 386, 1999.

Yajima M, Numano F, Park YB, and Sagar S, Comparative studies of patients with Takayasu arteritis in Japan, Korea and India — comparison of clinical manifestations, angiography and HLA-B antigen, *Jpn Circ J* 58, 9, 1994.

Yamagashi M, Miyatake K, Tamai J, et al., Intravascular ultrasound detection of atherosclerosis at the site of focal vasospasm in angiographically normal or minimally narrowed coronary segments, *J Am Coll Cardiol* 23, 352, 1994.

Yamaji K, Fujimoto S, Ikeda Y, et al., Apoptotic myocardial cell death in the setting of arrhythmogenic right ventricular cardiomyopathy, *Acta Cardiol* 60, 465, 2005.

Yamak B, Sener E, Kiziltepes U, et al., Low dose anticoagulation after St. Jude Medical prosthesis implantation in patients under 18 years of age, *J Heart Valve Dis* 4, 274, 1995.

Yamamoto A, Kamiya T, Yamamura T, et al., Clinical features of familial hypercholesterolemia, *Arteriosclerosis* 9 (Suppl), I66, 1989.

Yamamoto H, Tomoike H, Shimokawa H, et al., Development of collateral function with repetitive coronary occlusion in a canine model reduces myocardial reactive hyperemia in the absence of significant coronary stenosis, *Circ Res* 55, 623, 1984.

Yamori Y, Mano M, Nara Y, et al., Cathecolamine-induced polyploidization in vascular smooth muscle cells, *Circulation* 75 (Suppl 1), 1, 1987.

Yan GX and Antzelevitch C, Cellular basis for the Brugada syndrome and other mechanisms of arrhythmogenesis associated with ST-segment elevation, *Circulation* 100, 1660, 1999.

Yang BZ, Ding JH, Zhou C, et al., Identification of a novel mutation in patients with medium chain acyl-CoA dehydrogenase deficiency, *Mol Genet Metab* 69, 259, 2000.

Yang P, Kanki H, Drolet B, et al., Allelic variants in long-QT disease genes in patients with drug associated torsades de pointes, *Circulation* 105, 1943, 2002.

Yaoita H, Sakabe A, Maehara K, and Maruyama Y, Different effects of carvedilol, metoprolol, and propranolol on left ventricular remodelling after coronary stenosis or after permanent coronary occlusion in rats, *Circulation* 105, 975, 2002.

Yeager SB, Hougen TJ, and Levy AM, Sudden death in infants with chaotic atrial rhythm, *Am J Dis Child* 138, 689, 1984.

Yee ES and Khonsari S, Right-sided infective endocarditis: valvuloplasty, valvectomy or replacement, *J Cardiovasc Surg* 30, 744, 1989.

Yetman AT, Hamilton RM, Benson LN, and McCrindle BW, Long-term outcome and prognostic determinants in children with hypertrophic cardiomyopathy, *J Am Coll Cardiol* 32, 1943, 1998.

Yilmaz G, Ozme S, Ozer S, et al., Evaluation by exercise testing of children with mild and moderate valvular aortic stenosis, *Pediatr Int* 42, 48, 2000.

Yoshida NI, Ogura Y, and Wakasugi C, Myocardial lesions induced after trauma and treatment, *Foresic Sci Int,* 54, 181, 1992.

Yoshioka J and Lee RT, Cardiovascular genomics, *Cardiovasc Pathol* 12, 249, 2003.

Yoshioka K, Gao OW, Chin M, et al., Heterogeneous sympathetic innervation influences local myocardial repolarization in normally perfused rabbit hearts, *Circulation* 101, 1060, 2000.

Young JB, Winters WL, Bourge R, and Uretsky BF, Task force 4: function of the heart transplant recipient, *J Am Coll Cardiol* 22, 31, 1993.

Zack F, Terpe H, Hammer U, and Wegener R, Fibromuscular dysplasia of the coronary arteries as a rare cause of death, *Int J Legal Med* 108, 215, 1996.

Zack F and Wegener R, Zur Problematik der Diagnose rhythmoger Herztod durch histologische Untersuchungen des Erregungsbildungs- und -leitungssystems, *Rechtsmed* 5, 1–5, 1994.

Zamir M and Silver MD, Vasculature in the walls of human coronary arteries, *Arch Pathol Lab Med* 109, 659, 1985.

Zamora-Quezada JC, Dinerman H, Stadecker MJ, et al., Muscle and skin infarction after free basing cocaine (crack), *Ann Intern Med* 108, 564, 1988.

Zareba W, Moss AJ, Schwartz PJ, et al., Influence of genotype on the clinical course of the long-QT syndrome: International Long-QT syndrome Registry Research Group, *N Engl J Med* 339, 960, 1998.

Zehender M, Buchner C, Meinertz T, et al., Prevalence, circumstances, mechanisms, and risk stratification of sudden cardiac death in unipolar single-chamber ventricular pacing, *Circulation* 85, 596, 1992.

Zeltser D, Justo D, Halkin A, et al., Torsade de pointes due to noncardiac drugs: most patients have easily identifiable risk factors, *Medicine* 82, 282, 2003.

Zenker G, Erbel R, Kramer G, et al., Transesophogeal two-dimensional echocardiography in young patients with cerebral ischemic events, *Stroke* 19, 345, 1988.

Zerbini EJ, Coronary ligation in wounds of the heart: report of a case in which ligation of the anterior descending branch of the left coronary artery was followed by complete recovery, *J Thorac Surg* 12, 642, 1943.

Zhang H, Li Y, Peng T, Aasa M, et al., Localization of enteroviral antigen in myocardium and other tissues from patients with heart muscle disease by an improved immunohistochemical technique, *J Histochem Cytochem* 48, 579, 2000.

Zhang JM and Riddick L, Cytosckeleton immunohistochemical study of early ischemic myocardium, *For Sci Int* 80, 229, 1996.

Zhang L, Timothy KW, Vincent GM, et al., Spectrum of ST-T-wave patterns and repolarization parameters in congenital long-QT syndrome: ECG findings identify genotypes, *Circulation* 102, 2849, 2000.

Zhao M, Zhang H, Robinson TF, et al., Profound structural alterations of the extracellular collagen matrix in postischemic dysfunctional ("stunned") but viable myocardium, *J Am Coll Cardiol* 10, 1322, 1987.

Zheng ZJ, Croft BJ, Giles WH, et al., Sudden cardiac death in the United States, 1989 to 1998, *Circulation* 104, 2158, 2001.

Zimmermann ANE, Daems W, Hulsmann WC, et al., Morphopathological changes of heart muscle caused by successive perfusion with calcium-free and calcium containing solutions (calcium paradox), *Cardiov Res* 1, 201, 1967.

Zimring HJ, Fitzgerald RL, Engler RL, et al., Intracoronary versus intravenous effects of cocaine on coronary flow and ventricular function, *Circulation* 89, 1819, 1994.

Zipes DP, Electrophysiological mechanism involved in ventricular fibrillation, *Circulation* 51–52 (Suppl 3), 3, 1975.

Zipes DP and Jalife J, *Cardiac Electrophysiology: From Cell to Bedside*, W.B. Saunders, Philadelphia, 1995.

Zorec Karlovek M, Alibegovi A, and Balaûic J, Our experience with fatal ecstasy abuse (two case reports), *Forensic Sci Int* 147S, S77, 2005.

Index

A

A bands, 24
Abruptio placentae, 74
Abscess
    with endocarditis, 149, 151, 154
        infective, 150
        nonbacterial thrombotic (NBTE), 155
    myocarditis, 115
    pulmonary, 162
    tissue, 152
    traumatic injury, penetrating, 320
Accessory tracts, 219
    atrial tachyarrhythmias, miscellaneous, 234
    Lown-Ganong-Levine syndrome, 233
    oncocytic cardiomyopathy, 222
    sudden infant death (SID), 249, 250
    ventricular preexcitation syndromes, 231–232
Acetylcholine, 26
Acommissural aortic valve, 167
Acquired immunodeficiency syndrome (AIDS)
    cardiac arrest, 54
    colliquative myocytolysis, 48
    heart weight and left ventricular wall thickness, 137
    myocardial cell necrosis, frequency and extent of, 46
    myocardial disarray, 129
    myocarditis, 116–117, 124
        parasitosis, 121–122
        sudden and unexpected death, 124
        and sudden death, 124
    myofiber breakup in malignant arrhythmia/ventricular fibrillation, 54
    transplanted heart, 136
Acquired prolongation of Q-T interval, 227–228
Actin filaments, 24
Active plaque, 111
Activity, *see* Exercise/activity/exertion
Acute coronary syndromes
    embolism, extramural coronary arteries, 71
    functional occlusion-coronary spasm, 95–96
    reflow/reperfusion necrosis, 273
    thrombi, role of, 90
Acute myocardial infarct, *see* Infarct, acute
Acute necrotizing eosinophilic myocarditis, 119–120
Acute restrictive hypovolemia, 130
Adaptation
    lymphatics and, 75
    perivascular fibrosis and, 72
    vascular, 19–20
Adenocarcinoma, 155
Adenosine diphosphate, 92–93
Adenosine triphosphate, 24, 39
Adipose artery, left, 7
Adipose tissue, *see* Lipids/fatty tissue
Adolescents and young adults
    anomalous origin of coronary arteries and, 178
    congenital lesions of tricuspid valve and, 161–162
    dissecting aneurysm, 200
    fibromuscular dysplasia, 194
    hypertrophic cardiomyopathy, 211
Adrenal cortical carcinoma, 162
Adrenaline, cocaine-induced cardiac dysfunction, 343
Adrenergic control, and arterial wall thickness, 12–13
Adrenergic lesions, transplanted heart, 134
Adrenergic nerves
    innervation, 25
    reduction or denervation, 75
Adrenergic stress, 49
    anomalous origin of coronary arteries, 178
    and cardiac arrest, 53, 56
    and cardiac rupture, 104
    causal hypotheses, 281–282
    cocaine and, 345, 348
    congestive heart failure, 138
    contraction band necrosis, 47
    experimental coronary occlusion, 44
    infants and children, 224
    interpretation of pathological changes, 276–280
    MDMA and, 351
    myocarditis, 124
    obliterative intimal thickening, 72
    and ventricular fibrillation, 56, 272

M

X

Z